Essentials of
ECOLOGY
Second Edition

Colin R. Townsend
Department of Zoology
University of Otago
Dunedin, New Zealand

Michael Begon
Population Biology Research Group
School of Biological Sciences
The University of Liverpool
Liverpool, England

John L. Harper
Professor Emeritus in the
University of Wales
Visiting Professor in the
University of Exeter
Exeter, England

Blackwell
Publishing

© 2003 by Blackwell Science Ltd
a Blackwell Publishing company

350 Main Street, Malden, MA 02148-5018, USA
108 Cowley Road, Oxford OX4 1JF, UK
550 Swanston Street, Carlton South, Melbourne, Victoria 3053, Australia
Kurfürstendamm 57, 10707 Berlin, Germany

First published 2000 by Blackwell Science, Inc.
Second edition 2003

Library of Congress Cataloging-in-Publication Data

Townsend, Colin R.
    Essentials of ecology / Colin R. Townsend, Michael Begon, and John L. Harper. — 2nd ed.
        p.   cm.
    Includes bibliographical references and index.
    ISBN 1-40510-328-0
    1. Ecology. I. Begon, Michael. II. Harper, John L. III. Title.
    QH541 .T66 2002
    577—dc21

                                                            2002003801

A catalogue record for this title is available from the British Library.

Set in 9$\frac{1}{2}$/12pt Berkeley Old Style
by Graphicraft Limited, Hong Kong
Printed and bound in
Italy by G. Canale & C. S.p.A., Turin

For further information on
Blackwell Publishing, visit our website:
www.blackwellpublishing.com

Brief Contents

Preface   xi
Acknowledgments   xiv

Part I   Introduction   **1**
Chapter 1: Ecology and How To Do It   3
Chapter 2: The Ecology of Evolution   38

Part II   Conditions and Resources   **73**
Chapter 3: Physical Conditions and the Availability of Resources   75
Chapter 4: Conditions, Resources, and the World's Communities   111

Part III   Individuals, Populations, Communities, and Ecosystems   **149**
Chapter 5: Birth, Death, and Movement   151
Chapter 6: Interspecific Competition   188
Chapter 7: Organisms as Habitats   223
Chapter 8: Predation, Grazing, and Disease   256
Chapter 9: Population Processes—the Big Picture   293
Chapter 10: Patterns in Species Richness   330
Chapter 11: The Flux of Energy and Matter through Ecosystems   366

Part IV   Applied Issues in Ecology   **397**
Chapter 12: Sustainability   399
Chapter 13: Pollution   437
Chapter 14: Conservation   467

References   501
Index   513

Table of Contents

Preface   xi
Acknowledgments   xiv

Part I   Introduction   1

Chapter 1:   Ecology and How To Do It   3

1.1   **Introduction**   4
1.2   **Scales, diversity, and rigor**   7
    1.2.1   Questions of scale   8
    1.2.2   The diversity of ecological evidence   10
    1.2.3   Statistics and scientific rigor   11
1.3   **Ecology in practice**   18
    1.3.1   The brown trout in New Zealand—effects on individuals,
           populations, communities, and ecosystems   18
    1.3.2   Successions on old fields in Minnesota—a study in time
           and space   25
    1.3.3   Hubbard Brook—a long-term commitment of large-scale
           significance   28
    1.3.4   A model study: genetically modified crops—bad for
           biodiversity?   32
Summary   35
Review Questions   37

Chapter 2:   The Ecology of Evolution   38

2.1   **Introduction**   39
2.2   **Evolution by natural selection**   39
2.3   **Evolution within species**   43
    2.3.1   Geographical variation within species   43
    2.3.2   Variation within a species with man-made selection pressures 51
    2.3.3   Adaptive peaks and specialized abysses   53

2.4   **The ecology of speciation**   55
    2.4.1   What do we mean by a "species"?   55
    2.4.2   Islands and speciation   57
2.5   **The effects of climatic change on the evolution and distribution of species**   61
2.6   **The effects of continental drift on the ecology of evolution**   64
2.7   **Interpreting the results of evolution: convergent and parallel evolution**   67
Summary   70
Review Questions   71
Web Research Questions   72

Part II   Conditions and Resources   73

Chapter 3:   Physical Conditions and the Availability of Resources   75

3.1   **Introduction**   76
3.2   **Environmental conditions**   76
    3.2.1   What do we mean by "harsh", "benign", and "extreme"?   76
    3.2.2   Effects of conditions   77
    3.2.3   Conditions as stimuli   80
    3.2.4   The effects of conditions on interactions between organisms   81
    3.2.5   Responses by sedentary organisms   84
    3.2.6   Animal responses to environmental temperature   85
    3.2.7   Microorganisms in extreme environments   88
3.3   **Plant resources**   88
    3.3.1   Solar radiation   89
    3.3.2   Water   90
    3.3.3   Mineral nutrients   92
    3.3.4   Carbon dioxide   94
3.4   **Animals and their resources**   96
    3.4.1   Nutritional needs and provisions   98
    3.4.2   Defense   100
3.5   **The effect of intraspecific competition for resources**   103
3.6   **Conditions, resources, and the ecological niche**   106
Summary   108
Review Questions   110

Chapter 4:   Conditions, Resources, and the World's Communities   111

4.1   **Introduction**   112
4.2   **Geographical patterns at large and small scales**   112
    4.2.1   Large-scale climatic patterns   112
    4.2.2   Small-scale patterns in conditions and resources   114
    4.2.3   Patterns in conditions and resources in aquatic environments   118
4.3   **Temporal patterns in conditions and resources—succession**   118
4.4   **Terrestrial biomes**   120
    4.4.1   Describing and classifying biomes   120
    4.4.2   Tropical rain forest   127
    4.4.3   Savanna   130
    4.4.4   Temperate grasslands   130
    4.4.5   Desert   131
    4.4.6   Temperate forest   131
    4.4.7   Northern coniferous forest (taiga) grading into tundra   132

4.5   **Aquatic environments**   133
        4.5.1   Stream ecology   134
        4.5.2   Lake ecology   137
        4.5.3   Oceans   139
        4.5.4   Coasts   141
        4.5.5   Estuaries   142
Summary   145
Review Questions   146
Web Research Questions   147

Part III   Individuals, Populations, Communities, and Ecosystems   149

Chapter 5:   Birth, Death, and Movement   151

5.1   **Introduction**   152
        5.1.1   What is an individual?   152
        5.1.2   Counting individuals, births, and deaths   153
5.2   **Life cycles**   157
        5.2.1   Life cycles and reproduction   157
        5.2.2   Annual life cycles   159
        5.2.3   Longer life cycles   160
5.3   **Monitoring birth and death: life tables and fecundity schedules** 163
        5.3.1   Annual life tables   165
        5.3.2   Life tables for populations with overlapping generations   168
        5.3.3   A classification of survivorship curves   169
5.4   **Dispersal and migration**   170
        5.4.1   Dispersal determining abundance   172
        5.4.2   The role of migration   174
5.5   **The impact of intraspecific competition on populations**   174
        5.5.1   Patterns of population growth   176
5.6   **Life-history patterns**   181
Summary   185
Review Questions   187

Chapter 6:   Interspecific Competition   188

6.1   **Introduction**   189
6.2   **Ecological effects of interspecific competition**   189
        6.2.1   Competition between diatoms for silicate   189
        6.2.2   Coexistence and exclusion of competing salmonid fishes   190
        6.2.3   Some general observations   192
        6.2.4   Coexistence and exclusion of competing bumble bees   194
        6.2.5   Coexistence of competing diatoms   194
        6.2.6   Coexistence of competing rodents and ants   195
        6.2.7   The Competitive Exclusion Principle   197
        6.2.8   Environmental heterogeneity   201
6.3   **Evolutionary effects of interspecific competition**   203
        6.3.1   Character displacement and release in the Indian mongoose   204
        6.3.2   Canadian sticklebacks   206
6.4   **Interspecific competition and community structure**   207
        6.4.1   Limiting resources and the regulation of diversity in
                phytoplankton communities   207

6.4.2   Bumble bees in Colorado   208
6.4.3   Aggregation alters competitive interactions in plant
        communities   210
6.4.4   Niche differentiation amongst animals and amongst plants   211
6.5   How significant is interspecific competition in practice?   **213**
6.5.1   The prevalence of current competition   216
6.5.2   Competition or mere chance?   217
Summary   **220**
Review Questions   **222**

Chapter 7:   Organisms as Habitats   223

7.1   Introduction   **224**
7.2   Hosts as habitats   **228**
7.2.1   Life on the surface of another organism   228
7.2.2   Inhabitants of the body cavity   232
7.2.3   Inhabitants of host tissues and cells   236
7.3   Hosts as responsive habitats   **240**
7.3.1   Changes in the growth and form of hosts caused by their
        inhabitants   240
7.3.2   Changes in host behavior caused by their inhabitants   242
7.3.3   Immune and other defensive responses of the host   244
**7.4   Distribution and regulation of parasites and mutualists within hosts and
        their populations**   **245**
7.5   Life cycles and dispersal   **246**
7.6   Coevolution   **249**
Summary   **253**
Review Questions   **254**

Chapter 8:   Predation, Grazing, and Disease   256

8.1   Introduction   **257**
8.2   Prey fitness and abundance   **258**
8.3   The subtleties of predation   **261**
8.3.1   Interactions with other factors   261
8.3.2   Compensation and defense by individual prey   262
8.3.3   From individual prey to prey populations   264
8.4   Predator behavior—foraging and transmission   **267**
8.4.1   Foraging behavior   268
8.5   Population dynamics of predation   **273**
8.5.1   Underlying dynamics of predator–prey interactions:
        a tendency to cycle   274
8.5.2   Predator–prey cycles in practice   274
8.5.3   Disease dynamics and cycles   279
8.5.4   Crowding   282
8.5.5   Predators and prey in patches   284
8.6   Predation and community structure   **288**
Summary   **291**
Review Questions   **292**

Chapter 9:   Population Processes—the Big Picture   293

9.1	Introduction	294
9.2	Multiple determinants of the dynamics of populations	294
	9.2.1 Fluctuation or stability?	295
	9.2.2 Theories of species abundance	296
	9.2.3 Key-factor analysis	299
9.3	Dispersal, patches, and metapopulation dynamics	306
9.4	Temporal patterns in community composition	309
	9.4.1 Founder-controlled and dominance-controlled communities	309
	9.4.2 Community succession	312
9.5	Food webs	317
	9.5.1 Indirect and direct effects	317
	9.5.2 Top-down or bottom-up control of food webs?	320
	9.5.3 Community stability and food web structure	323
Summary		328
Review Questions		329

Chapter 10: Patterns in Species Richness   330

10.1	Introduction	331
10.2	A simple model of species richness	333
10.3	Spatially varying factors that influence species richness	335
	10.3.1 Productivity and resource richness	335
	10.3.2 Predation intensity	339
	10.3.3 Spatial heterogeneity	340
	10.3.4 Environmental harshness	340
10.4	Temporally varying factors that influence species richness	342
	10.4.1 Climatic variation	343
	10.4.2 Disturbance	343
	10.4.3 Environmental age: evolutionary time	345
10.5	Gradients of species richness	347
	10.5.1 Habitat area and remoteness—island biogeography	347
	10.5.2 Latitudinal gradients	352
	10.5.3 Gradients with altitude and depth	354
	10.5.4 Gradients during community succession	356
10.6	Patterns in taxon richness in the fossil record	358
10.7	Appraisal of patterns in species richness	360
Summary		363
Review Questions		365

Chapter 11: The Flux of Energy and Matter through Ecosystems   366

11.1	Introduction	367
11.2	Primary productivity	368
	11.2.1 Geographical patterns in primary productivity	368
	11.2.2 Factors limiting primary productivity	370
11.3	The fate of primary productivity	372
	11.3.1 Relationship between primary and secondary productivity	373
	11.3.2 The fundamental importance of energy transfer efficiencies	374

      11.3.3 Relative roles of the "live-consumer" and "decomposer"
              systems   376

11.4   **The process of decomposition**   378
      11.4.1 The decomposers: bacteria and fungi   378
      11.4.2 The detritivores and specialist microbivores   378
      11.4.3 Consumption of plant detritus   380
      11.4.4 Consumption of feces and carrion   383

11.5   **The flux of matter through ecosystems**   384
      11.5.1 Nutrient budgets in terrestrial ecosystems   386
      11.5.2 Nutrient budgets in aquatic communities   388

11.6   **Global biogeochemical cycles**   389
      11.6.1 The hydrological cycle   389
      11.6.2 The phosphorus cycle   390
      11.6.3 The nitrogen cycle   390
      11.6.4 The sulfur cycle   392
      11.6.5 The carbon cycle   392
      11.6.6 Human impacts on biogeochemical cycles   392

Summary   393
Review Questions   394
Web Research Questions   395

Part IV   **Applied Issues in Ecology**   **397**

Chapter 12: Sustainability   **399**

12.1   **Introduction**   400
12.2   **The human population "problem"**   400
      12.2.1 Introduction   400
      12.2.2 Population growth up to the present   402
      12.2.3 Predicting the future   404
      12.2.4 Two future inevitabilities   406
      12.2.5 A global carrying capacity?   407

12.3   **Harvesting living resources from the wild**   409
      12.3.1 Fisheries—maximum sustainable yields   409
      12.3.2 Obtaining MSYs through fixed quotas   411
      12.3.3 Obtaining MSYs through fixed effort   413
      12.3.4 Beyond MSY   414

12.4   **The farming of monocultures**   415
      12.4.1 Degradation and erosion of soil   418
      12.4.2 Sustainability of water as a resource   420

12.5   **Pest control**   423
      12.5.1 Aims of pest control: economic injury levels and action
              thresholds   423
      12.5.2 Problems with chemical pesticides—and their virtues   424
      12.5.3 Biological control   426

12.6   **Integrated farming systems**   428
12.7   **Forecasting agriculturally driven global environmental change**   433
Summary   434
Review Questions   436

Chapter 13: Pollution   437

13.1   Introduction   438
13.2   Urban pollution   438
    13.2.1 Burial and cremation   438
    13.2.2 Feces, urine, and other urban waste   439
13.3   Agricultural pollution   443
    13.3.1 Intensive livestock management   443
    13.3.2 Nitrate   444
    13.3.3 Pesticides   446
13.4   Atmospheric pollution   449
    13.4.1 Carbon dioxide—a major atmospheric pollutant   450
    13.4.2 The greenhouse effect   453
    13.4.3 Acid rain   455
    13.4.4 Thinning of the ozone layer   457
13.5   Nuclear radiation   457
13.6   Mining and quarrying   461
13.7   Restoration ecology   464
Summary   465
Review Questions   466

Chapter 14: Conservation   467

14.1   Introduction   468
14.2   Threats to species   469
    14.2.1 Overexploitation   474
    14.2.2 Habitat disruption   474
    14.2.3 Introduced species   475
    14.2.4 Possible genetic problems in small populations   476
    14.2.5 A review of risks   479
    14.2.6 The dynamics of small populations   480
14.3   Threats to communities   483
14.4   Conservation in practice   486
    14.4.1 Species management plans   486
    14.4.2 *Ex situ* conservation   490
    14.4.3 Protected areas   492
    14.4.4 Restoration ecology   494
    14.4.5 Finale—a healthy approach to conservation   496
Summary   496
Review Questions   498
Web Research Questions   500

References   501
Index   513

Preface

By writing this book we hope to share with you some of our wonder at the complexity of nature—the Rousseau painting on the front cover portrays the beauty of the natural world. But we are also familiar with a darker side: the shadows in Rousseau's jungle. Citizens need to be ecologically literate, so that they can take part in political debate and contribute to solving the ecological problems that we carry into the new millennium. We hope our book will contribute to this objective.

The genesis of this book can be found in the more comprehensive treatment of ecology in our big book *Ecology: Individuals, Populations and Communities* (Begon, Harper & Townsend—3rd edn, 1996). This is used as an advanced university text around the world, but many of our colleagues have called for a more succinct treatment of the essence of the subject. One recent reviewer of the 'big book' wondered whether one would need a wheelbarrow to carry around the next edition! Thus, we have been spurred into action to produce a distinctively different book, written with clear objectives for a different audience—those taking a semester-long beginners course in the essentials of ecology. We hope that some readers will be excited enough to go on to sample the big book and the rich literature of ecology that it can lead into.

In this second edition of *Essentials of Ecology* we have made the text, including mathematical topics, even more accessible. Ecology is a vibrant subject and this is reflected by our inclusion of fully 100 new studies. Some readers will be engaged most by the fundamental principles of how ecological systems work. Others will be impatient to focus on the ecological problems caused by human activities. We place heavy emphasis on both fundamental and applied aspects of ecology: there is no clear boundary between the two. However, we have chosen to deal first in a systematic way with the fundamental side of the subject, and we have done this for a particular reason. An understanding of the scope of the problems facing us (the unsustainable use of ecological resources, pollution, extinctions, and the erosion of natural biodiversity) and the means to counter and solve these problems depend absolutely on a proper grasp of ecological fundamentals.

The book is divided into four sections. In the introduction we deal with two foundations for the subject that are often neglected in texts. Chapter 1 aims to show not only what ecology is but also how ecologists do it—how ecological understanding is achieved, what we understand (and, just as important, what we do not yet understand), and how our understanding helps us predict and manage. We then introduce 'The ecology of evolution' and show that ecologists need a full understanding of the evolutionary biologist's discipline in order to make sense of patterns and processes in nature (Chapter 2).

What makes an environment habitable for particular species is that they can tolerate the physicochemical conditions there and find in it their essential resources. In the second section we deal with conditions and resources, both as they influence individual species (Chapter 3) and in terms of their consequences for the composition and distribution of multispecies communities, for example in deserts, rain forests, rivers, lakes, and oceans (Chapter 4).

The third section (Chapters 5–11) deals systematically with the ecology of individual organisms, populations of a single species, communities consisting of many populations, and ecosystems (where we focus on the fluxes of energy and matter between and within communities). To understand patterns and processes at each of these levels we need to know the behavior of the level below.

Finally, armed with knowledge and understanding of the fundamentals, the book turns to the applied questions of how to deal with pests and manage resources sustainably (whether wild populations of fish or agricultural monocultures) (Chapter 12), then to a diversity of pollution problems ranging from local enrichment of a lake by sewage to global climate change associated with the use of fossil fuels (Chapter 13), and lastly we develop an armoury of approaches that may help us to save endangered species from extinction and conserve some of the biodiversity of nature for our descendants (Chapter 14).

A number of pedagogical features have been included to help you.

- Each chapter begins with a set of key concepts that you should understand before proceeding to the next chapter.
- Marginal headings provide signposts of where you are on your journey through each chapter—these will also be useful revision aids.
- Unanswered questions in ecology are highlighted—we believe passionately that the awareness of ignorance is the beginning of knowledge. Ecologists do not know all the answers.
- Each chapter concludes with a summary and a set of review questions, some of which are designated challenge questions.
- You will also find three categories of boxed text:

  1 'History boxes' emphasize some landmarks in the development of ecology.
  2 'Quantitative boxes' set aside mathematical and quantitative aspects of ecology so they do not unduly interfere with the flow of the text and so you can consider them at leisure.

  3 'Topical ECOncerns' are boxes that highlight some of the applied problems in ecology, particularly those where there is a social or political dimension (as there often is). In these, you will be challenged to consider some ethical questions related to the knowledge you are gaining.

An important further feature of the book is the companion internet web site, **e.cology**, accessed through www.blackwellpublishing.com. This provides an easy-to-use range of resources to aid study and enhance the content of the book. A particularly novel feature is a set of web research questions that appear at the end of the four sections of the book. These explore current ecological issues in depth by exploiting the internet as an important new research medium. Each question is repeated on the web site, accompanied by a carefully chosen set of links to other relevant sites to guide the student in their own research. Other features of the web site include self-assessment multiple choice questions for each chapter in the book and an interactive tutorial to help students to understand the use of mathematical modeling in ecology.

Acknowledgments

It is a pleasure to record our gratitude to the people who helped with the planning and writing of this book. We thank Bob Campbell and Simon Rallison for getting the original enterprise off the ground and Nancy Hill-Whilton and Irene Herlihy for ably managing the project. We are also grateful to the following colleagues who provided insightful reviews of early drafts of one or more chapters: Tim Mousseau (University of South Carolina), Vickie Backus (Middlebury College), Kevin Dixon (Arizona State University, West), James Maki (Marquette University), George Middendorf (Howard University), William Ambrose (Bates College), Don Hall (Michigan State University), Clayton Penniman (Central Connecticut State University), David Tonkyn (Clemson University), Sara Lindsay (Scripps Institute of Oceanography), Saran Twombly (University of Rhode Island), Katie O'Reilly (University of Portland), Catherine Toft (UC Davis), Bruce Grant (Widener University), Mark Davis (Macalester College), Paul Mitchell (Staffordshire U., UK), and William Kirk (Keele U., UK).

Last but not least, we are glad to thank our wives and families for supporting us, listening to us, and ignoring us precisely as was required—thanks to Laurel and Dominic, to Borgny, and to Linda, Jessica, and Robert.

The publisher and the authors gratefully acknowledge the creative and scholarly input of Douglas Reed. He was instrumental in pulling together important parts of the website package.

For the second edition, we should additionally like to thank James Cahill (University of Alberta), Liane Cochrane-Stafira (Saint Xavier University), Hans deKroon (University of Nijmegen), Jake Weltzin (University of Tennessee at Knoxville) and Alan Wilmot (University of Derby, UK) for their comments on individual chapters, and especially Nathan Brown (Blackwell, US) and Rosie Hayden (Blackwell, UK) for making it so easy for us to take this book from manuscript into print.

Part I

Introduction

Chapter 1

Ecology and How To Do It

*Nowadays, ecology is a subject about which almost everyone has heard and most people consider to be important—even when they are unsure about the exact meaning of the term. There can be no doubt that it **is** important; but this makes it all the more critical that we understand what it is and how to do it.*

Chapter Contents

1.1 Introduction

1.2 Scales, diversity, and rigor

1.3 Ecology in practice

Key Concepts

In this chapter you will

- know how to define ecology and appreciate its development as both an applied and a pure science

- recognize that ecologists seek to describe and understand, and on the basis of their understanding to predict, manage, and control

- appreciate that ecological phenomena occur on a variety of spatial and temporal scales, and that patterns may be evident only at particular scales

- recognize that ecological evidence and understanding can be obtained by means of observations, field and laboratory experiments, and mathematical models

- understand that ecology relies on truly scientific evidence (and the application of statistics)

1.1 Introduction

The question "What is ecology?" could be translated into "How do we define ecology?" and answered by examining various definitions of ecology that have been proposed, and even choosing one of them as the best (Box 1.1). But while definitions have conciseness and precision, and they are good at preparing you for an examination, they are not so good at capturing the flavor, the interest, or the excitement of ecology. There is a lot to be gained by replacing that single question about definition with a series of more provoking ones: "What do ecologists *do*?" "What are ecologists *interested* in?" and "Where did ecology emerge from in the first place?"

the earliest ecologists   Ecology can lay claim to be the oldest science. If, as our preferred definition has it, "Ecology is the scientific study of the distribution and abundance of organisms and the interactions that determine distribution and abundance" (Box 1.1) then the most primitive humans must have been ecologists of sorts—driven by the need to understand

---

Box 1.1   **Historical Landmarks**

Definitions of Ecology

Ecology (originally in German: *Öekologie*) was first defined in 1866 by Ernst Haeckel, an enthusiastic and influential disciple of Charles Darwin. To him, ecology was "the comprehensive science of the relationship of the organism to the environment". The spirit of this definition is very clear in an early discussion of biological subdisciplines by Burdon-Sanderson (1893), in which ecology is "the science which concerns itself with the external relations of plants and animals to each other and to the past and present conditions of their existence", to be contrasted with physiology (internal relations) and morphology (structure). For many, such definitions have stood the test of time. Thus, Ricklefs (1973) in his textbook defined ecology as "the study of the natural environment, particularly the interrelationships between organisms and their surroundings".

In the years after Haeckel, plant ecology and animal ecology drifted apart. Influential works defined ecology as "those relations of *plants*, with their surroundings and with one another, which depend directly upon differences of habitat among plants" (Tansley, 1904), or as the science "chiefly concerned with what may be called the sociology and economics

of *animals*, rather than with the structural and other adaptations possessed by them" (Elton, 1927). The botanists and zoologists, though, have long since agreed that they belong together and that their differences must be reconciled.

There is, nonetheless, something disturbingly vague about the many definitions of ecology that seem to suggest that it consists of all those aspects of biology which are neither physiology nor morphology. In search of more focus, therefore, Andrewartha (1961) defined ecology as "the scientific study of the distribution and abundance of organisms", and Krebs (1972), regretting that the central role of "relationships" had been lost in this, modified it to "the scientific study of the *interactions* that determine the distribution and abundance of organisms", explaining that ecology was concerned with "*where* organisms are found, *how many* occur there, and *why*". This being so, it might be better still to define ecology as:

"the scientific study of the distribution and abundance of organisms and the interactions that determine distribution and abundance".

where and when not only their food but also their (non-human) enemies were to be found—and the earliest agriculturalists needed to be even more sophisticated: having to know how to manage their living but domesticated sources of food. These early ecologists, then, were *applied* ecologists, seeking to understand the distribution and abundance of organisms in order to apply that knowledge for their own collective benefit. They were interested in many of the sorts of things that applied ecologists are still interested in: how to maximize the rate at which food is collected from natural environments, and how this can be done repeatedly over time; how domesticated plants and animals can best be planted or stocked so as to maximize rates of return; how food organisms can be protected from their own natural enemies; how to control the populations of pathogens and parasites that live on us.

In the last century or so, however, since ecologists have been self-conscious enough to give themselves a name, ecology has consistently covered not only applied but also fundamental, "pure" science. A.G. Tansley was one of the founding fathers of ecology. He was concerned especially to understand, for understanding's sake, the processes responsible for determining the structure and composition of different plant communities. When, in 1904, he wrote from Britain about "The problems of ecology" he was particularly worried by a tendency for too much ecology to remain at the descriptive and unsystematic stage (i.e., accumulating descriptions of communities without knowing whether they were typical, temporary, or whatever), too rarely moving on to experimental or systematically planned, or what we might call a "scientific", analysis.

a pure and applied science

His worries were echoed across the Atlantic in the United States by another of ecology's founders, F.E. Clements, who in 1905 in his *Research Methods in Ecology* complained:

> The bane of the recent development popularly known as ecology has been a widespread feeling that anyone can do ecological work, regardless of preparation. There is nothing . . . more erroneous than this feeling.

On the other hand, the need in applied biology *for* ecology, and the contribution that applied biology can make *to* ecology, were clear in the introduction to Charles Elton's (1927) *Animal Ecology* (Figure 1.1):

> Ecology is destined for a great future . . . The tropical entomologist or mycologist or weed-controller will only be fulfilling his functions properly if he is first and foremost an ecologist.

In the intervening years, the coexistence of these pure and applied threads has been maintained and built upon. Many applied areas have contributed to the development of ecology and have seen their own development enhanced by ecological ideas and approaches. All aspects of food and fiber gathering, production, and protection have been involved: plant ecophysiology, soil maintenance, forestry, grassland composition and management, food storage, fisheries, and control of pests and pathogens. Each of these classic areas is still at the forefront of lots of good ecology and they have been joined by others. The biological control of pests (the use of pests' natural enemies to control them) has a history going back at least to the Ancient Chinese but has seen a

Figure 1.1
One of the great founders of ecology:
Charles Elton.

resurgence of ecological interest since the shortcomings of chemical pesticides began to be widely apparent in the 1950s. The ecology of pollution has been a growing concern from around the same time and expanded further in the 1980s and 1990s from local to global issues. The closing decades of the millenium also saw expansions both in public interest and ecological input into the conservation of endangered species and the biodiversity of whole areas, in the control of disease in humans as well as many other species, and in the potential consequences of profound alterations to the global environment.

unanswered questions   And yet, at the same time, many fundamental problems of ecology remain unanswered. To what extent does competition for food determine which species can coexist in a habitat? What role does disease play in the dynamics of populations? Why are there more species in the tropics than at the poles? What is the relationship between soil productivity and plant community structure? Why are some species more vulnerable to extinction than others? And so on. Of course, unanswered questions—if they are *focused* questions—are a symptom of the health not the weakness of any science. But ecology is not an easy science, and it has particular subtlety and complexity, in part because ecology has the distinction of being peculiarly confronted with uniqueness: millions of different species, countless billions of genetically distinct individuals, all living and interacting in a varied and ever-changing world. The beauty of ecology is that it challenges us to develop an understanding of very basic and apparent problems, in a way that recognizes the uniqueness and complexity of all aspects of nature but seeks patterns and predictions within this complexity rather than being swamped by it.

Summarizing this brief historical overview, it is clear that ecologists try to do a number of different things. First and foremost ecology is a science, and ecologists therefore try to *explain* and *understand*. There are two different classes of explanation in biology: "proximate" and "ultimate". For example, the present distribution and abundance of a particular species of bird may be "explained" in terms of the physical environment that the bird tolerates, the food that it eats, and the parasites and predators that attack it. This is a *proximate* explanation—an explanation in terms of what is going on "here and now". However, we may also ask how this species of bird has come to have these properties that now appear to govern its life. This question has to be answered by an explanation in evolutionary terms; the *ultimate* explanation of the present distribution and abundance of this bird lies in the ecological experiences of its ancestors (see Section 2.2).

In order to understand something, of course, we must first have a description of whatever it is we wish to understand. Ecologists must therefore *describe* before they explain. On the other hand, the most valuable descriptions are those carried out with a particular problem or "need for understanding" in mind. Undirected description, carried out merely for its own sake, is often found afterwards to have selected the wrong things and has little place in ecology—or any other science.

understanding, description, prediction, and control

Ecologists also often try to *predict* what will happen to a population of organisms under a particular set of circumstances, and on the basis of these predictions try to control or exploit them. We try to minimize the effects of locust plagues by predicting when they are likely to occur and taking appropriate action. We try to exploit crops most effectively by predicting when conditions will be favorable to the crop and unfavorable to its enemies. We try to preserve rare species by predicting the conservation policy that will enable us to do so. Some prediction and control can be carried out without deep explanation or understanding: there is no great difficulty in predicting that if a woodland is destroyed, all the woodland birds that live there will disappear from the area—and that their disappearance can be controlled by saving the woodland. But insightful predictions, precise predictions, and predictions of what will happen in unusual circumstances can be made only when we can also explain and understand what is going on.

This book is therefore about:

1 how ecological understanding is achieved;
2 what we do understand (but also what we do not understand—indeed, throughout the book, numbers of "unanswered questions" have been highlighted in the marginal notes);
3 how that understanding can help us predict, manage, and control.

1.2   Scales, Diversity, and Rigor

The rest of this chapter is about the two "hows" above: how understanding is achieved, and how that understanding can help us predict, manage, and control. Later in the chapter we illustrate three fundamental points about doing ecology by examining a limited number of examples in some detail (see Section 1.3). But first we elaborate on the three points, namely:

- ecological phenomena occur at a variety of scales;
- ecological evidence comes from a variety of different sources;
- ecology relies on truly scientific evidence and the application of statistics.

1.2.1 Questions of scale

Ecology operates at a range of scales: timescales, spatial scales, and "biological" scales. It is important to appreciate the breadth of these and how they relate to one another.

the "biological" scale

The living world is often said to comprise a biological hierarchy beginning with subcellular particles and continuing through cells, tissues, and organs. Ecology then deals with the next three levels:

- individual organisms;
- populations (consisting of individuals of the same species);
- communities (consisting of a greater or lesser number of populations).

At the level of the organism, ecology deals with how individuals are affected by (and how they affect) their environment. At the level of the population, ecology deals with the presence or absence of particular species, with their abundance or rarity, and with the trends and fluctuations in their numbers. Community ecology then deals with the composition or structure of ecological communities.

We can also focus on the pathways followed by energy and matter as these move among living and non-living elements of a fourth category of organization:

- *ecosystems* (comprising the community together with its physical environment).

With this level of organization in mind, Likens (1992) would extend our preferred definition of ecology (Box 1.1) to include "the interactions between organisms and the transformation and flux of energy and matter". However, we take energy/matter transformations as being subsumed in the "interactions" of our definition.

a range of spatial scales

Within the living world, there is no arena too small nor one so large that it does not have an ecology. Even the popular press talks increasingly about the "global ecosystem" and, while we may have doubts about the level of understanding of some media commentators, there is no question that several ecological problems can only be examined at this very large scale. These include the relationships between ocean currents and fisheries, or between climate patterns and the distribution of deserts and tropical rain forests, or between elevated carbon dioxide in the atmosphere (from burning fossil fuels) and global climate change.

At the opposite extreme, an individual cell may be the stage on which two populations of pathogens compete with one another for the resources that the cell provides. At a slightly larger spatial scale, a termite's gut is the habitat for bacteria, protozoans, and other species (Figure 1.2)—a community whose diversity may reasonably be compared with that of a tropical rain forest in terms of the richness of organisms living there, the variety of interactions in which they take part, and indeed the extent to which we remain ignorant about the species identity of many of the participants. Between these extremes, different ecologists, or the same ecologist at different times,

Figure 1.2
The diverse community of a termite's gut.
(After Breznak, 1975.)

may study the inhabitants of pools that form in small tree holes, the temporary watering holes of the savannas, or the great lakes and oceans; others may examine the diversity of fleas on different species of birds, the diversity of birds in different-sized patches of woodland, or the diversity of woodlands at different altitudes.

To some extent related to this range of spatial scales, and to the levels in the biological hierarchy, ecologists also work on a variety of timescales. "Ecological succession"—the successive and continuous colonization of a site by certain species populations, accompanied by the extinction of others—may be studied over a period from the deposition of a lump of sheep dung to its decomposition (a matter of weeks), or from the change in climate at the end of the last ice age to the present day and beyond (around 14,000 years and still counting). Migration may be studied in butterflies over the course of days, or in the forest trees that are still (slowly) migrating into deglaciated areas following that last ice age.

a range of timescales

Although it is undoubtedly the case that "appropriate" timescales vary, it is also true that many ecological studies are not as long as they might be. Longer studies cost more and require greater dedication and stamina. An impatient scientific community, and the requirement for concrete evidence of activity for career progression, both put pressure on ecologists, and all scientists, to publish their work sooner rather than later. Why are long-term studies potentially of such value? The reduction over a few years in the numbers of a particular species of wild flower, or bird, or butterfly might be a cause for conservation concern—but one or more decades of study may be needed to be sure that the decline is more than just an expression of the random ups and downs of "normal" population dynamics. Similarly, a 2-year rise in the abundance of a wild rodent followed by a 2-year fall might be part of a regular "cycle" in abundance, crying out for an explanation. But ecologists could not be sure until perhaps 20 years of study has allowed them to record four or five repeats of such a cycle.

the need for long-term studies

This does not mean that all ecological studies need to last for 20 years—nor that every time an ecological study is extended the answer changes. But it does emphasize the great value to ecology of the small number of long-term investigations that have been carried out or are ongoing.

1.2.2 The diversity of ecological evidence

Ecological evidence comes from a variety of different sources. Ultimately, ecologists are interested in organisms in their natural environments (though for many organisms, the environment which is "natural" for them now is itself man-made). Progress would be impossible, however, if ecological studies were limited to such natural environments. And even in natural habitats, unnatural acts (experimental manipulations) are often necessary in the search for sound evidence.

observations and field experiments

Many ecological studies involve careful *observation* and monitoring, in the natural environment, of the changing abundance of one or more species over time, or over space, or both. In this way, ecologists may establish patterns; for example, that red grouse (birds shot for "sport") exhibit regular cycles in abundance peaking every 4 or 5 years, or that vegetation can be mapped into a series of zones as we move across a landscape of sand dunes. But scientists do not stop at this point—the patterns require explanation. Careful analysis of the descriptive data may suggest some plausible explanations. But establishing what causes the patterns may well require *manipulative field experiments*: ridding the red grouse of intestinal worms, suggested to underlie the cycles, and checking if the cycles persist (they do not: Hudson et al., 1998), or treating experimental areas on the sand dunes with fertilizer to see whether the changing pattern of vegetation itself reflects a changing pattern of soil productivity.

laboratory experiments

Perhaps less obviously, ecologists also often need to turn to laboratory systems and even mathematical models. These have played a crucial role in the development of ecology, and they are certain to continue to do so. Field experiments are almost inevitably costly and difficult to carry out. Moreover, even if time and expense were not issues, natural field systems may simply be too complex to allow us to tease apart the consequences of the many different processes that may be going on. Are the intestinal worms actually capable of having an effect on the reproduction or mortality of an individual grouse? Which of the many species of sand dune plant are, in themselves, sensitive to changing levels of soil productivity and which are relatively insensitive? *Controlled, laboratory experiments* are often the best way to provide answers to such specific questions, which may themselves be key parts of any overall explanation of the complex situation in the field.

simple laboratory systems . . .

Of course, the complexity of natural ecological communities may simply make it inappropriate for an ecologist to dive straight into them in search of understanding. We may wish to explain the structure and dynamics of a particular community of 20 animal and plant species comprising various competitors, predators, parasites, and so on (relatively speaking, a community of remarkable simplicity). But we have little hope of doing so unless we already have some basic understanding of even simpler communities of just one predator and one prey species; or two competitors; or (especially ambitious) two competitors that also share a common predator. For this, it is usually most appropriate to construct, for our own convenience, *simple laboratory systems* that can act as benchmarks or jumping-off points in our search for understanding.

. . . and mathematical models

What is more, you have only to ask anyone who has tried to rear caterpillar eggs, or take a cohort of shrub cuttings through to maturity, to discover that even the simplest ecological communities may not be easy to maintain and keep free from invasion by other species, be they pathogens, predators, or competitors. Nor is it necessarily possible to construct precisely the particular, simple, artificial community that interests you; nor to subject it to precisely the conditions or the perturbation of interest. In many cases, therefore, there is much to be gained from the analysis of *mathematical models* of ecological communities: constructed and manipulated according to the ecologist's design.

On the other hand, although a major aim of science is to simplify, and thereby make it easier to understand the complexity of the real world, ultimately it is the real world that we are interested in, and the worth of models and simple laboratory experiments must always be judged in terms of the light they throw on the working of more natural systems. They are a means to an end—never an end in themselves. Like all scientists, ecologists need to "seek simplicity, but distrust it" (Whitehead, 1953).

1.2.3 Statistics and scientific rigor

It is never a good thing to overreact. For any scientist to take offence at some popular phrase or saying is to invite accusations of a lack of a sense of humor. It is nonetheless difficult to remain calm when phrases like "There are lies, damn lies and statistics" or "You can prove anything with statistics" are used by those who know no better, in order to justify continuing to believe what they wish to believe, whatever the evidence to the contrary. There is no doubt that statistics are sometimes *mis*-used to derive dubious conclusions from sets of data that actually suggest either something quite different or perhaps nothing at all. But these are not grounds for mistrusting statistics in general —rather for insuring that all those capable of doing so are educated in at least the principles of scientific evidence and its statistical analysis, so as to protect them from those who may seek to manipulate their opinions.

In fact, not only is it not true that you can prove anything with statistics, the contrary is the case: you cannot *prove* anything with statistics—that is not what statistics are for. Statistical analysis is essential, however, for attaching a level of confidence to conclusions that we may wish to draw; and ecology, like all science, is a search not for statements that have been "proved to be true" but for conclusions in which we can be confident.

ecology: a search for conclusions in which we can be confident

Indeed, what distinguishes science from other activities—what makes science "rigorous"—is that it is based, not on statements that are simply assertions, but (1) on conclusions that are the results of investigations (as we have seen, of a wide variety of types) carried out with the express purpose of deriving those conclusions, and (2) even more important, on conclusions to which a level of confidence can be attached, measured on an agreed scale. These points are elaborated in Boxes 1.2 (Interpreting probabilities) and 1.3 (Attaching confidence to results).

Statistical analyses are carried out after data have been collected, and they help us to interpret those data. There is no really good science, however, without forethought.

ecologists must think ahead

Box 1.2   *Quantitative Aspects*

Interpreting Probabilities

P-values

The term that is most often used, at the end of a statistical test, to measure the strength of conclusions being drawn is a *P*-value, or probability level. It is important to understand what these are. Imagine we are interested in establishing whether high abundances of a pest insect in the summer are associated with high temperatures in the previous spring, and imagine that the data we have to address this question consist of summer insect abundances and mean spring temperatures for each of a number of years. We may reasonably hope that statistical analysis of our data will allow us either to conclude, with a stated degree of confidence, that there is an association, or to conclude that there are no grounds for believing there to be an association (Figure 1.3).

Null hypotheses

To carry out a statistical test we first need a *null hypothesis*, which simply means, in this case, that there is *no* association: that is, no association between insect abundance and temperature. The statistical test (stated simply) then generates a probability (a *P*-value) of getting a data set like ours if the null hypothesis is correct.

Suppose the data were like those in Figure 1.3a. The probability generated by a test of association on these data is $P = 0.5$ (equivalently 50 percent). This means that, if the null hypothesis really was correct (no association), then 50 percent of studies like ours should generate just such a data set, or one even further from the null hypothesis. So, if there was no association, there would be nothing very remarkable in this data set, and we could have no confidence in any claim that there *was* an association.

Suppose, however, that the data were like those in Figure 1.3b, where the *P*-value generated is

$P = 0.001$ (0.1 percent). This would mean that such a data set (or one even further from the null hypothesis) could be expected in only 0.1 percent of similar studies if there was really no association. In other words, either something very improbable has occurred, or there *was* an association between insect abundance and spring temperature. Thus, since by definition we do not expect highly improbable events to occur, we can have a high degree of confidence in the claim that there *was* an association between abundance and temperature.

Significance testing

Both 50 percent and 0.01 percent, though, make things easy for us. Where, between the two, do we draw the line? There is no objective answer to this, and so scientists and statisticians have established a convention in *significance testing*, which says that if *P* is less than 0.05 (5 percent), written $P < 0.05$ (e.g Figure 1.3d), then the results are described as statistically significant and confidence can be placed in the effect being examined (in our case, the association between abundance and temperature), whereas if $P > 0.05$, then there is no statistical foundation for claiming the effect exists (e.g. Figure 1.3c). A further elaboration of the convention often describes results with $P < 0.01$ as "highly significant."

"Insignificant" results?

Naturally, some effects are strong (for example, there is a powerful association between people's weight and their height) and others are weak (the association between people's weight and their risk of heart disease is real but weak, since weight is only one of many

Box 1.2 (cont'd)

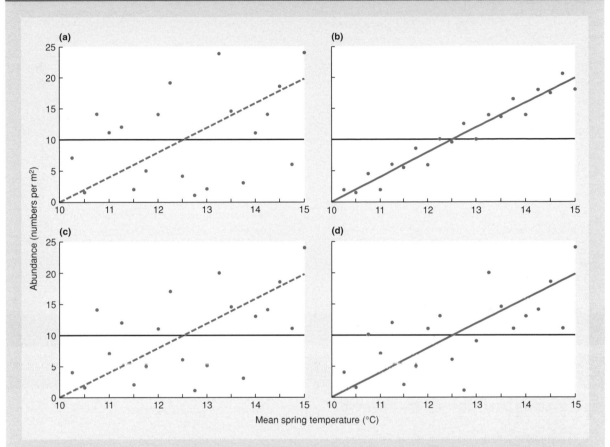

Figure 1.3
The results from four hypothetical studies of the relationship between the abundance of an insect pest in the summer and the mean temperature in the previous spring. In each case, the points are the data actually collected. Horizontal lines represent the *null hypothesis*—that there is no association between abundance and temperature, and thus the best estimate of expected insect abundance, irrespective of spring temperature, is the mean insect abundance overall. The angled line is the *line of best fit* to the data, which in each case offers some suggestion that abundance rises as temperature rises. However, whether we can be confident in concluding that abundance does rise with temperature depends, as explained in the text, on statistical tests applied to the data sets. (a) The suggestion of a relationship is weak ($P = 0.5$). There are no good grounds for concluding that the true relationship differs from that supposed by the null hypothesis and no grounds for concluding that abundance is related to temperature. (b) The relationship is strong ($P = 0.001$) and we can be confident in concluding that abundance increases with temperature. (c) The results are suggestive ($P = 0.1$) but it would not be safe to conclude from them that abundance rises with temperature. (d) The results are not vastly different from those in (c) but are powerful enough ($P = 0.04$, i.e., $P < 0.05$) for the conclusion that abundance rises with temperature to be considered safe.

important factors). More data are needed to establish support for a weak effect than for a strong one. A rather obvious but very important conclusion follows from this: a *P*-value in an ecological study greater than 0.05 (lack of statistical significance) may mean one of two things:

1 There really is no effect of ecological importance.
2 The data are simply not good enough, or there are not enough of them, to support the effect even though it exists, possibly because the effect itself is real but weak, and extensive data are therefore needed but have not been collected.

Box 1.2 (cont'd)

Quoting *P*-values

Furthermore, applying the convention strictly and dogmatically means that when $P = 0.06$ the conclusion should be "no effect has been established", whereas when $P = 0.04$ the conclusion is "there is a significant effect". Yet very little difference in the data is required to move a P-value from 0.04 to 0.06. It is therefore far better to quote exact P-values, especially when they exceed 0.05, and think of conclusions in terms of shades of grey rather than the black and white of "proven effect" and "no effect". In particular, P-values close to, but not less than, 0.05 suggest that something seems to be going on; they indicate, more than anything else, that more data need to be collected so that our confidence in the conclusions can be more clearly established.

Throughout this book, then, studies of a wide range of types are described, and their results often have P-values attached to them. Of course, as this is a textbook, the studies have been selected because their results *are* significant. Nonetheless, it is important to bear in mind that the repeated statements $P < 0.05$ and $P < 0.01$ mean that these are studies where (1) sufficient data have been collected to establish a conclusion with which we can be confident, (2) that confidence has been established by agreed means (statistical testing), and (3) confidence is being measured on an agreed and interpretable scale.

Box 1.3   *Quantitative Aspects*

Attaching Confidence to Results

Standard errors and confidence intervals

Following Box 1.2, another way in which the significance of results, and confidence in them, is assessed is through reference to standard errors. Again, simply stated, statistical tests often allow standard errors to be attached either to mean values calculated from a set of observations or to slopes of lines like those in Figure 1.3. Such mean values or slopes can, at best, only ever be estimates of the "true" mean value or true slope, because they are calculated from data that are only a sample of all the imaginable items of data that could be collected. The standard error, then, sets a band around the estimated mean (or slope, etc.) within which the true mean can be expected to lie with a given, stated probability. In particular, there is a 95 percent probability that the true mean lies within roughly two standard errors of the estimated mean; we call this the *95 percent confidence interval*.

Hence, when we have, say, two sets of observations, each with its own mean value (for instance, the number of seeds produced by plants from two sites—Figure 1.4) the standard errors allow us to assess whether the means are significantly different from one another, statistically. Roughly speaking, if each mean is more than two standard errors from the other mean, then the difference between them is statistically significant with $P < 0.05$. Thus, for the study illustrated in Figure 1.4a, it would not be safe to conclude that plants from the two sites differed in their seed production. However, for the similar study illustrated in Figure 1.4b, the means are roughly the same as they were in the first study and are roughly as far apart, but the standard errors are smaller. Hence, the difference between the means is significant ($P < 0.05$), and we can conclude with confidence that plants from the two sites differed.

Box 1.3 (cont'd)

When are standard errors small?

Note that the large standard errors in the first study, and hence the lack of statistical significance, could have been due to data that were, for whatever reason, more variable; but they may also have been due to a

sampling of fewer plants in the first study than the second. Standard errors are smaller, and statistical significance is easier to achieve, *both* (1) when data are more consistent (less variable), *and* (2) when there are more data.

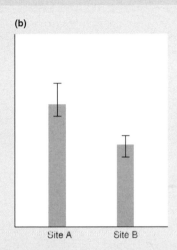

Figure 1.4

The results of two hypothetical studies in which the seed production of plants from two different sites was compared. In all cases, the heights of the bars represent the mean seed production of the sample of plants examined, and the lines crossing those means extend one standard error above and below them. (a) Although the means differ, the standard errors are relatively large and it would not be safe to conclude that seed production differed between the sites ($P = 0.4$). (b) The differences between the means are very similar to those in (a), but the standard errors are much smaller, and it can be concluded with confidence that plants from the two sites differed in their seed production ($P < 0.05$).

Ecologists, like all scientists, must know what they are doing, and why they are doing it, *while* they are doing it. This is entirely obvious at a general level: nobody expects ecologists to be going about their work in some kind of daze. But it is perhaps not so obvious that ecologists should know how they are going to analyze their data, statistically, not only after they have collected it, not only while they are collecting it, but even before they begin to collect it. Ecologists must plan, so as to be confident that they have collected the right kind of data, and a sufficient amount of data, to address the questions they hope to answer.

Ecologists typically seek to draw conclusions about groups of organisms overall: what is the birth rate of the bears in Yellowstone Park? What is the density of weeds in a wheat field? What is the rate of nitrogen uptake of tree saplings in a nursery? In doing so, we can only very rarely examine every individual in a group, or the entire

ecology relies on representative samples

sampling area, and we must therefore rely on what we hope will be *representative* samples from the group or habitat as a whole. Indeed, even if we examined a whole group (we might examine every fish in a small pond, say), we are likely to want to draw general conclusions from it: we might hope that the fish in "our" pond can tell us something about fish of that species in ponds of that type, generally. In short, ecology relies on obtaining *estimates* from representative samples. This is elaborated in Box 1.4.

Box 1.4   *Quantitative Aspects*

Sampling, Accuracy, and Precision

The discussion in Boxes 1.2 and 1.3 about when standard errors will be small or large, or when our confidence in conclusions will be strong or weak, has implications not only for the interpretation of data after they have been collected. It also carries a general message about planning the collection of data.

In undertaking a sampling program to collect data, the aim is to satisfy a number of criteria (Figure 1.5):

■ that the estimate should be accurate or unbiased: that is, neither systematically too high nor too low as a result of some flaw in the program;

Figure 1.5

The results of hypothetical programs to estimate weed density in a wheat field. (a) The three studies have equal precision (95% confidence intervals) but only the first (from a random sample) is accurate. (b) In the first study, individual samples from different parts of the field (southeast and southwest) fall into two groups (left); thus the estimate, although accurate, is not precise (right). In the second study, separate estimates for southeast and southwest are both accurate and precise—as is the estimate for the whole field obtained by combining them. (c) Following on from (b), most sampling effort is directed to the southwest, reducing the confidence interval there, but with little effect on the confidence interval for the southeast. The overall interval is therefore reduced: precision has been improved.

Box 1.4 (cont'd)

- that the estimate should have as narrow confidence limits (be as precise) as possible;
- that the time, money, and human effort invested in the program should be used as effectively as possible (because these are always limited).

Random and stratified random sampling

To understand these criteria, consider another hypothetical example. Suppose that we are interested in the density of a particular weed (say wild oat) in a wheat field. To prevent bias, it is necessary to insure that each part of the field has an equal chance of being selected for sampling. Sampling units should therefore be selected at random. We might, for example, divide the field into a measured grid, pick points on the grid at random, and count the wild oat plants within a 50 cm radius of the selected grid point. This unbiased method can be contrasted with a plan to sample only weeds from between the rows of wheat plants, giving too high an estimate, or within the rows, giving too low an estimate (Figure 1.5a).

Remember, however, that random samples are not taken as an end in themselves, but because random sampling is a means to truly representative sampling. Thus, randomly chosen sampling units may end up being concentrated, by chance, in a particular part of the field that, unknown to us, is not representative of the field as a whole. It is often preferable, therefore, to undertake *stratified random sampling* in which, in this case, the field is divided up into a number of equal-sized parts (*strata*) and a random sample taken from each. This way, the coverage of the whole field is more even, without our having introduced bias by selecting particular spots for sampling.

Separating subgroups and directing effort

Suppose now, though, that half the field is on a slope facing southeast and the other half on a slope facing southwest, and that we know that aspect (which way the slope is facing) affects weed density considerably. Random sampling (or stratified random sampling) ought still to provide an unbiased estimate of density for the field as a whole, but for a given investment in effort, the confidence interval for the estimate will be unnecessarily high. To see why, consider Figure 1.5b. The individual values from samples fall into two groups a substantial distance apart on the density scale: high from the southwest slope; low (mostly zero) from the southeast slope. The estimated mean density is close to the true mean (it is accurate), but the variation among samples leads to a very large confidence interval (it is not very precise).

If, however, we acknowledge the difference between the two slopes and treat them separately from the outset, then we obtain means for each that have much smaller confidence intervals. What is more, if we average those means and combine their confidence intervals to obtain an estimate for the field as a whole, then that interval too is much smaller than previously (Figure 1.5b).

But has our effort been directed sensibly, with equal numbers of samples from the southwest slope, where there are lots of weeds, and the southeast slope, where there are virtually none? The answer is no. Remember that narrow confidence intervals arise from a combination of a large number of data points *and* little intrinsic variability (Box 1.3). Thus, if our efforts had been directed mostly at sampling the southwest slope, the increased amount of data would have noticeably decreased the confidence interval (Figure 1.5c), whereas less sampling of the southeast slope would have made very little difference to that confidence interval because of the low intrinsic variability there. Careful direction of a sampling program can clearly increase overall precision for a given investment in effort. And generally, sampling programs should, where possible, identify biologically distinct subgroups (males and females, old and young, etc.) and treat them separately, but sample at random within subgroups.

1.3   Ecology in Practice

In the previous sections we have established in a general way how ecological under-standing can be achieved, and how that understanding can be used to help us predict, manage, and control ecological systems. However, the practice of ecology is easier said than done. To discover the real problems faced by ecologists and how they try to solve them, it is best to consider some real research programs in a little detail. While read-ing the following examples you should focus on how they illuminate our three main points—ecological phenomena occur at a variety of scales; ecological evidence comes from a variety of different sources; and ecology relies on truly scientific evidence and the application of statistics. Every other chapter in this book will contain descriptions of similar studies, but in the context of a systematic survey of the driving forces in eco-logy (Chapters 2–11) or of the application of this knowledge to solve applied problems (Chapters 12–14). For now, we content ourselves with seeking an appreciation of how four research teams have gone about their business.

1.3.1   The brown trout in New Zealand—effects on individuals, populations, communities, and ecosystems

It is rare for a study to encompass more than one or two of the four levels in the biolo-gical hierarchy (individuals, populations, communities, ecosystems). For most of the 20th century, physiological and behavioral ecologists (studying individuals), popula-tion dynamicists, and community and ecosystem ecologists have tended to follow separate paths, asking different questions in different ways. However, there can be little doubt that, ultimately, our understanding will be enhanced considerably when the links between all these levels are made clear—a point that can be illustrated by examining the impact of the introduction of an exotic fish to streams in New Zealand.

Prized for the challenge they provide to anglers, brown trout (*Salmo trutta*) have been transported from their native Europe all around the world; they were introduced to New Zealand beginning in 1867, and self-sustaining populations are now found in many streams, rivers, and lakes there. Until quite recently, few people cared about native New Zealand fish or invertebrates, so little information is available on changes in the ecology of native species after the introduction of trout. However, trout have colonized some streams but not others. We can therefore learn a lot by comparing the current ecology of streams containing trout with those occupied by non-migratory native fish in the genus *Galaxias* (Figure 1.6).

the individual level—consequences for invertebrate feeding behavior

Mayfly nymphs of various species commonly graze microscopic algae growing on the beds of New Zealand streams, but there are some striking differences in their activ-ity rhythms depending on whether they are in *Galaxias* or trout streams. In one experi-ment, nymphs collected from a trout stream and placed in small artificial laboratory channels were less active during the day than the night, whereas those collected from a *Galaxias* stream were active both day and night (Figure 1.7a). In another experiment, with another mayfly species, records were made of individuals visible in daylight on the surface of cobbles in artificial channels placed in a real stream. Three treatments were each replicated three times—no fish in the channels, trout present, and *Galaxias*

Figure 1.6
(a) A brown trout and (b) a galaxiid fish in a New Zealand stream—is the galaxiid hiding from the introduced predator? (Photographs by Angus McIntosh.)

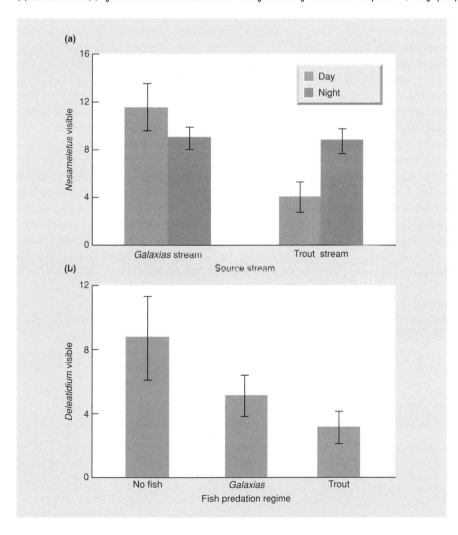

Figure 1.7
(a) Mean number (± standard error) of *Nesameletus ornatus* mayfly nymphs collected either from a trout stream or a *Galaxias* stream that were recorded by means of video as visible on the substrate surface in laboratory stream channels during the day and night (in the absence of fish). (After McIntosh & Townsend, 1994.) (b) Mean number (± standard error) of *Deleatidium* mayfly nymphs observed on the upper surfaces of cobbles during late afternoon in channels (placed in a real stream) containing no fish, trout, or *Galaxias*. (After McIntosh & Townsend, 1996.) In (a) nymphs from the trout stream show a statistically significantly greater tendency to be active at night, while those from a *Galaxias* stream do not. In (b) significantly fewer *Deleatidium* were visible during the day in the trout channels.

present. Daytime activity was significantly reduced in the presence of either fish species—but to a greater extent when trout were present (Figure 1.7b).

These differences in activity pattern reflect the fact that trout rely principally on vision to capture prey, whereas *Galaxias* rely on mechanical cues. Thus, invertebrates in a trout stream are considerably more at risk of predation during daylight hours. And these conclusions are all the more robust because they derive both from the readily controlled conditions of a laboratory experiment and from the more realistic, but more variable, circumstances of a field experiment.

the population level—brown trout and the distribution of native fish

In the Taieri River in New Zealand, 198 sites were selected in a stratified manner by choosing streams of similar dimensions at random in each of three tributaries from each of eight subcatchments of the river. Care was taken not to succumb to the temptation of choosing sites with easy access (near roads or bridges), in case this biased the results. The sites were classified as containing (1) no fish, (2) *Galaxias* only, (3) trout only, or (4) both *Galaxias* and trout. At every site a variety of physical variables was measured (stream depth, flow velocity, phosphorus concentration in the stream water, percentage of the stream bed composed of gravel, etc.). A statistical procedure called multiple discriminant functions analysis was then used to determine which physical variables, if any, distinguished one type of site from another. Means and standard errors of these key environmental variables are presented in Table 1.1.

Trout occurred almost invariably below waterfalls that were large enough to prevent their upstream migration; they tended to occur at low elevations because sites without waterfalls downstream tended to be at lower elevation. Sites containing *Galaxias* (or with no fish) were always upstream of one or several large waterfalls. The few sites that contained both trout and *Galaxias* were below waterfalls, at intermediate elevations, and in sites with cobble beds; the unstable nature of these stream beds may have promoted the coexistence (at low densities) of the two species. This descriptive study at the population level therefore takes advantage of a "natural" experiment (streams that happen to contain trout or *Galaxias*) to determine the effect of the introduction of trout. The most probable reason for the restriction of populations of *Galaxias* to sites upstream of waterfalls, which cannot be climbed by trout, is direct predation by trout on the native fish below the waterfalls (a single small trout in a laboratory aquarium has been recorded as consuming 135 *Galaxias* fry in a day).

Table 1.1
Means and, in brackets, standard errors for important discriminating variables for fish assemblage classes in 198 sites in the Taieri River. (From Townsend & Crowl, 1991.)

| Site type | Number of sites | Variables | | |
		Number of waterfalls downstream	Elevation (m above sea level)	% of the bed composed of cobbles
No fish	54	4.37 (0.64)	339 (31)	15.8 (2.3)
Brown trout only	71	0.42 (0.05)	324 (28)	18.9 (2.1)
Galaxias only	64	12.3 (2.05)	567 (29)	22.1 (2.8)
Trout + *Galaxias*	9	0.0 (0)	481 (53)	46.7 (8.5)

That an exotic predator such as trout has direct effects on *Galaxias* distribution or mayfly behavior is not surprising. However, we can ask whether these changes have community consequences that cascade through to other species. In the relatively species-poor stream communities in the south of New Zealand, the plants are mainly algae that grow on the stream bed. These are grazed by various insect larvae, which in turn are prey to predatory invertebrates and fish. As we have seen, trout have replaced *Galaxias* in many of these streams. An experiment involving artificial flow-through channels (several meters long, with mesh ends to prevent the escape of fish but to allow invertebrates to colonize naturally) placed into a real stream was undertaken to determine whether trout affect the stream food web differently from the displaced *Galaxias*. Three treatments were established (no fish, *Galaxias* present, trout present: at naturally occurring densities) in each of several randomized blocks located in a stretch of a stream with each block separated by more than 50 m. Algae and invertebrates were allowed to colonize for 12 days before introducing the fish. After a further 12 days, invertebrates and algae were sampled (Figure 1.8).

A significant effect of trout reducing invertebrate biomass was evident ($P = 0.026$), but the presence of *Galaxias* did not depress invertebrate biomass from the no-fish control. Algal biomass, perhaps not surprisingly then, achieved its highest values in the trout treatment ($P = 0.02$). It is clear that trout do have a more pronounced effect than *Galaxias* on invertebrate grazers and, thus, on algal biomass. The indirect effect of trout on algae occurs partly through a reduction in invertebrate density, but also because trout restrict the grazing behavior of the invertebrates that are present (see Figure 1.7b).

The sequence of studies above provided the impetus for a detailed energetics investigation of two neighboring tributaries of the Taieri River (with very similar physicochemical conditions), one being occupied by just trout and the other (because of a waterfall downstream) containing only *Galaxias*. No other fish were present in either stream. The hypothesis under examination was that the rate at which radiant energy was captured through photosynthesis by the algae would be greater in the trout stream because there would be fewer invertebrates and thus a lower rate of consumption of algae. Indeed, annual net "primary" production (the rate of production of plant, in this case algal, biomass) was six times greater in the trout stream than in the *Galaxias* stream (Figure 1.9a).

Moreover, the primary consumers (invertebrates that eat algae) produced new biomass in the trout stream at about 1.5 times the rate in the *Galaxias* stream, while trout themselves produced new biomass at roughly nine times the rate that *Galaxias* do (Figure 1.9b, c).

Thus, the algae, invertebrates, and fish are all "more productive" in the trout stream than in the *Galaxias* stream; but *Galaxias* consume only about 18% of available prey production each year (compared to virtually 100% consumption by trout); while the grazing invertebrates consume about 75% of primary production in the *Galaxias* stream (compared to only about 21% in the trout stream) (Figure 1.9). Thus, the initial hypothesis appears to be confirmed: it is strong control by trout of the invertebrates that releases algae to produce and accumulate biomass at a fast rate.

A further ecosystem consequence ensues. In the trout stream, the higher primary production is associated with a faster rate of uptake by algae of plant nutrients

the community—brown trout cause a cascade of effects

the ecosystem—trout and energy flow

(nitrate, ammonium, phosphate) from the flowing stream water (Simon & Townsend, in press).

This series of studies, therefore, illustrates some of the variety of ways in which ecological investigations may be pursued, and both the range of levels in the biological hierarchy that ecology spans and the way in which studies at different levels may complement one another. While it is necessary to be cautious when interpreting the results of an unreplicated study (only one trout and one *Galaxias* stream in the "ecosystem study"), the conclusion that a trophic cascade is responsible for the patterns observed at the ecosystem level can be made with some confidence because of the variety of other corroborative studies conducted at the individual, population, and community levels. Although brown trout are exotic invaders in New Zealand, and they have far-reaching effects on the ecology of native ecosystems, they are now considered a valuable part of the fauna, particularly by anglers, and generate millions of dollars for the nation. Many other invaders have dramatic negative economic impacts (Box 1.5).

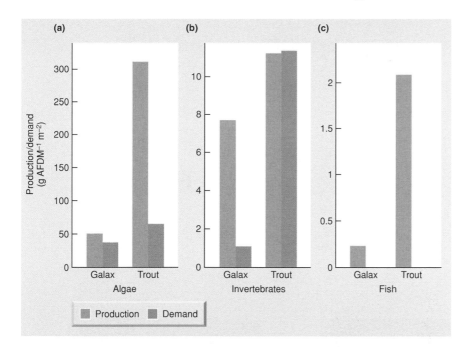

Figure 1.9
Annual estimates for "production" of biomass at one trophic level, and the "demand" for that biomass (the amount consumed) at the next trophic level for: (a) primary producers (algae), (b) invertebrates (which consume algae), and (c) fish (which consume invertebrates). Estimates are for a trout stream and a *Galaxias* stream (Galax). In the former, production at all trophic levels is higher, but because the trout consume essentially all of the annual invertebrate production (b), the invertebrates consume only 21% of primary production (a). In the *Galaxias* stream, these fish consume only 18% of invertebrate production, "allowing" the invertebrates to consume the majority (75%) of annual primary production. (After Huryn, 1998.)

Box 1.5   ***Topical ECOncerns***

Invasions and Homogenization of the Biota—
Does it Matter?

A recent analysis concluded that tens of thousands of invading exotic species in the United States cause economic losses totaling $137 billion each year (Pimentel et al., 2000). Table 1.2 breaks down the total into a variety of taxonomic groups.

Let us consider a few invaders with particularly dramatic consequences. The yellow star thistle (*Centaurea solstitalis*) now dominates more than 4 million hectares in California, resulting in the total loss of once productive grassland. Rats are estimated to destroy $19 billion of stored grains nationwide per year, as well as causing fires (by gnawing electric wires), polluting foodstuffs, spreading diseases, and preying on native species. Introduced carp reduce water quality by increasing turbidity, while 44 native fish are threatened or endangered by fish invaders. The red fire ant (*Solenopsis invicta*) kills poultry, lizards, snakes, and ground-nesting birds; in Texas alone, its estimated damage to livestock, wildlife, and public health is put at about $300 million per year, and a further $200 million is spent on control. The zebra mussel (*Dreissena polymorpha*), which arrived in Michigan's Lake St. Clair in ballast water released from ships from Europe, has reached most aquatic habitats in the eastern United States, and is expected to spread nationwide in the next 20 years. The large populations that develop threaten native mussels and other fauna, not only by reducing food and oxygen availability but by physically smothering them. The mussels also invade and clog water intake pipes, so that millions of dollars need to be spent clearing them from water filtration and hydroelectric-generating plants. Overall, pests of crop plants, including weeds, insects, and pathogens, engender the biggest economic costs. However, imported human disease organisms, particularly HIV and influenza viruses,

Box 1.5 (cont'd)

Table 1.2
Estimated annual costs (billions of dollars) associated with invaders in the United States. (After Pimentel et al., 2000.)

Type of organism	Number of invaders	Major culprit	Loss and damage	Control costs	Total costs
Plants	5,000	Crop weeds	24.4	9.7	34.1
Mammals	20	Rats and cats	37.2	NA	37.2
Birds	97	Pigeons	1.9	NA	1.9
Reptiles and amphibians	53	Brown tree snake	0.001	0.005	0.006
Fishes	138	Grass carp, etc.	1.0	NA	1.0
Arthropods	4,500	Crop pests	17.6	2.4	20.0
Mollusks	88	Asian clams	1.2	0.1	1.3
Microbes (pathogens)	>20,000	Crop pathogens	32.1	9.1	41.2

NA, not available

(a)

(b)

(c)

cost $6.5 billion to treat and result in 40,000 deaths per year. (See Pimentel et al., 2000, for further details and references.)

Globalization has been the prevalent economic ideology in recent times. Globalization of the biota, in which successful invaders are moved around the world, often driving local species extinct, can be expected to lead to a general homogenization of the world's biota. [Lövei (1997) has colorfully referred to this as "Macdonaldization" of the biosphere.] Does biotic homogenization matter? Why?

(a) Yellow star thistle (© Greg Hodson, Visuals Unlimited). (b) Red fire ants (© Visuals Unlimited/ARS). (c) Zebra mussels (© Visuals Unlimited/OMNR).

1.3.2 Successions on old fields in Minnesota—a study in time and space

"Ecological succession" is a concept that must be familiar to many who have simply taken a walk in open country—the idea that a newly created habitat, or one in which a disturbance has created an opening, will be inhabited, in turn, by a variety of species appearing and disappearing in some recognizably repeatable sequence. Widespread familiarity with the idea, however, does not mean that we understand fully the processes that drive or fine-tune successions. Yet developing such an understanding is important not just because succession is one of the fundamental forces structuring ecological communities, but also because human disturbance of natural communities has become ever more frequent and profound, and we need to know how communities may respond, and hopefully recover, from such disturbance, and how we may aid that recovery.

One particular focus for the study of succession has been the old agricultural fields of the eastern United States, abandoned as farmers moved west in search of "fresh fields and pastures new". One such site is now the Cedar Creek Natural History Area, roughly 50 km north of Minneapolis, Minnesota. The area was first settled by Europeans in 1856 and was initially subject to selective and clear-cut logging. Clearing for cultivation then began about 1885, and land was first cultivated between 1900 and 1910. Now there are agricultural fields that are still under cultivation and others that have been abandoned at various times since the mid-1920s. Cultivation led to depletion of nitrogen from what was already naturally nitrogen-poor soil.

In the first place, studies at Cedar Creek illustrate the value of "natural experiments". We particularly want to know the successional sequence of plants that occur in fields in the years following abandonment, and want to be able to account for that sequence. We could plan an artificial manipulation, under our control, in which a number of fields currently under cultivation were "forcibly" abandoned and the communities in them sampled repeatedly into the future. (We would need a *number* of fields because any single field might be atypical, whereas several would allow us to calculate the mean values for, say, "number of new species per year", and to place confidence intervals around those means.) But the results of this experiment would take decades to accumulate. The natural experiment alternative, therefore, is to use the fact that records already exist of when many of the old fields were abandoned. Thus, Figure 1.10 illustrates data from a group of 22 old fields, surveyed in 1983, having been abandoned at various times between 1927 and 1982 (i.e., between 1 and 56 years previously). Interpreted cautiously, these can be treated as 22 "snapshots" of the continuous process of succession in old fields at Cedar Creek in general, even though each field was itself only surveyed once.

A number of the shifting balances during succession are clear from the figures as statistically significant trends. Over the 56 years, the cover of "introduced" species (mostly agricultural weeds) decreased (Figure 1.10a) while the cover of species from nearby prairies increased (Figure 1.10b): the natives reclaimed their land. Of more general applicability, the cover of annual species decreased over time, while the cover of perennial species increased (Figure 1.10c, d). Annual species (those that complete a whole generation from seed to adult through to seeds again within a year) tend to be good at increasing in abundance rapidly in relatively empty habitats (the early stages of succession); whereas perennials (those that live for several or many years and may not reproduce in their early years) are slower to establish but more persistent once they do.

the use of natural experiments ...

Figure 1.10
Twenty-two fields at different stages in an old-field succession were surveyed to generate the following trends with succession age: (a) introduced species decreased, (b) prairie species increased, (c) annual species decreased, (d) perennial species increased, and (e) soil nitrogen content increased. (After Inouye et al., 1987.)

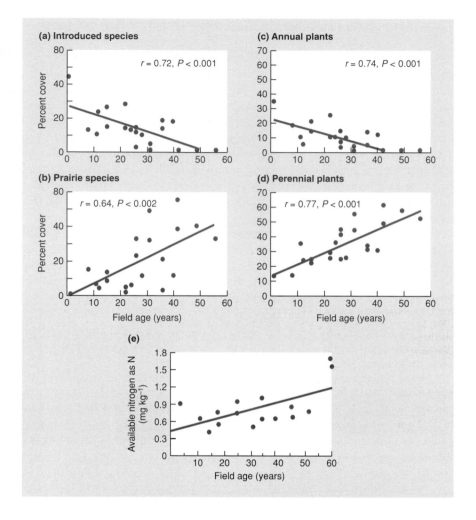

... in generating correlations

On the other hand, natural experiments like this, while frequently suggestive and stimulating (and too good an opportunity to miss) usually only generate *correlations*. They may therefore fail to establish what actually causes the observed patterns. In the present case, we can see the problem by noting, first, that field age is itself strongly correlated with nitrogen concentration in the soil—perhaps the single most important plant nutrient (Figure 1.10e). The question therefore arises: are the correlations in Figure 1.11a–d the result of an effect of field age itself? Or is the causal agent nitrogen, with which age is correlated?

artificial experiments: the search for causation

Can manipulative field experiments help support—or refute—what so far is no more than a plausible explanation based on correlation? It seems to follow from the proposed explanation (time matters) that nitrogen itself has little role to play in driving these successions, and that manipulating nitrogen should do little to alter the species sequences that these fields have followed. The results of an experiment are therefore of interest, where three of the old fields (last cultivated in 1934, 1957, and 1968) were

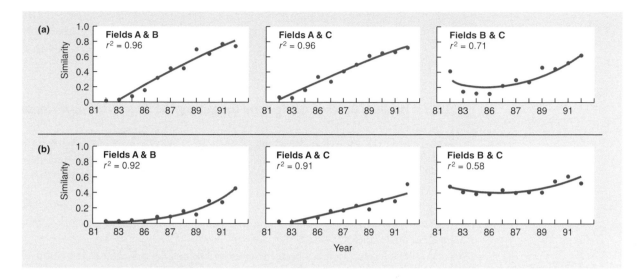

Figure 1.11
Results from an experiment in which three old fields from Figure 1.10 were given artificial nitrogen treatments starting in 1982: field A (abandoned 1968), field B (1957), and field C (1934). (a) Between 1982 and 1992, plots receiving 17 g of nitrogen m^{-2} yr^{-1} in pairs of fields became increasingly similar in composition. The similarity index measures the extent to which the species composition in pairs of fields are similar—identical compositions produce a similarity of 1. (b) Similar to (a) but with only 1 g of nitrogen m^{-2} yr^{-1} and less convergence. (After Inouye & Tilman, 1995.)

selected, and over an 11-year period starting in 1982, six replicate 4 × 4 m plots at each site were subjected to one of eight treatments: nitrogen added at rates ranging from 0 to 27.2 g m^{-2} yr^{-1} (Inouye & Tilman, 1995). Two questions in particular were being asked. Do patches receiving different supply rates of nitrogen become less similar in species composition over time? Do patches receiving similar supply rates of nitrogen become more similar in species composition over time?

The answer to the first question was clear: plots within a field were initially similar to one another but, 10 years later, plots receiving different amounts of nitrogen had diverged in species composition—and the greater the difference in nitrogen input, the greater the divergence (Inouye & Tilman, 1995). Furthermore, at the start of the experiment fields of different age tended to be very different in species composition, but 10 years later plots within them that had been subjected to similar rates of nitrogen input had become remarkably similar, despite there being, in one case, 34 years difference in their age (Figure 1.11). Thus, this experiment tends to refute the simplicity of our proposed explanation. Time itself is not the only cause of successional changes in species composition of these old fields. Differences in available nitrogen cause successions to diverge; similarities cause them to converge much more quickly than they would otherwise do. Time ("opportunity", "colonization") and nitrogen are clearly intimately intertwined and further experiments will be required to disentangle their web of cause and effect—just one of many unanswered ecological questions.

Finally, experimental manipulations over extended periods like these may also provide important insights into the possible effects of more chronic human disturbances to natural communities. The lowest rates of nitrogen addition in the experiment were similar to those experienced in many parts of the world as a result of increased

insight into the effects of nitrogen pollution

atmospheric deposition of inorganic nitrogen. Even these low levels apparently led to convergence of previously dissimilar communities over a 10-year period (Figure 1.11b). Experiments like this are crucial in helping us to predict the effects of pollutants, a point that is taken further in the next example.

1.3.3   Hubbard Brook—a long-term commitment of large-scale significance

The Cedar Creek study took advantage of a temporal pattern (a succession that takes decades to run its course) being reflected more or less accurately by a pattern in space (fields abandoned for different periods). The spatial pattern has the advantage that it could be studied within the time-bite of most research projects (3–5 years). It would have been better still to follow the ecological pattern through time but rather few researchers or institutions have risen to the challenge of designing research programs that last for decades.

A notable exception has been the work of Likens and associates at the Hubbard Brook Experimental Forest, an area of temperate deciduous forest drained by small streams in the White Mountains of New Hampshire. The researchers were pioneers with no precedents to follow. They decided to think big, and their work has shown the value of large-scale studies and long-term data records. The study commenced in 1963 and continues to the present. In the second edition of their classic book *Biogeochemistry of a Forested Ecosystem*, Likens and Bormann (1995) make poignant reference to three of their original collaborators who had died since the study began. Long-term indeed.

The research team developed an approach called "the small watershed technique" to measure the input and output of chemicals from individual catchment areas in the landscape. Because many chemical losses from terrestrial communities are channeled through streams, a comparison of the chemistry of stream water with that of incoming precipitation can reveal a lot about the differential uptake and cycling of chemical elements by the terrestrial biota. The same study can reveal much about the sources and concentrations of chemicals in the stream water, which in turn may influence the productivity of stream algae and the distribution and abundance of stream animals.

the catchment area as a unit of study

The catchment area (or watershed)—the extent of terrestrial environment drained by a particular stream—was taken as the unit of study because of the role that streams play in chemical export from the land. Six small catchments were defined and their outflows were monitored (Figure 1.12). A network of precipitation gauges recorded the incoming amounts of rain, sleet, and snow. Chemical analyses of precipitation and stream water made it possible to calculate the amounts of various chemical elements entering and leaving the system. In most cases, the output of chemicals in streamflow was greater than their input from rain, sleet, and snow (Table 1.3). The source of the excess chemicals was weathering of parent rock and soil, estimated at about 70 g m^{-2} yr^{-1}. The exception was nitrogen; less was exported in stream water than was added to the catchment in precipitation and by fixation of atmospheric nitrogen by microorganisms in the soil.

insights from a large-scale field experiment

Likens had the brilliant idea of performing a large-scale experiment in which all the trees were felled in one of Hubbard Brook's six catchments. In terms of experimental design, statistical purists might argue the study was flawed because it was unreplicated. However, the scale of the undertaking rather precluded replication. In

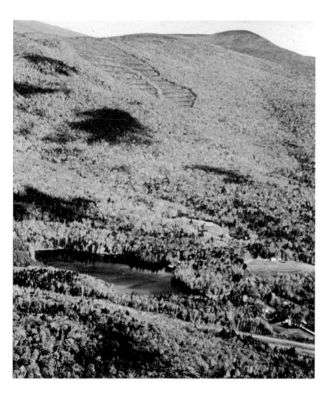

Table 1.3
Annual chemical budgets for forested watersheds at Hubbard Brook (kg ha^{-1} yr^{-1}). Inputs are for dissolved materials in precipitation or as dry fall (gases or associated with particles falling from the atmosphere). Outputs are losses in stream water as dissolved material plus particulate organic material. (After Likens et al., 1971.)

	NH_4^+	NO_3^-	SO_4^{2-}	K^+	Ca^{2+}	Mg^{2+}	Na^+
Input	2.7	16.3	38.3	1.1	2.6	0.7	1.5
Output	0.4	8.7	48.6	1.7	11.8	2.9	6.9
Net change*	+2.3	+7.6	−10.3	−0.6	−9.2	−2.2	−5.4

* Net change is positive when the catchment gains matter and negative when it loses it.

any case, it was the asking of a dramatically new question that made this study a classic rather than elegant statistical design.

Within a few months of felling all the trees in the drainage basin, the consequences were evident in the stream water. The overall export of dissolved inorganic substances from the disturbed catchment rose to 13 times the normal rate (Figure 1.13). Two phenomena were responsible. First, the enormous reduction in transpiring surfaces (leaves) led to 40% more precipitation passing through the ground water to be discharged to the streams, and this increased outflow caused greater rates of leaching of chemicals and weathering of rock and soil. Second, and more significantly, deforestation effectively broke the link between decomposition and nutrient uptake. In the

Figure 1.13
Concentrations of ions in stream water from
the experimentally deforested watershed 2
and the control (unmanipulated) watershed
6 at Hubbard Brook. The timing of
deforestation is indicated by arrows. Note
that the "nitrate" axis has a break in it.
(After Likens & Borman, 1975.)

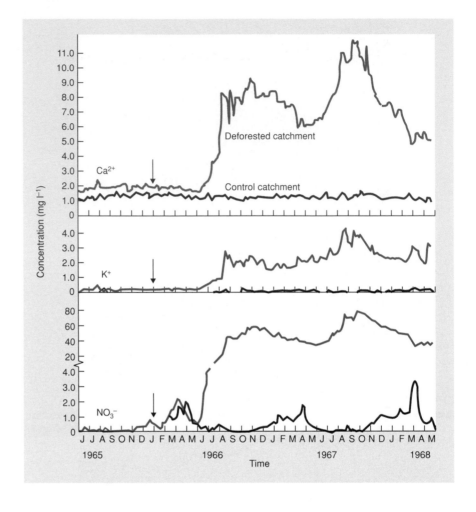

spring, when the deciduous trees would normally have started production and taken up inorganic nutrients released by decomposer activity, these were instead available to be leached in the drainage water.

statistically significant trends may require many years of data to become evident

Likens knew from the beginning that the rain and snow at Hubbard Brook were quite acid but it was some years before the widespread nature of acid rain in North America became clear. In fact, Hubbard Brook is more than 100 km from the nearest urban industrial area, yet precipitation and stream water were both markedly acid as a result of atmospheric pollution from fossil fuels. The long-term records kept so meticulously since 1963 at Hubbard Brook have proved invaluable in monitoring progress in the war against acid rain and its long-term consequences. The value of such records of stream water concentrations can be seen for hydrogen, sulfate, and nitrate, three ions associated with acid rain (which in simple terms is a mixture of dilute nitric and sulfuric acids; sulfuric acid is the dominant acid in the eastern United States). There have been statistically significant declines in average annual concentrations of H^+ and SO_4^{2-} since 1964–5, and also of NO_3^-, though the latter is subject to much greater year

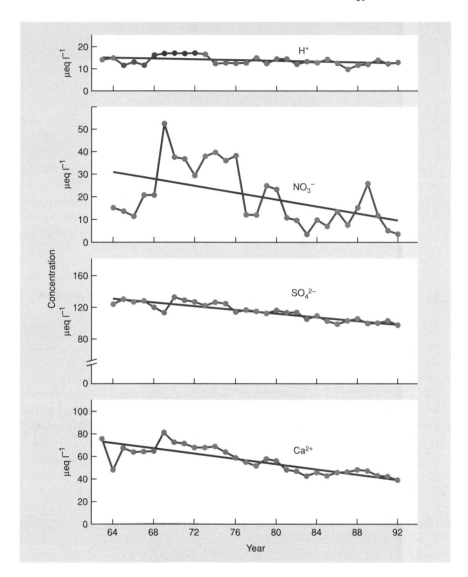

Figure 1.14
Long-term changes in concentrations (microequivalents per liter) of H^+, NO_3^-, SO_4^{2-}, and Ca^{2+} in stream water from Hubbard Brook watershed 6 from 1963–4 to 1992–3. The regression lines for all these ions have a probability of being significantly different from zero (no change) of $P < 0.05$; in other words there is a statistically significant pattern of decline in each. However, many years of data were needed before these patterns could be convincingly demonstrated. This is particularly marked for the hydrogen ion graph, where three periods of 4 years are highlighted in different colours. The first (in red) suggests an increasing trend, the second (in blue) no change, and the third (in green) a decreasing trend. (After Likens & Bormann, 1995.)

to year variation (Figure 1.14). Of note, however, is the fact that the results for shorter periods suggest quite different trends. Consider the hydrogen ion graph where three periods of 4 years are highlighted in different colours. The first suggests an increasing trend, the second no change, and the third a decreasing trend. In fact, no statistically significant, long-term trend was established until nearly two decades of data had been amassed (Likens 1989).

It is thought that acid rain began in the United States in the early 1950s (before monitoring began at Hubbard Brook). After the passage of the Clean Air Act in 1970, emissions of SO_2 and particulates were reduced and this has been clearly reflected in stream water chemistry (Figure 1.14). Additional reductions in emissions are expected as a result of the 1990 amendments to the Clean Air Act. However, critical questions

long data runs reveal the history of acid rain

UNANSWERED QUESTION: how long to recover from acid rain?

remain—will forest and aquatic ecosystems recover from the effects of acid rain and how long will it take (Likens et al., 1996)?

Using long-term data from Hubbard Brook and predictions of reductions to SO_2 emissions as a result of government legislation, Likens and Bormann (1995) estimated that by the turn of the millennium the sulfur loading in the atmosphere would still be three times higher than values recommended for the protection of sensitive forests and aquatic communities (many plants, fish, and aquatic invertebrates are intolerant of acid conditions). Moreover, declining inputs to Hubbard Brook of basic cations, such as calcium, may be causing the forests and streams to become even more sensitive to acidic inputs. Likens and Bormann (1995) hypothesize that a dramatic decline in forest growth rates during recent years may be related to a decline in calcium in the soil, a critical nutrient for tree growth. Acid rain may be responsible for the calcium deficiency. An associated reduction in bird populations in the forest may even be linked to this scenario. These unanswered questions are the subject of new phases of research at Hubbard Brook.

1.3.4 A model study: genetically modified crops—bad for biodiversity?

The progressive intensification of agriculture, and particularly increases in mechanization, field size, and pesticide use, have been linked with declines in bird, insect, and plant diversity in farm landscapes. A new wave of technological innovation involves genetic modification (GM) of crops—will the introduction of GM crops accelerate the trend in loss of biodiversity associated with farmland ecosystems? The value of mathematical models in ecology can be illustrated by examining the problem of predicting the effects of such GM crops on the population dynamics of crop weeds and the birds that eat their seeds.

genetic modification of sugar beet allows crop weeds to be more effectively controlled . . .

Sugar beet has been genetically modified to be resistant to the broad-spectrum herbicide glyphosate. This allows the herbicide to be used to effectively control weeds that normally compete with the crop without adverse affect on the sugar beet itself. Fat hen (*Chenopodium album*), a plant that occurs worldwide, is one weed that can be expected to be affected adversely by the farming of GM crops; but the seeds of fat hen are an important winter food source for farmland birds, including the skylark (*Alauda arvensis*). Watkinson et al. (2000) took advantage of the fact that the population ecologies of both fat hen and skylarks have been intensively studied to incorporate both, in a realistic manner, into a model of the impacts of GM sugar beet on farmland biodiversity.

. . . but the seeds of crop weeds are important as food for farmland birds

In the typical 5-year crop rotation, sugar beet is grown every fifth year with winter cereals grown in the other 4 years. Fat hen can only establish every fifth year, when sugar beet is grown. Between sugar beet crops, populations of fat hen persist as dormant seed banks (seeds in the soil). Survival from germination to flowering depends on weed control regimes (whether traditional or involving GM sugar beet and glyphosate), whilst seed production depends on competition for resources between weed and crop plants.

skylarks aggregate in weedy fields

Skylarks do not feed on fat hen alone, but this weed is a suitable "model" species to generate predictions of how GM sugar beet may affect the weed–bird interaction. Figure 1.15 shows how skylarks aggregate in fields in eastern England in response to seed density. They forage preferentially in weedy fields and aggregate locally in direct

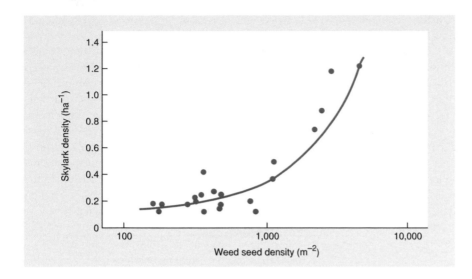

Figure 1.15
The relationship between density of skylarks (per hectare, *x*) in fields in Norfolk, England and weed seed density per square meter near the soil surface (*y*). The fitted equation is $y = 0.14 + 0.0002x$. (From Robinson & Sutherland, 1999.)

response to weed seed abundance. Hence, the impact of GM sugar beet on the birds will depend critically on the extent to which high-density patches of weeds are affected.

The steps in building a mathematical model of GM sugar beet/fat hen/skylark populations are as follows:

1 Construct a general model of the population dynamics of fat hen in a sugar beet crop (under a typical 5-year rotation).
2 Modify the model to simulate the effect of the introduction of GM sugar beet on the weed population—which suggests that densities of fat hen in GM sugar beet are likely to be less than 10% of those in conventional crops.
3 Incorporate into the model the predicted use of fields by skylarks according to weed seed density (essentially the equation shown in the legend to Figure 1.15).
4 Incorporate the possible effects of weed seed density on farming practice. The model assumes:
   - That before the introduction of GM technology, most farms have a relatively low density of weed seeds, with a few farms having very high densities (solid line in Figure 1.16).
   - Then, the probability of a farmer adopting GM crops is related to seed bank density through a parameter ρ. Positive values of ρ mean that farmers are more likely to adopt GM technology where seed densities are currently high and there is the potential to reduce yield losses to weeds. This leads to an increase in the relative abundance of low-density fields (dotted line in Figure 1.16). Negative values of ρ indicate that farmers are more likely to adopt the technology where seed densities are currently low (intensively managed farms), perhaps because a history of effective weed control is correlated with a willingness to adopt new technology. This leads to a decreased frequency of low-density fields (dashed line in Figure 1.16). Note that ρ is not an ecological parameter. Rather, it reflects a socioeconomic response to the introduction of new technology. The way that farmers will respond is not self-evident and needs to be included as a variable in the model.

Figure 1.16
Frequency distributions of mean seed densities across farms before the introduction of GM sugar beet (solid line), and in two situations where the technology has been adopted: where the technology is preferentially adopted on farms where weed density is currently high (dotted line) and where it is currently low (dashed line). (After Watkinson et al., 2000.)

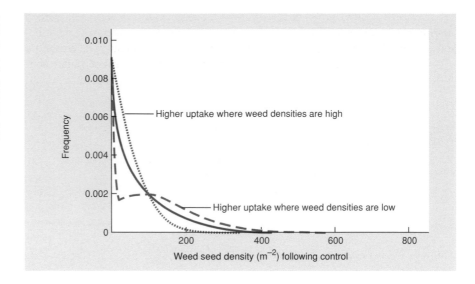

Figure 1.17
The relative density of skylarks in fields in winter (vertical axis; unity indicates field use before the introduction of GM crops) in relation to ρ (horizontal axis; positive values mean farmers are more likely to adopt GM technology where seed densities are currently high, negative values where seed densities are currently low), and to the approximate reduction in weed seed bank density due to the introduction of GM crops (Γ—third axis; realistic values are those less than 0.1). (After Watkinson et al., 2000.)

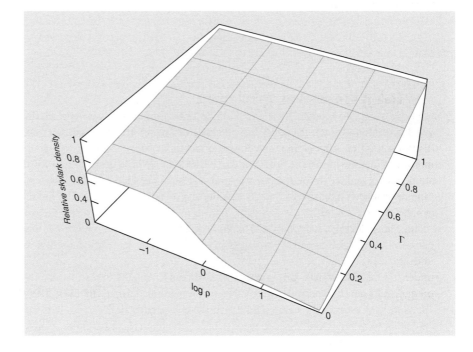

It turns out that the relationship between current weed levels and uptake of the new technology (ρ) is as important to bird populations as the direct impact of the technology on weed abundance. Note that the "parameter space" that real systems are expected to occupy in the three-dimensional graph in Figure 1.17 is the "slice" of the diagram nearest to you. In this region, small positive or negative values of ρ give quite different skylark densities.

Perhaps the most profound lesson of the model, therefore, is that predicting the impact of GM technology on biodiversity requires not only an understanding of how the ecological system operates but also of how the farming community will respond.

This example, then, has illustrated a number of important general points about mathematical models in ecology:

1 Models can be valuable for exploring scenarios and situations for which we do not have, and perhaps cannot expect to obtain, real data.
2 They can be valuable, too, for summarizing our current state of knowledge and generating predictions in which the connection between current knowledge, assumptions, and predictions is explicit and clear.
3 In order to be valuable in these ways, a model does not have to be (indeed, cannot possibly be) a full and perfect description of the real world it seeks to mimic—all models incorporate approximations.
4 Caution is therefore always necessary—all conclusions and predictions are provisional and can be no better than the knowledge and assumptions on which they are based.
5 Nonetheless, a model is inevitably applied with much more confidence once it has received support from real sets of data.

to predict impacts of GM technology we need to understand both ecology and socioeconomics

Summary

Ecology as a pure and applied science

We define ecology as the scientific study of the distribution and abundance of organisms and the interactions that determine distribution and abundance. From its origins in prehistory as an "applied science" of food gathering and enemy avoidance, the twin threads of pure and applied ecology have developed side by side, each depending on the other. This book is about how ecological understanding is achieved, what we do and do not understand, and how that understanding can help us predict, manage, and control.

Questions of scale

Ecology deals with four levels of ecological organization—individual organisms, populations (individuals of the same species), communities (a greater or lesser number of populations), and ecosystems (the community together with its physical environment). Ecology can be done at a variety of spatial scales, from the "community" within an individual cell to that of the whole biosphere. Ecologists also work on a variety of timescales. Ecological succession, for example, may be studied during the decomposition of animal dung (weeks), or during the period of climate change since the last ice age (millennia). The normal period of a research program (3–5 years) may often miss important patterns that occur over long timescales.

The diversity of ecological evidence

Many ecological studies involve careful observation and monitoring, in the natural environment, of the changing abundance of one or more species over time, or over space, or both. Establishing the cause(s) of any patterns observed often requires manipulative field experiments. For complex ecological systems (and most of them are) it will often be appropriate to construct simple laboratory systems that can act as jumping-off points in our search for understanding. Mathematical models of ecological communities also have an important role to play in unraveling ecological complexity. However, the worth of models and simple laboratory experiments must always be judged in terms of the light they throw on the working of natural systems.

Statistics and scientific rigor

What makes the science of ecology rigorous is that it is based not on statements that are simply assertions, but on conclusions that are the results of carefully planned investigations with well thought-out sampling regimes, and on conclusions, moreover, to which a level of statistical confidence can be attached. The term that is most often used, at the end of a statistical test, to measure the strength of conclusions being drawn is a "P-value" or probability level. The statements "$P < 0.05$" (significant) or "$P < 0.01$" (highly significant) mean that these are studies where sufficient data have been collected to establish a conclusion in which we can be confident.

Ecology in practice

Studies of the impacts of brown trout introduced to New Zealand last century have spanned all four ecological levels (individuals, populations, communities, ecosystems). Trout have replaced populations of native galaxiid fish below waterfalls. Laboratory and field experiments have established that grazing invertebrates in trout streams show an individual response, spending more time hiding and less time grazing. Trout cause a cascading community effect because the grazers impact less on the algae. Finally, a descriptive study revealed an ecosystem consequence—primary productivity by algae is higher in a trout than a galaxiid stream.

In the Cedar Creek Natural History Area are agricultural fields that are still under cultivation and others that have been abandoned at various times since the mid-1920s. This natural experiment was exploited to provide a description of the species sequence associated with succession on such abandoned fields. However, the fields differed not only in age but also in soil nitrogen. A set of field experiments, where soil nitrogen was augmented in a systematic way in fields of different age, showed that time and nitrogen interacted to cause the observed successional sequences.

The Hubbard Brook Experimental Forest study has been running since 1963. A large-scale experiment, involving the felling of all the trees in a single catchment area, resulted in a dramatic increase in chemical concentrations (particularly nitrate) in stream water. The loss of nitrate from the land and its increase in water can be expected to have consequences for the communities on both sides of the land–water interface. Monitoring of chemical concentrations for more than three decades in undisturbed catchments has revealed how acid rain has been diminishing as a result of the Clean Air Act. However, neither the forest nor the stream are immune from continuing effects of the pollution that caused acid rain.

Sugar beet has been genetically modified to be resistant to the broad-spectrum herbicide glyphosate. This allows the herbicide to be used to effectively control weeds that normally compete with the crop without adverse affect on the sugar beet itself. Genetic modification of crops promises benefits in terms of increased production. However, we need to beware unwanted consequences of such technological advances. The value of mathematical models in ecology is illustrated by examining the problem of predicting the effects of such GM crops on the population dynamics of crop weeds and the birds that eat their seeds. It turns out that predicting the impact of GM technology on biodiversity requires not only an understanding of how the ecological system operates, but also of the socioeconomic question of how the farming community will respond to the availability of the new technology.

Review Questions

1 ▲ Discuss the different ways that ecological evidence can be gained. How would you go about trying to answer one of ecology's unanswered questions, namely "Why are there more species in the tropics than at the poles?"

2 ▲ The variety of microorganisms that live on your teeth have an ecology like any other community. What do you think might be the similarities in the forces determining species richness (the number of species present) in your oral community as opposed to a community of seaweeds living on boulders along the shoreline?

3 Why do some temporal patterns in ecology need long runs of data to detect them, while other patterns need only short runs of data?

4 Discuss the pros and cons of descriptive studies as opposed to laboratory studies of the same ecological phenomenon.

5 What is a "natural field experiment"? Why are ecologists keen to take advantage of them?

6 Search the library for a variety of definitions of ecology: which do you think is most appropriate and why?

7 ▲ In a study of stream ecology, you need to choose 20 sites to test the hypothesis that brown trout have higher densities where the stream bed consists of cobbles. How might your results be biased if you chose all your sites to be easy to access because they are near roads or bridges?

8 How might the results of the Cedar Creek study of old-field succession have been different if a single field had been monitored for 50 years, rather than simultaneously comparing fields abandoned at different times in the past?

9 ▲ When all the trees were felled in a Hubbard Brook catchment area, there were dramatic differences in the chemistry of the stream water draining the catchment. How do you think stream chemistry would change in subsequent years as plants begin to grow again in the catchment area?

10 What are the main factors affecting the confidence we can have in predictions of a mathematical model?

▲ = Challenge Question

Chapter 2

The Ecology of Evolution

As the great Russian–American biologist Dobzhansky said, "Nothing in biology makes sense, except in the light of evolution". But equally, very little in evolution makes sense except in the light of ecology: ecology provides the stage directions through which the "evolutionary play" is performed. Ecologists and evolutionary biologists need a thorough understanding of each other's disciplines to make sense of key patterns and processes.

Chapter Contents

2.1 Introduction

2.2 Evolution by natural selection

2.3 Evolution within species

2.4 The ecology of speciation

2.5 The effects of climatic change on the evolution and distribution of species

2.6 The effects of continental drift on the ecology of evolution

2.7 Interpreting the results of evolution: convergent and parallel evolution

Key Concepts

In this chapter you will

- appreciate that Darwin and Wallace, who were responsible for the theory of evolution by natural selection, were both, essentially, ecologists

- understand that the populations of a species vary in their characteristics from place to place on both geographic and more local scales, and that some of the variation is heritable

- realize that natural selection can act very quickly on heritable variation—we can study it in action and control it in experiments

- understand that reciprocal transplanting of individuals of a species into each other's habitats can show a finely specialized fit between organisms and their environments

- appreciate that the origin of species requires the reproductive isolation of populations as well as natural selection forcing them to diverge

- realize that natural selection fits organisms to their past—it does not anticipate the future

- realize that the evolutionary history of species constrains what future selection can achieve

- understand that natural selection may produce similar forms from widely different ancestral lines (convergent evolution) or the same range of forms in populations that have become separated (parallel evolution)

2.1   Introduction

The Earth is inhabited by a multiplicity of types of organism. They are distributed neither randomly nor as a homogeneous mixture over the surface of the globe. Any sampled area, even on the scale of a whole continent, contains only a subset of the variety of species present on Earth. One of the greatest of all ecological generalizations is that all species are so specialized that they are always absent from almost everywhere. A great part of the science of ecology tries to explain why there are so many types of organism and why their distributions are so restricted. A proper answer to these ecological questions depends fundamentally on an understanding of the processes of evolution that have led to present-day diversity and distribution.

all species are so specialized that they are always absent from almost everywhere

Until relatively recently in the history of biology, the emphasis on diversity was to use it (e.g., for medicine, food, and fiber), to exhibit it in zoological and botanic gardens, and to catalogue it in museums (Box 2.1). Without an understanding of how this diversity developed, such catalogues are more like stamp collecting than science. The enduring contribution of Charles Darwin and Alfred Russell Wallace was to provide ecologists with the scientific foundations to comprehend patterns in diversity and distribution over the face of the Earth.

2.2   Evolution by Natural Selection

Darwin and Wallace (Figure 2.1) were both ecologists (although their seminal work was performed before the term was coined) who were exposed to the diversity of nature in the raw. Darwin sailed around the world as naturalist on the 5 year expedition of HMS *Beagle* (1831–6) recording and collecting in the enormous variety of environments that he explored on the way. He gradually developed the view that the natural diversity of nature was the result of a process of evolution in which *natural selection* favored some variants within species through a "struggle for existence". He developed this theme over the next 20 years through detailed study and an enormous correspondence with his friends as he prepared a major work for publication with all the evidence carefully marshaled. But he was in no hurry to publish.

Darwin and Wallace were both ecologists

In 1858, Wallace wrote to Darwin spelling out in all its essentials the same theory of evolution. Wallace was a passionate amateur naturalist. He had read Darwin's journal of the voyage of the *Beagle* and after a visit to the Jardin des Plantes in Paris and the insect room at the British Museum he wrote in 1847, "I should like to take some one family to study thoroughly, principally with a view to the theory of the origin of species." From 1847 to 1852, with his friend H.W. Bates, he explored and collected in the river basins of the Amazon and Rio Negro, and from 1854 to 1862 he made an extensive expedition in the Malay Archipelago. He recalled lying on his bed in 1858 "in the hot fit of intermittent fever, when the idea [of natural selection] suddenly came to me. I thought it all out before the fit was over, and . . . I believe I finished the first draft the next day."

Today, competition for fame and financial support would commonly lead to fierce conflict about priority—who had the idea first. Instead, in an outstanding example of selflessness in science (Darwin wrote to his friend Hooker, "It is miserable in me to care at all about priority"), sketches of Darwin's and Wallace's ideas were presented together

Box 2.1   *Historical Landmarks*

A Brief Catalogue of the Study of Diversity

An awareness of the diversity of living organisms, and of what lives where, is part of the knowledge that the human species accumulates and hands down through the generations. Hunter–gatherer peoples needed (and still need) detailed knowledge of the natural history of their environments to obtain food successfully and at the same time escape the hazards of being poisoned or eaten. The Arawaks of the South American equatorial forest know where to find and how to catch the species of large animals around them and also the names of trees and how they can be used.

The Chinese emperor Shen Nung had compiled what was perhaps the first written "herbal" of useful plants before 2000 BC, and by the first century AD Dioscorides had described 500 species of medicinal plants and illustrated many of them.

Collections of living specimens in zoos and gardens also have a long history—certainly back to Greece in the seventh century BC. The urge to collect from the diversity of nature developed in the West in the 17th century when some individuals made their living by finding interesting specimens for other people's collections. John Tradescant the father (died 1638) and John Tradescant the son (1608–1662) spent most of their lives collecting plants and importing live specimens for the gardens of royalty and the nobility. The father was the first botanist to visit Russia (1618), bringing back many living plants; his son made three visits (1637, 1642, and 1654) to the New World to collect specimens in the American colonies.

Wealthy individuals built up vast collections into personal museums and traveled or sent travelers in search of novelties from new lands as they were discovered and colonized. Naturalists and artists (often the same people) were sent to accompany the major voyages of exploration to report and take home, dead or alive, collections of the diversity of organisms and artefacts that they found. The study of taxonomy and systematics developed and flourished—taxonomy gave names to the various types of organism and systematics provided systems for classifying and pigeon-holing them.

When big national museums were established (the British Museum in 1759 and the Smithsonian in Washington in 1846), they were largely compiled from the gifts of personal collections. Like zoos and gardens, the museums' main role was to make a public display of the diversity of nature, especially the new and curious and rare.

There was no need to explain the diversity—the biblical theory of the 7-day creation of the world sufficed. However, the idea that the diversity of nature had "evolved" over time by progressive divergence from pre-existing stocks was beginning to be discussed in the early 19th century. In 1844 an anonymous publication, *The Vestiges of Creation*, put the cat among the pigeons with a popular account of the idea that animal species had descended from other species.

at a meeting of the Linnean Society in London. Darwin's *On the Origin of Species* was then hastily prepared and published in 1859 as an "abstract" of what was intended eventually to become his "big book". [In fact, most of what Darwin called his big book, with all its detail, footnotes, and references, was not published until 1975 (Stauffer, 1975).] *On the Origin of Species* may be considered the first major textbook of ecology, and aspiring ecologists would do well to read at least the third chapter.

influence of Malthus's essay on Darwin and Wallace

Both Darwin and Wallace had read *An Essay on the Principle of Population*, published by Malthus in 1798. Wallace commented, "The most interesting coincidence in

(b)

Figure 2.1
(a) Charles Darwin (lithograph by T.H. Maguire, 1849; courtesy of The Wellcome Library, London). (b) Alfred Russell Wallace, 1862 (courtesy of the Natural History Museum, London).

(a)

the matter, I think, is that I, as well as Darwin, was led to the theory itself through Malthus." Malthus's essay was concerned with the human population, which, if its intrinsic rate of increase remained unchecked, would, he calculated, be capable of doubling every 25 years and overrunning the planet. Malthus realized that limited resources slowed the growth of populations and placed absolute limits on their size, and that disease, wars, and other disasters also checked population growth. As experienced field naturalists, Darwin and Wallace realized that the Malthusian argument applied with equal force to the whole of the plant and animal kingdoms.

The living world is dominated by reproduction, overcrowding, and death. Darwin and Wallace were almost obsessed by this great truth. They appreciated that all organisms possess a potential to multiply that is impossible to realize. Darwin noted the great fecundity of some species—a single individual of the sea slug *Doris* may produce 600,000 eggs; the parasitic roundworm *Ascaris* may produce 64 million—and as an example of the absurd consequences of unimpeded population growth, he used a population of fish, each laying 2,000 eggs: in eight generations this "would cover like a sheet the whole globe, land and water". But he realized that every species "must suffer destruction during some period of its life, and during some season or occasional year, otherwise, on the principle of geometrical increase, its numbers would quickly become so inordinately great that no country could support the product" (Darwin, 1859, *On the Origin of Species*); and he observed that this was so. In one of the earliest examples of population ecology, Darwin counted all the seedlings that emerged from a plot of

the forces of reproduction, crowding, and death

cultivated ground 3 feet long and 2 feet wide: "Out of 357 no less than 295 were destroyed, chiefly by slugs and insects." Both authors, then, emphasized that most individuals die before they can reproduce and contribute nothing to future generations. Both, though, tended to ignore the important fact that those individuals that do survive in a population may leave different numbers of descendants.

The theory of evolution by natural selection, then, rests on a series of established truths:

1 Individuals that form a population of a species are not identical.

2 Some of the variation between individuals is heritable—that is, it has a genetic basis and is therefore capable of being passed down to descendants.

3 All populations could grow at a rate that would overwhelm the environment; but, in fact, most individuals die before reproduction and most (usually all) reproduce at less than their maximal rate. Hence, each generation, the individuals in a population are only a subset of those that "might" have arrived there from the previous generation.

4 Different ancestors leave different numbers of descendants (descendants, *not* just offspring): they do not all contribute equally to subsequent generations. Hence, those that contribute most have the greatest influence on the heritable characteristics of subsequent generations.

Evolution is the change, over time, in the heritable characteristics of a population or species. Given the above four truths, the heritable features that define a population will inevitably change. Evolution is inevitable.

But which individuals make the disproportionately large contributions to subsequent generations and hence determine the direction that evolution takes? The answer is: those that were best able to survive the risks and hazards of the environments in which they were born and grew; and those who, having survived, were left by their environments most capable of successful reproduction. Thus, interactions between organisms and their environments—the stuff of ecology—lie at the heart of the process of evolution by natural selection.

The philosopher Herbert Spencer described the process as "the survival of the fittest", and the phrase has entered everyday language—which is regrettable. First, we now know that survival is only part of the story: differential reproduction is often equally important. But more worryingly, even if we limit ourselves to survival the phrase gets us nowhere. Who are the fittest?—those that survive. Who survives?—those that are fittest. Nonetheless, the term *fitness* is commonly used to describe the success of individuals in the process of natural selection. An individual will survive better, reproduce more, and leave more descendants—it will be fitter—in some environments than in others. In a given environment, some individuals will survive better, reproduce more, and leave more descendants—they will be fitter—than other individuals.

Darwin had been greatly influenced by the achievements of plant and animal breeders—for example, the extraordinary variety of pigeons, dogs, and farm animals that had been deliberately bred by selecting individual parents with exaggerated traits. He and Wallace saw nature doing the same thing—"selecting" those individuals that survived from their excessively multiplying populations—hence the phrase "natural selection". But even this phrase can give the wrong impression. There is a great difference between human and natural selection. Human selection has an aim for the future—to breed a cereal with a higher yield, a more attractive pet dog, a better hunter, or a cow that will yield more milk. But nature has no aim. Evolution happens because some individuals have survived the death and destruction of the past and reproduced more successfully in the past, not because they were somehow chosen or selected by "Mother Nature" as improvements for the future.

Hence, past environments may be said to have selected particular characteristics of individuals that we see in present-day populations. Those characteristics are "suited" to present-day environments only because environments tend to remain the same, or at least change only very slowly. We shall see later in this chapter that when environments do change more rapidly, often under human influence, organisms can find themselves, for a time, left "high and dry" by the experiences of their ancestors.

Darwin and Wallace placed slightly different emphases on the forces that drive evolution. Wallace emphasized the killing forces of physical conditions such as frost, drought, and predators. Darwin laid more emphasis on competition for limited resources and the lethal effects of crowding that result from overpopulation. We pick up these powerful ecological forces in Chapters 3 (Physical Conditions and the Availability of Resources), 5 (Birth, Death, and Movement), 6 (Interspecific Competition), and 8 (Predation, Grazing, and Disease).

2.3   Evolution within Species

The natural world is not composed of a continuum of types of organism each grading into the next; we recognize boundaries between one sort of organism and another. In one of the great achievements of biological science, Linnaeus in 1789 devised an orderly system for naming the different sorts. Part of his genius was to recognize that there were features of both plants and animals that were not easily modified by the organisms' immediate environment, and that these "conservative" characteristics were especially useful for classifying organisms. In flowering plants, the form of the flowers was particularly stable, whereas the size of leaves and stems was much more readily affected by heat and cold, watering and drought, and the giving and withholding of fertilizers. Nevertheless, within what we recognize as species, there is often considerable variation, and some of this is heritable. It is on such intraspecific variation, after all, that plant and animal breeders work. In nature, some of this intraspecific variation is clearly correlated with variations in the environment and represents local specialization.

to understand the evolution of species we need to understand evolution within species

Darwin called his book *On the Origin of Species by Means of Natural Selection*, but evolution by natural selection does far more than create new species. Natural selection and evolution occur *within* species, and we now know that we can study them in action and within our own lifetime. Moreover, we need to study the way that evolution occurs within species if we are to understand the origin of new species.

2.3.1   Geographical variation within species

Since the environments experienced by a species in different parts of its range are themselves different (to at least some extent), we might expect natural selection to

the characteristics of a species may vary over its geographical range have favored different variants of the species at different sites. But evolution forces the characteristics of populations to diverge from each other (1) only if there is sufficient heritable variation on which selection can act, and (2) provided that the forces of selection favoring divergence are strong enough to counteract the mixing and hybridization of individuals from different sites. Two populations will not diverge completely if their members (or, in the case of plants, their pollen) are continually migrating between them, mating and mixing their genes.

The sapphire rockcress, *Arabis fecunda*, is a rare perennial herb restricted to calcareous soil outcrops in western Montana—so rare, in fact, that there are just 19 existing populations separated into two groups ("high elevation" and "low elevation") by a distance of around 100 km. Whether there is local adaptation here is of practical importance: four of the low elevation populations are under threat from spreading urban areas and may require reintroduction from elsewhere if they are to be sustained. Reintroduction may fail if local adaptation is too marked. Observing plants in their own habitats and checking for differences between them would not tell us if there was local adaptation in the evolutionary sense. Differences may simply be the result of immediate responses to contrasting environments made by plants that are essentially the same. Hence, high and low elevation plants were grown together in a "common garden" (Figure 2.2), eliminating any influence of contrasting immediate environments (McKay et al., 2001). The low elevation sites were more prone to drought: both the air and the soil were warmer and drier; and the low elevation plants in the common garden were indeed significantly more drought tolerant. For example, they had significantly better "water use efficiency" (their rate of water loss through the leaves was low compared to the rate at which carbon dioxide was taken in) as well as being much taller and "broader" (Figure 2.3).

variation over very short distances

Differentiation over a much smaller spatial scale was demonstrated at a site called Abraham's Bosom on the coast of North Wales. Here there was an intimate mosaic of very different habitats at the margin between maritime cliffs and grazed pasture, and a common species, creeping bent grass (*Agrostis stolonifera*) was present in many of the habitats. Figure 2.4 shows a map of the site and one of the transects from which plants were sampled; it also shows the results when plants from the sampling points along this transect were grown in a common garden. Lengths of shoot taken from this species readily form roots, so that a number of independent rooted plants can be cloned from a single plant taken from the field. Hence, in a careful statistical design, each of four plants taken from each sampling point was represented by five rooted clonal replicates of itself. The plants spread by sending out shoots along the ground surface (stolons), and the growth of the plants was compared by measuring the lengths of these. In the field, cliff plants formed only short stolons, whereas those of the pasture plants were long. In the experimental garden, these differences were maintained, even though the sampling points were typically only around 30 m apart—certainly within the range of pollen dispersal between plants. Indeed, the gradually changing environment along the transect was matched by a gradually changing stolon length, presumably with a genetic basis, since it was apparent in the common garden. Even here, the forces of selection seem to outweigh the mixing forces of hybridization.

On the other hand, it would be quite wrong to imagine that local selection always overrides hybridization—that all species exhibit geographically distinct variants with a genetic basis. For example, in a study of *Chamaecrista fasciculata*, an annual legume

(a) Common garden experiments

(b) Reciprocal transplant experiments

Figure 2.2
Common garden experiments (a) and reciprocal transplant experiments (b).

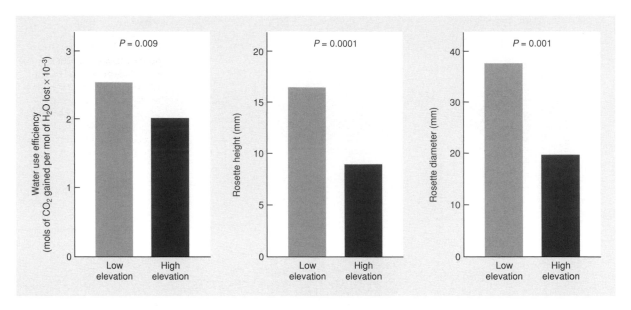

Figure 2.3
When plants of the rare sapphire rockcress from low elevation (drought prone) and high elevation sites were grown together in a common garden, there was local adaptation; those from the low elevation site had significantly better water use efficiency as well as having both taller and broader rosettes. (From McKay et al., 2001.)

Figure 2.4
(a) Map of Abraham's Bosom, the site chosen for a study of evolution over very short distances. The green area is grazed pasture; the pale brown area is cliffs falling to the sea. The numbers indicate sites from which the grass *Agrostis stolonifera* was sampled. Note that the whole area is only 200 m long. (b) A vertical transect across the study area showing a gradual change from pasture to cliff conditions. (c) The mean length of stolons produced in the experimental garden from samples taken from the transect. (From Aston & Bradshaw, 1966.)

from disturbed habitats in eastern North America, plants were grown in a common garden that were derived from the "home" site or were transplanted from distances of 0.1, 1, 10, 100, 1,000, and 2,000 km (Galloway & Fenster, 2000). The study was replicated three times—in Kansas, Maryland, and northern Illinois. Five characteristics

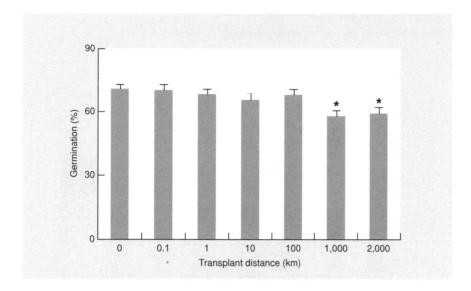

Figure 2.5
Percentage germination of local and transplanted *Chamaecrista fasciculata* populations to test for local adaptation along a transect in Kansas. Data for 1995 and 1996 have been combined because they do not differ significantly. Populations that differ from the home population at $P < 0.05$ are indicated by an asterisk. Local adaptation occurs at only the largest spatial scales. (From Galloway & Fenster, 2000.)

were measured: germination, survival, vegetative biomass, fruit production, and the number of fruit produced per seed planted; but for all characters in all replicates there was little or no evidence for local adaptation except at the very furthest spatial scales (Figure 2.5). There is "local adaptation"—but it's clearly not *that* local.

We can also test whether organisms have evolved to become specialized to life in their local environment in *reciprocal transplant* experiments (see Figure 2.2), comparing their performance when they are grown "at home" (i.e., in their original habitat) with their performance "away" (i.e., in the habitat of others).

It can be difficult to detect the local specialization of animals by transplanting them into each other's habitat: if they do not like it, most species will run away. But invertebrates like corals and sea anemones are sedentary, and some can be lifted from one place and established in another. The sea anemone *Actinia tenebrosa* is found in pools on headlands around the coast of New South Wales, Australia. Ayre (1985) chose three colonies on headlands within 4 km of each other on which the anemone was abundant. Within each colony, he selected three transplant sites (each 3–5 m long) and at each he set aside three 1 m wide strips—two to receive anemones from the away sites and one to receive "transplanted" individuals from the home site itself. Ayre cleared the experimental sites of all the anemones present and transplanted anemones into them. The anemone multiplies clonally by producing broods of asexual juveniles. The number of juveniles brooded per adult was used as a measure of the performance of the anemones in the various pools (home and away).

The proportion of adults that were found brooding 11 months later is shown in Table 2.1. Anemones originally sampled from Green Island were rather successful in brooding young after being transplanted both home and away and did not show any specialization to their home environment. However, in all the other transplant experiments a greater proportion of anemones brooded young at home than at away sites: strong evidence of evolved local specialization. In later experiments Ayre (1995) lifted anemones from a variety of sites as before, but he then kept them for a period to

reciprocal transplants test the match between organisms and their environment—sea anemones transplanted into each other's habitats

Table 2.1
A reciprocal transplant experiment of the sea anemone *Actinia tenebrosa*. (From Ayre, 1985.)

		Transplanted to sites at:		
Site of origin		Green Island	Salmon Point	Strickland Bay
Green Island	a	**0.42**	0.68	0.78
	b	**0.80**	0.63	0.75
	c	**0.67**	0.62	0.61
Salmon Point	a	0.11	**0.42**	0.13
	b	0.18	**0.43**	0.28
	c	0.00	**0.50**	0.40
Strickland Bay	a	0.11	0.06	**0.33**
	b	0.00	0.06	**0.27**
	c	0.04	0.20	**0.27**

a, b, and c are the three replicate sites in each colony. In each case the proportion of adults that were found brooding young is shown. Transplants back to the home sites are shown in bold print.

acclimate at a common site before transplanting them in a reciprocal experiment. This more severe test convincingly confirmed the results in Table 2.1.

a reciprocal transplant experiment involving a plant

Another reciprocal transplant experiment was carried out with white clover (*Trifolium repens*), which forms clones in grazed pastures. Individual clones differ in features such as the pattern of white markings on the leaves, ability to release hydrogen cyanide when damaged or bitten, and susceptibility to various diseases. To determine whether the characteristics of individual clones matched local features of their environment, Turkington and Harper (1979) removed plants from marked positions in the field and multiplied them into clones in the common environment of a greenhouse. They then transplanted samples from each clone into the place in the sward of vegetation from which it had originally been taken, and also to places from where all the others had been taken. The plants were allowed to grow for a year before they were removed, dried, and weighed. The mean weight of clover plants transplanted back into their home sites was 0.89 g, but at away sites it was only 0.52 g, a statistically highly significant difference.

The clover plants studied were not random samples but had been chosen from patches dominated by four different species of grass. Hence, in a second experiment, clonal samples of the different clovers were planted into dense experimental plots of the four grasses. Again the clovers were removed after 12 months, dried, and weighed; the results are shown in Figure 2.6. The mean yield of clovers grown with their original neighbor grass was 59.4 g; the mean yield with "alien" grasses was 31.9 g, again a highly significant difference. Thus, both clover experiments provide strong direct evidence that clover clones in the pasture had evolved to become specialized, such that they tended to perform best (make most growth) in their local environment and with their local neighbors.

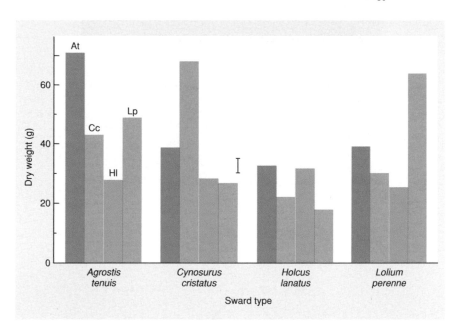

Figure 2.6
Plants of white clover (*Trifolium repens*) were sampled from a field of permanent grassland from local patches dominated by four different species of grass. The clover plants were multiplied into clones and transplanted (in all possible combinations) into plots that had been sown with seed of the four grass species. The histograms show the average weights of the transplanted clones after 12 months' growth. Clover types were sampled from patches dominated by *Agrostis tenuis* (At), *Cynosurus cristatus* (Cc), *Holcus lanatus* (Hl), and *Lolium perenne* (Lp). The vertical bar indicates the difference between the height of any pair of columns that is statistically significant at $P < 0.05$. (From Turkington & Harper, 1979.)

natural selection by predation—
a controlled field experiment in
fish evolution

In most of the examples so far, geographic variants of species have been identified, but the selective forces favoring them have not. This is not true of the next example. The guppy (*Poecilia reticulata*), a small freshwater fish from northeastern South America, has been the material for a classic series of evolutionary experiments. In Trinidad, many rivers flow down the northern range of mountains and are subdivided by waterfalls that isolate fish populations above the falls from those below. Guppies are present in almost all these water bodies, and in the lower waters they meet various species of predatory fish that are absent higher up the rivers. The populations of guppies in Trinidad differ from each other in almost every feature that biologists have examined. Forty-seven of these traits tend to vary in step with each other (they *covary*) and with the intensity of the risk from predators. This correlation suggests that the guppy populations have been subject to natural selection from the predators. But the fact that two phenomena are correlated does not prove that one causes the other. Only controlled experiments can establish cause and effect.

Where guppies have been free or relatively free from predators, the males are brightly decorated with different numbers and sizes of colored spots (Figure 2.7). Females are dull and dowdy and (at least, to us) inconspicuous. Whenever we study natural selection in action, it becomes clear that compromises are involved. For every selective force that favors change, there is a counteracting force that resists the change. Color in male guppies is a good example. Female guppies prefer to mate with the most gaudily decorated males—but these are more readily captured by predators because they are easier to see.

This sets the stage for some revealing experiments on the ecology of evolution. Guppy populations were established in ponds in a greenhouse and exposed to different intensities of predation. The number of colored spots per guppy fell sharply and rapidly when the population suffered heavy predation (Figure 2.8a). In a field

Figure 2.7
Male and female guppies (*Poecilia reticulata*) showing two flamboyant males, courting a typical dull-colored female. (Courtesy of Anne Magurran.)

Figure 2.8
(a) An experiment showing changes in populations of guppy *Poecilia reticulata*, exposed to predators in experimental ponds. The graph shows changes in the number of colored spots per fish in ponds with different populations of predatory fish. The initial population was deliberately collected from a variety of sites so as to display high variability and was introduced to the ponds at time 0. At time S weak predators (*Rivulus hartii*) were introduced to ponds R; a high intensity of predation by the dangerous predator *Crenicichila alta* was introduced into ponds C; while ponds K continued to contain no predators (the vertical lines show ± two standard errors). (b) Results of a field experiment showing changes in the size, number, and color diversity of guppy spots. A population of guppies originating in a locality with dangerous predators (c) was transferred to a stream having only a weak predator (*Rivulus hartii*) and, until the introduction, no guppies (x). Another stream nearby with guppies and *R. hartii* served as a control (r). The results shown are from guppies collected at the three sites 2 years after the introductions. Note how x and r have converged and changed dramatically from c. (After Endler, 1980.)

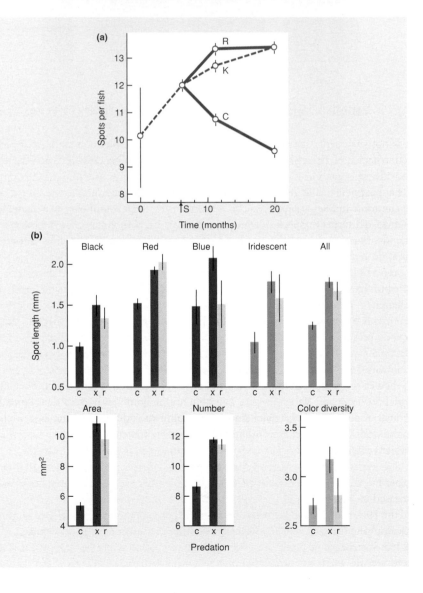

experiment, 200 guppies were moved from a site far down the Aripo River where predators were common and introduced to a site high up the river where there were neither guppies *nor* predators. The transplanted guppies thrived in their new site, and within just 2 years the males had more and bigger spots of more varied color (Figure 2.8b). The females' choice of the more flamboyant males had had dramatic effects on the gaudiness of their descendants, but this was only because predators did not reverse the direction of selection.

The speed of evolutionary change in this experiment in nature was as fast as that in artificial selection experiments in the laboratory. There was plenty of overpopulation (as many as 14 generations of fish occurred in the 23 months during which the experiment took place) and there was considerable genetic variation in the populations upon which natural selection could act. The guppies transplanted into the nearly predator-free environment evolved in other respects too. The females were larger and older at maturity, they produced fewer but bigger offspring, and they began to lose the habit of moving together in schools—behavior that defends them against predators (Endler, 1980; Magurran, 1998).

2.3.2 Variation within a species with man-made selection pressures

It is not surprising that some of the most dramatic examples of natural selection in action have been driven by the ecological forces of environmental pollution—these can provide rapid change under the influence of powerful selection pressures. Pollution of the atmosphere in and after the Industrial Revolution has left evolutionary fingerprints in the most unlikely places. *Industrial melanism* is the phenomenon in which black or blackish forms of species of moths and other organisms have come to dominate populations in industrial areas. In the dark individuals, a dominant gene is responsible for producing an excess of the black pigment melanin. Industrial melanism is known in most industrialized countries, including some parts of the United States (e.g., Pittsburgh), and more than a hundred species of moth have evolved forms of industrial melanism.

natural selection by pollution—the evolution of a melanic moth

The earliest recorded species to evolve in this way was the peppered moth (*Biston betularia*); the first black specimen was caught in Manchester (England) in 1848. By 1895, about 98 percent of the Manchester peppered moth population was melanic. Following many more years of pollution, a large-scale survey of pale and melanic forms of the peppered moth in Britain recorded more than 20,000 specimens between 1952 and 1970 (Figure 2.9). The winds in Britain are predominantly westerlies, spreading industrial pollutants (especially smoke and sulfur dioxide) toward the east. Melanic forms were concentrated toward the east and were completely absent from unpolluted western parts of England and Wales, northern Scotland, and Ireland.

The moths are preyed upon by insectivorous birds that hunt by sight. In a field experiment, large numbers of melanic and pale ("typical") moths were reared and released in equal numbers in a rural and largely unpolluted area of southern England. Of the 190 moths that were captured by birds, 164 were melanic and 26 were typicals. An equivalent study was made in an industrial area near the city of Birmingham. Twice as many melanics as typicals were recaught. This showed that a significant selection pressure was exerted through bird predation, and that moths of the typical form were

Figure 2.9
Sites in Britain where the frequencies of the pale (*forma typica*) and melanic forms of *Biston betularia* were recorded by Kettlewell and his colleagues. In all more than 20,000 specimens were examined. The principal melanic form (*forma carbonaria*) was abundant near industrial areas and where the prevailing westerly winds carry atmospheric pollution to the east. A further melanic form (*forma insularia*, which looks like an intermediate form but is due to several different genes controlling darkening) was also present but was hidden where the genes for *forma carbonaria* were present. (From Ford, 1975.)

clearly at a disadvantage in the polluted industrial environment (where their light color stood out against a sooty background), whereas the melanic forms were at a disadvantage in the pollution-free countryside (Kettlewell, 1955).

In the 1960s, however, industrialized environments in Western Europe and the United States started to change as oil and electricity began to replace coal and legislation was passed to impose smoke-free zones and to reduce industrial emissions of sulfur dioxide (Chapter 13). The frequency of melanic forms then fell back to near preindustrial levels with remarkable speed (Figure 2.10).

The forces of selection at work, first in favor and then against melanic forms, have clearly been related to industrial pollution, but the idea that melanic forms were favored simply because they were camouflaged against smoke-stained backgrounds

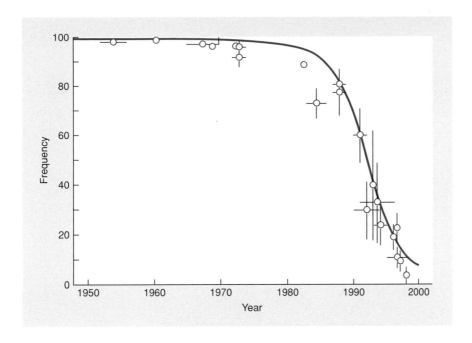

Figure 2.10
Change in the frequency of the *carbonaria* form of the peppered moth *Biston betularia* in the Manchester area since 1950. Vertical lines show the standard error and the horizontal lines show the range of years included. (After Cook et al., 1998.)

may be only part of the story. The moths rest on tree trunks during the day, and non-melanic moths are well hidden against a background of mosses and lichens. Industrial pollution has not just blackened the moths' background; atmospheric pollution, especially SO_2, has also destroyed most of the moss and lichen on the tree trunks. Indeed the distribution of melanic forms in Figure 2.9 closely fits the areas in which tree trunks were likely to have lost lichen cover as a result of SO_2 and so ceased to provide such effective camouflage for the nonmelanic moths. Thus SO_2 pollution may have been as important as smoke in selecting melanic moths.

Some plants are tolerant of another form of pollution: the presence of toxic heavy metals such as lead, zinc, and copper, which contaminate the soil after mining. Populations of plants on contaminated areas may be tolerant, while at the edge of these areas a transition from tolerant to intolerant forms can occur over very short distances (Figure 2.11). In some cases it has been possible to measure the speed of evolution. Zinc-tolerant forms of two species of grass *Agrostis capillaris* were found to have evolved under zinc-galvanized electricity pylons within 20–30 years of their erection (Al-Hiyaly et al., 1988).

natural selection by pollution— evolution of heavy-metal tolerance in plants

2.3.3 Adaptive peaks and specialized abysses

Natural selection changes the character of a population by sifting out and eliminating much of its variation and leaving behind a residue for future generations with a narrower range and more restricted potential. This is commonly pictured as a force that drives populations toward a peak of *adaptation*—of a perfect match between organism and environment. This is an optimist's view (Figure 2.12a). An alternative picture of

Figure 2.11
The grass *Anthoxanthum odoratum* colonizes land heavily contaminated with zinc (Zn) on an old mine. This is possible because the grass has evolved zinc-tolerant forms. Samples of the grass were taken along a transect from a mine (Trelogan) into surrounding grassland (zinc concentrations in the soil are shown as parts per million), and were tested for zinc tolerance by measuring the length of roots that they produced when grown in a culture solution containing excess zinc. The index of zinc tolerance falls off steeply over a distance of 2–5 m at the mine boundary. (After Putwain, in Jain & Bradshaw, 1966.)

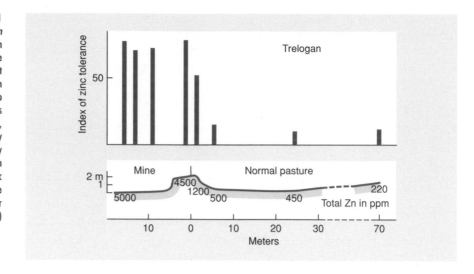

Figure 2.12
Two models of natural selection in action: (a) an optimistic and (b) a pessimistic interpretation of natural selection. In both (a) and (b) a horizontal plane is drawn to represent two dimensions of a range of environmental conditions. On both diagrams ellipses are drawn to show the range of this variation that is tolerated by four populations of organisms. (a) In the optimistic view, the vertical scale is a measure of the range of fitness of the organisms in the populations and natural selection is shown as arrows driving the population to ever higher "adaptive" peaks. Population 1 is highly variable and tolerates a wide range of conditions. Natural selection is relatively weak. Population 2 is a very uniform population, and natural selection is fierce and driving the population to a very high degree of specialization and local fitness. (b) In the pessimistic view, natural selection is shown as arrows driving the populations into ruts, troughs, and pits. The vertical downward axis is a measure of the intensity of specialization. The highly variable population with weak selection is rather safe if the environment changes but population 2 is at extreme risk of extinction. The two depictions should not be seen as one right and one wrong impression of natural selection in action, but rather as two views of the same truths.

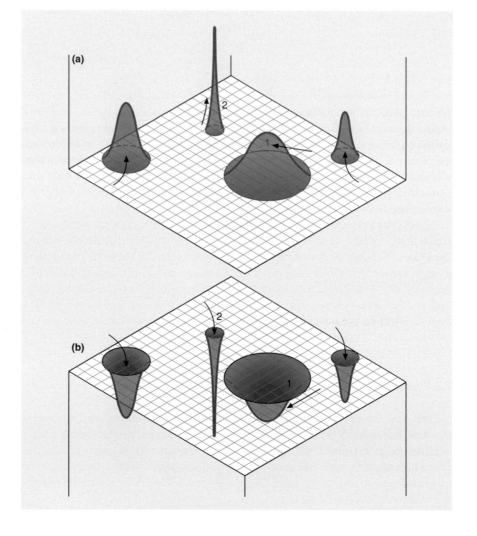

natural selection is that it forces populations into an ever-narrowing rut of overspecialization—an ever-deepening trap (Figure 2.12b). This pessimist's view emphasizes how the effects of natural selection are to limit and constrain, and that the specialization of species means that they risk extinction when the environment changes.

It is easy to see that a population of plants faced with repeated drought is likely to evolve a tolerance of water shortage, and an animal repeatedly faced with cold winters is likely to evolve habits of hibernation or a thick protective coat. But droughts do not become any less severe as a result, nor winters milder. Physical conditions are not heritable: they leave no descendants, and they are not subject to natural selection.

natural selection does not act on physical conditions . . .

But the situation is quite different when two species interact: predator on prey, parasite on host, competitive neighbor on neighbor. Natural selection may select from a population of parasites those forms that are more efficient at infecting their host. But this immediately sets in play forces of natural selection that favor more resistant hosts. As they evolve they put further pressure on the ability of the parasite to infect. Host and parasite are then caught in never-ending reciprocating selection. A result is that both host and parasite become increasingly specialized—caught in an ever-deepening rut. Eventually only a specialized form of the parasite can infect and can do so only on a highly specialized form of the host. We will find examples of this extreme form of natural selection—coevolution—when we consider organisms as habitats in Chapter 7.

. . . but parasites, predators, and competitors can all be both forces of natural selection and objects of selection

2.4 The Ecology of Speciation

We have seen that natural selection can force populations of plants and animals to change their character—to evolve. But none of the examples we have considered has involved the evolution of a new species. Indeed Darwin's *On the Origin of Species* is about natural selection and evolution but is not really about the origin of species! Those who studied the evolution of melanism in the peppered moth named the black and normal forms *forma carbonaria* and *forma typica*: they classified them as forms within a species, not as different species. Likewise the different growth forms of the grasses on the cliffs and pastures of Abraham's Bosom and the dull and flamboyant races of guppies are just local genetic classes. None qualifies for the status of distinct species. But when we ask just what criteria justify naming two populations as different species we meet real problems.

2.4.1 What do we mean by a "species"?

Cynics have said, with some truth, that a species is what a competent taxonomist regards as a species! Darwin himself regarded species (like genera) as "merely artificial combinations made for convenience". On the other hand, back in the 1930s, two American biologists, Ernst Mayr and Theodosius Dobzhansky, proposed an empirical test that could be used to decide whether two populations were part of the same or of two different species. They recognized organisms as being members of a single species if they could, at least potentially, breed together in nature to produce fertile offspring. They called a species tested and defined in this way a *biospecies*. In the examples that we have used earlier in this chapter we know that melanic and normal peppered moths

Figure 2.13
The role of isolation in the evolution of species. A uniform species with a large range (1) differentiates (2) into local forms, varieties, or subspecies, which (3) become genetically isolated from each other, for example, separated by geographical barriers or dispersed onto different islands. After evolution in isolation they may meet again (4) when they are unable to hybridize and have become true biospecies.

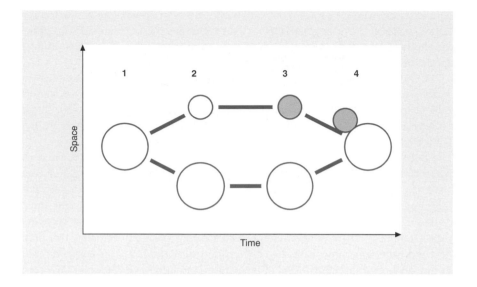

can mate and that the offspring are fully fertile; this is also true of colored and dull guppies and of plants from the different types of *Agrostis*. They are all variations within species—not separate species.

In practice, however, biologists do not apply the Mayr–Dobzhansky test before they recognize every species: there is simply not enough time and resources; but it is there to resolve arguments if they arise. What is more important is that the test recognizes a crucial element in the evolutionary process. If the members of two populations are able to hybridize and their genes are combined and reassorted in their progeny, then natural selection can never make them truly distinct. Although natural selection may tend to force a population to evolve into two or more distinct forms, sexual reproduction and hybridization mix them up again.

biospecies do not exchange genes   Two parts of a population can evolve into distinct species only if some sort of barrier prevents gene flow between them; they might, for example, be isolated on different islands. While isolated from each other they may then evolve and become so different that, if they meet again, they can no longer hybridize and their populations can no longer exchange genes. They are now different biospecies. Figure 2.13 illustrates this process.

Differences that are particularly effective in keeping newly evolved species distinct are different rituals of courtship, different signals of attraction between the sexes, and in flowering plants, different species of insect pollinator. It may sometimes happen that hybrids form between two evolving species but their parental chromosomes have become so different that they fail to pair at meiosis: the hybrids are then sterile (for example, the horse–donkey hybrid is the sterile mule).

evolution in sea gulls   The evolution of species and the balance between natural selection and hybridization are illustrated by the extraordinary case of two species of sea gull. The lesser black-backed gull (*Larus fuscus*) originated in Siberia and colonized progressively to the west, forming a chain or *cline* of different forms, spreading from Siberia to Britain and Iceland (Figure 2.14). The neighboring forms along the cline are distinctive, but

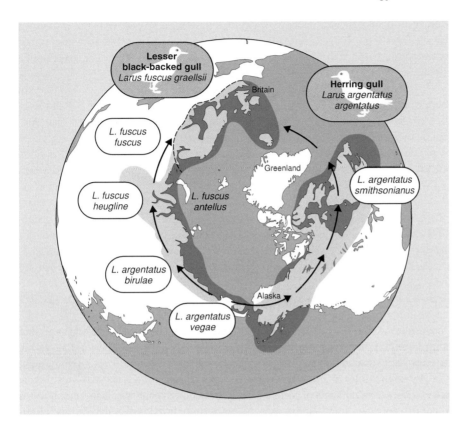

Figure 2.14
Two species of gull, the herring gull and the lesser black-backed gull, have diverged from a common ancestry as they have colonized and encircled the northern hemisphere. Where they occur together in northern Europe they fail to interbreed and are clearly recognized as two distinct species. However, they are linked along their ranges by a series of freely interbreeding races or subspecies. (After Brookes, 1998.)

they hybridize readily in nature. Neighboring populations are therefore regarded as part of the same species and taxonomists give them only "subspecific" status (e.g., *Larus fuscus graellsii*, *Larus fuscus fuscus*). Populations of the gull have, however, also spread east from Siberia, again forming a cline of freely hybridizing forms. Together, the populations spreading east and west encircle the northern hemisphere. They meet and overlap in northern Europe. There, the eastward and westward clines have diverged so far that it is easy to tell them apart, and they are recognized as two different *species*, the lesser black-backed gull (*Larus fuscus*) and the herring gull (*Larus argentatus*). Moreover, the two species do not hybridize: they have become true biospecies.

In this remarkable example we can see how two distinct species have evolved from one primal stock, and that the stages of their divergence remain frozen in the cline that connects them. But it is where a population becomes split into completely isolated populations, dispersed onto different islands for example, that they most readily diverge into distinct species.

2.4.2 Islands and speciation

The most celebrated example of evolution and speciation on islands is the case of Darwin's finches in the Galapagos archipelago. Darwin had been chided for

Darwin's finches

underestimating the importance of isolation in the evolution of species but responded (in a letter, 1876), "It would have been a strange fact if I had overlooked the importance of isolation, seeing that it was such cases as that of the Galapagos archipelago which chiefly led me to study the evolution of species."

The Galapagos are volcanic islands isolated in the Pacific Ocean about 1000 km west of Equador and 750 km from the island of Cocos, which is itself 500 km from Central America. At more than 500 m above sea level the vegetation is open grassland. Below this is a humid zone of forest that grades into a coastal strip of desert vegetation with some endemic species of prickly pear cactus (*Opuntia*). Fourteen species of finch are found on the islands, and there is every reason to suppose that these evolved from a single ancestral species that invaded the islands from the mainland of Central America.

In their remote island isolation, the Galapagos finches have radiated into a variety of species in groups with contrasting ecologies (Figure 2.15). Members of one group, including *Geospiza fuliginosa* and *G. fortis*, have strong bills and hop and scratch for seeds on the ground. *Geospiza scandens* has a narrower and slightly longer bill and feeds on the flowers and pulp of the prickly pears as well as on seeds. Finches of a third group have parrot-like bills and feed on leaves, buds, flowers, and fruits, and a fourth group with a parrot-like bill (*Camarhynchus psittacula*) has become insectivorous, feeding on beetles and other insects in the canopy of trees. A so-called woodpecker finch, *Camarhynchus* (*Cactospiza*) *pallida*, extracts insects from crevices by holding a spine or a twig in its bill. Yet a further group includes a species (*Certhidea olivacea*) that, rather like a warbler, flits around actively and collects small insects in the forest canopy and in the air. Populations of ancestor species became reproductively isolated, most likely after chance colonization of different islands within the archipelago, and evolved separately for a time. Subsequent movements between islands may have brought non-hybridizing biospecies together, and subsequently these have evolved to fill different niches. We will see in Chapter 6 that when individuals from different species compete, natural selection may act to favor those individuals that compete least with members of the other species. An expected consequence is that among a group of closely related species, such as Darwin's finches, differences in feeding and other aspects of their ecology are likely to become enhanced with time.

The entire process of evolutionary divergence of these species appears to have happened in less than 3 million years. Very rarely, hybridization occurs among species that have similar ecologies. However, these occasions are so rare that the species are true or emerging biospecies. We need to remember that the origin of a species is normally a process rather than an event. For the formation of a new species, like the boiling of an egg, there is some freedom to argue about when it is completed!

The evolutionary relationships among the various Galapagos finches have been traced by molecular techniques (analyzing variation in "microsatellite" DNA) (Petren et al., 1999) (Figure 2.15). These accurate modern tests confirm the long-held view that the family tree of the Galapagos finches radiated from a single trunk (i.e., was *monophyletic*) and also provides strong evidence that the warbler finch (*Certhidea olivacea*) was the first to split off from the founding group and is likely to be the most similar to the original colonist ancestors.

The flora and fauna of many other archipelagos show similar examples of great richness of species with many local *endemics* (i.e., species known only from one island or area). The Hawaiian Islands are home to an extraordinary diversity of picture-winged

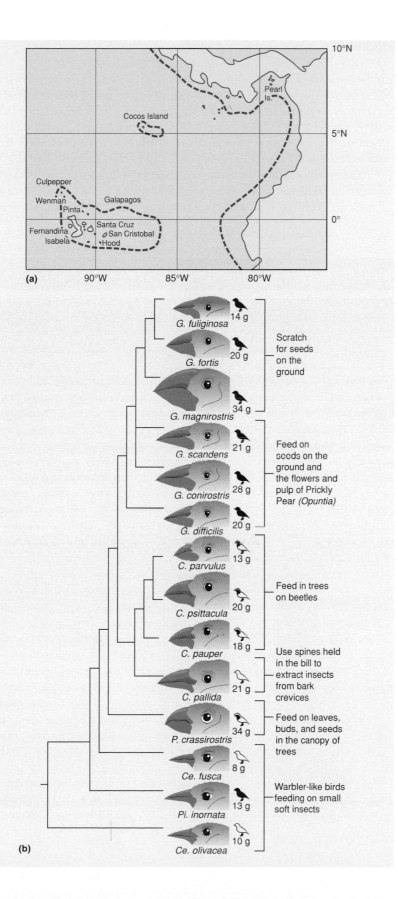

Figure 2.15
(a) Map of the Galapagos Islands showing their position relative to Central America; on the equator 5° equals approximately 560 km. (b) A reconstruction of the evolutionary history of the Galapagos finches based on variation in the length of microsatellite DNA. The feeding habits of the various species are also shown. Drawings of the birds are proportional to actual body size. The maximum amount of black coloring in the male plumage and the average body mass are shown for each species. The *genetic distance* (a measure of the genetic difference) between species is shown by the length of the horizontal lines. Notice the great and early separation of the warbler finch (*Certhidea olivacea*) from the others, suggesting that it may closely resemble the founders that colonized the islands. *C., Camarhynchus; Ce., Certhidea; G., Geospiza; P., Platyspiza; Pi., Pinaroloxias.* (After Petren et al., 1999.)

fruit flies (species of *Drosophila*) and closely related species of honey creeper that have diverged in feeding habit and bill shape remarkably like the Galapagos finches. Lizards of the genus *Anolis* have evolved a kaleidoscopic diversity of species on the islands of the Caribbean; and isolated groups of islands, such as the Canaries off the coast of North Africa, are treasure troves of endemic plants. These unusual and often rich communities may pose particular problems for the applied ecologist (Box 2.2).

Invaders onto marine islands may become isolated from other parts of their population and are then free to diverge under natural selection and become different species. But there are other kinds of "islands" in which colonists can also become genetically

Box 2.2   *Topical ECOncerns*

Deep Sea Vent Communities at Risk

Deep sea vents are islands of warmth in oceans (literally and metaphorically) that are otherwise cold and inhospitable. As a consequence, they support unique communities, rich in endemic species. One of the latest controversies to pit environmentalists against industrialists concerns these deep sea vents (see figure), which are also now known to be sites rich in minerals. This newspaper article by William J. Broad appeared in the *San Jose Mercury News*, January 20, 1998:

> With miners staking claim to valuable metals lying in undersea lodes in the South Pacific, questions surface about how to prevent

A deep sea vent community (© Whoi, J. Edmond, Visuals Unlimited).

disasters in these fragile, little understood ecosystems.

The volcanic hot springs of the deep sea are dark oases that teem with blind shrimp, giant tube worms and other bizarre creatures, sometimes in profusions great enough to rival the chaos of rain forests. And they are old.

Scientists who study them say these odd environments, first discovered two decades ago, may have been the birthplace of all life on Earth, making them central to a new wave of research on evolution.

Now, in a moment that diverse ranks of experts have feared and desired for years, miners are invading the hot springs, possibly setting the stage for the last great battle between industrial development and environmental preservation.

The undersea vents are rich not just in life but in valuable minerals such as copper, silver and gold. Indeed, their smoky chimneys and rocky foundations are virtual foundries for precious metals. . . . The fields of undersea gold have long fired the imaginations of many scientists and economists, but no mining took place, in part because the rocky deposits were hard to lift from depths of a mile or more.

Now, however, miners have staked the first claim to such metal deposits after finding the richest ores ever. The estimated value of copper, isolated from the rest of a population. Mountains isolate valleys from each other and valleys isolate mountains. A few individuals that chance to be dispersed to a habitable site in a mountain range can form the nucleus of an expanding new species. Its character will have been colored by the particular genes that were represented among the colonists—unlikely to be a perfect sample of the parent population. What natural selection can do with this *founder population* is limited by what is in its limited sample of genes (plus occasional rare mutations). Indeed much of the deviation among populations isolated on islands appears to be due to a *founder effect*—the chance composition of the pool of founder genes puts limits and constraints on what variation there is for natural selection to act upon.

The evolutionary biologist's understanding of island patterns depends on a thorough appreciation of ecological processes such as dispersal (Chapter 5) and interspecific competition (Chapter 6). Likewise, the ecologist's understanding of ecological specialization, species distributions, species diversity, and niche partitioning, among many other ecological phenomena, would be rudimentary indeed without the underpinning provided by the evolutionary processes discussed in this chapter.

2.5 The Effects of Climatic Change on the Evolution and Distribution of Species

Changes in climate, particularly during the ice ages of the Pleistocene (the past 2–3 million years), bear a lot of the responsibility for the present patterns of distribution of plants and animals. As climates have changed, species populations have advanced and retreated, have been fragmented into isolated patches, and may have then rejoined. Much of what we see in the present distribution of species represents phases in a recovery from past climatic change. Modern techniques for analyzing and dating biological remains (particularly buried pollen) are beginning to allow us to detect just how much of the present distribution of organisms is a precise,

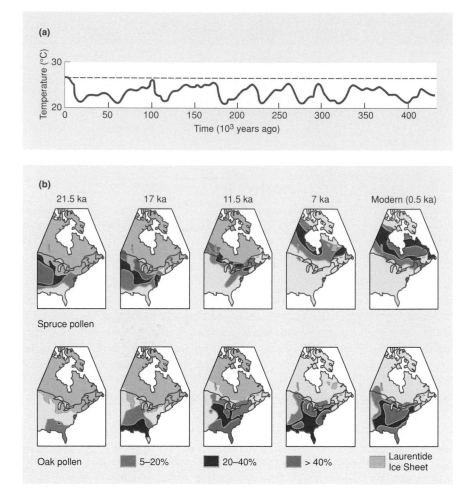

Figure 2.16
(a) An estimate of the temperature variations with time during glacial cycles over the past 400,000 years. The estimates were obtained by comparing oxygen isotope ratios in fossils taken from ocean cores in the Caribbean. The dashed line corresponds to the ratio 10,000 years ago, at the start of the present warming period. Periods as warm as the present have been rare events, and the climate during most of the past 400,000 years has been glacial. (After Emiliani, 1966; Davis, 1976.) (b) Ranges in eastern North America, as indicated by pollen percentages in sediments, of spruce species (above) and oak species (below) from 21,500 years ago to the present. (After Davis & Shaw, 2001.)

local-evolved match to present environments, and how much is a fingerprint left by the hand of history.

cycles of glaciation have occurred repeatedly

For most of the past 2–3 million years the Earth has been very cold. Evidence from the distribution of oxygen isotopes in cores taken from the deep ocean floor shows that there may have been as many as 16 glacial cycles in the Pleistocene, each lasting for up to 125,000 years (Figure 2.16a). Each cold (glacial) phase may have lasted for as long as 50,000–100,000 years, with brief intervals of only 10,000–20,000 years when the temperatures rose to, or above, those of today. In this case, present floras and faunas are unusual, having developed at the warm end of one of a series of unusual catastrophic warm periods.

the distribution of trees has changed gradually since the last glaciation

During the 20,000 years since the peak of the last glaciation, global temperatures have risen by about 8°C. The analysis of buried pollen—particularly of woody species, which produce most of the pollen—can show how vegetation has changed during this period (Figure 2.16b). As the ice retreated, different forest species advanced in different ways and at different speeds. For some, like the spruce of eastern North America,

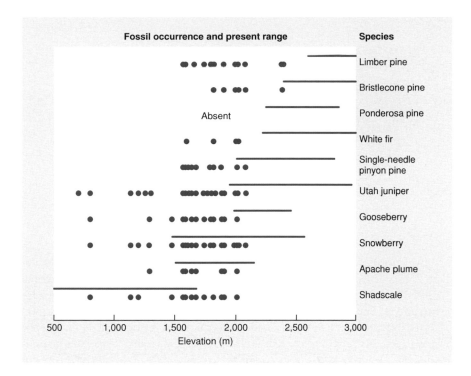

Figure 2.17
The elevation ranges of ten species of woody plant from the mountains of the Sheep Range, Nevada during the last glaciation (dots) and at present (solid line). (After Davis & Shaw, 2001.)

there was displacement to new latitudes; for others, like the oaks, the picture was more one of expansion.

We do not have such good records for the postglacial spread of the animals associated with the changing forests, but it is at least certain that many species could not have spread faster than the trees on which they feed. Some of the animals may still be catching up with their plants, and tree species are still returning to areas they occupied before the last ice age! It is quite wrong to imagine that our present vegetation is in some sort of equilibrium with (adapted to) the present climate.

Even in regions that were never glaciated, pollen deposits record complex changes in distributions: range shifts to new regions, large changes in population size, both up and down, and in the mountains of the Sheep Range, Nevada, shifts in elevation range varying in amount and even direction (Figure 2.17). The species composition of vegetation has continually been changing and is almost certainly still doing so.

The records of climatic change in the tropics are far less complete than those for temperate regions. Many believe, though, that during cooler, drier glacial periods, the tropical forests retreated to smaller patches, surrounded by a sea of savanna. Support for this comes from the present-day distribution of species in the tropical forests of South America (Figure 2.18). There, particular "hot-spots" of species diversity are apparent, and these are thought to be likely sites of forest refuges during the glacial periods, and sites too, therefore, of increased rates of speciation (Ridley, 1993). On this interpretation, the present distributions of species may again be seen as largely accidents of history (where the refuges were) rather than precise matches between species and their differing environments.

Figure 2.18
(a) The present-day distribution of tropical forest in South America. (b) The possible distribution of tropical forest refuges at the time when the last glaciation was at its peak, as judged by present-day hot-spots of species diversity within the forest. (After Ridley, 1993.)

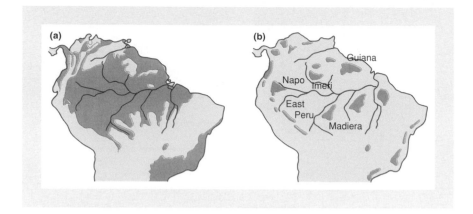

predicted global warming by the "greenhouse effect" is nearly a hundred times faster than postglacial warming

Evidence of changes in vegetation that followed the last retreat of the ice hint at the likely consequences of the global warming (maybe 3°C in the next 100 years) that is predicted to result from continuing increases in "greenhouse" gases in the atmosphere (Chapter 13). But the scales are quite different. Postglacial warming of about 8°C occurred over around 20,000 years, and changes in the vegetation failed to keep pace even with this. But current projections for the 21st century require range shifts for trees at rates of 300–500 km per century compared to typical rates in the past of 20–40 km per century (and exceptional rates of 100–150 km). It is striking that the only precisely dated extinction of a tree species in the Quaternary, that of *Picea critchfeldii*, occurred around 15,000 years ago at a time of especially rapid postglacial warming (Jackson & Weng, 1999). Clearly, future even more rapid change could result in extinctions of many additional species (Davis & Shaw, 2001).

2.6   The Effects of Continental Drift on the Ecology of Evolution

land masses have moved . . .

The patterns of species formation that occur on islands appear on an even larger scale in the evolution of genera and families across continents. Many curious distributions of organisms between continents seem inexplicable as the result of dispersal over vast distances. Biologists, especially Wegener (1915), met outraged scorn from geologists and geographers when they argued that it must have been the continents that had moved rather than the organisms that had dispersed. Eventually, however, measurements of the directions of the Earth's magnetic fields required the same, apparently wildly improbable, explanation and the critics capitulated. The discovery that the tectonic plates of the Earth's crust move and carry the migrating continents with them reconciles geologist and biologist (Figure 2.19). While major evolutionary developments were occurring in the plant and animal kingdoms, their populations were being split and separated, and land areas were moving across climatic zones. This was happening while changes in temperature were occurring on a vastly greater scale than the glacial cycles of the Pleistocene episode.

. . . and divided populations that have then evolved independently

The established drift of the continents answers many questions in the ecology of evolution. The curious world distribution of large flightless birds is one example

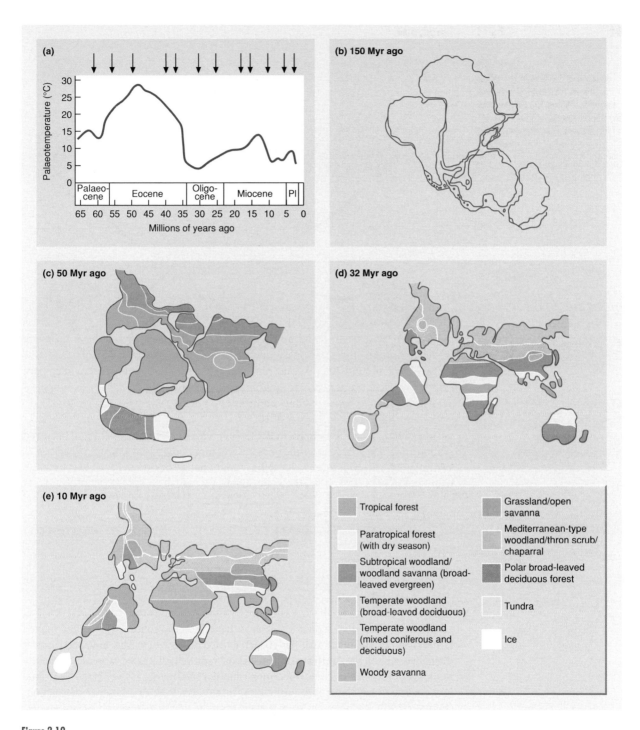

Figure 2.19

(a) Changes in temperature in the North Sea over the past 65 million years. During this period there were large changes in sea level that allowed the dispersal of both plants and animals between land masses. (b–e) Continental drift. (b) The ancient supercontinent of Gondwanaland began to break up about 150 million years ago. (c) About 50 million years ago (early Middle Eocene) recognizable bands of distinctive vegetation had developed, and (d) by 32 million years ago (early Oligocene) these had become more sharply defined. (e) By 10 million years ago (early Miocene) much of the present geography of the continents had become established but with dramatically different climates and vegetation from today: the position of the Antarctic ice cap is highly schematic. (After Norton & Sclater, 1979; Janis, 1993; and other sources.)

Figure 2.20
(a) The distribution of terrestrial flightless
birds. (b) The phylogenetic tree of the
flightless birds and the estimated times
(million years ago) of their divergence.
(After Diamond, 1983; from data of Sibley
& Ahlquist.)

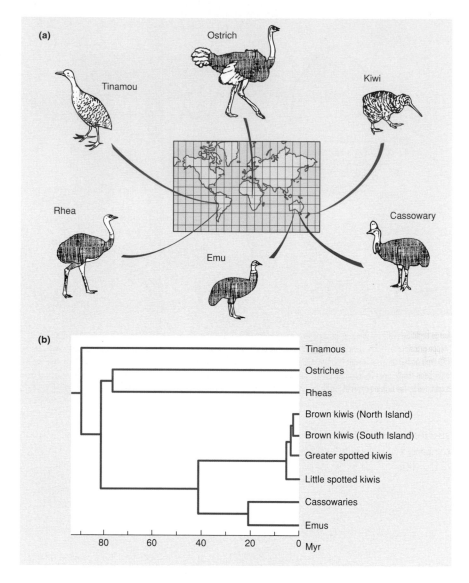

(Figure 2.20a). The presence of the ostrich in Africa, the emu in Australia, and the very similar rhea in South America could scarcely be explained by dispersal of some common flightless ancestor. Now, techniques of molecular biology make it possible to analyze the time at which the various flightless birds started their evolutionary divergence (Figure 2.20b). The tinamous seem to have been the first to diverge and became evolutionarily separate from the rest, the *ratites*. Australasia next became separated from the other southern continents, and from the latter, the ancestral stocks of ostriches and rheas were subsequently separated when the Atlantic opened up between Africa and South America. Back in Australasia, the Tasman Sea opened up about 80 million years ago and ancestors of the kiwi are thought to have made their way, by island hopping, about 40 million years ago across to New Zealand, where divergence into the present

(a)

(b)

(c)

Large flightless birds are found in three major continents. (a) The ostrich (*Struthio camelus*) is African and commonly occurs together with herds of zebra and antelope in savanna or steppe grasslands. (b) The rhea (*Rhea americana*) is found in similar grasslands in South America (e.g., Brazil and Argentina), commonly together with herds of deer and guanocos (© Walt Anderson, Visuals Unlimited). (c) The emu (*Dromaius novaehollandiae*) inhabits equivalent habitats in Australia. Many other species of these very large, mainly herbivorous birds have been sought after by humans for food and have become extinct. The presence of these evolutionarily related and ecologically similar species in three widely separated continents is explained by the drifting apart of the continents from the time (150 million years ago) when they were portions of the primitive continent of Gondwanaland (see Figure 2.19)

species happened relatively recently. The unraveling of this particular example implies the early evolution of the property of flightlessness and only subsequently the isolation of the different types between the emerging continents.

2.7 Interpreting the Results of Evolution: Convergent and Parallel Evolution

Flightlessness did not evolve independently on the different continents. However, there are many examples of organisms that have evolved in isolation from each other and then converged on remarkably similar forms or behavior. Such similarity is particularly striking when similar roles are played by structures that have quite different evolutionary origins—that is, when the structures are *analogous* (similar in superficial form or function) but not *homologous* (derived from an equivalent structure in a common ancestry). When this occurs, it is termed *convergent evolution*. Bird and bat wings are a classic example (Figure 2.21).

Such convergence is direct evidence of the power of evolutionary forces to shape the same form from quite different starting material. The French geneticist Jacob said that evolution was like "tinkering". It did not create ideal forms from ideal beginnings —rather it tinkered together what it could from what was available at the time (a good tinker can make a saucepan from a bicycle or a bicycle from saucepans).

convergent evolution

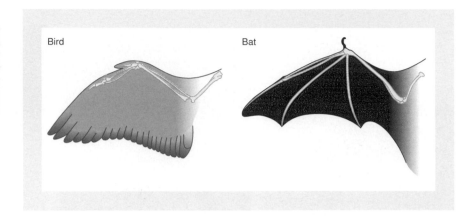

parallel evolution

Further examples show *parallels* in the evolutionary pathways of ancestrally related groups occurring after they were isolated from each other. The classic example is provided by the placental and marsupial mammals. Marsupials arrived on what would become the Australian continent in the Cretaceous period (around 90 million years ago; see Figure 2.19), when the only other mammals present were the curious egg-laying monotremes (now represented only by the spiny anteaters and the duck-billed platypus). An evolutionary process of radiation then occurred among the Australian marsupials that in many ways accurately paralleled what was occurring among the placental mammals on other continents (Figure 2.22). It is hard to escape the view that the environments of placentals and marsupials contained ecological pigeonholes (niches) into which the evolutionary process has neatly "fitted" ecological equivalents. In contrast to convergent evolution, however, the marsupials and placentals started to diversify from a common ancestral line, and both inherited a common set of potentials and constraints.

interpreting the match between organisms and their environment

When we marvel at the diversity of complex specializations by which organisms match their varied environments there is a temptation to regard each case as an example of evolved perfection. But there is nothing in the process of evolution by natural selection that implies perfection. For example, no population of organisms can contain all the possible genetic variants that might exist and might influence fitness. The evolutionary process works on the genetic variation that is available. It favors only those forms that are fittest *from among the range of variety available*, and this may be a very restricted choice. The very essence of natural selection is that organisms come to match their environments by being "the fittest available" or "the fittest yet"— they are not "the best imaginable".

It is particularly important to realize that past events on the Earth can have profound repercussions on the present. Our world has not been constructed by taking each organism in turn, testing it against each environment, and molding it so that every organism finds its perfect place. It is a world in which organisms live where they do for reasons that are often, at least in part, accidents of history. Moreover the ancestors of the organisms that we see around us lived in environments that were profoundly different from those of the present. Evolving organisms are not free agents—some of the features acquired by their ancestors hang like millstones around their necks, limiting and constraining where they can now live and what they might become. It is very easy

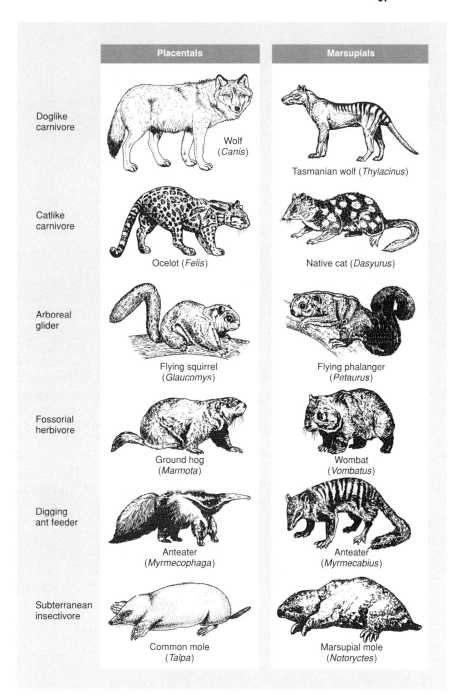

to wonder and marvel at how beautifully the properties of fish fit them to live in water —but just as important to emphasize that these same properties prevent them from living on land. Indeed it was the few that chanced to escape from their ancestral watery home that started the evolutionary lines that diversified into the dinosaurs and mammals.

Summary

The force of natural selection

Life is represented on Earth by a diversity of specialist species, each of which is absent from almost everywhere. Early interest in this diversity mainly existed among explorers and collectors, and the idea that the diversity had arisen by evolution from earlier ancestors over geological time was not seriously discussed until the first half of the 19th century. Charles Darwin and Alfred Russell Wallace (strongly influenced by having read Malthus's *An Essay on the Principle of Population*) independently proposed that natural selection constituted a force that would drive a process of evolution. The theory of natural selection is an ecological theory. The reproductive potential of living organisms leads them inescapably to compete for limited resources. Success in this competition is measured by leaving more descendants than others to subsequent generations. When these ancestors differ in properties that are heritable, the character of populations will necessarily change over time and evolution will happen.

Darwin had seen the power of human selection to change the character of domestic animals and plants and he recognized the parallel in natural selection. But there is one big difference: humans select for what they want in the future, but natural selection is a result of events in the past—it has no intentions and no aim.

Natural selection in action

We can see natural selection in action within species in the variation within species over their geographic range, and even over very short distances where we can detect powerful selective forces in action and recognize ecologically specialized races within species (*ecotypes*). Transplanting plants and animals between habitats reveals tightly specialized matches between organisms and their environments. The evolutionary responses of animals and plants to pollution demonstrate the speed of evolutionary change, as do experiments on the effects of predators on the evolution of their prey. The evolution of specialization can be interpreted as matching organisms more accurately with their environment or as forcing them into ever tighter straitjackets of form and behavior.

The origin of species

Natural selection does not normally lead to the origin of species unless it is coupled with the reproductive isolation of populations from each other—as occurs, for example, on islands and is illustrated by the finches of the Galapagos Islands. *Biospecies* are recognized when they have diverged enough to prevent them from forming fertile hybrids if and when they meet.

Climatic change and continental drift

Much of what we see in the present distribution of organisms is not so much a precise, local-evolved match to present environments as a fingerprint left by the hand of history. Changes in climate, particularly during the ice ages of the Pleistocene, bear a lot of the responsibility for the present patterns of the distribution of plants and animals. On a longer timescale, many distributions make sense only once we realize that while major evolutionary developments were occurring, populations were being split and separated, and land areas were moving across climatic zones.

Parallel and convergent evolution

Evidence of the power of ecological forces to shape the direction of evolution comes from *parallel evolution* (in which populations long isolated from common ancestors have followed similar patterns of diversification) and from *convergent evolution* (in which populations evolving from very different ancestors have converged on very similar forms and behaviors).

Review Questions

1 ▲ What do you consider to be the essential distinction between *natural selection* and evolution?

2 What was the contribution of Malthus to Darwin's and Wallace's ideas about evolution?

3 Why is "the survival of the fittest" an unsatisfactory description of natural selection?

4 What is the essential difference between natural selection and the selection practiced by plant and animal breeders?

5 What are *reciprocal transplants*? Why are they so useful in ecological studies?

6 Is sexual selection, as practiced by female guppies in choosing gaudy males, different from or just part of natural selection?

7 In what ways do the results of natural selection by parasites and predators differ from selection by physical conditions of the environment?

8 What is it about the Galapagos finches that has made them such ideal material for the study of evolution?

9 What is the difference between convergent and parallel evolution?

10 ▲ The process of evolution can be interpreted as optimizing the fit between organisms and their environment or as narrowing and constraining what they can do. Discuss whether there is a conflict between these interpretations.

▲ = Challenge Question

Web Research Questions

1 The words "ecology", "ecologist", "ecological", and so on have spread from the science of ecology to the press, to pressure groups, and to the general public. Survey and describe briefly the range of meanings these words are now being given and by whom. Some have suggested that the words have been hijacked and misused: do you see any evidence of that? Does it/would it matter if they were?

2 Discuss the pros and cons of long-term ecological research, using the Hubbard Brook Experimental Forest study as an example. This program is run by the Institute of Ecosystem Studies (IES). To get some idea of the scale of long-term ecosystem studies find out roughly how many scientists and related staff work at the IES. The Hubbard Brook Experimental Forest is one of a number of long-term ecological research (LTER) sites in North America. Compare and contrast the objectives of the Hubbard Brook study with three of the other sites. Should all long-term ecological programs go on indefinitely? If not, what criteria would you suggest to determine whether and when a long-term program should be terminated?

3 Like museums, the traditional role of zoos was to make a public display of the diversity of nature. Is this still the case? Or has the conservation of endangered species come to assume more importance? Visit the websites of local zoos, and others around the world, and discuss the relative roles of display and conservation in the different institutions. Describe the variety of endangered species that zoos are working with.

Part II

Conditions and Resources

Chapter 3

Physical Conditions and the Availability of Resources

For ecologists, organisms are really only worth studying where they can live. Perhaps the most fundamental features of an environment that make it habitable are that the organisms can tolerate the local conditions and that their essential resources are being provided. We cannot expect to go very far in understanding the ecology of any species without understanding its interactions with conditions and resources.

Chapter Contents

3.1 Introduction

3.2 Environmental conditions

3.3 Plant resources

3.4 Animals and their resources

3.5 The effect of intraspecific competition for resources

3.6 Conditions, resources, and the ecological niche

Key Concepts

In this chapter you will

- understand the nature of, and contrasts between, conditions and resources

- understand how organisms respond to the whole range of a condition like temperature, but also to "extreme" conditions and to the timing of both variations and extremes

- appreciate how a plant's responses to, and its consumption of, the resources of solar radiation, water, minerals, and carbon dioxide are intertwined

- appreciate the importance of contrasting body compositions in the consumption of plants by animals, and of overcoming defenses in the consumption of animals by other animals

- understand the effects of intraspecific competition for resources

- appreciate how responses to conditions and resources interact to determine ecological niches

3.1   Introduction

It is important to distinguish conditions and resources as two properties of environments that determine where organisms can live. Conditions are physicochemical features of the environment such as its temperature and humidity, and in aquatic environments, osmotic pressure or pH. Important features of conditions include their diurnal and annual cycles and the frequency of extreme events such as coldest nights and hottest days. The presence of an organism always alters the conditions in its immediate environment—sometimes on a very large scale (a tree, for example, maintains a zone of higher humidity on the ground beneath its canopy) and sometimes only on a microscopic scale (an algal cell in a pond alters the pH in the shell of water that surrounds it). But conditions are not consumed nor used up by the activities of organisms.

resources, unlike conditions, are consumed

Environmental resources, by contrast, *are* consumed by organisms in the course of their growth and reproduction. Green plants photosynthesize and obtain both energy and materials for growth and reproduction from inorganic materials. Their resources are solar radiation, carbon dioxide, water, and mineral nutrients. "Chemosynthetic" organisms like many of the primitive Archaebacteria obtain energy by oxidizing methane, ammonium ions, hydrogen sulfide, or ferrous iron; they live in environments like hot springs and deep sea vents using resources that were abundant during the early phases of life on Earth. All other organisms use the bodies of other organisms as their food resource. In each case, what has been consumed is no longer available to another consumer. The rabbit eaten by an eagle is no longer available to another eagle. The quantum of solar radiation absorbed and photosynthesized by a leaf is no longer available to another leaf. This has an important consequence: organisms may *compete* with each other to capture a share of a limited resource.

In this chapter we consider, first, examples of the ways in which environmental conditions limit the behavior and distribution of organisms. We draw most of our examples from the effects of temperature, which serve to illustrate many general effects of environmental conditions. We consider next the resources used by photosynthetic green plants, and then go on to examine the ways in which organisms that are themselves resources have to be captured, grazed, or even inhabited before they are consumed. Finally, we consider the ways in which organisms of the same species may compete with each other for limited resources.

3.2   Environmental Conditions

3.2.1   What do we mean by "harsh", "benign", and "extreme"?

It seems quite natural to describe environmental conditions as "extreme", "harsh", "benign", or "stressful". But these describe how we, human beings, feel about them. It may seem obvious when conditions are "extreme": the midday heat of a desert, the cold of an Antarctic winter, the salt concentration of the Great Salt Lake. What this means, however, is only that these conditions are extreme *for us*, given our particular physiological characteristics and tolerances. But to a cactus there is nothing extreme about the desert conditions in which cacti have evolved; nor are the icy fastnesses of Antarctica an

Penguins do not find the Antarctic in the least bit "extreme".

extreme environment for penguins. But a tropical rain forest *would* be a harsh environment for a penguin, though it is benign for a macaw; and a lake is a harsh environment for a cactus, though it is benign for a water hyacinth.

There is, then, a relativity in the ways organisms respond to conditions; and it is too easy and dangerous for the ecologist to assume that all other organisms sense the environment in the way we do. Rather, the ecologist should try to gain a "worm's-eye" (or "plant's-eye") view of the environment; to see the world as others see it. Emotive words like *harsh* and *benign*, even relativities such as hot and cold, should be used by ecologists only with care.

3.2.2 Effects of conditions

Temperature, relative humidity, and other physicochemical conditions induce a range of physiological responses in organisms, which determine whether the physical environment is habitable or not. There are three basic types of response curve (Figure 3.1a–c). In the first (Figure 3.1a), extreme conditions are lethal, but between the two extremes there is a continuum of more favorable conditions. Organisms are typically able to survive over the whole continuum, but can grow actively only over a more restricted range and can reproduce only within an even narrower band. This is a typical response curve for the effects of temperature or pH. In the second curve (Figure 3.1b), the condition is lethal only at high intensities. This is the case for poisons. At zero or low concentrations the organism is typically unaffected, but there is a threshold above which performance decreases rapidly: first reproduction, then growth, and finally survival. The third curve (Figure 3.1c), then, applies to conditions that are required by organisms at low concentrations but become toxic at high concentrations. This is the case for some minerals, such as copper and sodium chloride, that are essential resources for growth when they are present in trace amounts but become toxic conditions at higher concentrations.

Of these three, the first is the most fundamental. It is accounted for, in part, by changes in metabolic effectiveness. For each 10°C rise in temperature, for example, the rate of biological processes often roughly doubles (Figure 3.2). The increase is brought

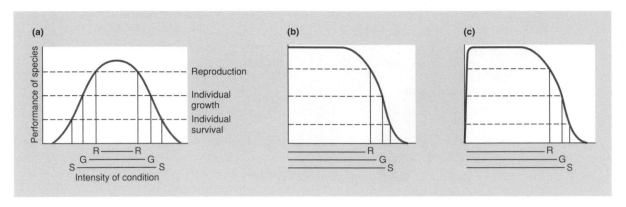

Figure 3.1

Response curves illustrating the effects of a range of environmental conditions on individual survival (S), growth (G), and reproduction (R). (a) Extreme conditions are lethal; less extreme conditions prevent growth; only optimal conditions allow reproduction. (b) The condition is lethal only at high intensities; the reproduction–growth–survival sequence still applies. (c) Similar to (b), but the condition is required by organisms, as a resource, at low concentrations.

Figure 3.2
The rate of oxygen consumption of the Colorado beetle (*Leptinotarsa decemlineata*), which doubles for every 10°C rise in temperature up to 20°C, but increases less fast at higher temperatures. (After Marzusch, 1952.)

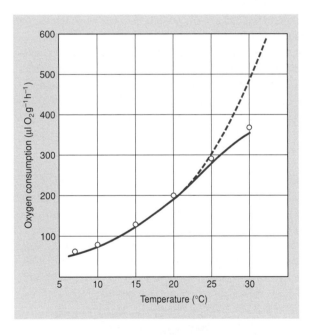

about because high temperature increases the speed of molecular movement and speeds up chemical reactions. Thus, at lower temperatures (though "lower" varies from species to species, as explained earlier) performance may be impaired simply as a result of metabolic inactivity.

high and low temperatures　　　At extremely high temperatures, on the other hand, enzymes and other proteins become unstable and break down, and the organism dies. But difficulties may set in before these extremes are reached. At high temperatures, terrestrial organisms are cooled by the evaporation of water (from open stomata on the surfaces of leaves, or through sweating or panting), but this may lead in turn to serious, perhaps lethal, problems of dehydration; or, as water reserves run low, body temperature may rise rapidly. However, even where loss of water is not a problem, for example among aquatic organisms, death is usually inevitable if temperatures are maintained for long above 60°C. The exceptions, *thermophiles*, are mostly specialized fungi and the primitive Archaebacteria. One of these, *Pyrodictium occultum*, can live at 105°C—something that is only possible because, under the pressure of the deep ocean, water does not boil at that temperature.

At temperatures a few degrees above zero, organisms may be forced into extended periods of inactivity and the cell membranes of sensitive species may begin to break down. This is known as *chilling injury*, which affects many tropical fruits—for example, bananas blacken at even a few degrees above 0°C. On the other hand, many species of both plants and animals can tolerate temperatures well below zero provided that ice does not form. If it is not disturbed, water can supercool to temperatures as low as −40°C without forming ice; but a sudden shock allows ice to form quite suddenly within plant cells, and this, rather than the low temperature itself, is then lethal, since ice formed within a cell is likely simply to disrupt and destroy it. If, however, temperatures fall slowly, ice can form between cells and draw water from within them. With dehydrated cells, the effects on plants are then very much like those of high-temperature drought.

The absolute temperature that an organism experiences is important. But the timing and duration of temperature extremes may be equally important. For example, unusually hot days in early spring may interfere with fish spawning or kill the fry but otherwise leave the adults unaffected. Similarly, a late spring frost might kill seedlings but leave saplings and larger trees unaffected. The duration and frequency of extreme conditions are also often critical. In many cases, a periodic drought or tropical storm

the timing of extremes

The saguaro cactus can only survive short periods at freezing temperatures.

may have a greater effect on a species' distribution than the average level of a condition. To take just one example: the saguaro cactus is liable to be killed when temperatures remain below freezing for 36 hours, but if there is a daily thaw it is under no threat. In Arizona, the northern and eastern edges of the cactus's distribution correspond to a line joining places where on occasional days it fails to thaw. Thus the saguaro is absent where there are occasionally lethal conditions—an individual need only be killed once.

3.2.3   Conditions as stimuli

Environmental conditions act primarily to modulate the rates of physiological processes. In addition, though, many conditions are important stimuli for growth and development and prepare an organism for conditions that are to come.

photoperiod is commonly used to time dormancy, flowering, or migration

The idea that animals and plants in nature can anticipate, and be used by us to predict, future conditions ("a big crop of berries means a harsh winter to come") is the stuff of folklore. But there are important advantages to an organism that can predict and prepare for repeated events such as the seasons. For this, the organism needs an internal clock that can be used to check against an external signal. The most widely used external signal is the length of day—the photoperiod. On the approach of winter —as the photoperiod shortens—bears, cats, and many other mammals develop a thickened fur coat, birds such as ptarmigan put on winter plumage, and very many insects enter a dormant phase (*diapause*) within the normal activity of their life cycle. Insects may even speed up their development as daylength decreases in the fall (as harsh winter conditions approach), but then speed up development again in the spring as daylength increases, once the pressure is on to have reached the adult stage by the start of the breeding season (Figure 3.3). Other photoperiodically timed events are the seasonal onset of reproductive activity in animals, the onset of flowering, and the seasonal migration in birds.

An experience of chilling is needed by many seeds before they will break dormancy. This prevents them from germinating during the moist warm weather immediately after ripening and then being killed by the winter cold. As an example, temperature

Figure 3.3
The effect of daylength on larval development time in the butterfly *Lasiommata maera* in the fall (third larval stage, before diapause) and spring. The arrows indicate the normal passage of time: daylength decreases through the fall (and development speeds up) but increases in the spring (development again speeds up). Bars are standard errors. (After Gotthard et al., 1999.)

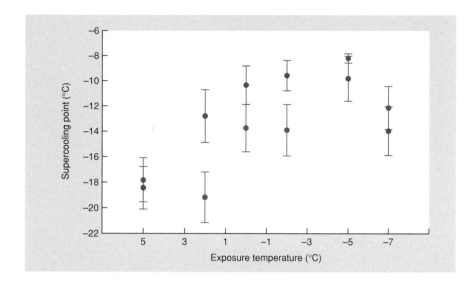

Figure 3.4
Acclimation to low temperatures. Samples of the Antarctic springtail *Cryptopygus antarcticus* were taken from field sites in the summer (ca. 5°C) on a number of days and their supercooling point (at which they froze) determined either immediately (controls, blue circles) or after a period of acclimation (red circles) at the temperatures shown. The supercooling points of controls themselves varied because of temperature variations from day to day, but acclimation at temperatures in the range +2 to −2°C (indicative of winter) led to a drop in the supercooling point, whereas no such drop was observed at higher temperatures (indicative of summer) or lower temperatures (too low for a physiological acclimation response). Bars are standard errors. (After Worland & Convey, 2001.)

and photoperiod interact to control the seed germination of birch (*Betula pubescens*). Seeds that have not been chilled need an increasing photoperiod (indicative of spring) before they will germinate; but if the seed has been chilled, it starts growth without the light stimulus. Either way, growth should be stimulated only after winter has passed. The seeds of lodgepole pine, on the other hand, remain protected in their cones until they are heated by forest fire. This stimulus is an indicator that the ground has been cleared and that new seedlings have a chance of becoming established.

Conditions may themselves trigger an altered response to the same or even more extreme conditions; for instance, exposure to relatively low temperatures may lead to an increased rate of metabolism at such temperatures and/or to an increased tolerance of even lower temperatures. This is the process of *acclimatization* (called *acclimation* when induced in the laboratory). Antarctic springtails (tiny arthropods), for instance, when taken from "summer" temperatures in the field (around 5°C in the Antarctic) and subjected to a range of acclimation temperatures, responded to temperatures in the range +2 to −2°C (indicative of winter) by showing a marked drop in the temperature at which they freeze (Figure 3.4). But at lower acclimation temperatures still (−5 and −7°C), they showed no such drop because the temperatures were themselves too low for the physiological processes required to make the acclimation response. One way in which such increased tolerance is achieved is by forming chemicals that act as antifreeze compounds: they prevent ice from forming within the cells and protect their membranes if ice does form (Figure 3.5). Acclimatization in some deciduous trees (frost hardening) can increase their tolerance of low temperatures by as much as 100°C.

acclimatization

3.2.4 The effects of conditions on interactions between organisms

Although organisms respond to each condition in their environment, the effects of conditions may be determined largely by the responses of other community members.

conditions may affect the availability of a resource

Figure 3.5
(a) Changes in the glycerol content of the larva of the goldenrod gall moth, which protect it from freezing. (b) Daily maximum and minimum temperatures. (After Rickards et al., 1987.)

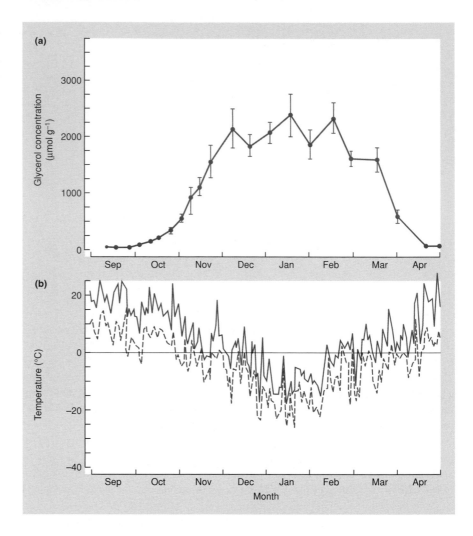

Temperature, for example, does not act on one species: it also acts on its competitors, prey, parasites, and so on. Most especially, an organism will suffer if its food is another species that cannot tolerate an environmental condition. This is illustrated by the distribution of the rush moth (*Coleophora alticolella*) in England. The moth lays its eggs on the flowers of the rush (*Juncus squarrosus*) and the caterpillars feed on the developing seeds. Above 600 m, the moths and caterpillars are little affected by the low temperatures, but the rush, although it grows, fails to ripen its seeds. This, in turn, limits the distribution of the moth, because caterpillars that hatch in the colder elevations will starve as a result of insufficient food (Randall, 1982).

the development of disease . . .   The effects of conditions on disease may also be important. Conditions may favor the spread of infection (winds carrying fungal spores), or favor the growth of the parasite, or weaken the defenses of the host. For example, during an epidemic of southern corn leaf blight (*Helminthosporium maydis*) in a corn field in Connecticut, the disease was more severe in local areas of the crop that were shaded by trees. The plants closest

Figure 3.6
The incidence of southern corn leaf blight (*Helminthosporium maydis*) on corn growing in rows at various distances from trees that shaded them. Wind-borne fungal diseases were responsible for most of this mortality (Harper, 1955). (From Lukens & Mullany, 1972.)

Figure 3.7
Changing temperature reverses the outcome of competition. At low temperature (6°C) on the left, the salmonid fish *Salvelinus malma* outsurvives cohabiting *S. leucomaenis*, whereas at 12°C, on the right, *S. leucomaenis* drives *S. malma* to extinction. Both species are quite capable, alone, of living at either temperature. (After Taniguchi & Nakano, 2000.)

to the trees that were shaded for the longest periods were the most heavily diseased (Figure 3.6).

Competition between species can also be profoundly influenced by environmental conditions, especially temperature. Two stream salmonid fishes, Dolly Varden and white-spotted char, *Salvelinus malma* and *S. leucomaenis*, coexist in sites at intermediate altitude (and therefore intermediate temperature) on Hokkaido Island, Japan, whereas only the former lives at higher altitudes (lower temperatures) and only the latter at lower altitudes. A reversal, by a change in temperature, of the outcome of competition between the species appears to play a key role in this. For example, in experimental streams supporting the two species maintained at 6°C over a 191-day period (a typical high altitude temperature), the survival of *S. malma* was far superior to that of *S. leucomaenis*; whereas at 12°C (a typical low altitude temperature), both species survived less well, but the outcome was so far reversed that by around 90 days all of the *S. malma* had died (Figure 3.7). Both species are quite capable, alone, of living at either temperature. (We return to this example in Section 6.2.2.)

... or competition

3.2.5   Responses by sedentary organisms

Motile animals have some choice over where they live: they can show preferences. They may move into shade to escape from heat or into the sun to warm up. Such choice of environmental conditions is denied to fixed or sedentary organisms. Plants are obvious examples, but so are many aquatic invertebrates such as sponges, corals, barnacles, mussels, and oysters.

form and behavior may change with the seasons

In all except equatorial environments, physical conditions follow a seasonal cycle. Indeed, there has long been a fascination with organisms' responses to these (Box 3.1). Morphological and physiological characteristics can never be ideal for all phases in the cycle and the jack of all trades is master of none. One solution is for the morphological and physiological properties of the organism to keep changing with the seasons (or even anticipating them, as in acclimatization). But continually changing may be costly:

Box 3.1   *Historical Landmarks*

Recording Seasonal Changes

Recording the changing behavior of organisms through the season (*phenology*) was essential before agricultural activities could be intelligently timed. The earliest phenological records were apparently the Wu Hou observations made in the Chou and Ch'in (1027–206 BC) dynasties. The date of the first flowering of cherry trees has been recorded at Kyoto, Japan, since AD 812.

A particularly long and detailed record was started in 1736 by Robert Marsham at his estate near the city of Norwich, England. He called these records "Indications of the Spring". Recording was continued by his descendants until 1947. Marsham recorded 27 phenological events every year: the first flowering of snowdrop, wood anemone, hawthorn, and turnip; the first leaf emergence of 13 species of tree; and various animal events such as the first appearance of migrants (swallow, cuckoo, nightingale), the first nest building by rooks, croaking of frogs and toads, and the appearance of the brimstone butterfly.

A long series of measurements of environmental temperature are not available for comparison with the whole period of Marsham's records, but they are available from 1771 for Greenwich, about 100 miles (160 km) away. There is surprisingly close agreement

between many of the flowering and leaf emergence events recorded by Marsham and the mean January–May temperature at Greenwich (Figure 3.8). However, not surprisingly, events such as the time of arrival of migrant birds bears little relationship to temperature.

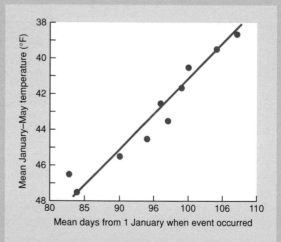

Figure 3.8
The relationships between mean January–May temperatures and the annual mean dates of ten flowering and leafing events from the classic Marsham records started in 1736. (From redrawn figures of Margary in Ford, 1982.)

Box 3.1 (cont'd)

Analysis of the Marsham data for the emergence of leaves on six species of tree indicates that the mean date of leafing is advanced by 4 days for every 1°F increase in the mean temperature from February to May (Figure 3.9). Similarly, for the eastern United States, Hopkins' *bioclimatic law* states that the indicators of spring such as leafing and flowering occur 4 days later for every 1° latitude northward, 5° longitude westward, or 120 m of altitude.

Collecting phenological records has now been transformed from the pursuit of gifted amateurs to sophisticated programs of data collection and analysis. At least 1,500 phenological observation posts are now maintained in Japan alone. The vast accumulations of data have suddenly become exciting and relevant as we try to estimate the changes in floras and faunas that will be caused by global warming.

Figure 3.9
The relationship between the mean temperature in the 4-month period February–May and the average date of six leafing events. The correlation coefficient is –0.81. (From redrawn figures of Kington in Ford, 1982.)

a deciduous tree may have leaves ideal for life in spring and summer but faces the costs of making new ones every year. Acclimatization may involve a change in biochemical characteristics to make costly antifreeze compounds. An alternative is to economize by having long-lasting leaves like those of pines, heathers, and the perennial shrubs of the desert and garrigue, but paying the cost of their more sluggish physiological processes.

3.2.6   Animal responses to environmental temperature

Most species of animals, like plants, are ectotherms: they rely on external sources of heat to determine the pace of their metabolism. This includes the invertebrates and also

ectotherms and endotherms

Figure 3.10
Changes in the body temperature of the
European hamster during a 3-day bout of
hibernation. (After Nedergaard & Cannon,
1990.)

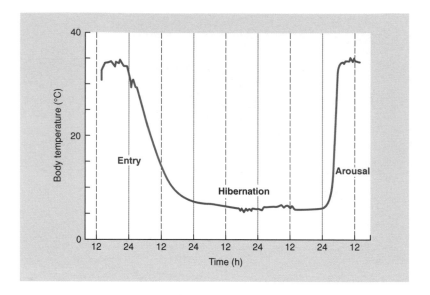

Figure 3.10
Changes in the body temperature of the European hamster during a 3-day bout of hibernation. (After Nedergaard & Cannon, 1990.)

fish, amphibians, and lizards. Others, mainly birds and mammals, are endotherms: they regulate their body temperature by producing heat within their body.

The distinction between ectotherms and endotherms is not absolute. Some typical ectotherms, some insects for example, can control body temperature through muscle activities (e.g., shivering flight muscles). Some fish and reptiles can generate heat for limited periods of time, and even some plants can use metabolic activity to raise the temperature of their flowers. Some typical endotherms, on the other hand, such as dormice, hedgehogs, and bats, allow their body temperature to fall and become scarcely different from that of their surroundings when they are hibernating (Figure 3.10).

Despite these overlaps, endothermy is inherently a different strategy from ectothermy. Over a certain narrow temperature range, an endotherm consumes energy at a basal rate. But at environmental temperatures further and further above or below that zone, endotherms expend more and more energy maintaining their constant body temperature. This makes them relatively independent of environmental conditions and allows them to stay longer at, or close to, peak performance. It makes them more efficient in both searching for food and escaping from predators. However, there is a cost—a high requirement for food to fuel this strategy.

Endotherms have morphological modifications that reduce their energetic costs. In cold climates most have low surface area to volume ratios (short ears and limbs), and this reduces heat loss through surfaces. Typically, endotherms that live in polar environments are insulated from the cold with extremely dense fur (polar bears, mink, foxes) or feathers and extra layers of fat. In contrast, desert endotherms often have thin fur, and long ears and limbs that help dissipate heat.

temperatures that vary seasonally pose special problems

Variability of conditions can set biological challenges as great as extremes. Seasonal cycles, for example, can expose an animal to summer heat close to its thermal maximum, and winter chill close to its thermal minimum. Responses to these changing conditions include the laying down of different coats in the fall (thick and underlain

The thick, white winter coat and the thinner, browner summer coat of the Arctic fox.

by a thick fat layer) and in the spring (a thinner coat and loss of the dense fat layer) (Figure 3.11). Some animals also take advantage of each other's body heat as a means to cope with cold weather by huddling together. Hibernation—relaxing temperature control—allows some vertebrates to survive periods of winter cold and food shortage (Figure 3.10) by *avoiding* the difficulties of finding sufficient fuel over these periods. Migration is another avoidance strategy: the Arctic tern, to take an extreme example, travels from the Arctic to the Antarctic and back each year, experiencing only the polar summers.

Despite the quite profound differences in the physiological characteristics of ectotherms and endotherms, they often have very similar ecological features. Large marine endotherms such as whales, porpoises, and dolphins survive broadly the same environmental conditions as large ectotherms such as sharks. Very small desert endotherms such as small seed-eating rodents share very similar conditions with small desert ectotherms such as lizards and snakes. On the other hand, there are no very large ectotherms on land that match the endothermic elephant, rhinoceros, or hippopotamus. Such ectotherms would have large volumes to heat up—but small surface areas, relative to that volume, over which to absorb heat. (There is still some dispute whether the large dinosaurs were endo- or ectotherms.) Similarly, there are plenty of species of very small fish but no comparably small aquatic endotherms (which would have a relatively large surface area over which to lose heat but only a small body volume from which to generate it).

3.2.7   Microorganisms in extreme environments

Microorganisms survive and grow in all the environments that are lived in or tolerated by animals and plants, and they show the same range of strategies—avoid, tolerate, or specialize. Many microorganisms produce resting spores that survive drought, high temperature, or cold. There are also some that are capable of growth and multiplication in conditions far outside the range of tolerance of higher organisms; they inhabit some of the most extreme environments on Earth. Temperatures maintained higher than 45°C are lethal to almost all plants and animals, but thermophilic ("temperature-loving") microbes grow at much higher temperatures. Although similar in many ways to heat-intolerant microbes, the enzymes of these thermophiles are stabilized by especially strong ionic bonds.

Microbial communities that not only tolerate but grow at low temperatures are also known; these include photosynthetic algae, diatoms, and bacteria that have been found on Antarctic sea ice. Microbial specialists have also been identified from other rare or peculiar environments: for example *acidophiles*, which thrive in environments that are highly acidic. One of them, *Thiobacillus ferroxidans*, is found in the waste from industrial metal-leaching processes and tolerates pH 1.0; *T. thiooxidans* can grow at a pH of 0! At the other end of the pH spectrum, the cyanobacterium *Plectonema nostocorum* from soda lakes can grow at pH 13. As noted previously, these oddities may be relicts from environments that prevailed much earlier in the Earth's history. Certainly, they warn us against being too narrow-minded when we consider the kind of organism we might look for on other planets.

3.3 Plant Resources

Resources may be either biotic or abiotic components of the environment: they are whatever an organism uses or consumes in its growth and maintenance, leaving less available for other organisms. When a photosynthesizing leaf intercepts radiation, it deprives some of the leaves or plants beneath it. When a caterpillar eats a leaf, there is less leaf material available for other caterpillars. By their nature, resources are both critical for survival, growth, and reproduction and also inherently a potential source of conflict and competition between organisms.

resource requirements of nonmotile organisms

If an organism can move about, it has the potential to search for its food. Organisms that are fixed and "rooted" in position cannot search. They must rely on growing toward their resources (like a shoot or root) or catching resources that move to them. The most obvious examples are green plants, which depend on (1) energy that radiates to them, (2) atmospheric carbon dioxide that diffuses to them, (3) mineral cations that they obtain from soil colloids in exchange for hydrogen ions, and (4) water and dissolved anions that their roots absorb from the soil. In the following sections, we concentrate on green plants. But it is important to remember that many of the nonmobile animals, like corals, sponges, and bivalve molluscs, depend on resources that are suspended in the watery environment and are captured by filtering the water or even just waiting for them open-mouthed.

3.3.1   Solar radiation

Solar radiation is a critical resource for green plants. We often refer to it loosely as "light", but green plants actually use only about 44 percent of that narrow part of the spectrum of solar radiation that is visible to us between infrared and ultraviolet. The rate of photosynthesis increases with the intensity of the radiation that a leaf receives, but with diminishing returns; and this relationship itself varies greatly between species (Figure 3.12), especially between those that usually live in shaded habitats (which reach saturation at low radiation intensities) and those that normally experience full sunlight and can take advantage of it.

sun and shade plants

The solar radiation that reaches a plant is forever changing. Its angle and intensity change annually and diurnally in a regular and systematic way. There are also irregular, unsystematic variations due to changes in cloud cover or shadowing by the leaves of neighboring plants. As light flecks pass over leaves lower in the canopy, they receive seconds or minutes of direct bright light and then plunge back into shade. The daily photosynthesis of a leaf integrates these various experiences; the whole plant integrates the diverse exposure of its various leaves.

There is enormous variation in the shapes and sizes of leaves that in a way parallels the diversity in the mouthparts of insects described later. But whereas insects search for widely different resources, all leaves garner the same two simple resources: radiation and carbon dioxide. Most of the heritable variation in the shapes of leaves has probably evolved under selection not primarily for high photosynthesis, but rather for optimal efficiency of water use (photosynthesis achieved per unit of water transpired) and minimization of the damage done by foraging herbivores.

Not all the variations in leaf shape are heritable, though; many are responses by the individual to its immediate environment. Many trees, especially, produce different

sun and shade leaves

Figure 3.12
The response of photosynthesis by the leaves of various types of green plant (measured as carbon dioxide uptake) to the intensity of solar radiation at optimal temperatures and with a natural supply of carbon dioxide. (After Larcher, 1980; and other sources.)

types of leaf in positions exposed to full sunlight ("sun leaves") and in places lower in the canopy where they are shaded ("shade leaves"). Sun leaves are thicker, with more densely packed chloroplasts (which process the incoming radiation) within cells, and more cell layers. The more flimsy shade leaves intercept diffused and filtered radiation low in the canopy but may nonetheless supplement the main photosynthetic activity of the sun leaves high in the canopy.

Among herbaceous plants and shrubs, specialist "sun species" or "shade species" are much more common than species with the ability to produce both sun and shade leaves. Leaves of sun plants are commonly exposed at acute angles to the midday sun and are typically superimposed into a multilayered canopy, where even the lower leaves may have a positive rate of net photosynthesis. The leaves of shade plants are typically arranged in a single-layered canopy and angled horizontally, maximizing their ability fully to capture the available radiation.

3.3.2   Water

water is lost from plants that photosynthesize

Most plant parts are largely composed of water. In some soft leaves and fruits, as much as 98 percent of the volume may be water. Yet this is a minute fraction of the water that passes from the soil through a plant to the atmosphere during plant growth. Photosynthesis depends on the plant absorbing carbon dioxide. This can only happen across surfaces that are wet—most notably the walls of the photosynthesizing cells in leaves. If a leaf allows carbon dioxide to enter, it is almost impossible to prevent water vapor from leaving. Likewise, any mechanism or process that slows down the rate of water loss, such as closing the stomata (pores) on the leaf surface, is almost bound to reduce the rate of carbon dioxide absorption and hence reduce the rate of photosynthesis.

wilting

Green plants serve as wicks that conduct water from the soil and release it to the atmosphere. If the rate of uptake falls below the rate of release, the body of the plant starts to dry out. The cells lose their turgidity and the plant wilts. This may just be temporary (though it may happen every day in summer), and they may recover and rehydrate at night. But if the deficit accumulates, the leaf is killed and the whole plant may die.

When a plant has absorbed water from the soil, less remains for others. Water that a plant absorbs from the soil may keep its leaves turgid and in a state in which it can photosynthesize—whereas a neighbor that it has deprived may have wilted and been unable to photosynthesize or even died. Water is certainly something for which plants may compete.

plant life in water deficit: avoiders and tolerators

Species of green plants differ in the ways in which they survive in dry environments. One strategy is to avoid the problem. Avoiders such as desert annuals, annual weeds, and most crop plants have a short life span: their photosynthetic activity is concentrated during periods when they can maintain a positive water balance. For the remainder of the year, they remain dormant as seeds, a stage that requires neither photosynthesis nor transpiration. Some perennial plants shed their photosynthetic tissues during periods of drought. Some species then replace them with new leaf forms that are less extravagant of water or spend the driest season with no leaves at all—just green stems.

Other plants, tolerators, have evolved a different compromise, producing long-lived leaves that transpire slowly (for example, by having few and sunken stomata).

They tolerate drought, but of course their photosynthesis is slower. These plants have sacrificed their ability to achieve rapid photosynthesis when water is abundant but gained the insurance of being able to photosynthesize at a low level throughout the seasons. This is not only a property of plants from arid areas but also of the pines and spruces that survive where water may be abundant but is usually frozen and therefore inaccessible.

The evaporation of water lowers the temperature of the body with which it is in contact. For this reason, if plants are prevented from transpiring they may overheat. This, rather than water loss itself, may be lethal. The desert honeysweet (*Tidestromia oblongifolia*) grows vigorously in Death Valley, California despite the fact that its leaves are killed if they reach 50°C—a temperature that is commonly reached in the surrounding air. Transpiration cools the surface of the leaf to a tolerable 40–45°C. Most desert plants bear hairs, spines, and waxes on the leaf surface. These reflect a high proportion of incident radiation and help to prevent overheating. Other more general modifications in desert plants include the characteristic "chunky" shape of succulents with few branches, giving a low surface area to volume ratio over which radiant heat is absorbed.

water and overheating

Specialized biochemical processes may increase the amount of photosynthesis that can be achieved per unit of water lost. The majority of plants on Earth photosynthesize using what is termed the *C3 pathway*. Although these plants are highly productive photosynthesizers, they are relatively wasteful of water, reach their maximum rates of photosynthesis at relatively low intensities of radiation, and are less successful in arid areas. Alternative pathways of photosynthesis—termed the *C4 pathway* and *CAM*—are more economical in their water use. C4 plants have a particularly high affinity for carbon dioxide—and so absorb more per unit of water lost. CAM plants open their stomata at night and absorb carbon dioxide and fix it as malic acid. They close their stomata during the day and release the cabon dioxide internally for photosynthesis. C4 and CAM plants are most common in arid and, in particular, hot arid areas. They are restricted in range because the associated costs of their systems apparently make them less competitive under less arid conditions. For example, the photosynthesis of C4 plants is inefficient at low radiation intensities (Figure 3.12) and so they are poor shade plants; while CAM plants must store their accumulated malic acid every night—most of them are succulents with extensive swollen water-storage tissues that cope with this problem.

increasing the efficiency of water use: C4 and CAM

Many of the problems in the life of terrestrial plants arise because the resources of radiation and water behave so differently. Solar radiation may be intercepted by leaves, and a small fraction may be fixed in photosynthesis. The products of photosynthesis are subsequently respired and the energy is released as heat. The remainder of the incident solar radiation is directly dissipated as heat. *It cannot be stored and can never be used again.* The behavior of water is in total contrast. Although plants may receive water as rain directly onto their leaves, they absorb scarcely any of it. Instead it passes to the soil. Some drains through the soil under the force of gravity, but much is held against gravity by capillary forces and as colloids. Plants obtain virtually all their water from this stored reserve. Sandy soils have wide pores: these do not hold much water but what is there is held with weak forces and plants can withdraw it rather easily. Clay soils have very fine pores. They retain more water against the force of gravity, but surface tension in the fine pores makes it more difficult for the plants to withdraw it.

obtaining water from the soil

The primary water-absorbing zone on roots is covered with root hairs that make intimate contact with soil particles (Figure 3.13). As water is withdrawn from the soil, the first to be released is from the wider pores, where capillary forces retain it only weakly. Subsequent water is withdrawn from narrower paths in which the water is more tightly held. Consequently, the more the soil around the roots is depleted of water, the more resistance there is to water flow. As a result of water withdrawal, roots create water depletion zones (or, more generally, resource depletion zones [RDZs]) around themselves. The faster the roots draw water from the soil, the more sharply defined the RDZs, and the more slowly water will move into that zone. In a soil that contains abundant water, rapidly transpiring plants may still wilt because water does not flow fast enough to replenish the RDZs around their root systems (or because the roots cannot explore new soil volumes fast enough).

The shapes of root systems are much less tightly programed than those of shoots. The root architecture that a plant establishes early in its life can determine its responsiveness to later events. Plants that develop under waterlogged conditions usually set down only a superficial root system. If drought develops later in the season these same plants can suffer because their root system did not tap the deeper soil layers. A deep tap root, however, will be of little use to a plant in which most water is received from occasional showers on a dry substrate. Figure 3.14 illustrates some characteristic differences between the root systems of plants from damp temperate and dry desert habitats.

3.3.3 Mineral nutrients

Roots extract water from the soil, but they also extract key minerals. Plants require mineral resources of nitrogen (N), phosphorus (P), sulfur (S), potassium (K), calcium

Figure 3.14
Profiles of root systems of plants from contrasting environments. (a–d) Northern temperate species of open ground: (a) *Lolium multiflorum*, an annual grass; (b) *Mercurialis annua*, an annual weed; (c) *Aphanes arvensis*; and (d) *Sagina procumbens*, both ephemeral weeds. (From Fitter, 1991.) (e–i) Desert shrub and semishrub species, Mid Hills, eastern Mojave Desert, California. (Redrawn from a variety of sources.)

(Ca), magnesium (Mg), and iron (Fe), together with traces of manganese (Mn), zinc (Zn), copper (Cu), and boron (B). All of these must be obtained from the soil (or directly from the water in the case of free-living aquatic plants). Soils are patchy and heterogeneous, and as roots grow through them, they may meet regions that vary

Figure 3.15
The root system developed by a young plant of wheat growing through a sandy soil with a layer of clay. Clays offer more nutrient resources and hold more water than sand and the roots respond by branching more intensively in the clay. (Courtesy of J.V. Lake.)

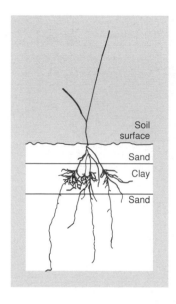

the architecture of roots determines their foraging efficiency

in nutrient and water content. They tend to branch profusely in the richer patches (Figure 3.15).

Root architecture is particularly important in this process because different nutrients behave differently and are held in the soil by different forces. Nitrate ions diffuse rapidly in soil water, and rapidly transpiring plants may bring nitrates to the root surface faster than they are accumulated in the body of the plant. However, other key nutrients such as phosphate are tightly bound in the soil (have low diffusion coefficients). The phosphate RDZs of two roots 0.2 mm apart scarcely overlap, and the parts of a finely branched root system scarcely compete with each other. Consequently, if phosphate is in short supply, a highly branched surface root will greatly improve phosphate absorbtion. A more widely spaced extensive root system, in contrast, will tend to maximize access to nitrate.

3.3.4 Carbon dioxide

Plants take in carbon dioxide through the stomatal pores on leaf surfaces and, using the energy of sunlight, capture the carbon atoms and release oxygen. Not only is carbon the base of all plant growth, and consequently the building block of every food chain, but life as we currently know it on Earth was not possible until photosynthetic organisms had substantially altered the atmosphere through the release of oxygen.

Carbon dioxide varies in its concentration at a variety of scales. In 1750, atmospheric carbon dioxide concentrations were approximately 280 μl l^{-1}. Currently, the figure is over 350 μl l^{-1} and is increasing by 0.4–0.5 percent per year, mainly as a result of the burning of fossil fuels (Box 3.2). Plants have responded to even larger fluctuations of carbon dioxide over geological history. During the Triassic, Jurassic, and Cretaceous periods, atmospheric concentrations of carbon dioxide were four to eight times greater than at present.

Box 3.2   *Topical ECOncerns*

Global Warming? Can We Risk it?

Carbon dioxide is one of several global "greenhouse gases" (see Section 13.4.2), whose increasing concentrations are believed by most scientists to be leading to rises in global mean temperatures, to a growth in the number of "extreme" and "record" weather events, and to the threat of the major biomes of the earth substantially changing their distribution (Box 4.3).

The Intergovernmental Panel on Climate Change (IPCC) was established in 1988 by the World Meteorological Organization (WMO) and the United Nations Environment Programme (UNEP). Each report produced by the IPCC is written by some 200 independent scientists and other experts in approximately 120 countries around the world and is reviewed by another 400 independent experts.

The most recent report (IPCC, 2001) describes the current state of understanding of the climate system, provides estimates of future change, and highlights areas of uncertainty. It concludes that an increasing body of observations point to a warming world—temperatures have risen during the past four decades in the lowest 8 km of the atmosphere, global average sea level has risen, and snow cover and ice extent have decreased. These changes have occurred at the same time as atmospheric greenhouse gases continue to increase due to human activities and the panel points to new and stronger evidence that most of the observed warming over the past half century is attributable to human activities. Scientists now have greater confidence in the ability of models to project future climate, and all reasonable scenarios point to a substantial temperature increase. Globally averaged surface temperature is expected to increase by between 1.4 and 5.8°C in the period 1990 to 2100, with complex consequences for weather patterns and sea level.

Policy-makers and law-makers are faced with different groups of scientific "experts" offering different projections into the future, and with many interest groups, including a number of industries, resisting attempts to force them to change their behavior in order to reduce emissions of greenhouse gases. Even though the majority of scientists believe the problem to be a very real one, the truth is that predictions of the future can never be made with absolute certainty.

*Put yourself in the position of a politician. Would it be reasonable of you to demand major changes of significant sectors of the national economy, in order to avert a disaster that may never happen in any case? Or, since the consequences of the "worst case" and even some of the "middle-of-the-road" scenarios are so profound, is the only responsible course of action to minimize risk—to behave as if disaster **is** certain if we do not change our collective behavior, even though it is not? One alternative might be to wait for better data. But suppose that by the time better data are available it is too late . . .*

Increasing carbon dioxide above atmospheric levels in greenhouses can increase crop yields. It might therefore be thought that global increases in atmospheric carbon dioxide will lead to global increases in crop production. The reality seems to be much more complicated, and it is not easy to see a general trend in the behavior of either crop plants or natural vegetation (Bazzaz, 1996).

In a terrestrial community, carbon dioxide increases and moves upward through the canopy at night as plants respire and there is no photosynthesis. The release of carbon dioxide from decomposing organic matter in the soil also contributes to the balance sheet. During the day, carbon dioxide is absorbed by plants within the canopy and the flux is downward as plants actively remove carbon dioxide from the air. It is very doubtful whether plants in the field ever compete with each other for carbon dioxide. Gaseous diffusion and mixing are extremely rapid, and active uptake of carbon dioxide by one plant is very unlikely to create resource depletion zones that overlap with those of its neighbors.

3.4 Animals and their Resources

autotrophs and heterotrophs

Green plants are *autotrophs*: their resources are quanta of radiation, ions, and simple molecules. Plants assemble them into complex molecules (carbohydrates, fats, and proteins) and then package them into cells, tissues, organs, and whole organisms. It is these packages that form the food resources for virtually all other organisms, the *heterotrophs* (decomposers, predators, grazers, and parasites). These consumers unpack the packages, metabolize and excrete some of the contents, and reassemble the remainder into their own bodies. They in turn may be consumed, unpacked, and reconstituted in a chain of events in which each consumer becomes, in turn, a resource for some other consumer.

Heterotrophs can generally be grouped as follows:

1 *Decomposers*, which feed on already dead plants and animals.
2 *Parasites*, which feed on one or a very few host plants or animals while they are alive but do not (usually) kill their hosts, at least not immediately.
3 *Predators*, which, during their life, eat many prey organisms, typically (and in many cases always) killing them.
4 *Grazers*, which, during their life, consume parts of many prey organisms, but do not (usually) kill their prey, at least not immediately.

The usual mental image of a predator–prey relationship is something akin to a lion eating a gazelle, but the relationship encompasses a much wider array of consumer–resource interactions. For example, a squirrel is a predator when it eats an acorn (it kills the acorn embryo); a whale is a predator as it feeds on krill; a fungus can be regarded as a predator when it feeds on and kills a growing seedling; and carnivorous plants are of course predators of their (usually) insect prey. In each case, the predator kills its food resource as it consumes all or part of it. Here, we concentrate on animal consumers (and take the subject further still in Chapter 8).

monophagy and polyphagy

An important distinction between animal consumers is whether they are specialized or generalized in their diet. Generalists (*polyphagous* species) take a wide variety of prey species though they very often have clear preferences and a rank order of what they will choose when there are alternatives available. Specialists, on the other hand, may specialize on particular parts of their prey but range over a number of species. This is most common among herbivores because, as we shall see, different parts of plants are quite different in their composition. Thus, many birds specialize on eating seeds though they are seldom restricted to a particular species. Finally, a consumer may specialize on a single species or a narrow range of closely related species (when it is said to be *monophagous*). Examples are caterpillars of the cinnabar moth (which eat leaves, flower buds, and very young stems of species of ragwort, *Senecio*) and many species of host-specific parasites.

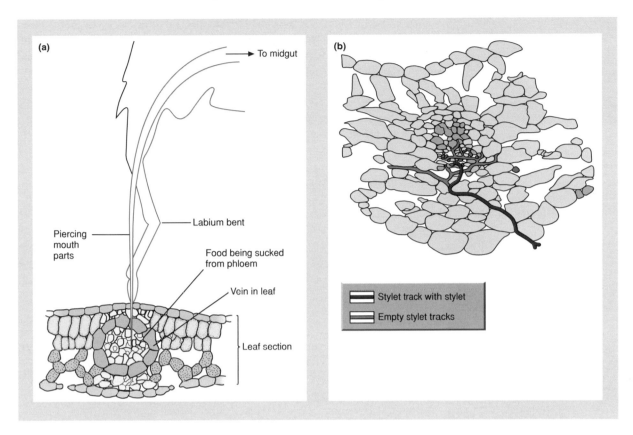

Figure 3.16
The stylet of an aphid penetrating the host tissues and reaching the sugar-rich phloem cells in the leaf veins. (a) Aphid mouthparts and cross-section of a leaf. (b) A stylet, showing its circuitous path through a leaf. (After Tjallingii & Hogen Esch, 1993.)

Many of the resource-use patterns found among animals reflect the different life spans of the consumer and what it consumes. Individuals of long-lived species are likely to be generalists: they cannot depend on one food resource being available throughout their life. Specialization is increasingly likely if a consumer has a short life span. Evolutionary forces can then shape the timing of the consumer's food demands to match the timetable of its prey. Specialization also allows the evolution of structures that make it possible to deal very efficiently with particular resources—this is especially the case with mouthparts. As we saw in Chapter 2, a structure like the stylet of the aphid (Figure 3.16) can be interpreted as an exquisite product of the evolutionary process that has given the aphid access to a valuable food resource—or as an example of the ever-deepening rut of specialization that has constrained what aphids can feed on. The more specialized the food resource required by an organism, the more it is constrained to live in patches of that resource *or* to spend time and energy in searching for it among a mixture of resources. This is one of the costs of specialization.

A number of these points can be illustrated by reference to a population of wild raspberries in a temperate woodland. Flowering brings only a short period of nectar production. Bees feed from this but also make use of nectar from many other species

the importance of life spans

and hence continue to feed throughout the summer. By contrast, the raspberry beetle lays eggs in, and its larvae feed on, the flowers of *only* this species, and the larva completes its life cycle within the developing fruit. It then remains, inactive, as a pupa until the next raspberry flowering season in 10–11 months' time. The larva of the raspberry moth, on the other hand, has a much longer active period because it feeds on a consistently present resource—the pith within the woody stems. Thus, a single resource (the raspberry plant) can be used by a variety of types of consumer, or, to look at it another way, a single species may provide many different resources.

3.4.1   Nutritional needs and provisions

plants as (a variety of) foods

The various parts of a plant have very different compositions (Figure 3.17a) and so offer quite different resources. Bark, for example, is largely composed of dead cells with corky and lignified walls, packed with defensive phenolics, and is quite useless as a food for most herbivores (even species of "bark beetle" specialize on the nutritious cambium layer just beneath the bark, rather than on the bark itself). The richest

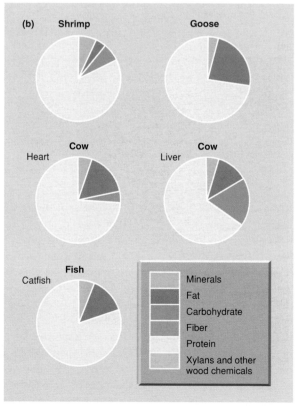

Figure 3.17
The composition of various plants and animals that may serve as food resources for herbivores or carnivores. Note that the different parts of a plant (a) have very different composition, whereas different species of animal (and their parts) (b) are remarkably similar.

concentrations of plant proteins (and hence of nitrogen) are in the meristems in the buds at shoot apices and in leaf axils. Not surprisingly, these are usually heavily protected with bud scales and defended from herbivores by prickles and spines. Seeds are usually dried, packaged reserves rich in starch or oils as well as specialized storage proteins. And the very sugary and fleshy fruits are resources provided by the plant as "payment" to animals that disperse the seeds. Very little of the plant's nitrogen is "spent" on these rewards.

The diversity of different food resources offered by plants is matched by the diversity of specialized mouthparts and digestive tracts that have evolved to consume them. This diversity is especially developed in the beaks of birds and the mouthparts of insects (Figure 3.18).

For a consumer, the body of a plant is a quite different package of resources from the body of an animal. First, plant cells are bounded by walls of cellulose, lignin, and other structural carbohydrates that give plants their high fiber content and contribute to their high ratio of carbon to other elements. These large amounts of fixed carbon mean that plants are potentially rich sources of energy. Yet the overwhelming majority of animal species lack cellulolytic and other enzymes that can digest these compounds; they are quite useless as a direct energy resource for most herbivores. Moreover, the cell wall material of plants hinders the access of digestive enzymes to the plant cell contents. The acts of chewing by the grazing mammal, cooking by humans, and grinding

from plants into animals

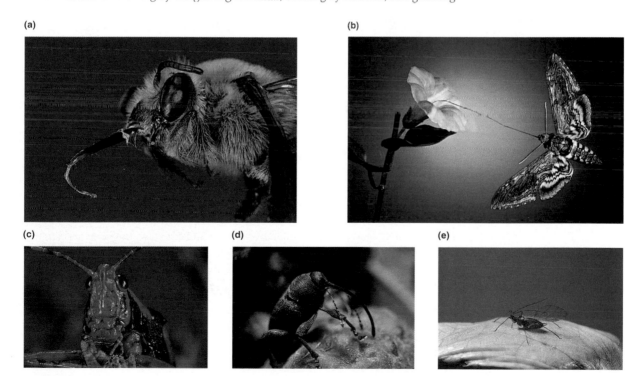

(a)   **(b)**

(c)   **(d)**   **(e)**

Figure 3.18
Examples of the variety of specialized mouthparts in herbivorous insects. (a) Honeybee with a proboscis for sucking (© Doug Sokell, Visuals Unlimited). (b) Hawkmoth with an even longer proboscis (© Visuals Unlimited). (c) Leichhardt's grasshopper with large plate-like chewing mandibles (© Mantis Wildlife Films, Oxford Scientific Films). (d) Acorn weevil with chewing mouthparts at the very end of its long rostrum (© Oxford Scientific Films). (e) Rose aphid with a piercing stylet (© Oxford Scientific Films).

in the gizzard of birds are necessary precursors to digestion of plant food because they allow digestive enzymes access to the cell contents. The carnivore, by contrast, can more safely gulp its food.

Many herbivores have made up for their own lack of cellulolytic enzymes by entering into a *mutualistic* (beneficial to both parties) association with cellulolytic bacteria and protozoa in their guts which do have the appropriate enzymes. The rumen (or sometimes the cecum) of many herbivorous mammals is a temperature-regulated culture chamber for these microbes into which already partially fragmented plant tissues flow continually. The microbes receive a home and a supply of food. The herbivorous "host" benefits by absorbing many of the major by-products of this microbial fermentation, especially fatty acids.

Unlike plants, animal tissues contain no structural carbohydrate or fiber component but are rich in fat and protein. The C : N ratio of plant tissues commonly exceeds 40 : 1, in contrast to ratios that rarely exceed 10 : 1 in bacteria, fungi, and animals. Thus herbivores, which undertake the first stage of making animal bodies out of plants, are involved in a massive burning off of carbon as the C : N ratio is lowered. The main waste products of herbivores are therefore carbon-rich compounds (carbon dioxide and fiber). Carnivores, on the other hand, get most of their energy from the protein and fats of their prey, and their main excretory products are consequently nitrogenous.

Even if the cell wall is excluded from consideration, the C : N ratio is high in plants compared with other organisms. Aphids, which gain direct access to cell contents by driving their stylets into the phloem transport system (Figure 3.16), gain a resource that is rich in soluble sugars (Figure 3.17a). In their search for valuable nitrogen, they use only a fraction of this energy resource and excrete the rest in sugary-rich honeydew that may drip as a rain from an aphid-infested tree. For most herbivores and decomposers, the body of a plant is a superabundant source of energy and carbon; it is other components of the diet, especially nitrogen, that are more usually limiting.

animals as food

The bodies of different species of animal have remarkably similar composition (Figure 3.17b). In terms of protein, carbohydrate, fat, water, and minerals per gram, there is very little to choose between a diet of caterpillars or cod, or of earthworms, shrimps, or venison. The packages may be differently packaged (and the taste may be different), but the contents are essentially the same. Moreover, the different parts of an animal have very similar nutritional content. Unlike herbivores, carnivores are not faced with difficult problems of digestion (and they vary very little in their digestive apparatus) but rather with difficulties in finding, catching, and overcoming the defenses of their prey.

3.4.2 Defense

The value of a resource to a consumer is determined not only by what it contains but by how well its contents are defended. Not surprisingly, organisms have evolved physical, chemical, morphological, and behavioral defenses against being attacked. These serve to reduce the chance of an encounter with a consumer and/or increase the chance of survival in such encounters. The spiny leaves of holly are not eaten by larvae of the oak

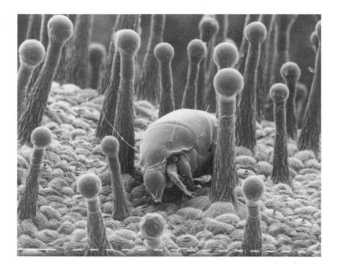

Figure 3.19
A mite trapped in the protective trichomes (hairs) on the surface of a *Primula* leaf. The trichomes themselves support capsules of irritant volatile oils at their tip. Each white bar towards the foot of the image represents 10 μm. (Courtesy of C.J. Veltkamp.)

eggar moth, but if the spines are removed the leaves are eaten quite readily. No doubt similar results would be achieved in equivalent experiments with foxes as predators and de-spined hedgehogs or porcupines as prey. On a smaller scale, many plant surfaces are clothed in epidermal hairs (trichomes) that may keep the smaller predators (such as thrips and mites) away from the leaf surface (Figure 3.19).

Any feature of an organism which increases the energy spent by a consumer in discovering or handling it is a defense if, as a consequence, the consumer eats less of it. The thick shell of a nut increases the time that an animal spends extracting a unit of effective food, and this may reduce the number of nuts that are eaten. We have already seen that most green plants are relatively overprovided with energy resources in the form of cellulose and lignin. It may therefore be cheap to build husks and shells around seeds (and woody spines on stems) if these defense tissues contain rather little protein, and if what is protected are the real riches.

some resources are protected . . .

Both plants and animals have a battery of chemical defenses. The plant kingdom, in particular, is very rich in "secondary" chemicals that apparently play no role in normal plant biochemical pathways. A defensive function is generally ascribed to these chemicals and a defensive role has been demonstrated unequivocally in some cases. Populations of white clover, for example, commonly contain some individuals that release hydrogen cyanide when their tissues are attacked (*cyanogenic* forms) and others that do not. Those that cannot are eaten by slugs and snails. The cyanogenic forms, however, are nibbled but then rejected (Table 3.1). Noxious plant chemicals have been broadly classified into two types: toxic (or qualitative) chemicals, which are poisonous even in small quantities, and quantitative chemicals, such as tannins, which act by binding proteins and make the tissues that contain them, such as mature oak leaves, relatively indigestible. The growth of caterpillars of the winter moth decreases with an increase in the concentration of tannin included in the diet.

. . . or defended

Animals have more options than plants when it comes to defending themselves, but some still make use of chemicals. For example, defensive secretions of sulfuric acid of pH 1 or 2 occur in some marine gastropod groups, including the cowries. Other

chemical defense in animals

Table 3.1
Slugs (*Agriolimax reticulatus*) graze on the leaves of clover (*Trifolium repens*). There are forms of clover that release hydrogen cyanide when the cells are damaged. Slugs nibble clover leaves and reject cyanogenic forms but continue to consume the leaves of noncyanogenic forms. Two plants, one of each form, were grown together in plastic containers and slugs were allowed to graze for seven successive nights. The table shows the numbers of leaves in different conditions after slug grazing. +/− indicates deviation from random expectation; the difference from random expectation is significant at $P < 0.001$. (After Dirzo & Harper, 1982.)

		Conditions of leaves after grazing		
	Not damaged	Nibbled	Up to 50% of leaf removed	More than 50% of leaf removed
Cyanogenic plants	160 (+)	22 (+)	38 (−)	9 (−)
Noncyanogenic plants	87 (−)	7 (−)	30 (+)	65 (+)

animals, which can tolerate the chemical defenses of their plant food, actually store the plant toxins and use them in their own defense. A classic example is the monarch butterfly, whose caterpillars feed on milkweeds, which contain cardiac glycosides, which are poisonous to mammals and birds. These caterpillars can store the poison, and it is still present in the adult. Thus, a bluejay will vomit violently after eating one, and, once it recovers, will reject all others on sight. In contrast, monarchs reared on cabbage are edible.

Chemical defenses are not equally effective against all consumers. Indeed, what is unacceptable to some animals may be the chosen, even unique, diet of others. Many herbivores, particularly insects, specialize on one or a few plant species whose particular defense they have overcome. For example, females of the cabbage root fly, with eggs to lay, home in on a brassica crop from distances as far as 15 m downwind of the plants. It is probably hydrolyzed glucosinolates (toxic to many other species) that provide the attractive odor.

crypsis, aposematism, and mimicry   An animal may be less obvious to a predator if it matches its background, or possesses a pattern that disrupts its outline, or resembles an inedible feature of its environment. A straightforward example of such *crypsis* is the green coloration of many grasshoppers and caterpillars. Cryptic animals may be highly palatable, but their morphological traits and color (and their choice of the appropriate background) reduce the likelihood that they will be used as a resource. In contrast, noxious or dangerous animals often seem to advertize the fact by bright, conspicuous colors and patterns (*aposematism*). The monarch butterfly (see earlier), for example, is aposematically colored. One attempt by a bird to eat an adult monarch is so memorable that others are subsequently avoided for some time. The adoption of memorable body patterns by distasteful prey, moreover, immediately opens the door for deceit by other species—there will be a clear advantage to a palatable prey if it *mimics* an unpalatable species. Thus, the palatable viceroy butterfly mimics the distasteful monarch, and a bluejay that has learned to avoid monarchs will also avoid viceroys.

behavior   By living in holes (e.g., millipedes, moles), animals may avoid stimulating the sensory receptors of predators, and by "playing dead" (e.g., opossum, African ground

Lepidopterous caterpillars illustrate a range of defense strategies. (a) the irritating hairs of the gypsy moth; (b) aposematism (advertising distastefulness) in the black swallowtail; (c) a cryptic (camouflaged) noctuid (it looks like bark); and (d) another swallowtail rearing and hence possibly startling a potential predator.

squirrel) animals may fail to stimulate a killing response. Animals that withdraw to a prepared retreat (e.g., rabbits and prairie dogs to their burrows, snails to their shells) or roll up and protect their vulnerable parts by a tough exterior (e.g., armadillos, hedgehogs) reduce their chance of capture. Other animals seem to try to bluff themselves out of trouble by threat displays. The startle response of moths and butterflies, which suddenly exposes eye spots on their wings, is one example. No doubt the most common behavioral response of an animal in danger of becoming a used resource is to run away.

3.5   The Effect of Intraspecific Competition for Resources

Resources are consumed. The consequence is that there may not be enough of a resource to satisfy the needs of a whole population of individuals. Individuals may then compete with each other for the limited resource. *Intraspecific competition* is competition between individuals of the same species.

In many cases, competing individuals do not interact with one another directly. Rather, they deplete the resources that are available to each other. Grasshoppers may compete for food, but a grasshopper is not directly affected by other grasshoppers so much as by the level to which they have reduced the food supply. Two grass plants may compete, and each may be adversely affected by the presence of close neighbors, but this is most likely to be because their resource depletion zones overlap—each may shade its neighbors from the incoming flow of radiation, and water or nutrients may be

exploitation: competitors depleting each other's resources

Figure 3.20
A population of the freshwater diatom
Asterionella formosa was grown in flasks of
culture medium. The diatom consumes
silicate during growth and the population of
diatoms stablizes when the silicate has been
reduced to a very low concentration.
(After Tilman et al., 1981.)

less accessible than they would otherwise be around the plants' roots. The data in Figure 3.20, for example, show the dynamics of the interaction between a single-celled aquatic plant, a diatom, and one of the resources it requires, silicate. As diatom density increases over time, silicate concentration decreases: there is then less available for the many than there had previously been for the few. This type of competition—in which competitors interact only indirectly, through their shared resources—is termed *exploitation.*

direct interference

On the other hand, competing individual vultures may fight one another over access to a newly found carcass. Individuals of other species may fight for ownership of a "territory" and access to the resources that a territory brings with it. A barnacle that settles on a rock denies the space to another barnacle. This is called *interference* competition.

competition and vital rates

Whether competition occurs through exploitation, interference, or a combination of the two, its ultimate effect is on the *vital rates* of the competitors—their survival, growth, and reproduction—compared with what they would have been if resources had been more abundant. Competition typically leads to decreased rates of resource intake per individual, and thus to decreased rates of individual growth or development, and perhaps to decreases in amounts of stored reserves or to increased risks of predation. Figure 3.21a shows how the mortality rate of steelhead trout increases as the number of competing fish rises, at a range of food levels. Figure 3.21b shows how the birth rate of the sand dune grass *Vulpia* declines as individuals become increasingly crowded.

In practice, intraspecific competition is often a very one-sided affair: a strong early seedling will shade and suppress a stunted, late one; a large vulture is likely to fight off a smaller one. Some of the competitive strength of individuals is related to timing (the early seedling) or to random events (one seed may germinate in a depression where it obtains more water than its neighbors). Sometimes the winner and loser may be genetically different and then competition will be playing a role in natural selection.

density dependence

The effects of intraspecific competition on any individual are typically greater the more crowded the individual is by its neighbors—the more the resource depletion zones of other individuals overlap its own. This often translates into saying that the greater the density of a population of competitors the greater is the effect of

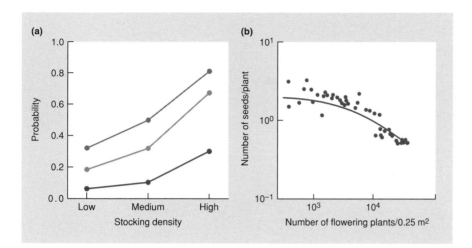

Figure 3.21
(a) The probability of mortality among steelhead trout (*Oncorynchus mykiss*) reared at a range of densities (32, 63 and 127 per m^2) and at a range of food levels (1.4, 2.9 and 5.8 g of food pellets per day: orange, mauve, and blue lines, respectively). (After Keeley, 2001.) (b) The average number of seeds produced per plant of the dune grass *Vulpia fasciculata* growing at a range of densities. (After Watkinson & Harper, 1978.)

competition. Hence, the effects of intraspecific competition are often said to be *density dependent*. But it is doubtful that any organism has a way of detecting the density of its population! Rather, it responds to the effects of being crowded.

On the other hand, at low densities in the case of *Vulpia* (Figure 3.21b), the *per capita* birth rate or fecundity was *independent* of density (where *per capita* means literally "per head" or "per individual"). That is, the fecundity was effectively the same at a density of 1,000 plants/0.25 m^2 as it was at a density of 500/0.25 m^2. Thus, there is no evidence at these densities that individuals are affected by the presence of other individuals and hence no evidence of intraspecific competition. But as density increases further, the per capita birth rate progressively decreases. These effects are now density dependent, and this may be taken as an indication that at these densities, individuals are suffering as a result of intraspecific competition.

The patterns in Figure 3.21 make the point that as crowding (or density) increases, the fecundity per individual is likely to decline and the mortality per individual likely to increase (which would mean that the survival rate per individual would *decrease*). But what can we expect to happen to the *total* number of seeds or eggs produced by populations at different densities—or to the total number of survivors? In some cases, although the rate per individual declines with increasing density, the total fecundity or total number of survivors in the population continues to increase ("undercompensation"). This can be seen (Figure 3.22a) to have been the case for the plant populations in Figure 3.21b—at least over the range of densities examined. In other cases, the rate per individual declines so rapidly with increasing density that the total fecundity or total number of survivors in the population actually gets smaller the greater the number of contributing individuals ("overcompensation"). This can be seen in Figure 3.22b for the highest densities of a bacterial parasite of the planktonic crustacean *Daphnia magna*.

In yet further cases, the mortality risk or fecundity per individual declines with increasing density such that the total number of survivors or total fecundity is the same irrespective of the number of contributing individuals. This is referred to as *exactly compensating density dependence*, and the competition leading to it is sometimes referred to as "contest-like", since this is the pattern you would expect to see if there were a fixed number of winners and all the other competitors were doomed to lose. Examples are

competition and the total number of survivors

Figure 3.22
Undercompensating, overcompensating, and exactly compensating effects of intraspecific competition. (a) An undercompensating effect on fecundity: the total number of seeds produced by *Vulpia fasciculata* continues to rise as density increases. (After Watkinson & Harper, 1978.) (b) When the planktonic crustacean *Daphnia magna* was infected with varying numbers of spores of the bacterium *Pasteuria ramosa*, the total number of spores produced per host in the next generation was independent of density (exactly compensating) at the lower densities, but declined with increasing density (overcompensating) at the higher densities. Standard errors are shown. (After Ebert et al., 2000.) (c) An exactly compensating effect on mortality: the number of surviving trout fry is independent of initial density at higher densities. (After Le Cren, 1973.) (d) The total number of eggs of the parasitic nematode *Marshallagia marshalli* produced by infected reindeer (eggs per gram of feces) increased in direct proportion to the number of adult nematodes in the reindeer: there was no evidence of competition between the nematodes. (After Irvine et al., 2001. Reprinted with the permission of Cambridge University Press.)

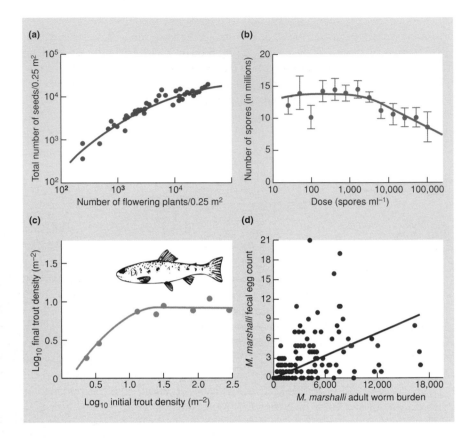

shown for fecundity in Figure 3.22b (at the lower densities) and for survivors in Figure 3.22c. Finally, of course, birth or mortality rates may be density independent (no competition) throughout the range examined, in which case the total number of births or survivors will simply continue to rise in direct proportion to the original density (e.g., Figure 3.22d).

3.6   Conditions, Resources, and the Ecological Niche

Finally, many of the ideas in this chapter can be brought together in the concept of the ecological niche. The term *niche*, though, is frequently misunderstood and misused. It is often used loosely to describe the sort of place in which an organism lives, as in the sentence "Woodlands are the niche of woodpeckers." However, more strictly, where an organism lives is its *habitat*. A niche is not a place but an idea: a summary of the organism's tolerances and requirements. The habitat of a gut microorganism would be an animal's alimentary canal; the habitat of an aphid might be a garden; and the habitat of a fish could be a whole lake. Each habitat, however, provides many different niches: many other organisms also live in the gut, the garden, or the lake—and with quite different lifestyles. The word *niche* began to gain its present scientific meaning when Charles Elton wrote in 1933 that the niche of an organism is its mode of life "in the sense that we speak of trades or jobs or professions in a human community". The niche of an organism started to be used to describe how, rather than just where, an organism lives.

The modern concept of the niche was proposed by Evelyn Hutchinson in 1957 to address the ways in which tolerances and requirements interact to define the conditions and resources needed by an individual (or a species) in order to practice its way of life. Temperature, for instance, is a condition that limits the growth and reproduction of all organisms, but different organisms tolerate different ranges of temperature. This range is one *dimension* of an organism's ecological niche. Figure 3.23a shows how

the niche of an organism is defined by its needs and tolerances

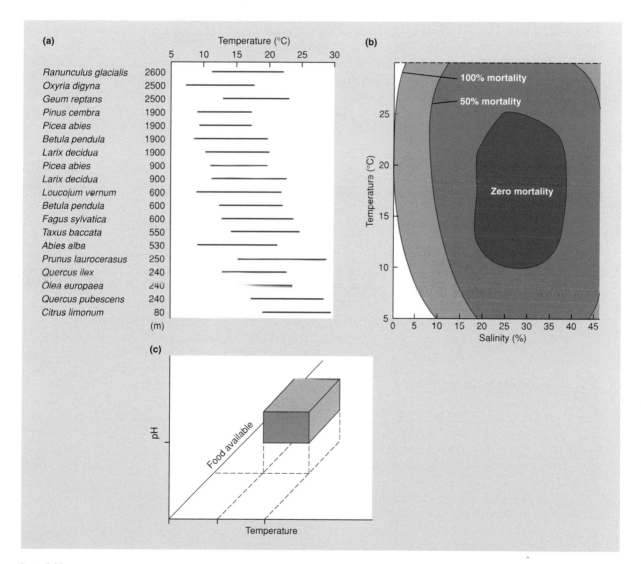

Figure 3.23
(a) A niche in one dimension. The range of temperatures at which a variety of plant species from the European Alps can achieve net photosynthesis of low intensities of radiation (70 W m^{-2}). (After Pisek et al., 1973.) (b) A niche in two dimensions for the sand shrimp (*Crangon septemspinosa*) showing the fate of egg-bearing females in aerated water at a range of temperatures and salinities. (After Haefner, 1970.) (c) A diagrammatic niche in three dimensions for an aquatic organism showing a volume defined by the temperature, pH, and availability of food.

species of plants vary in the temperature dimension of their niche. But there are many such dimensions for the niche of a species: its tolerance of various other conditions (relative humidity, pH, wind speed, water flow, and so on), and its need for various resources (nutrients, water, food, and so on). Clearly the real niche of a species must be multidimensional.

It is easy to visualize the early stages of building such a multidimensional niche. Figure 3.23b illustrates the way in which two niche dimensions (temperature and salinity) together define a two-dimensional area that is part of the niche of a sand shrimp. Three dimensions, such as temperature, pH, and the availability of a particular food, may define a three-dimensional niche volume (Figure 3.23c). It is hard to imagine (and impossible to draw) a diagram of a more realistic, multidimensional niche. (Technically, we now consider a niche to be an *n-dimensional hypervolume*, where *n* is the number of dimensions that make up the niche.) But the simplified three-dimensional version nonetheless captures the idea of the ecological niche of a species. It is defined by the boundaries that limit where it can live, grow, and reproduce, and it is very clearly a concept rather than a place. The concept has become a cornerstone of ecological thought, as we shall see in later chapters.

Summary

Conditions and resources

Conditions are physicochemical features of the environment such as its temperature and humidity. They may be altered but are not consumed. Environmental resources are consumed by living organisms in the course of their growth and reproduction.

Environmental conditions

There are three basic types of response curve to conditions. Extreme conditions may be lethal, with, between the two extremes, a continuum of more favorable conditions; or a condition may be lethal only at high intensities; or a condition may be required by organisms at low concentrations but becomes toxic at high concentrations.

These responses are accounted for, in part, by changes in metabolic effectiveness. But at extremely high temperatures, for example, enzymes and other proteins become unstable and break down, and the organism dies; and at high environmental

temperatures, terrestrial organisms may encounter serious, perhaps lethal, problems of dehydration. At temperatures a few degrees above zero, organisms may be forced into extended periods of inactivity, or ice may form between cells and draw water from within them. The timing and duration of temperature extremes, however, may be as important as absolute temperatures.

In practice, the effects of conditions may be determined largely by the responses of other community members, through food consumption, disease, or competition.

Many conditions are important stimuli for growth and development and prepare an organism for conditions that are to come.

Plant resources

Solar radiation, water, minerals, and carbon dioxide are all critical resources for green plants. The shape of the curve that relates the rate of photosynthesis to the intensity of radiation varies greatly among species. The radiation that reaches a plant is forever changing;

the plant integrates the diverse exposures of its various leaves.

Most variation in leaf shape has probably evolved under selection to optimize the photosynthesis achieved per unit of water transpired. Any mechanism or process that slows the rate of water loss, such as closing of the stomata, reduces the photosynthetic rate. If the rate of water uptake falls below the rate of release, the body of the plant starts to wilt. If the deficit accumulates, the whole plant may die. Plants may avoid or tolerate water shortage. Specialized biochemical processes may increase the amount of photosynthesis that can be achieved per unit of water lost in C4 and CAM (as opposed to C3) plants.

The primary water-absorbing zone on roots is covered with root hairs that make intimate contact with soil particles. Roots create water depletion zones around themselves. Root architectures are much less tightly programed than those of shoots, and those established early in a plant's life can determine its responsiveness to later events. Roots also extract key minerals from the soil. Root architecture is particularly important here because different nutrients are held in the soil by different forces.

Animals and their resources

Green plants are autotrophs. Decomposers, predators, grazers, and parasites are heterotrophs. The various parts of a plant have very different compositions and so offer quite different resources. This diversity is matched by the diversity of mouthparts and digestive tracts that have evolved to consume them. The body of a plant is a quite different package of resources from the body of an animal. To make better use of plant material, many herbivores enter into a mutualistic association with cellulolytic bacteria and protozoa in their alimentary canal.

The C : N ratio of plant tissues greatly exceeds those of bacteria, fungi, and animals. Thus, herbivores typically have a superabundant source of energy and carbon, but nitrogen is often limiting; their main waste products are carbon dioxide and fiber. The bodies of different species of animal have remarkably similar compositions. Carnivores are not faced with problems of digestion, but rather with difficulties in finding, catching, and overcoming the defenses of their prey. Carnivores' main excretory products are nitrogenous.

The effect of intraspecific competition for resources

Individuals may compete indirectly, via a shared resource, through exploitation, or directly, through interference. The ultimate effect of competition is on survival, growth, and reproduction of individuals. Typically, the greater the density of a population of competitors, the greater is the effect of competition (density dependence). As a result, though, the total number of survivors, or of offspring, may increase, decrease, or stay the same as initial densities increase.

Conditions, resources, and the ecological niche

Where an organism lives is its habitat. A niche is a summary of an organism's tolerances and requirements. The modern concept, proposed by Hutchinson in 1957, is an n-dimensional hypervolume.

Review Questions

1 ▲ Explain, referring to a variety of specific organisms, how the amount of water in different organisms' habitats may define either the conditions for these organisms, or their resource level, or both.

2 Discuss whether you think the following statement is correct: "A layperson might describe Antarctica as an extreme environment, but an ecologist should never do so."

3 In what ways do ectotherms and endotherms differ, and in what ways are they similar?

4 ▲ Drawing examples from a variety of both animals and plants, contrast the responses of tolerators and avoiders to seasonal variations in environmental conditions and resources.

5 Describe how plants' requirements to increase the rate of photosynthesis and to decrease the rate of water loss interact. Describe, too, the strategies used by different types of plants to balance these requirements.

6 ▲ Describe and account for the differences in both root and shoot architecture exhibited by different plants.

7 Account for the fact that the tissues of plants and animals have such contrasting carbon-to-nitrogen ratios. What are the consequences of these differences?

8 Describe the various ways in which animals use color to defend themselves against attacks by predators.

9 Explain, with examples, what exploitation and interference intraspecific competition have in common and how they differ.

10 What is meant when an ecological niche is described as an n-dimensional hypervolume?

▲ = Challenge Question

Chapter 4

Conditions, Resources, and the World's Communities

The interplay of conditions and resources profoundly influences the composition of the world's communities. At the global scale, patterns of climate circulation are largely responsible for distinctive terrestrial biomes, such as deserts and rain forests, with their characteristic assemblages of plants and animals. Distinct types of marine and freshwater communities can sometimes also be identified at a broad geographic scale. Within each biome or aquatic category, however, there are enormous variations in conditions and resources that are reflected in community patterns viewed at a smaller scale.

Chapter Contents

4.1 Introduction

4.2 Geographical patterns at large and small scales

4.3 Temporal patterns in conditions and resources—succession

4.4 Terrestrial biomes

4.5 Aquatic environments

Key Concepts

In this chapter you will

- understand that conditions and resources interact to help determine the composition of whole communities
- appreciate that climatic patterns over the surface of the Earth are responsible for the large-scale pattern of distribution of terrestrial biomes (such as tropical rain forest, desert, and tundra)
- recognize that biomes are not homogeneous because local topography, geology, and soil influence the communities of plants and animals that occur
- understand that in most aquatic environments it is difficult to recognize anything comparable to terrestrial biomes: communities tend to reflect local conditions and resources rather than global patterns in climate
- appreciate that conditions and resources at a location may change over timescales ranging from hours to millennia, leading to parallel temporal patterns in the composition of communities

4.1   Introduction

Having examined in Chapter 3 the ways in which individual organisms are affected by conditions and resources, we now turn to the larger question of how the interplay of conditions and resources influences whole communities (the assemblages of species that occur together). The answer to this question depends fundamentally on the scale at which we choose to study communities; this will be a pervasive theme throughout the chapter.

scale and patchiness—central themes of this chapter

Not surprisingly, because of its influence on both conditions and resources, climate plays a major role in determining the large-scale pattern of distribution of different types of community across the face of the Earth. However, local factors, such as soil type in terrestrial environments and water chemical composition in aquatic environments, are responsible for patchiness in community composition on much smaller scales. We discuss some of the causes of spatial patterns in community distribution in Section 4.2 before turning, in Section 4.3, to temporal patterns in conditions and resources whose consequences are changes in community composition over timescales from days to millennia. Section 4.4 describes the characteristics of the Earth's major terrestrial biomes; and Section 4.5 deals with the diversity of aquatic community types.

4.2   Geographical Patterns at Large and Small Scales

4.2.1   Large-scale climatic patterns

solar radiation, . . .

At the largest scale, the geography of life on Earth is mainly a consequence of the planet's movement through space. The tilt of the Earth as it makes its annual orbit around the sun causes solar radiation to strike the Earth's surfaces with different intensities at different latitudes (Figure 4.1). Because the equator is tilted toward the sun, equatorial and tropical areas receive more direct sunlight and are warmer than other latitudes. Warm air holds more moisture than cold air, increasing the water-holding capacity of air around the tropics. Solar radiation draws water from the vegetation by evaporation, but because the air is so moist, much of the water condenses and falls back as rain. Thus, the air that cycles to the atmosphere from the tropics is relatively dry, having lost most of its moisture as local rainfall before it ascends to the lower atmosphere.

The rotation of the Earth causes the air masses from the tropics to curve to the north and south. Air that was warmed at the tropics (and which lost moisture as local rain) cools in the atmosphere and descends again at a latitude of approximately 30° (north and south). The air mass warms as it descends, increasing its capacity to hold water and causing the descending air mass to "soak up" available water from the land. As a result, most of the major deserts, including the Sahara, Kalahari, Mojave, and Sonoran, are found at approximately this latitude. Another smaller evaporation/precipitation system occurs between 30° and 60° latitude, as warm air, now moist, rises and is blown further north or south, respectively. As it cools, the air descends again and rains, producing wetter environments.

. . . ocean currents, . . .

Ocean currents have further powerful effects on climatic patterns. Southern waters circulate counterclockwise; they carry cold Antarctic waters up along the western

Figure 4.1
The tilt of the Earth on its axis and its rotation around the sun define the amount of radiation striking the atmosphere around the Earth's surface. This, in combination with the daily spin of the Earth on its axis, is responsible for the large-scale patterns of rainfall and solar radiation that define the pattern of global climates. This diagram shows winter in the northern hemisphere with radiation falling almost vertically south of the equator, but the same amount of radiation is spread over greater areas north of the equator; less is therefore received, and there is less heating per unit area. (After Audesirk & Audesirk, 1996.)

Figure 4.2
The movements of the major ocean currents. (After Audesirk & Audesirk, 1996.)

coasts of continents and distribute warmer waters from the tropics along their eastern coasts (Figure 4.2). In the northern hemisphere, currents circulate clockwise, carrying cold Arctic waters along the western coasts of continents and returning warm tropical currents along the eastern coasts. The cool, dry climate of western South America is an

Figure 4.3
The typical influence of topography on rainfall (histogram bars) in the northern hemisphere. Moisture-laden westerlies are forced higher by a mountain range. As they rise they become cooler and release the moisture as rain or snow. This leaves a drier rain shadow on the eastern slopes. (After Audesirk & Audesirk, 1996.)

effect of the Antarctic Humboldt Current; the relatively dry climate of California is a result of Arctic currents. Conversely, on the eastern side of North America the strong tropical Gulf Stream carries with it warm and moist air far into the Atlantic Ocean, affecting even the climate of Western Europe.

. . . and mountain ranges . . .

The topography of the land has consequences at an intermediate scale for the pattern of terrestrial climates. As winds meet mountain ranges they are forced up and become cooler as they rise. The cooler air holds less moisture so that water is released (as rain and snow) on the windward slopes of the mountains (the Rockies and Himalayas provide striking examples of this effect). As the air passes over to the leeward sides of the mountains it descends, becomes warmer, and now absorbs water. This produces a desiccating effect and causes a *rain shadow* along the leeward slopes (Figure 4.3).

. . . produce a mosaic of dry, wet, cool, and warm climates over the face of the Earth . . .

The variety of influences on climatic conditions over the surface of the globe has given rise to a mosaic of dry, wet, cool, and warm climates. In the patches of this mosaic, distinctive terrestrial associations of vegetation and animals have formed. A world traveler sees repeatedly what can be recognized as characteristic types of vegetation, which ecologists call *biomes* (such as coniferous forest, savanna, and rain forest). Figure 4.4 provides an example of a set of biomes that can be recognized and used to construct a global map. Figure 4.5 shows the ranges of rainfall and mean monthly minimum temperature recorded within some major biomes of the world. We describe the characteristics of the communities inhabiting biomes in Section 4.4.

. . . that, in turn, are responsible for the large-scale distribution of terrestrial biomes

4.2.2   Small-scale patterns in conditions and resources

We need to be wary of seduction by cartographers who draw sharp lines on maps to show geographic boundaries. Neat pigeon-holes, sharp categories, and tidy boundaries are a convenience, not a reality of nature. Moreover, biomes are not homogeneous within their hypothetical boundaries; every biome has gradients of physicochemical

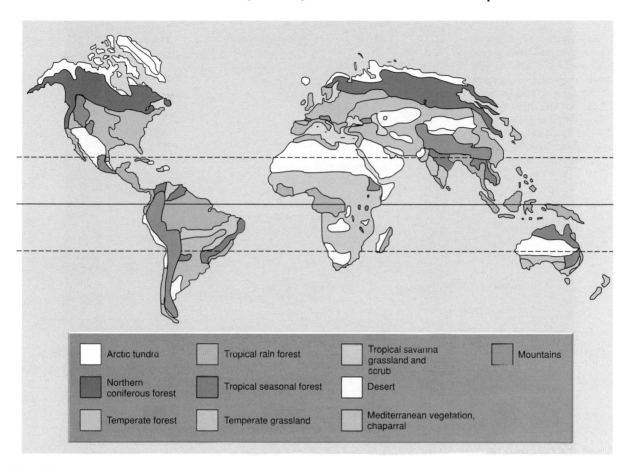

Figure 4.4
World distribution of the major biomes of vegetation. (After Audesirk & Audesirk, 1996.)

Legend:
- Arctic tundra
- Northern coniferous forest
- Temperate forest
- Tropical rain forest
- Tropical seasonal forest
- Temperate grassland
- Tropical savanna grassland and scrub
- Desert
- Mediterranean vegetation, chaparral
- Mountains

conditions related to features of local topography and geology. The communities of plants and animals that occur in these different regions may be quite distinct.

local topography

Local variations in topography can override elements of the broad climatic pattern described in Section 4.2.1. For example, temperature falls with increasing elevation and one effect is that vegetation high on a mountain in the tropics tends to resemble vegetation at low altitudes in northern latitudes. Traveling up a mountain in the tropics involves passing along a very similar ecological gradient as traveling northward (Figure 4.6).

local geology and soil

It is worth remembering that the Earth's surface would consist of a mosaic of different environments even if the climate were identical everywhere. Geological history has provided a variety of rocks that differ in their mineral composition. When the surfaces of these rocks are decomposed by heat, frost, and thaw they give rise to a variety of types of soil that reflect their geological origin. Without soil, it is impossible for significant terrestrial vegetation to grow. Soils provide a source of stored water, a reserve of mineral nutrients, a medium in which atmospheric nitrogen can be fixed

Figure 4.5
The variety of environmental conditions experienced in terrestrial environments can be described in terms of their annual rainfall and mean monthly minimum temperatures. The range of conditions experienced in (a) tropical rain forest, (b) savanna, (c) temperate deciduous forest, (d) northern coniferous forest (taiga), and (e) tundra. (After Heal et al., 1993, © UNESCO.)

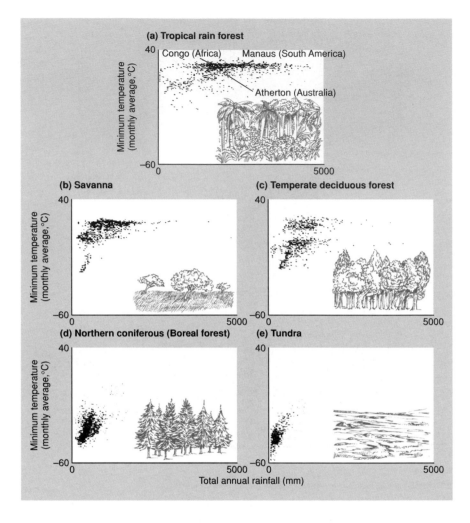

for plant use, and the support that allows plants to stand up and expose their leaves to the sunlight.

acidic and calcareous soils bear very different vegetation

Limestone rocks and chalk have formed as marine deposits of calcium carbonate, often containing some magnesium and other carbonates. Where these deposits have been raised and exposed as land surfaces they become the basis for neutral or slightly alkaline *calcareous* soils, which bear a characteristic *calcicole* (calcium-loving) flora. On the other hand, plants normally found on more acid soils (*calcifuges*), such as *Rhododendron*, *Azalea*, and species of *Kalmia* (sheep laurel and calico bush), commonly suffer nutrient deficiency on calcareous soils. Strict calcicoles, in contrast, suffer on acidic soils, where they are intolerant of aluminum ions released at low pH. In the United States, for example, yellow poplar (*Liriodendron tulipifera*) and northern white cedar (*Thuja occidentalis*) are found only on neutral or alkaline soils, and balsam fir (*Abies balsamea*) and eastern hemlock (*Thuja canadensis*) are usually confined to highly acidic soils.

Figure 4.6
The effect of altitude and latitude on the distribution of biomes. (After Audesirk & Audesirk, 1996.)

Organic matter accumulates at different rates in different soils; local variations in the balance between mineral and organic material in the soil contribute to the complexity of environmental mosaics. In extreme conditions, especially where the rocks are acidic, the temperatures are low, and/or the soil is waterlogged, the decomposition of organic matter may be seriously impeded. Then, peat bogs may form on the only semidecomposed organic matter, bearing very specialized plants and animals.

variations in the rate of decomposition of organic matter

To an ecologist, a *patch* in a community is an area in which a single variable distinguishes it from its surroundings. Thus, a fallen tree in a forest leaves a gap in the canopy and a patch on the forest floor where sufficient radiation may penetrate to allow seedlings to grow and eventually fill the gap. A tide pool is a patch on a rocky shore, but within that pool snails may graze and clear a patch free of algae. It is often useful to think of patches as the scale at which particular organisms experience the environment around them. For an aphid in a forest, an individual leaf of a particular species of tree is a patch—it provides both the conditions and the resources necessary for the insect. For a warbler feeding on caterpillars, the canopies of individual trees are patches that it encounters in its daily life. But owls or hawks hunt over a large part of the forest, and for them a patch may be the territory that each bird defends or perhaps even the whole forest over which it ranges.

patchiness is in the eye of the observer

all communities are patchy

4.2.3   Patterns in conditions and resources in aquatic environments

In most aquatic environments it is difficult to recognize anything comparable to terrestrial biomes. The exceptions occur at the ocean's edge; mangrove swamps, coral reefs, and kelp forests have a flora and fauna that are as distinctive as any of the various terrestrial biomes, but this is largely due to their link with major terrestrial climates. In contrast, the open oceans form a continuum in which there is a flow of water and dissolved chemicals across the globe. We have seen how variation in the intensity of solar radiation from place to place and between the seasons has dramatic effects on the temperature and water relations of terrestrial environments. But this is not the case in the oceans. The high thermal capacity of water makes the oceans slow to heat and slow to cool. One effect is that the temperature of the water at one point on the globe is a better reflection of where the water has come from than of the heat exchanged locally.

The world's large lakes can be distinguished and classified according to their physical conditions. For example, large lakes in lowland equatorial regions generally experience permanent stratification (distinct layers of water at particular temperatures), whereas seasonal patterns of stratification and mixing are the rule in temperate regions. Within the polar circles, permanent ice cover with no mixing is characteristic of large lakes. However, local geological conditions and basin size and shape have strong influences on conditions and resources in lakes, particularly relating to water chemical composition. Consequently, a broad geographic classification of lake communities has only limited merit. In the cases of streams, rivers, and estuaries, we will see that local conditions and resources are paramount in determining community patterns (Section 4.5).

4.3   Temporal Patterns in Conditions and Resources—Succession

The composition of communities can change over timescales ranging from hours to millennia, as conditions and resources themselves change. For example, the microbial community that colonizes and decomposes a dead worm or a fragment of a leaf may change from hour to hour (like a newly inoculated culture of yogurt). At the other extreme, we can trace patterns in community composition over tens of thousands of years. Thus, changes in climate during the Pleistocene ice ages bear much of the responsibility for the present patterns of distribution of plants and animals (Figure 4.7). In the 20,000 years since the peak of the last glaciation, global temperatures have risen by about 8°C. Many species continue, even today, to migrate northward, following the retreat of the glaciers. Climatic changes that occurred 3,200 years ago are still evident in persistent patches of hemlocks among the dominant hardwood species in the forests of the Upper Peninsula of Michigan state (Davis et al., 1994).

plant successions—"raised" peat bogs . . .

At intermediate temporal scales, predictable successions of plant species may occur over periods ranging from years to centuries. For example, the so-called raised peat bogs display a succession of species of moss of the genus *Sphagnum* that grow on the peat. As their dead parts accumulate, they form hummocks, raised above the

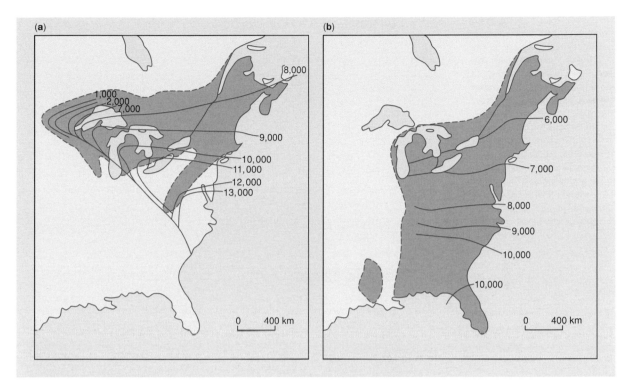

Figure 4.7
A map showing the spread of two species of forest tree in eastern North America after the retreat of the last ice age glaciation. Note that the two species (a) eastern white pine (*Pinus strobus*) and (b) beech (*Fagus grandifolia*) have not followed the same invasion path. The lines on the maps (isochrones) define the time of arrival of each species at 1,000-year intervals. The numbers on the maps refer to thousands of years before present. The shaded areas show their present distributions. (From Davis, 1976.)

surface of the peat. Each species of *Sphagnum* occupies a zone at a particular height on the hummocks and they replace each other through the decades as the hummocks grow. Ultimately the hummocks are invaded by heather, *Calluna vulgaris*, and when this dies the whole hummock breaks down and the sequence of *Sphagnum* species starts again (Figure 4.8). The peat bog as a whole is a mosaic of patches at various stages in this successional process. To the pilot of an aircraft, the peat bog may look very uniform, but observers on their knees see a finely structured and repetitive patchwork.

Old field successions occur over periods of decades or centuries and are particularly associated with eastern parts of the United States (see Section 1.3.2). Farmers moved west when new frontiers were opened up in the 19th century, and cultivated land in the east was abandoned. In the typical succession, annual weeds were first on the scene but were later replaced by herbaceous perennials that, in turn, were succeeded by shrubs, by early successional trees (e.g., eastern red cedar *Juniperus virginiana*), and finally by late successional species (e.g., sugar maple *Acer saccharum*). Similar successions also occur within smaller patches of established communities: for example, after the death of a single tree. The important topic of community succession is discussed in more detail in Section 9.4.

... and old field successions

Figure 4.8
The structure of the hummock–hollow cycle on an actively growing raised bog. The moss *Sphagnum cuspidatum* (C) invades small pools of water, and its slowly decaying parts accumulate in them, creating a habitat suitable for *S. pulchrum* (Pu). This in its turn is replaced by *S. papillosum* (A), and in its growth this species builds up hummocks of its dead parts. The hummocks rise above the level of standing water and become invaded by small woody shrubs such as *Calluna vulgaris* and then by the lichen *Cladonia arbuscula*. When the *Calluna* dies the hummocks erode and break down and new pools are formed—starting the whole sequence again.

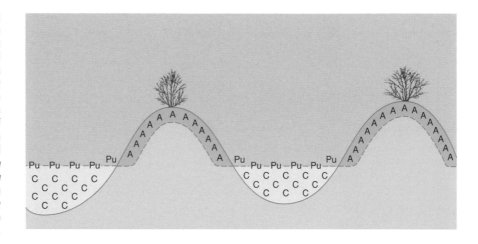

4.4 Terrestrial Biomes

patterns that we recognize in nature depend on how we focus our attention

Different geographers recognize different numbers of biomes; some make do with just five biomes and others find they need many more. The perspective of the scientist is as important as the system being studied; "splitters" tend to distrust broad generalizations and emphasize the diversity of the natural world, whereas "lumpers" force diversity into a minimum of easily mapped categories. Seven will be adequate for our purposes —tropical rain forest, savanna, temperate grassland, desert, temperate deciduous forest, northern or boreal coniferous forest (taiga), and tundra. This section is largely about global patterns that can be represented, without too many frills, on a map of the world. If we were dealing with California and not the whole world, we would be able to recognize more subtle variations and classify vegetation and map the land on a finer scale: biomes would be far too crude. What we see depends on the way we look; we can recognize patterns in nature whether we look through a telescope or a microscope!

The topographical and geological features of a landscape superimpose their own details on the broader patterns that geographers can display on maps. Nowhere is this more dramatic than on major mountain ranges in the tropics. The journey up a mountain in Kenya or in Mexico can pass from rain forest or savanna through grasslands, deciduous forest, coniferous forest, and tundra to snow-capped peaks. More generally, and on a smaller scale, the vegetation of the real world is a patchy and mottled carpet, and the fauna largely follows suit. Southern slopes on hillsides within the tundra biome, for example, often carry tussocks of leafy flowering plants and the occasional stunted tree. Similarly, north-facing hillsides within a northern conifer forest may, in fact, consist of pure tundra, bare of all but moss and lichens.

4.4.1 Describing and classifying biomes

We pointed out in Chapter 2 the crucial importance of geographic isolation in allowing populations to diverge under selection. The geographical distributions of species, genera, families, and even higher taxonomic categories of plants and animals often reflect this geographic divergence. All species of lemurs, for example, are found on the island of Madagascar and nowhere else. Similarly, 230 species in the genus *Eucalyptus* (gum tree) occur naturally in Australia (and two or three in Indonesia and Malaysia). The lemurs and the gum trees occur where they do because they evolved there—not because these are the only places where they could survive and prosper. Indeed, many *Eucalyptus* species grow with great success and spread rapidly when they have been introduced to California or to Kenya. A map of the natural world distribution of lemurs tells us quite a lot about the evolutionary history of this group. But as far as its relationship with a biome is concerned, the most we can say is that lemurs happen to be one of the constituents of the tropical rain forest biome in Madagascar.

Another theme of Chapter 2 concerned the way species with quite different evolutionary origins have been selected to converge in their form and behavior. There were also examples of taxonomic groups that have radiated into a range of species with strikingly similar form and behavior (parallel evolution, as in the marsupial and placental mammals). Examples like these reveal much about the ways in which organisms have evolved to match and restrict them to the conditions and resources in their environments. But the different species need not characterize different biomes. Thus, particular biomes in Australia include certain marsupial mammals, while the *same* biomes in other parts of the world are home to their placental counterparts.

A map of biomes, then, is not usually a map of the distribution of species. Instead, it shows where we could find areas of land dominated by plants with characteristic shapes, forms, and physiological processes. These are the types of vegetation that can be recognized from an aircraft passing over them or from the windows of a fast car or train. It does not require a botanist to identify them—indeed, by nature and training the botanist may not be able to see the wood for the trees.

describing and classifying vegetation

The scrubby maquis or garrigue vegetation of countries around the coast of the Mediterranean Sea provides a striking example. The spectrum of plant forms that gives this vegetation its distinctive nature also occurs in similar environments in California (as chaparral) and in Australia—but the species and genera of plants are quite different. We recognize different biomes and different types of aquatic community from the *types* of organisms that live in them. How can we describe their similarities so that we can classify, compare, and map them?

In addressing this question, the Danish biogeographer Raunkiaer developed, in 1907 (Raunkiaer, 1934), his idea of "life forms", a deep insight into the ecological significance of plant form (Figure 4.9). He then used the spectrum of life forms present in different types of vegetation as a means of describing their ecological character. Plants grow by developing new shoots from the buds that lie at the apices (tips) of existing shoots and in the leaf axils. Within the buds, the meristematic cells are the most sensitive part of the whole shoot—the "Achilles' heels" of plants. Raunkiaer argued that the ways in which these buds are protected in different plants are powerful indicators of the hazards in their environments and may be used to define the different plant forms (Figure 4.9).

Raunkiaer's plant life forms and life-form spectrums

Thus, trees expose their buds high in the air, fully exposed to wind, cold, and drought; Raunkiaer called them *phanerophytes*: Greek *phanero* = "visible"; *phyte* = "plant". The least protected tree buds are those of tropical rain forest, where the buds are freely exposed and produce flushes of new shoots throughout the year; the buds of trees in almost all other forests are dormant during the coldest or driest seasons of the

(a)

(b)

(c)

(d)

(e)

(f)

(g)

Each biome is illustrated with two photographs, one focusing on the detail of the vegetation and the other providing a distant view and emphasizing the great structural variation to be found among the world's terrestrial communities. The animals found in each of these biomes also cannot be ignored; they are obvious in the savanna photo, but invertebrate and vertebrate animals are busy behind the scenes in all the biomes. (a) Top: Carrizo Badlands, Anza-Bonnego Desert State Park, California (© Doug Sokell); bottom: Red Rock Canyon, Las Vegas, Nevada (© Mark E. Gibson); (b) top: Ozark Forest and Current River, Ozark National Scenic Riverways, Missouri (© Richard Thom); bottom: mature eastern deciduous forest (© Bill Beatty); (c) top: fir tree forest, Jasper National Park, Alberta, Canada (© Mark E. Gibson); bottom: foggy coniferous forest, Sierras (© Joe McDonald); (d) top: Maasai Mara Game Preserve at dawn (© Joe McDonald); bottom: African savanna with zebra and buffalo (© John Cunningham); (e) top: rain forest, western slope of Andes, Ecuador (© C.P. Hickman); bottom: lake in mixed dipterocarp forest, Mulu National Park, Sarawak, Borneo (© Brian Rogers); (f) top: a long pronghorn antelope looks tiny in this vast mixed-grass prairie, Stanley County, central South Dakota (© Ron Spomer); bottom: view of prairie in flower with blazing star and black-eyed Susan (© Ann B. Swengel); (g) top: green tundra with glacial morraine and Alaska mountain range, Denali National Park, Alaska (© Patrick J. Endres); bottom: wet summer tundra (© Doug Sokel). (All photos courtesy of Visuals Unlimited.)

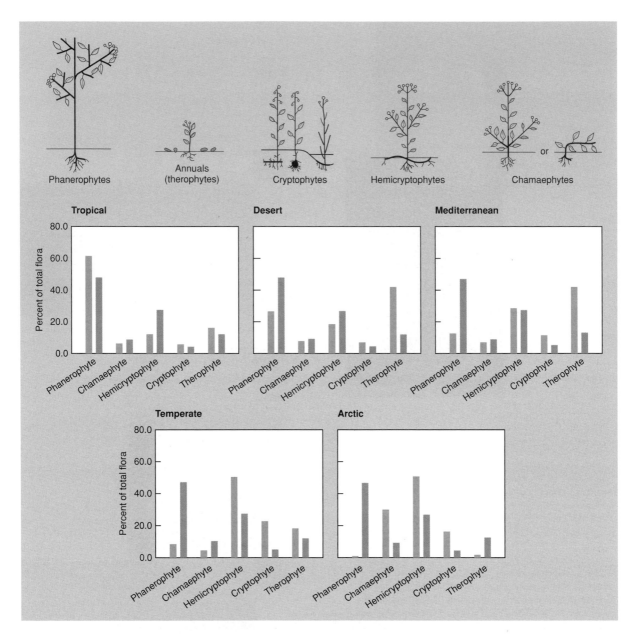

Figure 4.9
Above, drawings depict the variety of plant forms distinguished by Raunkiaer on the basis of where they bear their buds (shown in bold). Below are life-form spectrums for five different biomes. The brown bars show the percentage of the total flora that is composed of species with each of the five different life forms. The blue bars are the proportions of the various life forms in the world flora for comparison. (From Crawley, 1986.)

year and are strongly protected by sheaths of bud scales. By contrast, many perennial herbs form cushions or tussocks in which buds are borne above ground but are protected from drought and cold in the dense mass of old leaves and shoots (*chamaephytes*: "on the ground plants"). Buds are even better protected when they are formed at or in the soil surface (*hemicryptophytes*: "half-hidden plants") or on buried dormant storage organs (bulbs, corms, and rhizomes—*cryptophytes*: "hidden plants" or *geophytes*: "earth plants"). These allow the plants to make rapid growth and flower before they die back to a dormant state. A final major category consists of annual plants that depend wholly on dormant seed to carry their populations through seasons of drought and cold (*therophytes*: "summer plants").

Therophytes are the plants of deserts (they make up nearly 50 percent of the flora of Death Valley), sand dunes, and repeatedly disturbed habitats. They also include the annual weeds of arable lands, gardens, and urban wastelands—the botanical "camp followers" of humanity. But there is, of course, no vegetation that consists entirely of one growth form. All vegetation contains a mixture, a spectrum, of Raunkiaer's life forms. The spectrum on a tropical island, for example, is predominantly of phanerophytes, but these are completely absent from the spectrum of Arctic islands.

The composition of the spectrum in any particular habitat is as good a shorthand description of its vegetation as ecologists have yet managed to devise. Thus, we recognize a *Mediterranean* type of vegetation when we see it in Chile, Australia, California, or Crete because the life-form spectrums are similar, though the lists of the species present do not tell us that these communities have anything in common. Similarly, a Raunkiaer spectrum of the plant life forms in the Amazon rain forest would tell us how ecologically similar they are to the spectrum of life forms in rain forests of the Congo or Borneo. The detailed taxonomy of their floras and faunas would only emphasize how different they are.

Faunas are bound to be closely tied to floras—if only because most herbivores are choosy about their diet. Terrestrial carnivores range more widely than their herbivore prey, but the distribution of herbivores still gives the carnivores a broad vegetational allegiance (Figure 4.10). Plant scientists have tended to be keener on classifying floras than animal scientists on classifying faunas. Nevertheless, it is worth asking whether

describing and classifying faunas

Figure 4.10
The distribution of bird species along a gradient of plant succession in the piedmont region of Georgia, United States. The variation in intensity of shading indicates the relative abundance of the species. As a bare field is taken over by grassland and then by scrub and eventually forest, the community of associated birds also changes. (After Gathreaux, 1978.)

Figure 4.11
A comparison of the forest mammals present
in various communities of tropical forest.
The mammals were classified according to
their diet and according to the zones in
which they forage. The dietary categories
are: C, carnivores; HF, herbivores that are
leaf and fruit feeders; I, insect feeders; M,
mixed feeders or omnivores. The foraging
categories are: arboreal, aerial, ground
feeders, and scansorial (see key). (a) One
hundred and sixty-one species present in
forested areas of Malaysia; (b) 70 species
in Panama dry forest; (c) 50 species in
Cape York forest, Australia; (d) 96 species
in Irangi forest, Zaire, Africa.
(After Andrews et al., 1979.)

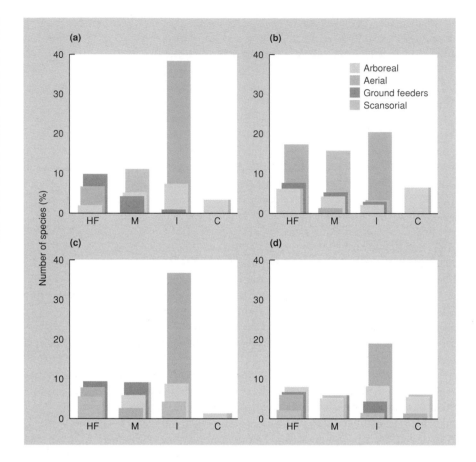

there is, for the fauna of an area, any one feature (or set of features) with a role comparable to the protection and position of buds in vegetation. An interesting attempt to classify faunas used a particularly animal characteristic, the mode of locomotion, and combined this with dietary habits, to compare the mammals of forests in Malaya, Panama, Australia, and Zaire (Andrews et al., 1979). Mammals were classified into carnivores, herbivores including fruit eaters, insectivores, and mixed feeders, and these categories were subdivided into those that were aerial (mainly bats and flying foxes), arboreal (tree dwellers), scansorial (climbers), and small ground mammals (Figure 4.11). The comparison reveals some strong contrasts and similarities. For example, the ecological diversity spectrums for the Australian and Malayan forests were very similar despite the fact that their faunas are taxonomically very distinct—the Australian mammals are marsupials and the Malaysian mammals are placentals.

There are splendid questions to be asked and answered about comparisons of birds or insects and indeed about whole faunas. Would the categories appropriate for mammals be appropriate for these other groupings? What groupings might be appropriate for comparisons of tropical, Mediterranean, temperate, and Arctic faunas? Perhaps the presence and character of dormancy (e.g., hibernation) might be as valuable for an ecological classification of faunas as they were for Raunkiaer in plants.

UNANSWERED QUESTION: how can ecologists most effectively define life-form spectrums of faunas?

Perhaps it will be spectrums of relative abundances of body sizes, endotherms versus ectotherms, or life-history traits that will form the basis for some future comparison and classification.

4.4.2   Tropical rain forest

We have chosen to discuss tropical rain forest in greater depth than the other biomes because it represents the global peak of evolved biological diversity: all the other biomes suffer from a relative poverty of resources or more strongly constraining conditions.

Tropical rain forest is the most productive of the Earth's biomes with a photosynthetic productivity that can exceed 1,000 g of carbon fixed per square meter per year (see Section 11.2.1). Such exceptional productivity results from the coincidence of high solar radiation received throughout the year and regular and reliable rainfall (illustrated in Figure 4.5). The production is achieved, overwhelmingly, high in the dense forest canopy of evergreen foliage (i.e., Raunkiaer's phanerophytes). It is dark at ground level except where fallen trees create gaps. A characteristic of this biome is that often many tree seedlings and saplings remain in a suppressed state from year to year and only leap into action if a gap forms in the canopy above them.

Almost all the action in a rain forest (not just photosynthesis but also flowering, fruiting, predation, and herbivory) happens high in the canopy. Apart from the trees, the vegetation is largely composed of plant forms that reach up into the canopy vicariously; they either climb and then scramble in the tree canopy (vines and lianas, including many species of fig) or grow as epiphytes, rooted on the damp upper branches. The epiphytes depend on the sparse resources of mineral nutrients that they extract from crevices and pockets of humus on the tree branches. The rich floras and faunas of the canopy are not easy to study; even to gain access to the flowers in order to identify the species of tree is difficult without the erection of tree walks. It is a measure of the problems of doing research in rain forest that botanists have trained monkeys to collect and throw down flowers and research teams have used hot air balloons to move over the canopy and work in it.

Most species of both animals and plants in tropical rain forest are active throughout the year, though the plants may flower and ripen fruit in sequence. In Trinidad, for example, the forest contains at least 18 trees in the genus *Miconia*, whose combined fruiting seasons extend throughout the year (Figure 4.12).

Dramatically high species richness is the norm for tropical rain forest (see Section 10.5.2), and communities rarely if ever become dominated by one or a few species—a very different situation from the biomonotony of northern coniferous forests. This raises some fundamental questions that have proved very difficult to resolve. First, what is it about the evolutionary history of tropical rain forest that has allowed such diversity to evolve? Part of the answer relates to the comparative stability of patches of rain forest during the ice ages. It is thought that during these periods, drought forced tropical rain forests to contract into "islands" (in a "sea" of savanna), and these expanded and coalesced again as wetter periods returned. This would have promoted genetic isolation of populations, a phenomenon that is important for speciation to occur (see Section 2.4). We may also ask why it is that among the diversity of species

UNANSWERED QUESTION: what permitted high species richness to evolve in tropical rain forests?

Figure 4.12
The seasonal patterns of fruit or seed
production by a rain forest flora and by the
flora of a temperate deciduous forest.
(a) The fruiting seasons of 18 species of
Miconia in the rain forest of Arima valley,
Trinidad, are spread throughout the year.
(b) The seasonal production of fruit and
seeds by herbs in a deciduous temperate
forest (lime and hornbeam) at Bialowieza,
Poland, are concentrated in a relatively
brief season. (After Harper, 1977.)

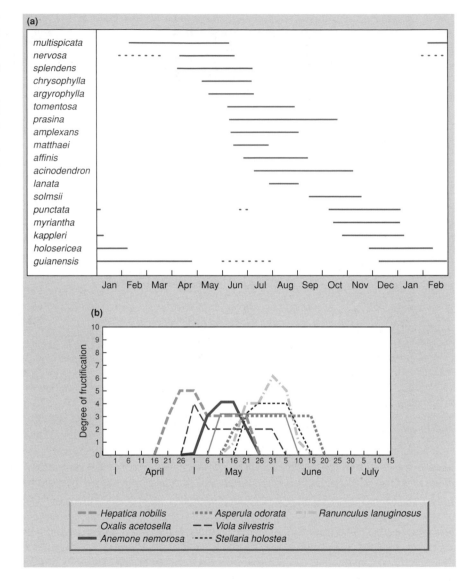

in tropical rain forest, a few have not dominated and suppressed the rest in a struggle for existence. We will see later (see Section 10.5.2) that at least part of the answer is that populations of specialized pathogens and herbivores develop near mature trees and attack new recruits of the same tree species nearby. Thus, the chance that a new seedling will survive can be expected to increase with its distance from a mature tree of the same species, reducing the likelihood of dominance by one or a few species in the forest.

tropical rain forest is also associated with high animal diversity . . .   The diversity of rain forest trees provides for a corresponding diversity of resources for herbivores (Figure 4.13). A variety of fresh young leaves is available throughout the year, and a constant procession of seed and fruit production provides reliable food for

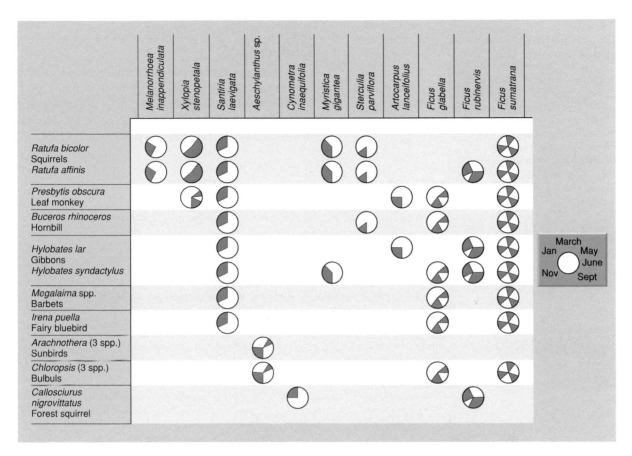

Figure 4.13
Animals (listed vertically) that feed on the fruit of trees (listed horizontally) at various times of the year at Selangor, Malaysia. Each circle is a calendar in which the feeding season is shown in purple. Each plant produces fruit only at certain times in the year, but there is fruit available for specialist fruit-eaters throughout the year. (After Harper, 1977.)

specialists such as fruit-eating bats. Moreover, a diversity of flowers, such as epiphytic orchids with their specialized pollinating mechanisms, require a parallel specialized diversity of pollinating insects. Rain forests are the center of diversity for ants—43 species have been recorded on a single tree in a Peruvian rain forest. And there is even more diversity among the beetles; Erwin (1982) estimated that there are 18,000 species of beetle in 1 ha of Panamanian rain forest (compared with only 24,000 in the whole of the United States and Canada!).

There is intense biological activity in the soil of tropical rain forests. Leaf litter decomposes faster than in any other biome and as a result the soil surface is often almost bare. The mineral nutrients in fallen leaves are rapidly released, and, as rainfall seeps down the soil profile, nutrients may be carried well below the levels at which roots can recover them. Almost all the mineral nutrients in a rain forest are held in the plants themselves, where they are safe from leaching. When such forests are cleared for agriculture, or the timber is felled or destroyed by fire, the nutrients are released and leached or washed away; on slopes the whole soil may go too. The full regeneration of

... and intense soil activity

soil and of a nutrient budget in a new body of forest biomass may take centuries. Evidence of cultivated patches within rain forest can still be seen clearly from the air 40 years or more after they have been deserted.

All the other terrestrial biomes can be seen as the poor relations of tropical rain forest. They are all colder or drier and all are more seasonal. They have had prehistories that prevented the evolution of a diversity of fauna and flora that approaches the remarkable species richness of tropical rain forest. Moreover, they are generally less suited to the lives of extreme specialists, both plant and animal.

4.4.3   Savanna

The vegetation of savanna characteristically consists of grassland with scattered small trees, but extensive areas have no trees. Herds of grazing herbivores (e.g., zebra and wildebeest in Africa) have a profound influence on the vegetation, favoring grasses (which protect their regenerating meristems in buds at or just below ground level) and hindering the regeneration of trees (which expose their meristems to browsing animals and to fire). Fire is a common hazard in the dry season and, like grazing animals, tips the balance in the vegetation against trees and favors grassland.

seasonal glut and food shortage are characteristic of savanna

Seasonal rainfall places the most severe restrictions on the diversity of plants and animals in savanna. Plant growth is limited for part of the year by drought, and there is a seasonal glut of food, alternating with shortage; as a consequence, the larger grazing animals suffer extreme famine (and mortality) in drier years. The strong seasonality of savanna ecology is well illustrated by its bird populations. An abundance of seeds and insects supports large populations of migrating birds, but only a few species can find sufficiently reliable resources to be resident year round.

4.4.4   Temperate grasslands

Temperate grassland is the natural vegetation over large areas in every continent—the steppes of Asia, the prairies of North America, the pampas of South America, and the veldt of South Africa. Typically, these grasslands experience seasonal drought, but the role of climate in determining vegetation is usually overridden by the effects of grazing animals. Populations of invertebrates, such as grasshoppers, are often very large and their biomass may exceed that of grazing vertebrates.

of all the biomes, temperate grassland has been most transformed by humans

Many of these natural grasslands have been cultivated and replaced by arable annual "grasslands" of wheat, oats, barley, rye, and corn. Such annual grasses of temperate regions, together with rice in the tropics, provide the staple food of human populations worldwide. In fact, the vast increase in the size of the human population in historical times (see Section 12.2) has depended on the domestication of grasses for human food or feed for domestic animals. At the drier margins of the biome, where cultivation is not economical, many of the grasslands are "managed" for meat or milk production, sometimes requiring a nomadic human lifestyle. The natural populations of grazing animals, especially bison and pronghorn antelope in North America and ungulates in Africa, have been driven back in favor of cattle, sheep, and goats. Of all the biomes, this is the one most coveted, used, and transformed by humans.

4.4.5 Desert

In their most extreme form, the hot deserts are too arid to bear any vegetation; they are as bare as the cold deserts of Antarctica. Where there is sufficient rainfall to allow plants to grow in arid deserts, its timing is always unpredictable.

Desert vegetation falls into two sharply contrasted patterns of behavior. Many species have an opportunistic lifestyle, stimulated into germination by the unpredictable rains (physiological clocks are useless in this environment). They grow fast and complete their life history by starting to set new seed after a few weeks. These are the species that can occasionally make a desert bloom; the ecophysiologist Fritz Went called them "belly plants" because only someone lying on the ground can appreciate their individual charm.

contrasting patterns of behavior of desert plants

A different pattern of behavior of arid desert plants is to be long-lived with sluggish physiological processes. Cacti and other succulents, and small shrubby species with small, thick, and often hairy leaves, can close their stomata (pores through which gas exchange takes place) and can tolerate long periods of physiological inactivity. In arid deserts, freezing temperatures are common at night and tolerance of frost is almost as important as tolerance of drought.

The relative poverty of animal life in arid deserts reflects the low productivity of the vegetation and the indigestibility of much of it. Desert perennials including species of wormwood (*Artemisia*) and creosote plant (*Larrea mexicana*) in the southwestern United States, and mallee species of *Eucalyptus* in Australia, carry high concentrations of chemicals that are repellent to herbivores. Ants and small rodents rely on seeds as a relatively reliable perennial resource, whereas bird species are largely nomadic, driven by the need to find water. Only desert carnivores can survive on the water they obtain from their food. In the deserts of Asia and Africa, camels, donkeys, and sheep are managed for transport and food by migrant groups of humans.

animal diversity is low in deserts

4.4.6 Temperate forest

Like all biomes, temperate forest includes, under one name, a variety of types of vegetation. At its low latitude limits in Florida and New Zealand, winters are mild, frosts and droughts are rare, and the vegetation largely consists of broad-leaved evergreen trees. At its northern limits in the forests of Maine and the upper midwest of the United States, the seasons are strongly marked, winter days are short, and there may be 6 months of freezing temperatures. Deciduous trees, which dominate in most temperate forests, lose their leaves in the fall and become dormant after transferring much of their mineral content to the woody body of the tree. On the forest floor, diverse floras of perennial herbs often occur, particularly those that grow quickly in the spring (Raunkiaer's geophytes), before the new tree foliage has developed.

All forests are patchy because old trees die, providing open environments for new colonists. This patchiness is on an especially large scale after hurricanes fell the older and taller trees or after fire kills the more sensitive species. In temperate forests the canopies are often composed of a mixture of long-lived species, such as red oaks in the midwest of the United States, and colonizers of gaps, such as sugar maple.

Temperate forests provide food resources for animals that are usually very seasonal in their occurrence (compare Figure 4.12a and b), and only species with short life cycles, such as leaf-eating insects, can be dietary specialists. Many of the birds of temperate forests are migrants that return in spring but spend the remainder of the year in warmer biomes.

temperate forest soils are rich in organic matter

Soils are usually rich in organic matter that is continually added to, decomposed, and churned by earthworms and a rich community of other detritivores (organisms that feed on dead organic matter). Only waterlogging and low pH inhibit the decomposition of organic matter and force it to accumulate as peat or "raw humus".

Large swathes of deciduous forest in Europe and the United States have been cut down to provide for agriculture, but these have sometimes been allowed to regenerate as farmers abandoned the land (a conspicuous feature in New England).

4.4.7   Northern coniferous forest (taiga) grading into tundra

Northern coniferous forest (also known as *taiga*) and tundra occur in regions where the short growing season and the cold of winter limit the vegetation and its associated fauna.

Coniferous forest consists of a very limited tree flora. In areas with less severe winters, the forests may be dominated by pines (*Pinus* species, which are all evergreens) and deciduous trees such as larch (*Larix*), birch (*Betula*), or aspens (*Populus*), often as mixtures of species. Farther north, these species give way to monotonous single-species forests of spruce (*Picea*) over immense areas of North America, Europe, and Asia. This is *biomonotony*, an extreme contrast to the biodiversity of tropical rain forests.

The areas of vegetation now occupied by tundra and northern coniferous forests (and much of the northern deciduous forests) were occupied by the ice sheet during the last ice age, which only started to withdraw 20,000 years ago. Temperatures are now as high as they have ever been since that time, but the vegetation has not yet caught up with the changing climate and the forests are still spreading north. The very low diversity of northern floras and faunas is in part a reflection of a slow recovery from the catastrophes of the ice ages.

the low diversity of northern coniferous forest provides ideal conditions for pest outbreaks

Biomonotonous communities provide ideal conditions for the development of disease and epidemics of pests. For example, the spruce budworm (*Choristoneura fumiferana*) lives at low densities in immature northern forests of spruce. As the forests mature, the budworm populations explode in devastating epidemics. These wreck the old forest, which then regenerates with young trees. This cycle takes about 40 years to run its course.

The overriding environmental constraint in northern spruce forests is the presence of permafrost: the water in the soil remains frozen throughout the year, creating permanent drought except when the sun warms the actual soil surface. The root system of spruce can develop in the superficial soil layer, from which the trees derive all their water during the short growing season.

To the north of the spruce forest, the vegetation changes to tundra, and the two often form a mosaic in the Low Arctic. In the colder areas, grasses and sedges disappear, leaving nothing rooted in the permafrost. High winds exaggerate the aridity of the environment, and ultimately vegetation that consists only of lichens and mosses gives way, in its turn, to the polar desert. The number of species of higher plants (i.e., excluding mosses and lichens) that form the vegetation of the Arctic decreases from the Low Arctic (600 species in North America) to 100 species in the High Arctic (north of 83°) of Greenland and Ellesmere Island. In contrast, the flora of Antarctica contains only two native species of vascular plant and some lichens and mosses that support a few small invertebrates. The biological productivity and diversity of Antarctica are concentrated at the coast and depend almost entirely on resources harvested from the sea.

The faunas of northern coniferous forests and tundra have intrigued ecologists because populations of lemmings, mice, voles, and hares (herbivores), and the fur-bearing carnivores (e.g., lynx and ermine) that feed on them, pass through remarkable cycles of expansion and collapse (see Section 8.5.2). Lemmings (*Lemmus*) are famous for their population cycles and the role they play in the tundra. When the snow melts during a period when the lemming cycle is at a high point, the animals are exposed and they support large migratory populations of predatory birds (owls, skuas, and gulls) and mammals, such as weasels. Reindeer and caribou (they are the same species, *Rangifer tarandus*) occur in migrant herds capable of foraging on lichens of the tundra, which they can reach through snow cover.

dramatic animal population cycles are characteristic of northern biomes

The leaves of conifers are hard and resistant to decomposition; low temperatures further slow their rate of decomposition. Fallen leaves accumulate in layers on the forest floor as an acid humus, and there is little animal activity in the soil to churn and mix the organic matter into the mineral soil. In the tundra, the decomposition of organic matter is severely limited by the low temperature of the soil and the absence of any forest protection allows ferocious winds to develop so that dead organic matter is blown from place to place.

4.5 Aquatic Environments

The dominating characteristics of aquatic environments result from the physical properties of water. A water molecule is composed of an oxygen atom, which is slightly negatively charged, bonded with two hydrogen atoms, which are slightly positively charged. This *dipolar structure* enables water molecules to attract and dissolve more substances than any other liquid on Earth. Consequently, water can hold mineral ions in solution, providing the nutrient resources required for the growth of algae and higher plants.

On the other hand, the solubility of oxygen, an essential resource for both plants and animals, decreases rapidly with increasing temperature, and oxygen diffuses only slowly in water. This problem can place major limits on life in water. Oxygen is rapidly used up when dead organic matter decomposes. In places where tree leaves accumulate or untreated sewage is discharged into a river or lake, decomposition can create anaerobic conditions, which are lethal for fish and other animals that have a high biological oxygen demand. Many aquatic animals maintain access to oxygen by forcing a continual flow of water over their respiratory surfaces (e.g., the gills of fish) or have very large surface areas relative to their body volume.

special properties of water as a medium in which to live

Water is viscous, and moving water transports whole living organisms, such as small plants and animals. It offers resistance to the movement of motile animals such as fish, otters, and aquatic birds; not surprisingly, many motile aquatic animals are streamlined. Many plants that live in moving water depend on rooting in the sub-stratum to hold them against water currents, and many smaller animals are attached to the plants or hide in crevices or under rocks where they are protected from the drag of moving water.

Water is unusual in remaining liquid over a wide range of temperatures. It requires a lot of energy to heat it (i.e., it has *high thermal capacity*), but retains heat efficiently. One consequence is that the temperature of large bodies of water (oceans and large lakes) varies little over the seasons. A further peculiar physical property of water is that it is less dense when frozen than when liquid. Like most liquids, water becomes denser and sinks as it cools. However, at temperatures below 4°C, water becomes less dense and when ice forms (at 0°C), it floats. Ice on a water surface insulates the water beneath; lakes and streams can remain liquid, free flowing, and inhabitable under a layer of ice.

4.5.1   Stream ecology

Streams and rivers contain a minute portion of the world's water (0.006 percent), but an enormous proportion of the fresh water that can be used by people. Consequently, they have been tapped, dammed, straightened, rerouted, dredged, and polluted since the beginning of civilization. Understanding the impacts and sustainability of some of these practices begins with understanding the basics of stream ecology.

Streams and rivers are characterized by their linear form, unidirectional flow, fluctuating discharge, and unstable beds. The narrow nature of river channels means that they are very intimately connected to the surrounding terrestrial environment. Thus, a proper understanding of river ecology requires us to consider the river and its drainage basin as a unit.

importance of oxygen concentration

Oxygen concentration is often high in turbulent, upstream locations and low farther downstream, where higher temperatures cause reduced solubility. This is reflected in river fish communities, with active upstream species such as brown trout (*Salmo trutta*) having a high oxygen demand, whereas more sluggish species such as pike (*Esox lucius*) can tolerate the lower concentrations in their habitats downstream.

pH and temperature . . .

A variety of other chemical and physical conditions vary from stream to stream, or down the length of a given river. Figure 4.14 illustrates how the species composition of stream invertebrate communities varies with conditions. There were 30–40 species at each site, with much overlap in the list of species present. The data were subjected to an analysis called *community classification*, which is conceptually similar to taxonomic classification. In taxonomy, similar individuals are grouped together in species, similar species in genera, and so on. In community classification, communities with similar species compositions are grouped together in sets. These sets are then, in turn, grouped into more inclusive sets, and so on. In this case, the conditions that were most influential in determining the pattern of grouping—and thus were most influential in determining community composition—were pH, stream temperature, and the volume of water flowing per unit time (discharge).

. . . and disturbance of the streambed

Because stream discharge responds to events such as thunderstorms and snowmelt, streams are highly disturbed systems. Stream ecologists have recently been looking at ways in which different regimes of disturbance of the streambed are reflected

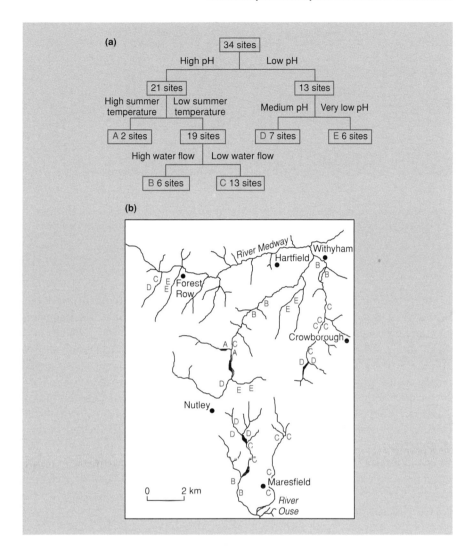

Figure 4.14
The species composition of stream invertebrate communities varies with conditions such as pH, summer temperature, and water flow. (a) Classification of 34 stream communities. (b) Distribution of community classes A to E, derived from community classification in (a). (After Townsend et al., 1983.)

in the composition of the community. For example, the disturbance regimes of 54 stream sites in New Zealand were assessed by painting particles (pebbles, cobbles, boulders) representative of their bed and determining the percentage that moved during several periods; this varied from 10 to 85 percent. The insect inhabitants of the streams were categorized according to properties that might help them deal with highly disturbed conditions, including small size (small species generally have short life cycles and their populations can rapidly rebuild), a streamlined or flattened body (less prone to being dislodged), and good powers of flight of the adult insects that emerge from the stream to mate (more likely to recolonize after a disturbance). The representation of these traits was higher in the more disturbed streams, testifying to the ecological importance of disturbance regimes (Figure 4.15). This relationship between animal traits and a predominant condition in their stream environment parallels the

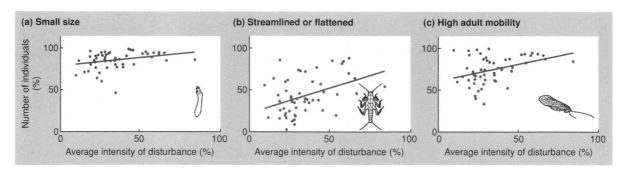

Figure 4.15
Disturbances are important determinants of stream ecology and, in particular, of communities of insects. Disturbed streams contained proportionately more larval insects that (a) were small, (b) had streamlined bodies, and (c) became adults that were strong fliers: characteristics that would enable these organisms to withstand a disturbance and recolonize afterward. $P < 0.001$ in each case. (After Townsend et al., 1997.)

interactions between the stream and surrounding land . . .

relationships between plant form and conditions in their terrestrial environments identified by Raunkiaer (see Figure 4.9).

The terrestrial vegetation surrounding a stream (the *riparian* vegetation) has two influences on the resources available to its inhabitants. First, by shading the streambed it may reduce primary production of attached algae and other plants. Second, by shedding leaves it can contribute directly to the food supply of animals and micro-organisms. Rivers that begin their course in forested regions are often dominated by the external supply of organic matter, and many of the invertebrates have mouthparts that can handle large particles (*shredders*) (Vannote et al., 1980). Farther downstream, where the stream is wider and where shading is less intense, invertebrates that graze or scrape algae from stones (grazer–scrapers) may be more abundant. As a result of the shredding of large particles into small organic particles (and also physical processes that break up leaves), food for *collector–gatherers* and *collector–filterers* may also increase downstream (Figure 4.16).

. . . may be disrupted by human activities

When riparian vegetation is changed, for example when forest is converted to agriculture, there can be far-reaching effects. Less particulate organic matter enters the stream, but there is less shading and more nutrient runoff from farmland. Results are an increase in productivity of stream plants and a corresponding change in the stream food web. There may also be effects on discharge (increased when trees are removed), water temperature (higher if shading is removed), and streambed characteristics (increased input of fine mineral particles). The more specific consequences of one particular interaction between human activity and stream ecology are described in Box 4.1.

The intimate relationship between land and water is also obvious on the floodplains of rivers such as the Amazon, where seasonal floods inundate huge areas of surrounding forest and provide massive inputs of nutrients and organic matter to the river. Many of the world's floodplains have been deliberately drained or cut off from their associated river channels. The effects of this dislocation of spatial continuity are only beginning to be understood (Townsend, 1996).

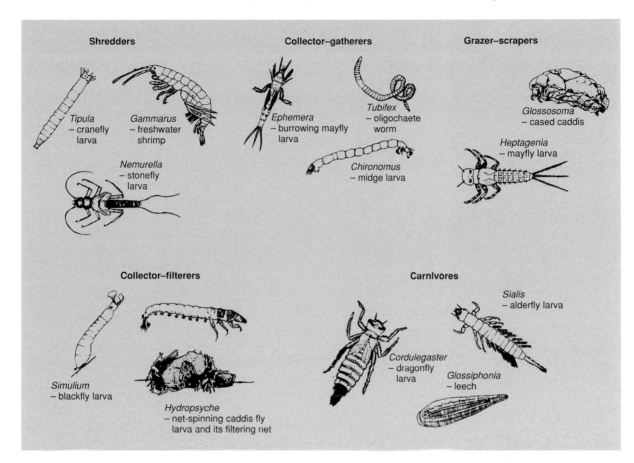

Figure 4.16
Examples of the various categories of invertebrate consumers in stream environments.

4.5.2   Lake ecology

Just as river ecology is defined by the unidirectional flow of water, lake ecology is defined by the relatively stationary nature of water within its basin. A critical component of lake ecology is the way in which water can stratify vertically in response to temperature (as mentioned in Section 4.2.3). As water sits in a lake basin, the upper layer is exposed to the sun and heats up. Because warm water is less dense than cold (and therefore tends to rise), and because there is resistance of water to heating, the top layer stratifies—that is to say, it forms a layer that is quite separate from the colder water beneath. This layer, the *epilimnion*, is warm and well illuminated and has a high oxygen content because surface waters exchange oxygen with the atmosphere. It is usually extremely productive, with high densities of plant and animal life.

The importance of vertical spatial scales becomes apparent in deeper lakes, where two further layers may form. Below the epilimnion is a transitional layer, the *thermocline*, in which temperature, oxygen concentration, and light all decrease. The deepest

lakes may become thermally stratified, with major consequences for their ecology

Box 4.1   *Topical ECOncerns*

A Tiny Stream Fish with Big Consequences for Property Development

Because streams are so intimately connected with their terrestrial catchment areas, human activity in their vicinity can have negative consequences for stream ecology. For example, landscaping or the construction of roads and buildings in the vicinity of waterways increases soil erosion and leads to silt runoff into streams. The Cherokee darter lives in clear streams with beds made up of pebbles and cobbles. Stream beds covered in silt deny this species the ability to forage and spawn; it is now restricted to just a few streams.

The following article by Clint Williams appeared in the *Atlanta Journal* on July 2, 2001.

CHEROKEE DARTER: Tiny fish forces changes to project

While barely 2 inches long, the Cherokee darter has the power to move roads and redesign a golf course.

The tiny fish, protected under the federal Endangered Species Act, swims in the small, gravel-bottomed streams that wind through a planned 730-acre gated community straddling the Cobb-Paulding county line. And it's forcing the developer to reshape his plans in order to protect the fish.

"We have fine-tuned our layout in order to minimize our impact on the Cherokee darter," said Joe Horton, developer of the Governor's Club, a high-dollar golf course development. "We're now on our sixth-generation site plan," Horton said.

The Cherokee darter, pale straw yellow with dark olive markings, was listed as threatened by the US Fish and Wildlife Service in 1994, not long after it was identified as a species distinct from the Coosa darter. The Cherokee darter is found only in roughly 20 small tributary systems of the Etowah River, according to a Fish and Wildlife report. But just a few streams have healthy populations.

"It's in a number of streams but that number is declining rapidly," said Seth Winger, a conservation ecologist at the Institute of Ecology at the University of Georgia. . . .

The creeks running through the Governor's Club property are tributaries of Pumpkinvine Creek, which flows into the Etowah River below Allatoona Dam. There are 8,000 feet of creeks on the tract, Horton said. A biological survey conducted before purchasing the property found four Cherokee darters. . . .

"We're proud we have them," Horton said. Having them will be a bit costly, however.

Is it reasonable that a small population of a species that occurs in about 20 other streams should disrupt economic development?

More specifically, how widespread would the species have to be (in how many streams, in how many States or countries) before developers should be allowed to ignore it?

Do you think it should be the responsibility of ecologists, such as the one quoted, simply to inform the public of the facts (as in this article)? Or is it reasonable for them to become involved in advocacy for a conservation cause?

layer, the *hypolimnion*, is cold and often poor in oxygen. It is here that sunken dead organic matter is decomposed and its mineral nutrients released. In temperate regions of the Earth, stratification of lake water breaks down in the fall when the upper layer cools. Currents then mix the water layers and the minerals released in the hypolimnion become available at the lake surface.

Lake ecologists are increasingly turning their attention to the larger spatial scale of whole lake districts. Lakes high in a landscape (such as those in northern Wisconsin) receive a greater proportion of their water from direct precipitation, whereas lakes at lower altitude receive more water as an input from ground water (Figure 4.17). This is reflected in the higher concentrations of important ions in lakes low in the landscape. The contrasting ion concentrations can be expected, for example, to affect the ecology and distribution of freshwater sponges, whose skeletons require silica, and crayfish and snails, which have a particular need for calcium.

Nutrient-rich lakes may support a rich flora of microscopic floating phytoplankton (microscopic plants), together with a diversity of invertebrates and fish species, but a rooted flora of flowering plants is confined to shallow waters near the shore, the *littoral* zone. This zone is usually rich in oxygen, light, food resources, and hiding places. However, some fish and invertebrates specialize in the deeper colder waters of lakes. Lake trout and walleye are two popular sport fish whose habitat is restricted to the colder regions of lakes.

Many lakes in arid regions, lacking a stream outflow, lose water only by evaporation and become rich in sodium and other ions. These saline lakes should not be considered as oddities; globally, they are just as abundant as freshwater lakes. They are usually very fertile and have dense populations of blue–green algae, and some, such as Lake Nkuru in Kenya, support huge aggregations of plankton-filtering flamingoes.

saline lakes are common in some parts of the world

4.5.3   Oceans

The oceans cover the major part of the Earth's surface and receive most of the Earth's income of solar radiation. However, much of this radiation is reflected at the water surface or absorbed by water itself and by particles in suspension. Even in clear water the intensity of radiation falls off exponentially with depth and photosynthesis is mainly restricted to the upper 100 m—the *euphotic zone*. In most waters the euphotic zone is much shallower, especially where water is more turbid close to coasts and estuaries.

The green plants that photosynthesize in the open oceans are planktonic, mainly single-celled algae that we know from experimental cultures are capable of using solar radiation very efficiently. But, in the real world, many areas of ocean that receive the greatest intensity of solar radiation have the lowest biological activity—because they are limited by shortage of mineral nutrients. The great tropical parts of the Atlantic and Pacific oceans have a biological productivity of less than 35 g C m^{-2} yr^{-1}. This compares with more than 800 g C m^{-2} yr^{-1} in terrestrial communities at the same latitudes.

The areas of greatest marine productivity (exceeding 90 g C m^{-2} yr^{-1}) occur where there is a reliable supply of minerals (especially nitrogen and phosphorus and perhaps iron). This occurs via leaching from the land through rivers and estuaries or where deep currents in the oceans well up to the surface and bring dissolved nutrients into the

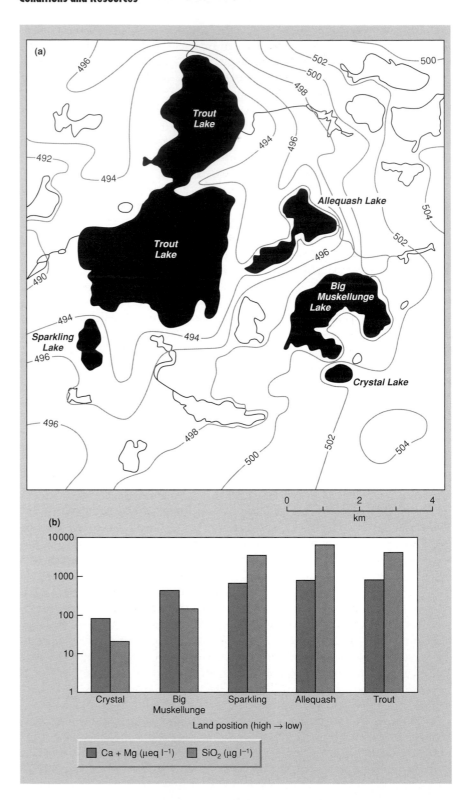

Figure 4.17
Lakes at different positions in the landscape differ in the source of their water and the concentrations of chemicals important to their inhabitants. (a) Map of Wisconsin Lake District: study lakes are shaded black and contours are shown (meters above sea level). (b) Relationship between landscape position and concentrations of calcium and magnesium (Ca + Mg) and silica (SiO_2) in the five lakes. (Based on Kratz et al., 1997.)

euphotic zone (see Section 11.2.2). In areas where upwellings occur, the desert of the ocean becomes transformed to a productive environment, as, for example, off the coast of Peru. Dense populations of planktonic algae support small crustacea, which in turn are eaten by schools of anchovies (*Engraulis ringens*). The fish support bonitos, sea lions, and flocks of cormorants, pelicans, and gannets.

We saw earlier in this chapter that the distribution of terrestrial communities depends largely on the intensity of solar radiation and its effects on temperature and water availability. In complete contrast, variations between oceanic communities are ruled mainly by the availability of mineral nutrients.

Below the euphotic zone is increasing darkness, and the ocean floor is in total darkness, intensely cold, and under great pressure. This abyssal environment supports the very slow biological activity of a community of extraordinary biological diversity (including worms, crustaceans, mollusks, and fish found nowhere else), which depends on the rain of dying and dead organisms falling from the euphotic zone above. Many of the invertebrate animals are tiny, have very low metabolic rates, and possess a life span that may last for decades. Yet further diversity has been discovered recently in hydrothermal vents that occur at a number of isolated places 2,000–4,000 m deep (see Box 2.2). In these remarkable environments, there are high sulfide concentrations and very high temperatures, up to 350°C, where superheated fluid emerges from "chimneys", and there is a sharp gradient down to 2°C, the temperature normally encountered in abyssal depths close by. The vent areas are inhabited by productive thermophilic (heat-loving) bacteria and a unique fauna of polychaete worms, crabs, and very large mollusks.

unique communities occur in the abyssal depths of the oceans

4.5.4 Coasts

Marine environments change dramatically near to coasts. Not only are they enriched by nutrients from the land, but they are also affected by waves and tides that bring new physical forces to bear. In particular, there are now surfaces to which organisms can attach; indeed, if they do not do so they are liable to be washed out to sea or stranded on the shore. At a broad scale, coastal communities are strongly influenced by waves and tides and the topography of the coast. Within a single stretch of coast, we can recognize a zonation in the flora and fauna marked by high and low tide marks and differing between areas with heavy or light wave action (Figure 4.18).

The extent of the intertidal zone depends on the height of tides and the slope of the shore. Away from the shore, the tidal rise and fall are rarely greater than 1 m, but closer to shore the shape of the land mass can funnel the ebb and flow of the water to produce extraordinary spring tidal ranges of, for example, nearly 20 m in the Bay of Fundy (between Nova Scotia and New Brunswick, Canada). In contrast, the shores of the Mediterranean Sea experience scarcely any tidal range.

waves and tides are key influences in coastal ecology

On steep shores and rocky cliffs the intertidal zone is very short and zonation is compressed. Both plants and animals are profoundly affected by the physical force of wave action. Anemones, barnacles, and mussels attach themselves securely and permanently to the substrate and filter planktonic plants and animals from the water when the tides cover them. Other animals, such as limpets, move to graze, and crabs move with the tides and use rock crevices as refuges. The flora in a rocky subtidal

Figure 4.18
A general zonation scheme for the seashore determined by relative lengths of exposure to the air and to the action of waves. (After Raffaelli & Hawkins, 1996.)

zone is usually dominated by the large brown seaweeds (kelps), which fix themselves to the rock with specialized "holdfasts".

Environments are quite different on shallow sloping shores on which the tides deposit and stir up sand and mud. Here, the dominant animals are mollusks and polychaete worms, living buried in the substrate and feeding by filtering the water when they are covered by the tides. This environment is completely free of large seaweeds, whose holdfasts can find no anchorage. Flowering plants are almost, but not completely, absent from intertidal environments. The exceptions occur where it is possible for them to be anchored by their roots and this requirement limits them to the more stable and muddy areas colonized by "sea grasses" such as *Zostera* and *Posidonia* or tussocks of *Spartina*. In the tropics, mangroves occupy this kind of habitat, adding a shrubby, woody dimension to the marine littoral zone.

4.5.5 Estuaries

Estuaries occur at the confluence of a river (fresh water) and a tidal bay (salt water). They provide an intriguing mix of the conditions normally experienced in rivers, shallow lakes, and tidal communities. Salt water, more dense than fresh water, tends to enter along the bottom of an estuary as a salt wedge. As it mixes with the outflowing fresh water, a brackish middle layer is created, then returns downstream. The shape of the wedge of salt water is determined in large part by the size of the discharge of the river flowing into the estuary; high discharge tends to create a smaller wedge of salt water and less mixing. The strong gradients in salinity, in both space and time, are reflected in a specialized estuarine fauna; some animals cope through particular physiological mechanisms while many avoid the variable salt concentrations by burrowing, closing protective shells, or moving away.

We have already noted that the broad-scale boundaries of biomes are not as clear-cut as maps imply. Nowadays, we are also acutely aware that these boundaries are not immutable, and that they may be on the move. Predicted changes in global climate over the next few decades can be expected to result in dramatic changes to the distribution of biomes over the face of the Earth (Box 4.2). But the exact nature of these changes remains profoundly uncertain.

UNANSWERED QUESTION:
biomes over the face of the Earth—where next?

Box 4.2 *Topical ECOncerns*

Predicted Changes in the Distribution of Biomes as a Result of Global Climate Change

As a result of human activities, the atmosphere contains increasing concentrations of gases, particularly carbon dioxide, but also nitrous oxide, methane, ozone, and chlorofluorocarbons (CFCs). These changes are predicted to lead to increased temperatures and altered patterns of climate over the face of the Earth (see Section 13.4.2). Given the controlling influence of climate on the distribution of biomes, ecologists expect the biome map of the world to change significantly as carbon dioxide concentrations double over the next 60–70 years.

It is no easy matter to predict the precise details of future climate or its consequences for biome distribution. Scientists have come up with a number of feasible scenarios, which differ according to the basic assumptions included in their models. The details of these need not concern us here; it is enough to note that the simulations shown in Figures 4.19 and 4.20 are based on a climate change model that assumes an effective doubling of carbon dioxide concentrations and takes into account the coupling of atmosphere and ocean in determining changes in

patterns of temperature and rainfall. This is translated into patterns in the distribution of biomes (using the biogeographic model MAPSS) by simulating the potential mature vegetation that could live under the "average" seasonal climate prevailing (see Neilson et al., 1998, for further details).

The distribution of biomes shown in Figure 4.19 is as simulated by the model for current climate (Neilson et al., 1998). In other words, it is the model's picture of the way biomes are distributed now (and reflects reality well; note that the biome categories are not exactly the same as those we discuss elsewhere in the chapter). The map in Figure 4.20, by contrast, is the predicted distribution of biomes in 60–70 years' time (Neilson et al., 1998).

This model predicts a reduction in area of the northern biomes of tundra and taiga/tundra (the open woodland that occurs between the treeless taiga and the dense northern coniferous forest). It also predicts a decrease in arid lands and an increase in temperate forest. These conclusions are in broad agreement with a variety of models that incorporate different starting assumptions.

(Thanks to Dr Ron Neilson for help in compiling this box.)

Box 4.2 (cont'd)

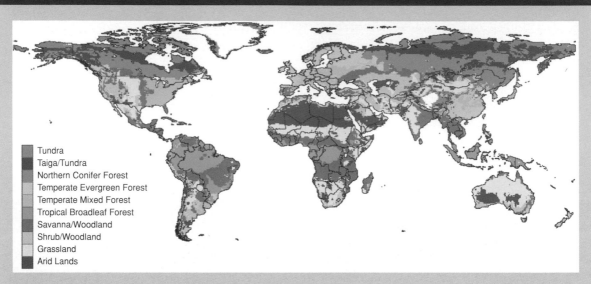

Figure 4.19
The distribution of major biome types under the current climate, as simulated by the MAPSS biogeography model. (After Neilson et al., 1998.)

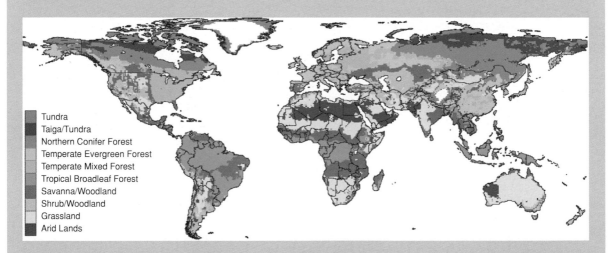

Figure 4.20
The potential distribution of major biomes resulting from climate changes associated with an effective doubling of carbon dioxide concentration, as simulated by the MAPSS biogeography model. (After Neilson et al., 1998.)

Summary

Geographical patterns at large and small scales

The variety of influences on climatic conditions over the surface of the globe causes a mosaic of climates. This, in turn, is responsible for the large-scale pattern of distribution of terrestrial biomes. However, biomes are not homogeneous within their hypothetical boundaries; every biome has gradients of physicochemical conditions related to local topographical, geological, and soil features. The communities of plants and animals that occur in these different locations may be quite distinct.

In most aquatic environments it is difficult to recognize anything comparable to terrestrial biomes; the communities of streams, rivers, lakes, estuaries, and open oceans reflect local conditions and resources rather than global patterns in climate.

The composition of local communities can change over timescales ranging from hours, through decades, to millennia.

Terrestrial biomes

A map of biomes is not usually a map of the distribution of species. Instead, it shows where we could find areas of land dominated by plants with characteristic shapes, forms, and physiological characteristics. The objective desciption of communities according to these gross features, such as Raunkiaer's "life forms", poses a difficult challenge for ecologists.

Tropical rain forest represents the global peak of evolved biological diversity. Its exceptional productivity results from the coincidence of high solar radiation received throughout the year and regular and reliable rainfall.

Savanna consists of grassland with scattered small trees. Seasonal rainfall places the most severe restrictions on the diversity of plants and animals in savanna; grazing herbivores and fire also influence the vegetation, favoring grasses and hindering the regeneration of trees.

Temperate grassland occurs in the steppes, prairies, pampas, and veldt. Typically, it experiences seasonal drought, but the role of climate in determining vegetation is usually overridden by the effects of grazing animals. Humans have transformed temperate grassland more than any other biome.

Many desert plants have an opportunistic lifestyle, stimulated into germination by the unpredictable rains; others, such as cacti, are long-lived and have sluggish physiological processes. Animal diversity is low in deserts, reflecting the low productivity of the vegetation and the indigestibility of much of it.

Temperate forests at lower latitudes experience mild winters, and the vegetation consists of broad-leaved evergreen trees. Nearer the poles, the seasons are strongly marked, and the vegetation is dominated by deciduous trees. Soils are usually rich in organic matter.

Northern coniferous forest often consists of a *biomonotony* that contrasts strongly with the biodiversity of tropical rain forests, reflecting a slow recovery from the catastrophes of the ice ages, and the overriding local constraint of frozen soil. Nearer the poles, the vegetation changes to tundra, and the two often form a mosaic in the Low Arctic. The mammal populations of the northern biomes often pass through remarkable cycles of expansion and collapse.

Aquatic environments

Streams and rivers are characterized by their linear form, unidirectional flow, fluctuating discharge, and unstable beds. The terrestrial vegetation surrounding a stream has strong influences on the resources available to its inhabitants; the conversion of forest to agriculture can have far-reaching effects.

Lake ecology is defined by the relatively stationary nature of its water. Some lakes stratify vertically in response to temperature, with consequences for the availability of oxygen and plant nutrients. Lakes higher in a landscape may receive more of their water from rainfall while those at lower altitude receive more from ground water. Saline lakes in arid regions lack a stream outflow and lose water only by evaporation.

The oceans cover the major part of the Earth's surface and receive most of the solar radiation. However, many areas have very low biological activity because of a shortage of mineral nutrients. Below the surface zone is increasing darkness, but at the ocean floor there may be an abyssal environment that supports a diverse community with very slow biological activity.

Coastal communities are enriched by nutrients from the land, but they are also affected by waves and tides. Within a single stretch of coast, there is a zonation in the flora and fauna that differs between areas with heavy or light wave action.

Estuaries occur at the confluence of a river (fresh water) and a tidal bay (salt water). Strong gradients in salinity, in both space and time, are reflected in a specialized estuarine fauna.

Review Questions

1 Describe the various changes in climate that occur with changing latitude, including an explanation of why deserts are more likely to be found at around 30° latitude than at other latitudes.

2 How would you expect the climate to change as you crossed from west to east over the Rocky Mountains?

3 ▲ Biomes are differentiated by gross differences in the nature of their communities, not by the species that happen to be present. Discuss possible schemes for the objective classification of gross differences among floras and faunas that can be used to distinguish among biomes.

4 The tropical rain forest is a diverse community supported by a nutrient-poor soil. Account for this.

5 ▲ Which of the Earth's biomes do you think have been most strongly influenced by people? How and why have some biomes been more strongly affected by human activity than others?

6 What is meant by the "stratification" of water in lakes? How does it occur? And what are the reasons for variations in stratification from time to time and from lake to lake?

7 Describe how the logging of a forest may influence the community of organisms inhabiting a stream running through the affected area.

8 Why is much of the open ocean, in effect, a "marine desert"?

9 Discuss some reasons why community composition changes as one moves (1) up a mountain and (2) down the continental shelf into the abyssal depths of the ocean.

10 ▲ Why are broad geographic classifications of aquatic communities less feasible than broad geographic classifications of terrestrial communities? What characteristics of aquatic ecosystems buffer the effects of climate?

▲ = Challenge Question

Web Research Questions

1 Global warming is expected to cause dramatic changes to patterns of rainfall and temperature. How could these changes impact on the boundaries of the world's biomes? Do different models of global warming give different results?

2 Hydrothermal vents in the depths of the oceans are home to bizarre and diverse arrays of creatures ("from inner space"). How do these ecosystems differ from those in terrestrial and shallow water locations in terms of their energy base and trophic stucture? Evaluate the idea that such vents may be where life originated on Earth.

Part III

Individuals, Populations, Communities, and Ecosystems

Chapter 5

Birth, Death, and Movement

All questions in ecology—however scientifically fundamental, however crucial to immediate human needs and aspirations—can be reduced to attempts to understand the distributions and abundances of organisms, and the processes—birth, death, and movement—that determine distribution and abundance. In this chapter, these processes, methods of monitoring them, and their consequences are introduced.

Chapter Contents

5.1 Introduction

5.2 Life cycles

5.3 Monitoring birth and death: life tables and fecundity schedules

5.4 Dispersal and migration

5.5 The impact of intraspecific competition on populations

5.6 Life-history patterns

Key Concepts

In this chapter you will

- appreciate the difficulties of counting individuals, but the necessity of doing so for understanding the distribution and abundance of organisms and populations
- appreciate the range of life cycles and patterns of birth and death exhibited by different organisms
- understand the nature and the importance of life tables and fecundity schedules
- understand the role and the importance of dispersal and migration in the dynamics of populations
- understand the impact of intraspecific competition on birth, death, and movement, and hence on populations
- appreciate that life-history patterns linking types of organism to types of habitat can be constructed but also recognize the limitations of those patterns

5.1   Introduction

Ecologists try to describe and understand the distribution and abundance of organisms. They may do so because they wish to control a pest or conserve an endangered species, or simply because they are fascinated by the world around them and the forces that govern it. A major part of their task, therefore, involves studying changes in the size of populations. It is usual to use the term *population* to describe a group of individuals of one species under investigation. What actually constitutes a population, though, varies from species to species and from study to study. In some cases, the boundaries of a population are readily apparent: the sticklebacks occupying a small lake are *the stickleback population of the lake*. In other cases, boundaries are determined more by an investigator's purpose or convenience. Thus, it is possible to study the population of lime aphids inhabiting one leaf, one tree, one stand of trees, or a whole woodland. What is common to all uses of *population* is that it is defined by the number of individuals that compose it and hence grows or declines by changes in those numbers.

The processes that change the size of populations are birth, death, and movement into and out of that population. Trying to understand the causes of changes in populations is a major preoccupation of ecologists because the science of ecology is not just about understanding nature but often also about predicting or controlling it. We might, for example, wish to reduce the size of a population of rabbits that can do serious harm to crops. We might do this by increasing the death rate by introducing the myxomatosis virus to the population, or by decreasing the birth rate by offering them food that contains a contraceptive. We might encourage their emigration by bringing in dogs, or prevent their immigration by fencing.

Similarly, a nature conservationist may wish to increase the population of a rare endangered species. In the 1970s, the numbers of bald eagles, ospreys, and other birds of prey in the United States began a rapid decline. This might have been because their birth rate had fallen, or their death rate had risen, or because the populations were normally maintained by immigration and this had fallen, or because individuals had emigrated and settled elsewhere. Eventually the decline was traced to reduced birth rates. The insecticide DDT was widely used at the time (it is now banned in the United States) and had been absorbed by fish and other prey of the birds. It accumulated in the bodies of the birds and affected their physiological processes so that the shells of their eggs became so thin that the chicks often died in the egg before they were able to hatch successfully. Conservationists charged with restoring the bald eagle population had to find a way to increase the birds' birth rate. The banning of DDT achieved this end.

5.1.1   What is an individual?

A population is a number of individuals, but for some kinds of organism it is not always clear what we mean by an individual. For many organisms there is no problem. An individual rabbit has two ears, two eyes, and four legs. An individual spider has eight legs. A spider that lived a long life would not grow more legs and a rabbit would not grow a third eye. The whole form of such organisms, and their program of development from the moment when a sperm fuses with an egg, is predictable and *determinate*. They are called *unitary* organisms.

Birds, insects, reptiles, and mammals are all unitary organisms. We could (in theory, anyway) determine the number of sheep in a flock by counting all their legs and dividing by four. But there is no way we could do anything similar with a population of *modular* organisms such as trees, shrubs and herbs, corals, sponges, and very many other marine invertebrates. These grow by the repeated production of modules (leaves, polyps, etc.) and almost always form a branching structure. Such organisms have an architecture: most are rooted or fixed, not motile (Figure 5.1). Both their structure and their precise program of development are not predictable but *indeterminate*. We could count all the leaves in a forest, but this would not allow us to determine the number of trees. We might count all the polyps in a piece of coral reef, but this could not tell us the number of fertilized eggs (zygotes) from which that piece of reef was derived.

In modular organisms, then, we need to distinguish between the genet, the genetic individual, and the module. The *genet* is the individual that starts life as a single-celled zygote and is considered dead only when all its component modules have died. A *module* starts life as a multicellular outgrowth from another module and proceeds through a life cycle to maturity and death even though the form and development of the whole genet are indeterminate. We usually think of unitary organisms when we write or talk about populations, perhaps because we ourselves are unitary, and there are certainly many more species of unitary than of modular organisms. But modular organisms are not rare exceptions and oddities. Most of the living matter (biomass) on Earth and a large part of that in the sea is of modular organisms: the forests, grasslands, coral reefs, and peat-forming mosses. In all of these, and in many other cases, distinguishing individuals can be fraught with difficulty.

modular organisms are themselves populations of modules

5.1.2 Counting individuals, births, and deaths

Even with unitary organisms, ecologists face enormous technical problems when they try to count what is happening to populations in nature. A great many ecological questions remain unanswered because of these problems. For example, resources can only be focused on controlling a pest effectively if it is known when its birth rate is highest. But this can only be known by monitoring accurately either births themselves or rising total numbers—neither of which is ever easy.

If we want to know how many fish there are in a pond we might obtain an accurate count by putting in poison and counting the dead bodies. But apart from the questionable morality of doing so, ecologists usually want to continue studying a population after they have counted it. Occasionally it may be possible to trap alive all the individuals in a population, count them, and then release them. With birds, for example, it may be possible to mark nestlings with leg rings and ultimately recognize every individual (except immigrants) in the population of a small woodland. It is not too difficult to count the numbers of large mammals such as deer on an isolated island. But it is very much more difficult to count the numbers of lemmings in a patch of tundra because they spend a large part of the year (and may reproduce) under thick snow cover! Even a census of the number of humans in a town is difficult when some individuals hide to escape tax collectors or immigration officials!

the difficulties of counting

(a)

(b)

Figure 5.1
Modular plants (on the left) and animals (on the right), showing the underlying parallels in the various ways they may be constructed. (a) Modular organisms that fall to pieces as they grow: duckweed (*Lemna* sp.) (© John D. Cunningham, Visuals Unlimited) and *Hydra* sp. (© Larry Stepanowicz, Visuals Unlimited). (b) Freely branching organisms in which the modules are displayed as individuals on "stalks": a vegetative shoot of a higher plant (*Lonicera japonica*) with leaves (feeding modules) and a flowering shoot (© Visuals Unlimited) and a hydroid colony (*Obelia*) bearing both feeding and reproductive modules (© L.S. Stepanowicz, Visuals Unlimited).

estimates from representative samples

Ecologists, therefore, are almost always forced to use indirect measures of the number of individuals in a population—they estimate rather than count. They may estimate the numbers of aphids on a crop, for example, by counting the number on a representative sample of leaves, then estimating the number of leaves per square meter of ground, and from this estimating the number of aphids per square meter. Sometimes more complex methods are used (Box 5.1).

(c)

(d)

(e)

Figure 5.1 (*cont'd*)
(c) Stoloniferous organisms in which colonies sperad laterally and reamin joined by "stolons" or rhizomes: a single plant of strawberry (*Fragaria*) spreading by means of stolons
(© Science VU, Visuals Unlimited) and a colony of the hydroid *Tubularia crocea* (© John D. Cunningham, Visuals Unlimited). (d) Tightly packed colonies of modules: a tussock of the
spotted saxifrage (*Saxifraga bronchialis*) (© Gerald and Buff Corsi, Visuals Unlimited) and a segment of the hard coral *Turbinaria reniformis* (© Dave B. Fleetham, Visuals Unlimited).
(e) Modules accumulated on a long persisent, largely dead support: an oak tree (*Quercus robur*) in which the support is mainly the dead woody tissues derived from previous modules
(© Silwood Park, Visuals Unlimited) and a gorgonian coral in which the support is mainly heavily calcified tissues from earlier modules (© Daniel W. Gotshall, Visuals Unlimited).

Box 5.1   *Quantitative Aspects*

Mark–Recapture Methods for Estimating Population Size

An estimate of the size of a population can sometimes be made by capturing a sample of individuals, marking them in some way (paint spots, leg rings), and then releasing them. Later, another sample is captured, and the proportion that is marked gives some estimate of the size of the whole population (see figure). For example, we might capture and mark 100 individuals from a population of sparrows and release them back into the population. If we later sample a further 100 individuals from the population and find half are marked, we could argue in the following way: half the sample are marked; the sample is representative of the whole population; therefore half the population are marked; 100 individuals were given a mark; therefore the whole population is composed of about 200 individuals. But this technique of mark and recapture is far less straightforward than it appears at first sight. There are many pitfalls in the sampling process and in interpretation of the data. Suppose, for example, that many of the individuals we marked died between our first and second visits. Modifications of the method would be needed to take account of this. For many motile organisms, however, it is the only technique that we have to estimate the size of a population.

The mark and recapture technique for estimating the size of a population of mobile organisms (in simplified form). (a) On a first visit to a population of unknown total size, N, a representative sample is caught (r individuals) and given a harmless mark. (b) These are released back into the population, where they remix with the unknown number of unmarked individuals. (c) On a second visit, a further representative sample is caught. Because it is representative, the proportion of marks in the sample (m out of a total sample of n) should, on average, be the same as that in the whole population (r out of a total of N). Hence N can be estimated.

UNANSWERED QUESTION:
modular organisms: what should we count?

It is *relatively* easy to count the number of plants in a pasture or barnacles on a rock face, because they remain in position, almost waiting to be counted. But what do we count in a forest? When is a tree a tree? Do we include saplings and seedlings? Do we include live seeds in the soil? Do we include trees of which all the parts are dead except for one or two shoots that still bear a live leaf or two? A commercial forester may make an arbitrary decision to count as trees only those above a certain size: for example, those more than 20 cm in diameter measured at breast height. To an entomologist interested in a pest that feeds on the trees, however, it may be the number of leaves rather than the number of trees that he needs to know.

And how are births to be counted? Students of genetics usually regard the formation of the zygote as the starting point in the life of an individual. This is when the genetic blueprint for development is first assembled. It is also the stage at which organisms start to die, but it is a stage that is often hidden and extremely hard to study. We simply do not know, for most animals and plants, how many embryos die before "birth", though in the rabbit at least 50 percent of embryos are thought to die in the womb, and in many higher plants it seems that about 50 percent of embryos abort before the seed is fully grown and mature.

For ecologists, though, it is almost always impossible in practice to treat the formation of a zygote as the time of birth. Some other, less logical starting point is usually forced upon us. A student of birds may count the moment that an egg hatches as the time of birth. A student of mammals may view birth as when an individual ceases to be supported within the mother on her placenta and starts to be supported outside her as a suckling. The botanist may regard the germination of a seed as the birth of a seedling, although it is really only the moment at which a developed embryo restarts into growth after a period of dormancy. We need to remember that half or more of a population will probably have died before it can be recorded as born!

Counting deaths in a population poses as many problems as counting births. Dead bodies do not linger long in nature. Only the skeletons of large animals persist long after death. Seedlings may be counted and mapped one day and gone without trace the next. Mice, voles, and soft-bodied animals such as caterpillars and worms are digested by predators or rapidly consumed by decomposing organisms (Chapter 11). They leave no carcasses to be counted and no forensic evidence of the cause of death.

Only a tiny fraction of the population of any species completes the full cycle from birth to reproduction to senescence and death. Those individuals that do so are anomalies—highly unusual exceptions. There are many more caterpillars than butterflies, many more tadpoles than frogs.

5.2 Life Cycles

5.2.1 Life cycles and reproduction

If we wish to understand the forces determining the abundance of a population of organisms, we need to know the important phases of those organisms' lives: that is, the phases when these forces act most significantly. For this, we need to understand the sequences of events that occur in those organisms' life cycles.

There is a point in the life of any individual when, if it survives that long, it will start to reproduce and leave progeny. A highly simplified, generalized life history (Figure 5.2) comprises birth, followed by a prereproductive period, a period of reproduction, a postreproductive period, and then death as a result of senescence (though of course other forms of mortality may intervene at any time). The life histories of all unitary organisms can be seen as variations around this simple pattern, though a postreproductive period (as seen in humans) is probably rather unusual.

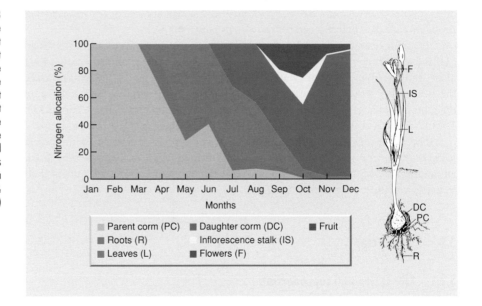

the conflict between growth and
reproduction

Some organisms fit several or many generations within a single year, some have just one generation each year (annuals), and others (perennials) have a life cycle extended over several or many years. For all organisms, though, a period of growth occurs before there is any reproduction, and growth usually slows down (and in some cases stops altogether) when reproduction starts. Growth and reproduction both require resources and there is clearly some conflict between them. Thus, as the perennial plant *Sparaxis grandiflora* enters its reproductive stage in the southwestern Cape of South Africa, flowers, flower stalks, and fruit can be seen to have been produced at the expense of roots and leaves (Figure 5.3). There are also many plants (e.g., foxgloves) which spend their first year in vegetative growth, and then flower and die in the second or a later year (called "biennial" plants). But if the flowers of these species are removed before their seeds begin to set, the plants usually survive to the following year, when they flower again and set seed even more vigorously. It seems to be the cost of provisioning the offspring (seeds) rather than flowering itself that is lethal. Similarly, pregnant women are advised to increase their caloric intake by as much as half their normal consumption: when nutrition is inadequate, pregnancy can harm the health of the mother.

Among both annuals and perennials, there are some—*iteroparous species*—that breed repeatedly, devoting some of their resources during a breeding episode not to breeding itself, but to survival to further breeding episodes (if they manage to live that long). We ourselves are examples. There are others, *semelparous species*, like the biennial plants already described, in which there is a single reproductive episode, with no resources set aside for future survival, so that reproduction is inevitably followed quickly by death.

iteroparous and semelparous species

5.2.2 Annual life cycles

In strongly seasonal, temperate latitudes, most annuals germinate or hatch as temperatures start to rise in the spring, grow rapidly, reproduce, and then die before the end of summer. The European common field grasshopper *Chorthippus brunneus* is an example of an annual species that is iteroparous. It emerges from its egg in late spring and passes through four juvenile stages of nymph before becoming adult in midsummer and dying by mid-November. During their adult life, the females reproduce repeatedly, each time laying egg pods containing about 11 eggs, and recovering and actively maintaining their bodies between the bursts of reproduction.

Many annual plants, by contrast, are semelparous: they have a sudden burst of flowering and seed set, and then they die. This is commonly the case among the weeds of arable crops. Others, such as groundsel, are iteroparous: they continue to grow and produce new flowers and seeds through the season until they are killed by the first lethal frost of winter. They die with their buds on.

Most annuals spend part of the year dormant as seeds, spores, cysts, or eggs. In many cases these dormant stages may remain viable for many years; there are reliable records of seeds of the annual weeds *Chenopodium album* and *Spergula arvensis* remaining viable in soil for 1,600 years. Similarly, the dried eggs of brine shrimps remain viable for many years in storage. This means that if we measure the length of life from the time of formation of the zygote, many so-called "annual" animals and plants live very much longer than a single year. Large populations of dormant seeds form a *seed bank* buried in the soil. As many as 86,000 viable seeds per square meter have been found in cultivated soils. Species of annuals that seem to have become locally extinct may suddenly reappear after the soil is disturbed and these seeds germinate.

seed banks

Dormant seeds, spores, or cysts are also necessary to the many ephemeral plants and animals of sand dunes and deserts that complete most of their life cycle in less than 8 weeks. They then depend on the dormant stage to persist through the remainder of the year and survive the hazards of low temperatures in winter and the droughts of summer. In desert environments, in fact, the rare rains are not necessarily seasonal, and it is only in occasional years that sufficient rain falls and stimulates the germination of the characteristic and colorful floras of very small ephemeral plants.

the ephemeral "annuals" of deserts

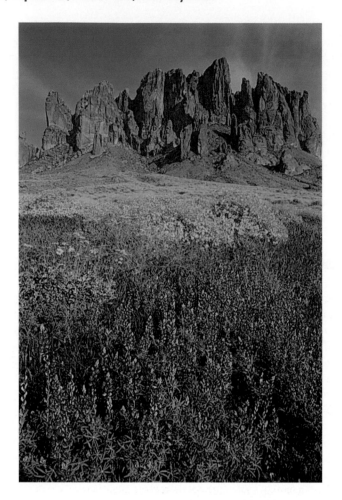

Desert in bloom. In desert areas where rainfall is rare and seasonally unpredictable, a dense and spectacular flora of very short-lived annuals commonly develops after rain storms. They often complete their life cycle from germination to seed set in little more than a month. (© Doug Sokell, Visuals Unlimited.)

5.2.3 Longer life cycles

repeated, seasonal breeders

There is a marked seasonal rhythm in the lives of many long-lived plants and animals, especially in their reproductive activity: a period of reproduction once per year (Figure 5.4a). Mating (or the flowering of plants) is commonly triggered by the length of the *photoperiod*—the light phase in the daily light–dark cycle, which varies continuously through the year—and usually makes sure that young are born, eggs hatch, or seeds are ripened when seasonal resources are likely to be abundant.

In populations of perennial species, the generations overlap and individuals of a range of ages breed side by side. The population is maintained in part by the survival of adults and in part by new births. A study of the great tit *Parus major*, for example, showed that of 50 eggs that were laid by a breeding population of ten birds in one season, only 30 hatchlings survived to become fully fledged, and only three of these survived to adulthood the following year. These three 1-year-old birds were joined in

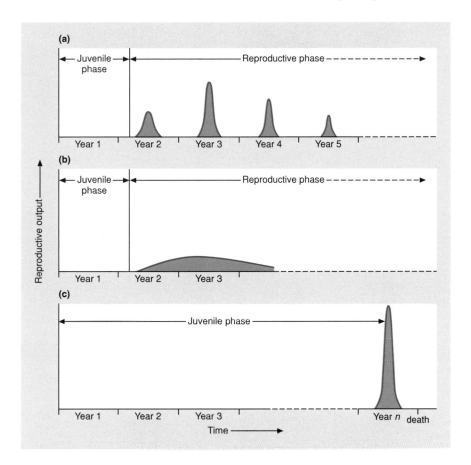

Figure 5.4
Simplified life histories for organisms living for more than 1 year. (a) An iteroparous species breeding seasonally once per year. Deaths tends not to occur predictably after any given time, though a decline toward senescence is often observed. (b) An iteroparous species breeding continuously throughout the year. The pattern of death and decline is similar to that in (a). (c) A semelparous species passing several or many years in a prereproductive juvenile phase, followed by a burst of reproduction, followed in turn by inevitable death.

that second year, though, by a further five birds aged between 2 and 5 years—the survivors of the previous year's ten (Figure 5.5).

In wet equatorial regions, on the other hand, where there is very little seasonal variation in temperature and rainfall and scarcely any variation in photoperiod, we find species of plant that are in flower and fruit throughout the year—and continuously breeding species of animal that subsist on this resource (Figure 5.4b). There are several species of fig (*Ficus*), for instance, that bear fruit continuously and form a reliable year-round food supply for birds and primates. In more seasonal climates, humans are unusual in also breeding continuously throughout the year, though numbers of other species, cockroaches, for example, do so in the stable environments that humans have created.

continuous breeders

In a contrasting program of lifetime fecundity, a plant or animal may spend almost all its life in a long nonreproductive (juvenile) phase and then have one lethal burst of reproductive activity (Figure 5.4c). We saw such semelparity earlier in biennial plants, but it is also characteristic of some species that live much longer than 2 years. The Pacific salmon is a familiar example. Salmon are spawned in rivers. They spend the first phase of their juvenile life in fresh water and then migrate to the sea, often traveling thousands of miles. At maturity they return to the stream in which they were hatched.

semelparous species like salmon and bamboo

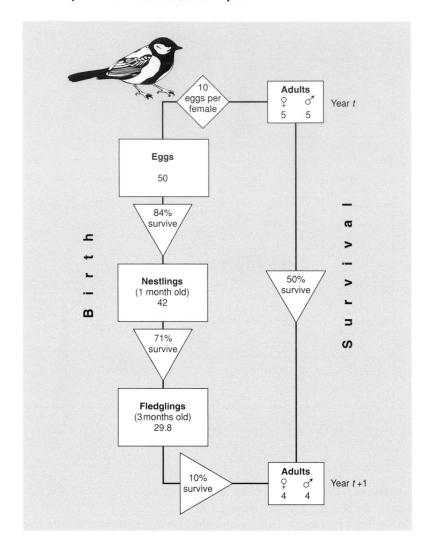

Some mature and return to reproduce after only 2 years at sea; others mature more slowly and return after 3, 4, or 5 years. At the time of reproduction the population of salmon is composed of overlapping generations of individuals. But all are semelparous: they lay their eggs and then die; their bout of reproduction is terminal.

There are even more dramatic examples of species that have a long life but reproduce just once. Many species of bamboo form dense clones of shoots that remain vegetative for many years: in some species, 100 years. The whole population of shoots then flowers simultaneously in a mass suicidal orgy. Even when shoots have become physically separated from each other, the parts still flower synchronously.

size is important   Organisms of long-lived species that are the same age, however, are not necessarily the same size—especially in modular organisms. Some individuals may be very old but have been suppressed in their growth and development by predators or by competition. Age is often a particularly poor predictor of fecundity. Charles Darwin

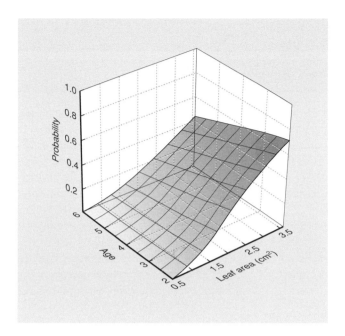

Figure 5.6
The effect of plant age (years) and plant size (as measured by leaf area) on the probability of *Rhododendron lapponicum* shoots entering their reproductive phase. The relationships have been "smoothed" by a statistical technique called "logistic regression". Note that the probability of reproduction increases with plant size at all ages. Also, older shoots are overall more likely to enter their reproductive phase because they tend to be bigger. However, at any given size, the probability of reproduction tends to *decrease* with age, making age itself a much poorer predictor of shoot-fate than size. (After Karlsson & Jacobson, 2001.)

examined a population of 32 dwarfed fir trees in a square yard of heathland and found one that was 26 years old—repeatedly grazed by cattle, still alive, but far too small to reproduce. An analysis that classifies the members of a population according to their size is often more useful in suggesting whether they will survive, reproduce, or whatever (Figure 5.6) than one that classifies them by their age.

5.3   Monitoring Birth and Death: Life Tables and Fecundity Schedules

The previous sections have outlined the different patterns of births and deaths in different species. But patterns are just a start. What are the *consequences* of these patterns in specific cases in terms of their effects on how a population might grow to pest proportions, say, or shrink to the brink of extinction? To determine these consequences, we need to monitor the patterns in a quantitative way.

There are different ways of doing so. To monitor and quantify survival, we may follow the fate of individuals from the same *cohort* within a population; that is, all individuals born within a particular period. A *cohort life table* then records the survivorship of the members of the cohort over time (Box 5.2). A different approach is necessary when we cannot follow cohorts but we know the ages of all the individuals in a population. We can then, at one time, describe the numbers of survivors of different ages in what is called a *static life table* (Box 5.2).

The fecundity of individuals also changes with their age, and to understand properly what is going on in a population we need to know how much individuals of different ages contribute to births in the population as a whole: these can be described in *age-specific fecundity schedules*.

Box 5.2   *Quantitative Aspects*

The Basis for Cohort and Static Life Tables

In the figure below, a population is portrayed as a series of diagonal lines, each line representing the life "track" of an individual. As time passes, each individual ages (moves from bottom left to top right along its track) and eventually dies (the dot at the end of the track). In some cases, individuals are classified by their age (as here), but in others it is more appropriate to split the life of each individual into the different developmental stages.

In the present case, three individuals are born prior to the time period t_0, four during t_0, and three during t_1. To construct a *cohort life table*, we direct our attention to a particular cohort (in this case,

those born during t_0) and monitor the subsequent development of the cohort. The life table is constructed by noting the number surviving to the start of each time period. Here, we note that two of the four individuals have survived to the beginning of t_1; only one of these is alive at the beginning of t_2; none survives to the start of t_3. The first data column of the cohort life table thus comprises the series of declining numbers in the cohort: 4, 2, 1, 0.

A different approach is necessary when we cannot follow cohorts but we know the ages of all the individuals in a population (perhaps from some clue such as the condition of the teeth in a species of deer).

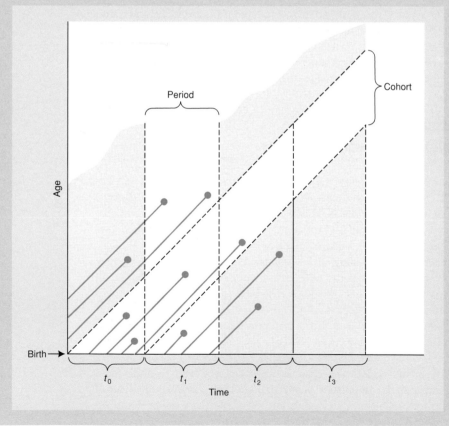

Box 5.2 (cont'd)

We can then, as the figure shows, direct our attention to the whole population during a single period (in this case, t_1) and note the numbers of survivors of different ages in the population. These may be thought of as entries in a life table if we assume that rates of birth and death are, and have previously been, constant—a very big assumption. What results is called a *static life table*. Here, of the seven individuals alive during t_1, three were actually born during t_1, two were born in the previous time interval, and two in the interval

before that. The first data column of the life table would thus comprise the series 3, 2, 2, 0. Considering this the first column of a life table implies, first, that a single life table can typify the survival of individuals during the whole period shown (i.e., birth and death rates remained roughly constant over that time), and second, that the typical cohort will have started with three and declined over successive time intervals to two, then two again, then zero.

5.3.1 Annual life tables

Probably the most straightforward life table to construct is a cohort life table for annuals, because with nonoverlapping generations it is indeed often possible to follow a single cohort from the first birth to the death of the last survivor. A simplified life table is shown in Table 5.1 for the common field grasshopper. The first column is a list of the stages of the grasshopper's life. The second is then the raw data from the study, collected in the field. It reports the number of individuals surviving to the beginning of each stage.

a grasshopper life table . . .

Since ecologists will typically be interested not just in examining populations in isolation but in comparing the dynamics of two or more perhaps rather different

Table 5.1
A simplified cohort life table for the common field grasshopper *Chorthippus brunneus*. The columns are explained in the text. (After Richards & Waloff, 1954.)

Stage (x)	Number observed at start of each stage a_x	Proportion of original cohort surviving to start of each stage l_x	Eggs produced in each stage F_x	Eggs produced per surviving individual in each stage m_x	Eggs produced per original individual in each stage $l_x m_x$
Eggs (0)	44,000	1.000	–	–	–
Instar I (1)	3,513	0.080	–	–	–
Instar II (2)	2,529	0.058	–	–	–
Instar III (3)	1,922	0.044	–	–	–
Instar IV (4)	1,461	0.033	–	–	–
Adults (5)	1,300	0.030	22,617	17	0.51

$$R_0 = \Sigma l_x m_x = \frac{\Sigma F_x}{a_0} = 0.51.$$

populations (in the presence and absence of a pollutant, for instance), it is necessary to standardize the raw data so that comparisons can be made. This is done in the third column of the table, which is said to contain l_x values, where l_x is defined as the proportion of the original cohort surviving to the start of stage x (or in other cases, age class x). The first value in the third column, l_0 (spoken: L zero), is therefore the proportion surviving to the beginning of this original stage. Obviously, here and in every life table, l_0 is 1.00 (the whole cohort is there at the start). In fact, the grasshopper population had an initial number of 44,000 eggs. The l_x values for subsequent stages are therefore expressed as proportions of this number. Only 3,513 individuals survived to reach the first nymphal stage. Thus, the second value in the third column, l_1, is the proportion $3{,}513/44{,}000 = 0.08$ (that is, only 0.08 or 8 percent of the original cohort survived this first step), and so on. In a full life table, subsequent columns would then use these same data to calculate the proportion of the original cohort that died at each stage and also the mortality rate for each stage, but for brevity these columns have been omitted here.

. . . and fecundity schedule . . .

Table 5.1 also includes a fecundity schedule for the grasshopper (columns 4 and 5). Column 4 shows the total number of eggs produced by each stage F_x (though in this case only the adult stage produced eggs). In fact, the 1,300 surviving adults produced 22,617 eggs. The fifth column is then said to contain m_x values: the mean number of eggs produced per surviving individual. In the present case there is only one such value, for the adults: $m_5 = 22{,}617/1{,}300 = 17$.

. . . combined to give the basic reproductive rate

In the final column of a life table, the l_x and m_x columns are brought together to express the overall extent to which a population increases or decreases over time—reflecting the dependence of this on both the survival of individuals (the l_x column) and the reproduction of those survivors (the m_x column). In this case, 0.03 of the original cohort survived to the adult stage, and those survivors produced 17 eggs on average. Hence, the mean number of eggs produced at the end of the cohort per egg present at the beginning was $0.03 \times 17 = 0.51$ eggs. This is therefore a measure of the overall extent by which this population has (in this case) decreased in a generation. We call this the *basic reproductive rate* and denote it by R. In this case, then, $R = 0.51$.

a life table for an annual plant

Data from annual plants can be treated in the same way. Thus an initial cohort of 996 seeds of the annual plant *Phlox drummondii* was followed from seed germination, with the life cycle broken down into successive periods of 30–60 days (Table 5.2). In contrast to the grasshoppers, whose population clearly declined over the annual cycle, this population set approximately two and a half times more seed at the end of the season than was present at the beginning. Notice, though, that these seeds were produced not by just one stage (as among the grasshoppers) but by several. It is apparent therefore that R (= 2.41 in this case) is generally the sum of all the $l_x m_x$ values, denoted by $\Sigma l_x m_x$, where the symbol Σ means "the sum of".

logarithmic survivorship curves

It is also possible to study the detailed pattern of decline in the *Phlox* cohort. Figure 5.7a, for example, shows the numbers surviving relative to the original population —the l_x values—plotted against the age of the cohort. However, this can be misleading. If the original population is 1,000 individuals, and it decreases by half to 500 in one time interval, then this decrease looks more dramatic on a graph like Figure 5.7a than a decrease from 50 to 25 individuals later in the season. Yet the risk of death to individuals is the same on both occasions. If, however, l_x values are replaced by $\log(l_x)$ values, that is, the logarithms of the values, as in Figure 5.7b (or, effectively the same thing, if l_x values are plotted on a log scale), then it is a characteristic of logs that the

Table 5.2
A simplified cohort life table for the annual plant *Phlox drummondii*. The columns are explained in the text. (After Leverich & Levin, 1979.)

Age interval (days) x–x'	Number surviving to day x a_x	Proportion of original cohort surviving to day x l_x	F_x	m_x	$l_x m_x$
0–63	996	1.000	–	–	–
63–124	668	0.671	–	–	–
124–184	295	0.296	–	–	–
184–215	190	0.191	–	–	–
215–264	176	0.177	–	–	–
264–278	172	0.173	–	–	–
278–292	167	0.168	–	–	–
292–306	159	0.160	53.0	0.33	0.05
306–320	154	0.155	485.0	3.13	0.49
320–334	147	0.148	802.7	5.42	0.80
334–348	105	0.105	972.7	9.26	0.97
348–362	22	0.022	94.8	4.31	0.10
362–	0	0.000	–	–	–
Total			2,408.2		2.41

$$R_0 = \Sigma l_x m_x = \frac{\Sigma F_x}{a_0} = 2.41.$$

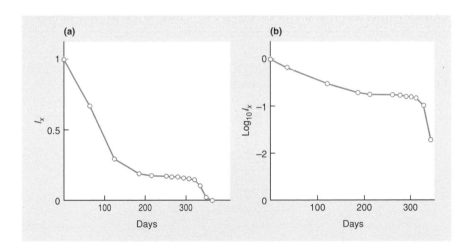

Figure 5.7
Following the survival of a cohort of *Phlox drummondii* (Table 5.2). (a) When l_x is plotted against cohort age, it is clear that most individuals are lost earlier in the life of the cohort, but there is no clear impression of the risk of mortality at different ages. (b) By contrast, a survivorship curve plotting $\log(l_x)$ against age shows that an initial 6 months of moderate survivorship was followed by an extended period of higher survivorship (less risk of mortality) and then by very low survivorship in the final weeks of the annual cycle.

reduction of a population to half its original size will always look the same. *Survivorship curves* are, therefore, conventionally plots of log(l_x) values against cohort age. Figure 5.7b shows that there was a relatively rapid decline in the size of the cohort over the first 6 months, but that the death rate thereafter remained steady and rather low until the very end of the season, when the survivors all died.

It is possible to see, therefore, even from these two examples, how life tables can be useful in characterizing the "health" of a population—the extent to which it is growing or declining—and identifying where in the life cycle, and whether it is survival or birth, that is apparently most instrumental in determining that rate of increase or decline. Either or both of these may be vital in determining how best to conserve an endangered species or control a pest.

5.3.2   Life tables for populations with overlapping generations

Many of the species for which we have important questions, and for which life tables may provide an answer, have repeated breeding seasons (or continuous breeding, as in the case of humans), but constructing life tables here is more complicated, largely because these populations have individuals of many different ages living together. Building a cohort life table is sometimes possible, but this is relatively uncommon. Apart from the mixing of cohorts in the population, it can be difficult simply because of the longevity of many species.

a static life table—useful if used with caution

Another approach is to construct a "population snapshot" in a static life table (Box 5.2). Superficially, the data look like a cohort life table: a series of different numbers of individuals in different age classes. But great care is required; they can only be treated and interpreted in the same way if patterns of birth and survival in the population have remained much the same since the birth of the oldest individuals—and this will happen only rarely. Nonetheless, useful insights can sometimes be gained by combining the data from a static life table (an *age structure*: the numbers in different age classes) with corresponding background information. This is illustrated by a study of two populations of the long-lived tree *Acacia burkittii* in South Australia (Figure 5.8). Although differences in age structure between the populations are obvious, the reasons are not. Fortunately, background information provides important clues.

Figure 5.8
Age structures (and hence static life tables) of *Acacia burkittii* populations at two sites in South Australia. (a) South Lake Paddock populations had been grazed by sheep from 1865 to 1970 and by rabbits from 1885 to 1970; (b) whereas the Reserve population had been fenced in 1925 to exclude sheep (but did not exclude rabbits). The effect of grazing from 1865 onward is evident in the decreased numbers of new recruits to both populations. However, the effects of fencing after 1925 are equally obvious in the Reserve population, where the proportion of new recruits increased dramatically. The effects of rabbit grazing on recruitment after fencing in the Reserve population can, however, still be detected, since, for example, the 1925–40 age class was much smaller than the (pregrazing) 1845–60 class, even though the latter had survived an additional 75 years. (After Crisp & Lange, 1976.)

5.3.3   A classification of survivorship curves

Life tables provide a great deal of data on specific organisms. But ecologists search for generalities—patterns of life and death that we can see repeated in the lives of many species. Ecologists conventionally divide survivorship curves into three types, in a scheme that goes back to 1928, generalizing what we know about the way in which the risks of death are distributed through the lives of different organisms (Figure 5.9).

■ Type I describes the situation in which mortality is concentrated toward the end of the maximum life span. It is perhaps most typical of humans in developed countries and their carefully tended zoo animals and pets.
■ Type II is a straight line that describes a constant mortality rate from birth to maximum age. It describes, for instance, the survival of buried seeds in a seed bank.
■ Type III indicates extensive early mortality, but a high rate of subsequent survival. This is typical of species that produce many offspring. Few survive initially, but once individuals reach a critical size, their risk of death remains low and more or less constant. This appears to be the most common survivorship curve among animals and plants in nature.

These types of survivorship curve are useful generalizations, but in practice, patterns of survival are usually more complex. Thus, in a population of *Erophila verna*, a very short-lived annual plant inhabiting sand dunes, survival can follow a type I curve when the plants grow at low densities; a type II curve, at least until the last stages of life, at medium densities; and a type III curve in the early stages of life at the highest densities (Figure 5.10).

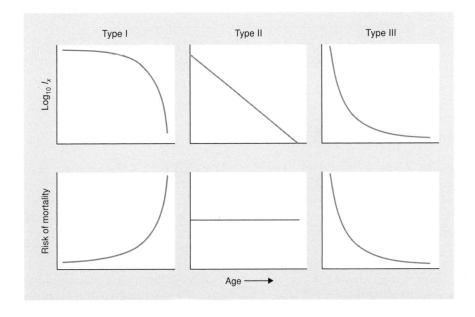

Figure 5.9
A classification of survivorship curves plotting log(l_x) against age, above, with corresponding plots of the changing risk of mortality with age, below. The three types are discussed in the text. (After Pearl, 1928; Deevey, 1947.)

Figure 5.10
Survivorship curves (l_x, where $l_0 = 1,000$) for the sand dune annual plant *Erophila verna* monitored at three densities: high (initially 55 or more seedlings per 0.01 m² plot); medium (15–30 seedlings per plot); and low (1–2 seedlings per plot). The horizontal scale (plant age) is standardized to take account of the fact that each curve is the average of several cohorts, which lasted different lengths of time (around 70 days on average). (After Symonides, 1983.)

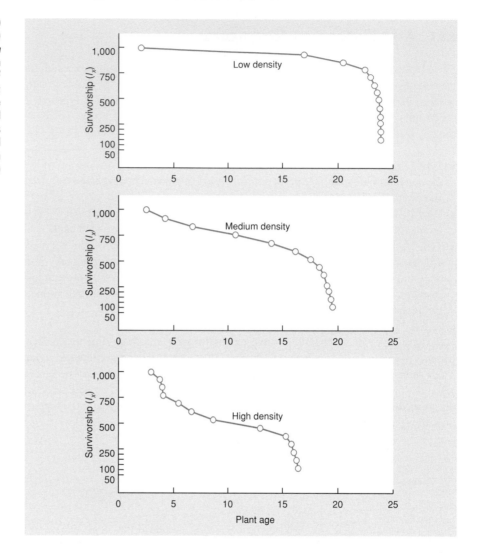

5.4   Dispersal and Migration

patterns of distribution

Birth is only the beginning. If we were to stop there in our studies, many crucial ecological questions would remain unanswered. From their place of birth, all organisms move to locations where we eventually find them. Plants grow where their seeds fall, but seeds may be moved by the wind, water, animals, or shifting soil. Animals move in search of food and safe havens, whether it is only to move 1 cm along a leaf from where their egg was deposited, or to move halfway around the globe. The effects of those movements are varied. In some cases they aggregate members of a population into clumps; in others they continually redistribute and shuffle them; and in still others they spread the individuals out and "dilute" their density. Three generalized spatial patterns

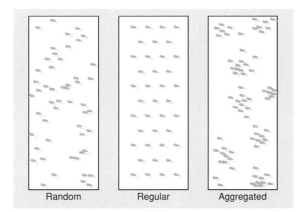

Figure 5.11
Three generalized spatial patterns that may
be exhibited by organisms across their
habitat.

Random   Regular   Aggregated

that result from this movement—aggregated (clumped), random, and regular (evenly spaced)—are illustrated in Figure 5.11. Clearly, movement and spatial distribution (the latter sometimes confusingly called "dispersion") are intimately related.

Technically, the term *dispersal* describes the way individuals spread away from each other, such as when seeds are carried away from a parent plant or young lions leave the pride in search of their own territory. *Migration* refers to the mass directional movement of large numbers of a species from one location to another. Migration therefore describes the movement of locust swarms but also includes the smaller scale movements of intertidal organisms, back and forth twice a day as they follow their preferred level of immersion or exposure.

Our view of dispersal and migration, and of the resulting distributions, is determined by the scale on which we are working. For example, consider the distribution of an aphid living on a particular species of tree in a woodland. On a large scale, the aphids appear to be aggregated in the woodlands and nonexistent in the open fields. If the samples we took were smaller, and taken only in woodlands, the aphids would still appear to be aggregated, but now aggregated on their host trees rather than on trees in general. However, if samples were collected at an even smaller scale—the size of a leaf within a canopy—the aphids might appear to be randomly distributed over the tree as a whole. And on the scale experienced by the aphid itself (1 cm^2), the distribution might appear regular as individuals on a leaf spread out to avoid one another (Figure 5.12).

perception of pattern depends on the spatial scale

This example also illustrates the difference between the "average density" and the crowding experienced by individuals in a population. The *average density* is simply the total number of individuals divided by the total size of the habitat—but it depends very much on how we define the habitat. For the aphids, if it includes everything, woodland and nonwoodland, then average density will be low. It will be higher, but still quite low, if we include only woodland but every species of tree. It will be much higher, however, if we include only the aphids' host trees.

density and crowding

The average density of individuals in the United States is about 75 persons km^{-2}. Yet there are vast areas of the United States—rural and wilderness areas—within which the density is low, but also crowded cities and towns within which the density is much higher. And because the majority of people live in urban and suburban settings, the

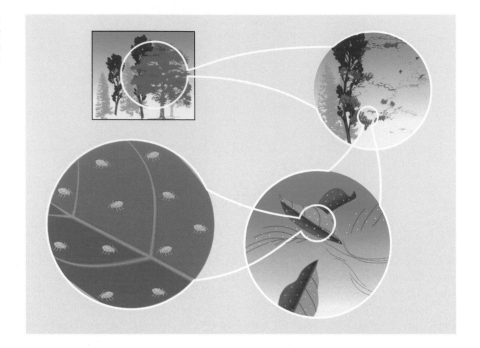

density actually experienced by people, on average, has been calculated at 3,630 persons km^{-2}. There may be little impetus for dispersal, or migration, at the relatively low population pressure of 75 individuals km^{-2}. At 3,630 people km^{-2}, however, individuals are much more likely to find ways to escape from their neighbors. Real measures of crowding as experienced by individuals are likely to be more important forces driving dispersal and migration than some average value of population density.

5.4.1 Dispersal determining abundance

a sand dune plant example

Dispersal can have a profound effect on the dynamics of a population. This was seen in a study of *Cakile edentula*, a summer annual plant growing on the sand dunes of Martinique Bay, Nova Scotia. The population was concentrated in the middle of the dunes and declined toward both the sea and the land. Only in the area toward the sea, however, was seed production high enough and mortality rate sufficiently low for the population to maintain itself year after year. At the middle and landward sites, mortality rate exceeded seed production rate. Hence, one might have expected the population to become extinct (Figure 5.13). But the distribution of *Cakile* did not change over time. Instead, large numbers of seeds from the seaward zone dispersed to the middle and landward zones. Indeed, more seeds were dispersed into and germinated in these two zones than were produced by the residents. The distribution and abundance of *Cakile* were directly due to the dispersal of seeds in the wind and the waves.

There are few populations in which the dynamics are weighted so heavily by immigration as they are in *Cakile*. Yet most populations are more affected by immigration and emigration than is commonly imagined. Within the United States, for

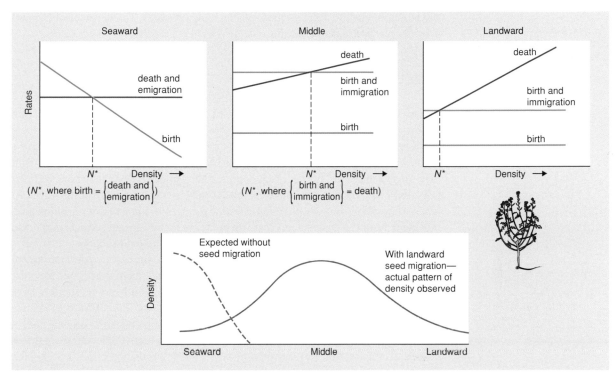

Figure 5.13
Diagrammatic representation of variations in mortality and seed production rate of *Cakile edentula* in three areas along an environmental gradient from open sand beach (seaward) to densely vegetated dunes (landward). In contrast to that in other areas, seed production was prolific at the seaward site. Births, however, declined with plant density, and where births were equal to {deaths + emigration}, an equilibrium population density can be envisaged (N^*) (see also Section 5.5). In the middle and landward sites, deaths always exceeded births resulting from local seeds, but populations persisted there because of immigration—the landward drift of the majority of seed produced by plants on the beach (seaward site). Thus, the sum of local births plus immigrating seeds can balance mortality in the middle and landward sites, resulting in equilibria at appropriate densities. (After Keddy, 1982; Watkinson, 1984.)

example, over 40 percent of US residents, over 100 million people, can trace their roots to the 12 million immigrants who entered the United States through the Ellis Island port from 1870 to 1920.

It is hard to count the births and deaths in a population. It is even harder to determine the number of individuals that immigrate or emigrate. It is possible to determine that of 5,000 seeds produced by a ragwort plant, 3,000 fell within a square meter of the parent, but it is next to impossible to determine the fate of the other 2,000 seeds. Yet the importance of dispersal cannot be overemphasized. In some cases, such as the *Cakile* example, it is repeated dispersal that maintains the abundance of a population. In other cases, it may be the extreme distances dispersed by a very few individuals that are responsible for the distribution of a species. It takes only one seed, carried to a new area, to begin a new population. Often ecologists have made the rash assumption that the individuals that leave a population are balanced by those that join it. We now know that this can be grossly misleading. Moreover, immigrants and emigrants not only influence the numbers in a population—they also affect its composition. A higher proportion of individuals usually escape from a crowded population than from a sparse one, and those that leave are not a random sample. The dispersers are often the young,

UNANSWERED QUESTION: where do all the dispersers go?

and males frequently do more moving about than females. The behavior of humans provides plenty of examples to illustrate these generalizations!

5.4.2   The role of migration

The mass movements of populations that we call migration are almost always from regions where the food resource is declining to regions where it is abundant (or where it will be abundant for the progeny). By day, planktonic plants live in the upper layers of the water in lakes where the light needed for photosynthesis is brightest. At night they migrate to lower, nutrient-rich depths. Crabs migrate along the shore with the tides, following the movement of their food supply as it is washed up in the waves. At longer timescales, some shepherds still follow the ages-old practice of transhumance, moving their flocks of sheep and goats up to mountain pastures in summer and down again in the fall to track the seasonal changes in climate and food supply.

The long-distance migrations of terrestrial birds in many cases involve movement between areas that supply abundant food, but only for a limited time. They are areas where seasons of comparative glut and famine alternate and cannot support large all-year-round resident populations. For example, swallows (*Hirundo rustica*) migrate seasonally from northern Europe in the fall, when flying insects start to become rare, to South Africa when they are becoming common. In both areas the food supply that is reliable throughout the year can support only a small population of resident species. The seasonal glut supports the populations of invading migrants, which make a large contribution to the diversity of the local fauna.

5.5   The Impact of Intraspecific Competition on Populations

The concept of intraspecific competition was introduced in Section 3.5 because its intensity is typically dependent on resource availability. It re-emerges here because its effects are expressed through the focal topics of this chapter—rates of birth, death, and movement. Competing individuals that fail to find the resources they need may grow more slowly or even die; survivors may reproduce later and less; or, if they are mobile, they may move farther apart or migrate elsewhere. Examples in which the dynamics of a species can be understood without a firm grasp of the effects of competition are exceedingly rare.

crowding not density—especially in modular organisms

The intensity of competition for limiting resources is often related to the density of a population, though, as we have seen, the straightforward density need not be a good measure of the extent to which its individuals are crowded. Modular sessile organisms are particularly sensitive to competition from their immediate neighbors: they cannot withdraw from each other and space themselves more evenly or escape by dispersal or migration. Thus, when silver birch trees (*Betula pendula*) were grown in small groups, there were more suppressed and dying branches on the sides of individual trees where their branches shaded each other, whereas on the sides away from neighbors there was more vigorous growth (Figure 5.14).

density-dependent birth and death and the carrying capacity

The inexact relationship between density and crowding notwithstanding, we saw in Section 3.5 that, over a sufficiently large density range, as density increases,

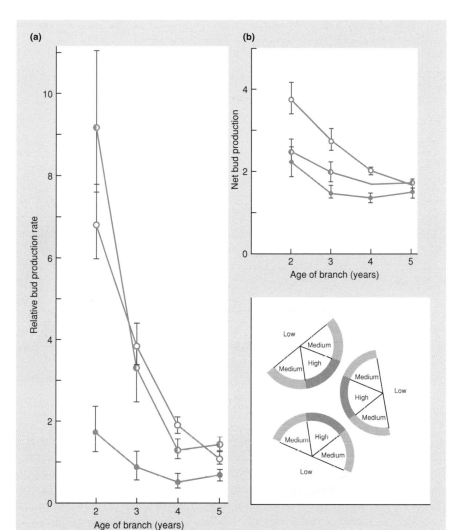

Figure 5.14
Mean relative bud production (new buds per existing bud) for silver birch trees (*Betula pendula*), expressed (a) as gross bud production and (b) as net bud production (birth minus death), in different interference zones. These zones are themselves explained in the inset. ●, high interference; ◑, medium; ○, low. Bars represent standard errors. (After Jones & Harper, 1987.)

competition between individuals generally reduces the per capita birth rate and increases the death rate, and that this effect is described as *density dependent*. Thus, when birth and death rate curves are plotted against density on the same graph, and either or both are density dependent, the curves must cross (Figure 5.15a–c). They do so at the density at which birth and death rates are equal, and because they are equal, there is no overall tendency at this density for the population either to increase or to decrease (ignoring, for convenience, both emigration and immigration). The density at the crossover point is called the *carrying capacity* and is denoted by the symbol K. At densities below K, births exceed deaths and the population increases. At densities above K, deaths exceed births and the population decreases. There is therefore an overall tendency for the density of a population under the influence of intraspecific competition to settle at K.

Figure 5.15
Density-dependent birth and mortality rates lead to the regulation of population size. When both are density dependent (a), or when either of them is (b, c), their two curves cross. The density at which they do so is called the *carrying capacity* (K). However, the real situation is closer to that shown by thick lines in (d), where mortality rate broadly increases, and birth rate broadly decreases, with density. It is possible, therefore, for the two rates to balance not at just one density, but over a broad range of densities, and it is toward this broad range ("K") that other densities tend to move.

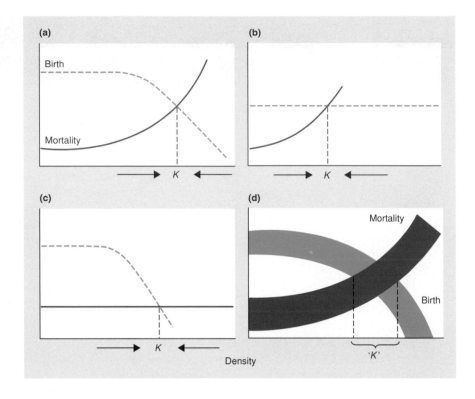

population regulation by competition—but not to a single carrying capacity

In fact, because of the natural variability within populations, the birth rate and death rate curves are best represented by broad lines, and K is best thought of not as a single density, but as a range of densities (Figure 5.15d). Thus, intraspecific competition does not hold natural populations to a single, predictable and unchanging level (K), but it may act upon a very wide range of starting densities and bring them to a much narrower range of final densities. It therefore tends to keep density within certain limits, and may thus be said to play a part in *regulating* the size of populations.

Of course, graphs like those in Figure 5.15 are generalizations on a grand scale. Many organisms, for example, have seasonal life cycles. For part of the year births vastly outnumber deaths, but later, after the period of peak births, there is likely to be a period of high juvenile mortality. Most plants, for example, die as seedlings soon after germination. Thus, although births may balance deaths over the year, a population that is "stable" from year to year will often change dramatically over the seasons.

5.5.1 Patterns of population growth

When populations are sparse and uncrowded they may grow rapidly (and this can cause real problems—even with species that were previously endangered as in Box 5.3)

Box 5.3 *Topical ECOncerns*

Sea Otter Populations on the Rise

It is estimated that as many as 300,000 sea otters once populated the North Pacific, from Russia to Mexico. But hunting caused the population to plummet to a few thousand by 1911. Since then, numbers have shot back up to more than 100,000, although the animals have not returned everywhere. The following newspaper article by Craig Welch concerns the situation along the Washington coastline in the northwest United States. It appeared in the *Philadelphia Inquirer* on March 4, 2001.

Sea otters colliding with fishing industry

Sea otters are rebounding in dramatic fashion along Washington's coast, and that is forcing marine biologists and wildlife managers to prepare for a potentially uncomfortable collision between the charismatic critters and some coastal fisheries.

"It's a classic recipe for political polarization," said Glenn VanBlaricom, an associate professor of marine ecology at the University of Washington. "People love sea otters, but they could run right into shellfish harvesters whose livelihoods depend on their food sources." Wiped out of Washington waters in the 19th century by pelt-hungry hunters, otters have staged a comeback since being reintroduced to the western shores of the Peninsula in the late 1960s.

The population has grown 30-fold in as many years, and their range is expanding so far and fast that some scientists suspect groups of otters may someday—for the first time—make Puget Sound home.

. . . While sea otters remain protected under Washington state law as an endangered species, their numbers are increasing by 10 percent a year. The population now hovers at 600 animals, roughly a quarter of what marine experts think the environment can sustain.

But such a healthy return comes with complications. Because they lack blubber, otters eat a quarter of their weight each day to fuel their supercharged metabolisms. Their munchies of choice include the seafood humans crave—sea urchins, Dungeness crabs, clams, abalone. And their recent travels toward rich harvest areas such as the Dungeness Spit put them on a direct route toward multimillion-dollar commercial, recreational and tribal shellfisheries.

Steven Jeffries, who heads marine-mammal investigations for the State Department of Fish and Wildlife, said it was tough to determine whether it would be a few years or a few decades before conflicts begin.

Shellfisheries are of considerable importance to commercial, recreational, and tribal fishers. How would you weigh up the competing demands of conservation and fishing? Should the sea otters remain absolutely protected or is there a case for culling or some other form of control of their spread?

The story in Washington is very different from that in parts of Alaska, where otter numbers are declining, or Los Angeles, where recent efforts have been made to reintroduce it. Suggest some plausible reasons for the different population trajectories in different areas.

Box 5.4   *Quantitative Aspects*

The Exponential and Logistic Equations of Population Growth

In this box, simple mathematical models are derived for populations first in the absence of, and then under the influence of, intraspecific competition. These and other mathematical models play an important part in ecology (Chapter 1). They help us to follow through the consequences of assumptions we may wish to make, and to explore the behavior of ecological systems that we may find it hard to observe in nature or construct in the laboratory. The particular models in this box themselves form the basis for more complex models of *interspecific* competition and predation: they are important building blocks. It is essential to appreciate, however, that a pattern generated by such a model—for example, the S-shaped pattern of population growth under the influence of intraspecific competition—is not of interest, or important, because it is generated by the model. There are, no doubt, many other models that could generate very similar (indistinguishable) patterns. Rather, the point about the pattern is that it reflects important, underlying ecological processes—and the model is useful in that it appears to capture the essence of these processes.

We start with a model of a population in which there is no intraspecific competition, and then incorporate that competition later. Our models are in the form of *differential equations*, describing the net rate of increase of a population, which will be denoted by dN/dt (spoken: dN by dt). This represents the speed at which a population increases in size, N, as time, t, progresses.

The increase in size of the whole population is the sum of the contributions of the various individuals within it. Thus, the average rate of increase per individual, or the per capita rate of increase (*per capita* means per head) is given by $dN/dt \cdot (1/N)$. In the absence of intraspecific competition (or any other force that increases the death rate or reduces the birth rate) this rate of increase is a constant and as high as it can be for the species concerned. It is called the *intrinsic rate of natural increase* and is denoted by r. Thus:

$$dN/dt(1/N) = r$$

and the net rate of increase for the whole population is therefore given by:

$$dN/dt = rN.$$

This equation describes a population growing *exponentially* (Figure 5.16).

Intraspecific competition can now be added. This is done by deriving the *logistic equation*: a simple mathematical model developed by Pierre-François Verhulst (1804–1849). It was almost completely ignored until it was rediscovered in 1920 by Raymond Pearl, who used it to model the growth of the population of the United States.

The equation can be derived by the method set out in the figure. The net rate of increase per individual is unaffected by competition when N is very close to zero. It is still therefore given by r (point A). When N rises to K (the carrying capacity) the net rate of increase per individual is zero (point B). For simplicity, we assume a straight line between A and B; that is, we assume a linear reduction in the per capita rate of increase, as a result of intensifying intraspecifuc competition, between $N = 0$ and $N = K$.

Thus, on the basis that the equation for any straight line takes the form $y = $ intercept $+$ slope x,

Box 5.4 (cont'd)

where x and y are the variates on the horizontal and vertical axes, here we have:

$$dN/dt(1/N) = r - (r/K)N$$

or, rearranging,

$$dN/dt = rN\{1 - (N/K)\}.$$

This is the logistic equation, and a population increasing in size under its influence is shown in Figure 5.16. It describes a sigmoidal growth curve approaching a stable carrying capacity, but it is only one of many reasonable equations that do this. Its major advantage is its simplicity. Nevertheless, it has played a central role in the development of ecology.

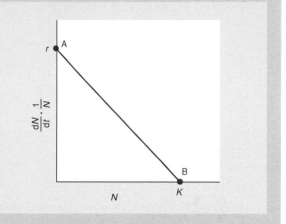

An ideal linear decline in the net rate of increase per individual with increasing population (N) (see text).

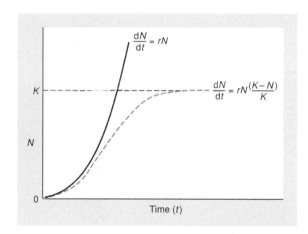

Figure 5.16
Exponential and S-shaped or *sigmoidal* increase in the size of a population over time. These patterns describe the growth to be expected in general in populations in the absence (exponential) and under the influence (sigmoidal) of intraspecific competition but are also generated, specifically, by the exponential and logistic equations shown (see also Box 5.4).

—it is only as crowding increases that density-dependent changes in birth and death rates start to take effect. In essence, populations at these low densities grow by simple multiplication over successive intervals of time. This is *exponential* growth (Figure 5.16 and Box 5.4) and the rate of increase is the population's *intrinsic rate of natural increase* (denoted by r). Of course, any population that behaved in this way would soon run out of resources, but as we have seen, the rate of increase tends to become reduced by competition as the population grows, and it falls to zero when

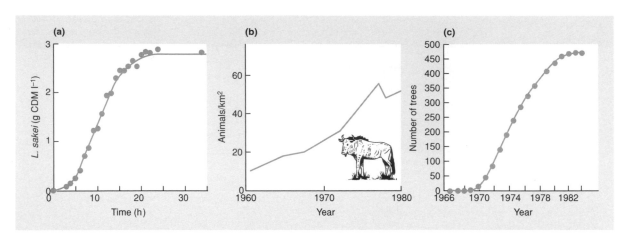

Figure 5.17
Real examples of S-shaped population increase. (a) The bacterium *Lactobacillus sakei* (measured as grams of cell dry mass [CDM] per liter) grown in nutrient broth. (After Leroy & de Vuyst, 2001.) (b) The population of wildebeest *Connochaetes taurinus*, of the Serengeti region of Tanzania and Kenya, seems to be leveling off after rising from a low density caused by the disease rinderpest. (After Sinclair & Norton-Griffiths, 1982; Deshmukh, 1986.) (c) The population of the willow tree (*Salix cinerea*) in an area of land after myxomatosis had effectively prevented rabbit grazing. (After Alliende & Harper, 1989.)

the population reaches its carrying capacity (since birth rate then equals death rate). A steady reduction in the rate of increase as densities move toward the carrying capacity gives rise to population growth that is not exponential but S-shaped (Figure 5.16). The pattern is also often called *logistic* growth after the so-called logistic equation (Box 5.4).

The S-shaped curve can best be seen in action in laboratory studies of microorganisms or animals with very short life cycles (Figure 5.17a). In these kinds of experiment it is easy to have experimental control of environmental conditions and resources. In the real world, outside the laboratory and the mind of the mathematician, the world is less simple. The complex life cycles of organisms, changing conditions and resources through the seasons, and the patchiness of habitats introduce many complications. In nature, populations often follow a very bumpy ride along the path of perfect logistic growth (Figure 5.17b), though not always (Figure 5.17c).

Another way to summarize the ways in which intraspecific competition affects populations is to look at *net recruitment*—the number of births minus the number of deaths in a population over a period of time. When densities are low, net recruitment will be low because there are few individuals available either to give birth or to die. Net recruitment will also be low at much higher densities as the carrying capacity is approached. Net recruitment will be at its peak, then, at some intermediate density. The result is a "humped" or "dome-shaped" curve (Figure 5.18). Again, of course, as with the ideal logistic curve, real data from nature never fall on a single line. But the dome-shaped curve reflects the essence of net recruitment patterns when density-dependent birth and death are the result of intraspecific competition.

Wildebeest (also know as gnu) are African antelopes of the genus *Connochaetes* that form large herds that graze on open plains and savannas, and are continually on the move as they search for new areas to graze.

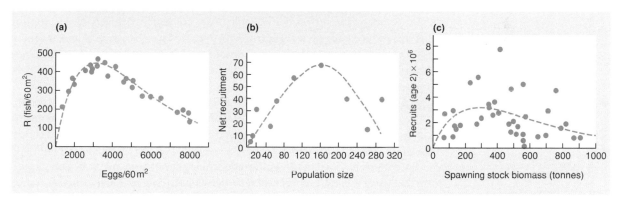

Figure 5.18
Some dome-shaped net recruitment curves. (a) Six-month old brown trout, *Salmo trutta*, in Black Brows Beck, England between 1967 and 1989. (After Myers, 2001; following Elliott, 1994.) (b) An experimental population of the fruit fly *Drosophila melanogaster*. (After Pearl, 1927.) (c) "Blackwater" herring, *Clupea harengus*, from the Thames estuary, England between 1962 and 1997. (After Fox, 2001. © Crown copyright, reproduced with permission of CEFAS, Lowestoft, UK.)

5.6   Life-History Patterns

One of the ways in which we can try to make sense of the world around us is to search for repeated patterns. In doing so, we are not pretending that the world is simple or that all categories are watertight, but we can hope to move beyond a description that is no more than a series of unique special cases. This final section of the chapter describes some simple, useful, though by no means perfect, patterns linking different types of life history and different types of habitat.

First, though, we return to a point made earlier: that in any life history there is a limited total amount of energy or some other resource available to an organism for growth and reproduction. Some trade-off may therefore be necessary: either grow more and reproduce less, or reproduce more and grow less. Specifically, there may be an observable cost of reproduction in that when reproduction starts, or increases,

the "cost" of reproduction— a life-history trade-off

Figure 5.19
The negative correlation between cone crop size and annual growth increment for a population of Douglas fir *Pseudotsuga menziesii*. (After Eis et al., 1965.)

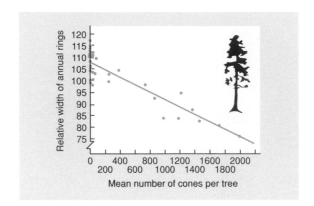

growth may slow or stop completely, as resources are diverted from growth to reproduction. In many forest trees, for example, there are years when very heavy crops of seeds are produced: *mast years*. The trunks of trees are formed by annual increments or rings of growth. In mast years these are conspicuously narrower (Figure 5.19). We can, of course, look at this trade-off between growth and reproduction the other way around: an organism that makes vigorous growth, and so thrives in competition with its neighbors, may do so by delaying or reducing reproductive activity. Furthermore, the diversion of resources to present reproduction may also jeopardize subsequent survival (as in the salmon and foxgloves described earlier), or simply reduce the capacity for future reproduction.

But early reproduction can yield some striking rewards, particularly because the progeny themselves start reproduction earlier. Populations of individuals that reproduce early in their life can grow extremely fast—even if this means producing many fewer total offspring over their life than they would otherwise. The effect is shown by considering the life cycle of fruit flies (*Drosophila*). The number of eggs produced by a female in her lifetime is about 780. Doubling that number would clearly boost the intrinsic rate of increase, but such a massive increase in reproductive output is asking a great deal of an individual. So, what other changes in the life history of *Drosophila* would have a similar effect? In fact, the same rise in the rate of increase would be attained by shortening the juvenile period by just 1.5 days in a total period of around 10 days (reproducing sooner, rather than growing longer). Conversely, the rate of growth of populations can be slowed by delaying the onset of reproduction. One way in which the growth rate of human populations can be slowed down, for example (Chapter 12), is by discouraging early marriage and childbearing.

r and *K* species   We can now turn to the life-history patterns themselves. The potential of a species to multiply rapidly—producing large numbers of progeny early in the life cycle—is advantageous in environments that are short-lived, allowing the organisms to colonize new habitats quickly and exploit new resources. This rapid multiplication is a characteristic of the life cycles of terrestrial organisms that invade disturbed land (for example, many annual weeds), or colonize newly opened habitats such as forest clearings, and of the aquatic inhabitants of temporary puddles and ponds. These are species whose populations are usually found expanding after the last disaster or exploiting the new opportunity. They have the life-cycle properties that are favored by natural selection in such

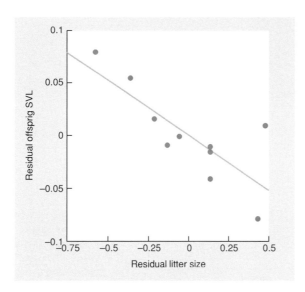

Figure 5.20
Evidence for a trade-off between the number of offspring produced in a clutch by a parent and the individual fitness of those offspring: a negative correlation between the size of offspring (as measured by their snout–vent length, SVL) and the number of them in a litter in the Australian highland copperhead snake, *Austrelaps ramsayi* ($r^2 = 0.63$; $P = 0.006$). "Residual" offspring and litter sizes have been used: these are the values arrived at after variations in maternal size have been allowed for, since both increase with maternal size. (After Rohr, 2001.)

conditions. They have been called *r* species, because they spend most of their life in the near-exponential, *r*-dominated phase of population growth (Box 5.4), and the habitats in which they are likely to be favored have been called *r*-selecting.

Organisms with quite different life histories survive in habitats where there is often intense competition for limited resources. The individuals that are successful in leaving descendants are those that have captured, and often held on to, the larger share of resources. Their populations are usually crowded and those that win in a struggle for existence do so because they have grown faster (rather than reproducing) or have spent more of their resources in aggression or some other activity that has favored them in competition with others. They are called *K* species because their populations spend most of their lives in the *K*-dominated phase of population growth (Box 5.4)—"bumping up" against the limits of environmental resources—and the habitats in which they are likely to be favored have been called *K*-selecting.

A further common distinction between *r* and *K* species is whether they produce many small progeny (characteristic of *r* species) or few large progeny (characteristic of *K* species). This is another example of a life-history trade-off: an organism has limited resources available for reproduction, and natural selection will influence how these are packaged. In environments where rapid population growth is possible, those individuals that produce large numbers of small progeny will be favored. The size of progeny can be sacrificed because they will usually not be in competition with others. However, in environments in which the individuals are crowded and there is competition for resources, those progeny that are well provided with resources by the parent will be favored. Producing progeny that are well endowed requires the trade-off of fewer progeny (see, for example, Figure 5. 20).

r, K, and progeny size and number

The *r/K* concept can certainly be useful in describing some of the general differences among different organisms. For instance, among plants it is possible to describe a number of very broad and general relationships (Figure 5.21). Trees in a forest are splendid examples of *K* species. They compete for light in the canopy, and survivors are

evidence for the r/K scheme?

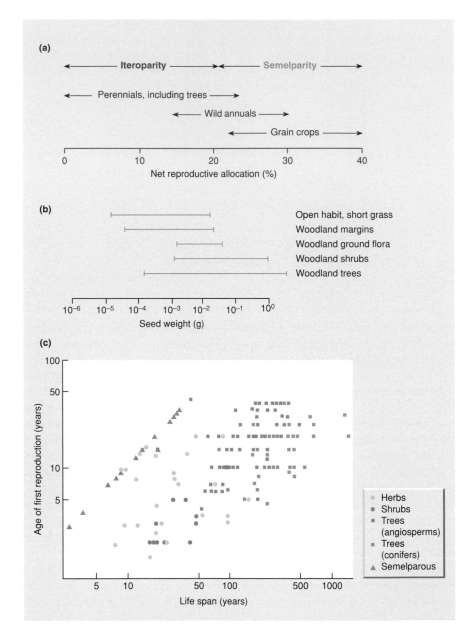

those that put their resources into early growth and overtopping their neighbors. They usually delay reproduction until their branches have an assured place in the canopy of leaves. Once established they hold on to their position and usually have a very long life, with a relatively low allocation to reproduction overall but large individual seeds. By contrast, in more disturbed, open, r-selecting habitats, the plants tend to conform to the general syndrome of r characteristics: a greater reproductive allocation, but smaller seeds, smaller size, earlier reproduction, and a shorter life (Figure 5.21).

On the other hand, there seem to be about as many examples that fail to fit the *r/K* scheme as examples that correspond. One might regard this as a damning criticism of the *r/K* concept, since it undoubtedly demonstrates that the explanatory powers of the scheme are limited. But it is equally possible to regard it as very satisfactory that a relatively simple concept can help make sense of a large proportion of the multiplicity of life histories. Nobody, though, can regard the *r/K* scheme as the whole story. Like all attempts to classify species and their characteristics into pigeonholes, the distinction between *r* and *K* species has to be recognized as a convenient (and useful) human creation rather than an all-encompassing statement about the living world.

Summary

Counting individuals, births, and deaths

Ecologists try to describe and understand the distribution and abundance of organisms. The processes that change the size of populations are birth, death, and movement. A population is a number of individuals, but for some kinds of organism, especially *modular* organisms, it is not always clear what we mean by an individual.

Ecologists face enormous problems when they try to count what is happening to populations in nature. They almost always estimate rather than count. There are particular problems in counting modular organisms and the numbers of births and deaths.

Life cycles and reproduction

The life histories of all unitary organisms can be seen as variations around a simple, sequential pattern. Some organisms fit several or many generations within a single year, some breed predictably just once each year (annuals), and others (perennials) have a life cycle extended over several or many years. Some, *iteroparous* species, breed repeatedly; others, *semelparous* species, have a single reproductive episode followed quickly by death.

Most annuals germinate or hatch in spring, grow rapidly, reproduce, and then die before the end of summer. Most spend part of the year dormant. There is a marked seasonal rhythm in the lives of many long-lived species. Where there is very little seasonal variation, some reproduce throughout the year; others have a long nonreproductive phase and then one lethal burst of reproductive activity.

Monitoring birth and death: life tables and fecundity schedules

Life tables can be useful in identifying what in a life cycle is apparently most instrumental in determining the rates of increase or decline. A cohort life table records the survivorship of members of a single cohort. When we cannot follow cohorts, it may be possible to construct a static life table, but great care is required. The fecundity of individuals also changes with age, described in age-specific fecundity schedules.

Ecologists search for patterns of life and death that we can see repeated in the lives of many species. A useful set of survivorship curves (types I–III) has been developed, but in practice patterns of survival are usually more complex.

Dispersal and migration

Dispersal is the way individuals spread away from each other. Migration is the mass directional movement of large numbers of a species from one location to another. Movement and spatial distribution are intimately related. Dispersal and migration can have a profound effect on the dynamics of a population and on its composition.

The impact of intraspecific competition on populations

Over a sufficiently large density range, competition between individuals generally reduces the birth rate as density increases and increases the death rate (i.e., is density dependent). Intraspecific competition therefore tends to keep density within certain limits and may thus be said to play a part in regulating the size of populations.

When populations are sparse and uncrowded they tend to exhibit *exponential* growth, but the rate of increase tends to become reduced by competition as the population grows, giving rise to population growth that is not exponential but S-shaped or *logistic*. Intraspecific competition also affects *net recruitment*, typically resulting in a humped curve.

Life-history patterns

There is typically a limited total amount of energy or some other resource available to an organism for growth and reproduction. There may be an observable cost of reproduction. But populations of individuals that reproduce early in their life can grow extremely fast.

The potential of a species to multiply rapidly is favored by natural selection in environments that are short-lived, allowing the organisms to colonize new habitats quickly and exploit new resources. Such species have been called r species. Where there is often intense competition for limited resources, the individuals that are successful in leaving descendants are those that have captured the larger share of resources, often because they were born larger and/or have grown faster (rather than reproducing): so-called K species. The r/K concept can be useful in interpreting many of the differences in form and behavior of organisms, but of course it is not the whole story.

Review Questions

1 Contrast the meaning of the word *individual* for unitary and modular organisms.

2 In a mark–recapture exercise during which a population of butterflies remained constant in size, an initial sample provided 70 individuals, each of which was marked and then released back into the population. Two days later, a second sample was taken, totaling 123 individuals of which 47 bore a mark from the first sample. Estimate the size of the population. State any assumptions that you have had to make in arriving at your estimate.

3 ▲ Define *annual*, *perennial*, *semelparous*, and *iteroparous*. Try to give an example of both an animal and a plant for each of the four possible combinations of these terms. In which cases is it difficult (or impossible) to come up with an example and why?

4 Contrast the derivation of *cohort* and *static life tables* and discuss the problems of constructing and/or interpreting each.

5 The following is an outline life table and fecundity schedule for a cohort of a population of sparrows. Fill in the missing values (wherever there is a question mark).

6 Describe what are meant by *aggregated*, *random*, and *regular* distributions of organisms in space, and outline, with actual examples where possible, some of the behavioral processes that might lead to each type of distribution.

7 ▲ Why is the average density of people in the United States lower than the density experienced by people, on average, in the United States? Is a similar contrast likely to apply to most species? Why? Under what conditions might it not apply?

8 ▲ Compare unitary and modular organisms in terms of the effects of intraspecific competition both on individuals and on populations.

9 What is meant by the *carrying capacity* of a population? Describe where it appears, and why, in (1) S-shaped population growth, (2) the logistic equation, and (3) dome-shaped net recruitment curves.

10 Explain why an understanding of life-history trade-offs is central to an understanding of life-history evolution. Explain the contrasting trade-offs expected to be exhibited by *r*-selected and *K*-selected species.

▲ = Challenge Question

Stage (x)	Numbers at start of stage (a_x)	Proportion of original cohort alive at start of stage (l_x)	Mean no. of eggs produced per individual in stage (m_x)
Eggs	173	?	0
Nestlings	107	?	0
Fledgelings	64	?	0
1-year-olds	31	?	2.5
2-year-olds	23	?	3.7
3-year-olds	8	?	3.1
4-year-olds	2	?	3.5
		R = ?	

Chapter 6

Interspecific Competition

Interspecific competition is one of the most fundamental phenomena in ecology, affecting not only the current distribution and success of species but also their evolution. Yet the existence and effects of interspecific competition are often remarkably difficult to establish and demand an armoury of observational, experimental, and modeling techniques.

Chapter Contents

6.1 Introduction

6.2 Ecological effects of interspecific competition

6.3 Evolutionary effects of interspecific competition

6.4 Interspecific competition and community structure

6.5 How significant is interspecific competition in practice?

Key Concepts

In this chapter you will

- appreciate the difficulty of distinguishing between the power and importance of interspecific competition in principle and in practice
- distinguish between fundamental and realized niches
- define the Competitive Exclusion Principle and understand its limitations
- appreciate the potential role of the evolutionary effects of competition in species coexistence and the difficulty of proving that role
- understand the nature of niche complementarity and the importance of scale in communities structured by interspecific competition
- appreciate the difficulties of determining the prevalence of current competition in nature, and of distinguishing between the effects of competition and mere chance

6.1   Introduction

Having been introduced to *intra*specific competition in previous chapters, it is not difficult to deduce what *inter*specific competition is. Its essence is that individuals of one species suffer a reduction in fecundity, survivorship, or growth as a result of exploitation of resources or interference by individuals from another species. These competitive effects on individuals are likely to affect the population dynamics of the competing species. These, in turn, can influence the species' distributions, and also their evolution. The distributions and abundances of species, of course, determine the compositions of the biological communities of which they are part. And evolution, in *its* turn, can itself influence the species' distributions and dynamics.

This chapter, then, is about both the ecological and the evolutionary effects of interspecific competition on individuals, on populations, and on communities. But it also addresses a more general issue in ecology and indeed in science—that there is a difference between what a process can do and what it does do: a difference between what, in this case, interspecific competition is capable of doing and what it actually does in practice. These are two separate questions, and we must be careful to keep them separate.

two separate questions—the possible and actual consequences of competition

The way these different questions can be asked and answered will be different, too. To find out what interspecific competition is capable of doing is relatively easy. Species can be *forced* to compete in experiments, or they can be examined in nature in pairs or groups chosen precisely because they seem most likely to compete. On the other hand, to discover how important interspecific competition is, actually, is much more difficult. It will be necessary to ask how realistic our experiments were, how typical they were of the way species interact in nature, and how typical of pairs and groups of species in general were those singled out for special attention. As will become evident, this is just one of those many areas of ecology, and of science in general, where clear-cut answers are as yet unavailable.

We begin, though, with some examples of what interspecific competition *can do*.

6.2   Ecological Effects of Interspecific Competition

6.2.1   Competition between diatoms for silicate

Competition was investigated in the laboratory between two species of freshwater diatoms (single-celled plants), *Asterionella formosa* and *Synedra ulna*, both of which require silicate in the construction of their cell walls (see Section 3.5). The population densities of the diatoms were monitored, but at the same time their impact on their limiting resource (silicate) was also being recorded. When either species was grown alone in a liquid medium to which resources were continuously being added, it established a steady population density while reducing the silicate to a constant low concentration (Figure 6.1a, b). However, in exploiting this resource, *Synedra* reduced the silicate concentration to a lower level than did *Asterionella*. Hence, when the two species were grown together, *Synedra* maintained the concentration at a level that was too low for the survival and reproduction of *Asterionella* and only *Synedra* survived (Figure 6.1c).

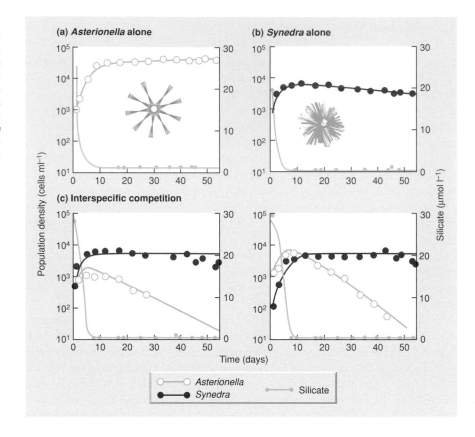

more effective exploiters exclude
less effective ones

Thus, although both species were capable of living alone in the laboratory habitat, when they competed *Synedra* excluded *Asterionella*, because it was the more effective exploiter of their shared, limiting resource. A similar result has been obtained for the nocturnal, insectivorous gecko *Hemidactylus frenatus*, an invader of urban habitats across the Pacific basin, where it is responsible for population declines of the native gecko *Lepidodactylus lugubris*. Petren and Case (1996) have established that the diets of the two geckos overlap substantially and that insects are a limiting resource for both. The invader is capable of depleting insect resources in experimental enclosures to lower levels than the native gecko, and the latter suffers reductions in body condition, fecundity, and survivorship as a result.

6.2.2 Coexistence and exclusion of competing salmonid fishes

Salvelinus malma (Dolly Varden charr) and *S. leucomaenis* (white-spotted charr) are morphologically similar and closely related salmonid fishes. The two species are found together in many streams on Hokkaido Island in Japan, but Dolly Varden are distributed at higher altitudes (further upstream) than white-spotted charr, with a zone of overlap at intermediate altitudes. In streams where one species is absent, the other

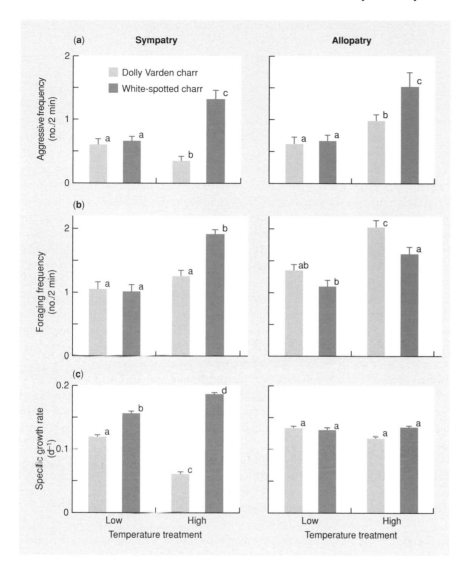

Figure 6.2
(a) Frequency of aggressive encounters initiated by individuals of each fish species during a 72-day experiment in artificial stream channels with two replicates each of 50 Dolly Varden (pale histograms) or 50 white-spotted charr (dark histograms) alone (allopatry) or 25 of each species together (sympatry); (b) foraging frequency; and (c) specific growth rate in length. Different letters indicate means are significantly different from each other. (From Taniguchi & Nakano, 2000.)

expands its range. Water temperature, an abiotic factor with profound consequences for fish ecology, increases downstream. Taniguchi and Nakano (2000) used experiments in laboratory streams to test the hypothesis that at higher temperatures (12°C as compared to 6°C) white-spotted charr would dominate in aggressive interactions, secure more profitable microhabitats (from which to forage for stream insects), and grow faster than Dolly Varden. They argued that such growth differentials would eventually force the subordinate Dolly Varden to emigrate or die. They further hypothesized that the reverse would hold at low temperature.

When the species were tested alone, higher temperatures led to increased aggression in both; however, this effect was reversed for Dolly Varden when in the presence of white-spotted charr (Figure 6.2a). Reflecting this, Dolly Varden were suppressed

competitive advantage determined by temperature-dependent aggressive behavior from obtaining favorable foraging positions when white-spotted charr were present (Figure 6.2b).

When alone, neither species' growth rates were influenced by temperature. However, when both species were present, growth of Dolly Varden decreased with increasing temperature, whereas that of white-spotted charr increased (Figure 6.2c). Furthermore, Dolly Varden grew more slowly in the presence of white-spotted charr at both temperatures, but the difference was much greater at the higher temperature.

Size-dependent mortality was evident in both species at each temperature. Fish that died were smaller than survivors (Figure 6.3). By day 72, densities of surviving fish were much reduced in experiments where both species were present. However, at the higher temperature more white-spotted charr survived to this date, and these were much larger than the few surviving Dolly Varden. Slight size advantages were probably gained by white-spotted charr early in the experiment and these were subsequently amplified as a result of interspecific competition. At the higher temperature, 2–10% of individuals of all populations survived to day 191, except for Dolly Varden which went extinct in the presence of white-spotted charr.

These results are consistent with the hypothesis that the lower altitudinal boundary of Dolly Varden in the Japanese streams was due to temperature-mediated competition favoring white-spotted charr. On the other hand, the results do not support the contention that the upper boundary of white-spotted charr is also due to temperature-mediated competitive differences (that is, Dolly Varden did not out-compete white-spotted charr in any of the experiments, even at the lower temperatures). Further work will be needed to test hypotheses that this upper limit may be due, for example, to effects at even lower temperatures (stream temperatures dip below the experimental temperature of 6°C during the colder parts of the year) or to poorer starvation resistance of white-spotted charr during long periods of food shortage at cold temperatures.

6.2.3   Some general observations

This example, and the previous one, illustrate several points of general importance.

- Competing species often coexist at one spatial scale but are found to have distinct distributions at a finer scale of resolution. Here, the fishes coexisted in the same stream, but each was more or less confined to its own altitudinal zone.
- Species are often excluded by interspecific competition from locations at which they could exist perfectly well in the absence of interspecific competition. Here, Dolly Varden charr can live in the white-spotted charr zone—but only when there are no white-spotted charr there. (Similarly *Asterionella* can live in laboratory cultures—but only when there were no *Synedra* there.)

fundamental and realized niches

- We can describe this by saying that the conditions and resources provided by the white-spotted charr zone are part of the *fundamental* ecological niche of Dolly Varden (see Section 3.6 for an explanation of ecological niches) in that the basic requirements for the existence of Dolly Varden are provided there. But the white-spotted charr zone does not provide a realized niche when white-spotted charr are present. (And the laboratory cultures provided the requirements of the fundamental niches of both *Synedra* and *Asterionella*, but those of the realized niche for only *Synedra*.)

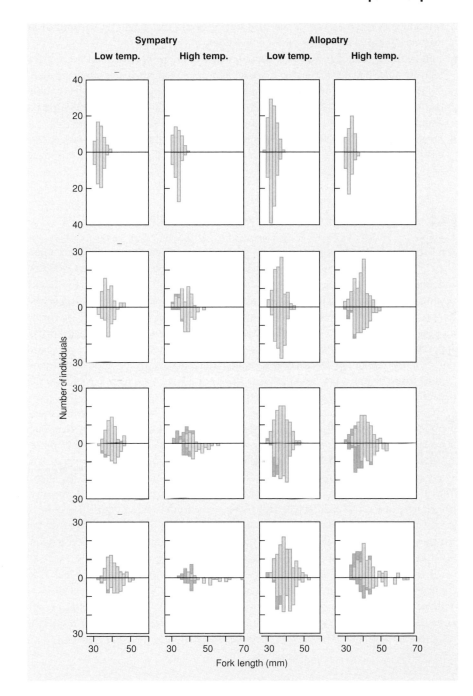

Figure 6.3
Frequency distributions of length of individuals (fork length) when species were grown alone or together, at low or high temperature for four time periods. Results for Dolly Varden are shown above the zero line, and for white-spotted charr below the line. The dark shaded areas denote individuals that died during the experimental period. (From Taniguchi & Nakano, 2000.)

- Thus, a species' fundamental niche is the combination of conditions and resources that allow that species to exist, grow, and reproduce when considered in isolation from any other species that might be harmful to its existence; whereas its realized niche is the combination of conditions and resources that allow it to exist, grow, and reproduce in the presence of specified other species that might be harmful to its existence—especially interspecific competitors.

■ Competing species can therefore coexist when both are provided with a realized niche by their habitat (in the present case, the stream as a whole provided a realized niche for both fishes); but even in locations that provide a species with the requirements of its fundamental niche, that species may be excluded by another, superior competitor that denies it a realized niche there.

■ Finally, the fish study illustrates the importance of experimental manipulation if we wish to discover what is really going on in a natural population—"nature" needs to be prodded to reveal its secrets.

6.2.4   Coexistence and exclusion of competing bumble bees

A related study, but one that focuses on species competing not through direct interference (as in the case of the salmonid fishes) but by exploitation of shared resources, concerns two bumble bee species, *Bombus appositus* and *B. flavifrons*, in the Rocky Mountains of Colorado (Inouye, 1978). The bees live in the same area, but *B. appositus* forages primarily from the flowers of larkspur, *Delphinium barbeyi*, and *B. flavifrons* forages primarily from monkshood, *Aconitum columbianum*. However, each species appears to specialize only in the presence of the other species. When one or other was removed, the remaining species quickly modified its behavior to include much more feeding from the previously less-preferred flower and it stayed longer at each visit (an indication of increased foraging success) (Figure 6.4).

The experiment therefore demonstrated that each bee's fundamental niche includes both species of flower: it is the presence of the competitor that restricts each bee's range to a realized niche of one flower type. Once again, therefore, the species coexist in the habitat as a whole, but exclude one another from particular parts of it. This reciprocal competitive effect is almost certainly related to differences in the length of each species' proboscis. Although capable of foraging on either flower, *B. appositus* has a longer proboscis, making it better able to forage from larkspur whose flower has a longer corolla. *B. appositus* therefore diminishes the nectar in larkspur to a level that makes it unattractive for *B. flavifrons*; *B. flavifrons* consequently specializes on monkshood, making it less attractive for *B. appositus*.

coexisting competitors holding different resources at levels too low for effective exploitation by others

In the case of the diatoms exploiting silicate, described above, *Synedra*, which held the resource at the lower level when exploiting it alone, outcompeted *Asterionella*. Here with the bumble bees, we see that two species exploiting two resources (nectar in each of two flower species) can compete but still coexist when each species holds one of the resources at a level that is too low for effective exploitation by the other species.

6.2.5   Coexistence of competing diatoms

In fact, a further experimental study of diatoms also shows two competing species coexisting on two shared, limiting resources. The two species were *Asterionella formosa* (again) and *Cyclotella meneghiniana*, and the resources, which were both capable of limiting the growth of both diatoms, were silicate and phosphate. However, whereas

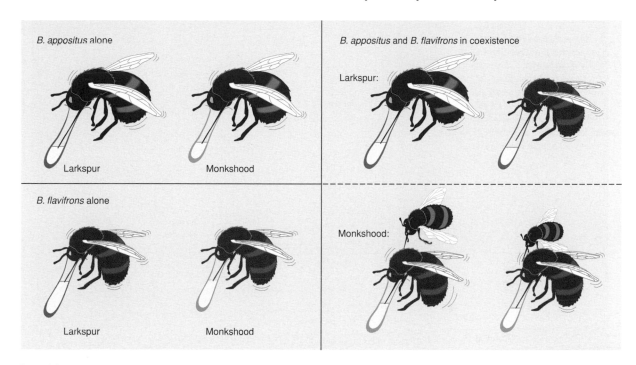

Figure 6.4
Bombus appositus and *B. flavifrons* both forage from both larkspur and monkshood if they are the only bumble bee present; but when they coexist, *B. appositus*, with a longer proboscis, excludes *B. flavifrons* from larkspur, whereas *B. flavifrons* specializes on monkshood, making it less attractive for *B. appositus*.

Cyclotella was the more effective exploiter of silicate, *Asterionella* was the more effective exploiter of phosphate. Thus, as we might now predict, in cultures with relatively balanced supplies of silicate and phosphate, the two diatoms coexisted; whereas *Cyclotella* excluded *Asterionella* when there were especially low supplies of silicate, *Asterionella* excluded *Cyclotella* when there were especially low supplies of phosphate (Figure 6.5).

6.2.6 Coexistence of competing rodents and ants

The examples described so far have all involved pairs of closely related species—diatoms, salmonid fish, or bumble bees. This is potentially misleading in at least two important respects. First, competition may occur amongst larger groups of species than just a pair—where it is sometimes, therefore, called "diffuse" competition. And second, competition may occur between completely unrelated species.

competition among groups of unrelated species

   Both points are illustrated by a study of interspecific competition involving seed-eating ants and seed-eating rodents in deserts of the southwestern United States. At the study sites, only two *guilds* (groups of species that feed on similar foods in a similar fashion) fed on seeds: the rodents and the ants. By studying the size of the seeds harvested by each guild, it was apparent that the two exhibited significant overlap in the size of the seeds they ate (Figure 6.6). Ants did eat a larger proportion of the

Figure 6.5
Asterionella formosa and *Cyclotella meneghiniana* coexist when there are roughly balanced supplies of silicate and phosphate, but *Asterionella* excludes *Cyclotella* when there are especially low supplies of phosphate (PO_4), whereas *Cyclotella* excludes *Asterionella* when there are especially low supplies of silicate (SiO_2). (After Tilman, 1982.)

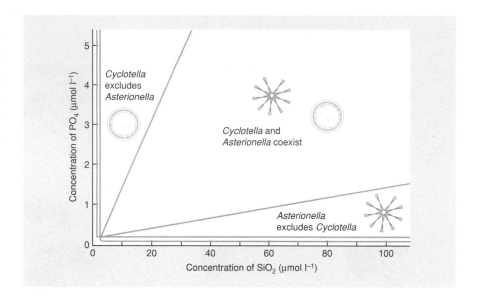

Figure 6.6
The diets of ants and rodents overlap: sizes of seeds harvested by coexisting ants and rodents near Portal, Arizona. (After Brown & Davidson, 1977.)

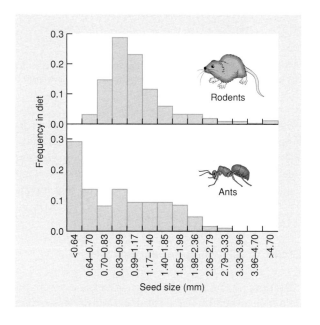

smallest seeds, but overall the potential for resource competition between them was very high.

As indicated earlier, however, the only true test for whether competition occurs between them would be to manipulate the abundance of each competitor and observe the response of its counterpart. Consequently, eight plots were established in similar habitats. In two, rodents were trapped and excluded by fencing, to insure that only ants now had access to the seeds within. In another two, ants were eliminated by repeated applications of pesticides. In two further plots both ants and rodents were excluded, and finally two plots were maintained as unmanipulated controls.

When either rodents or ants were removed, there was a statistically significant increase in the numbers of the other guild: the depressive effect of interspecific competition from each guild on the abundance of the other was apparent. Also, when rodents were removed, the ants ate as many seeds as the rodents and ants had previously eaten between them—as did the rodents when the ants were removed; only when both were removed did the amount of resource increase. In other words, under normal circumstances both guilds eat less and achieve lower levels of abundance than they would do if the other guild were absent. This clearly indicates that rodents and ants, although they coexist in the same habitat, compete interspecifically with one another.

6.2.7   The Competitive Exclusion Principle

The patterns that are apparent in these examples have also been uncovered in many others, and have been elevated to the status of a principle: the "Competitive Exclusion Principle" or "Gause's Principle" (named after an eminent Russian ecologist). It can be stated as follows:

- If two competing species coexist in a stable environment, then they do so as a result of niche differentiation, i.e., differentiation of their realized niches.
- If, however, there is no such differentiation, or if it is precluded by the habitat, then one competing species will eliminate or exclude the other.

Although the principle has emerged here from a contemplation of patterns evident in real sets of data, its establishment was—and many modern discussions of inter specific competition still are—bound up with a simple mathematical model of interspecific competition, usually known by the names of its two (independent) originators: Lotka–Volterra (Box 6.1).

There is no question that there is some truth in the principle that competitor species *can* coexist as a result of niche differentiation, and that one competitor species *may* exclude another by denying it a realized niche. But it is crucial also to be aware of what the Competitive Exclusion Principle does *not* say.

It does *not* say that whenever we see coexisting species with different niches it is reasonable to jump to the conclusion that this is the principle in operation. Each species, on close inspection, has its own unique niche. Niche differentiation does not prove that there are coexisting competitors. The species may not be competing at all and may never have done so in their evolutionary history. We require *proof* of interspecific competition. In the examples above, this was provided by experimental manipulation—remove one species (or one group of species) and the other species increases its abundance or its survival. But most of even the more plausible cases for competitors coexisting as a result of niche differentiation have not been subjected to experimental proof. So just how important is the Competitive Exclusion Principle in practice? We return to this question in Section 6.5.

Part of the problem is that although species may not be competing now, their ancestors may have competed in the past, so that the mark of interspecific competition

The Competitive Exclusion Principle —what it does and does not say

Box 6.1   *Quantitative Aspects*

The Lotka–Volterra Model of
Interspecific Competition

The most widely used model of interspecific competition is the Lotka–Volterra model (Volterra, 1926; Lotka, 1932). It is an extension of the logistic equation described in Box 5.4. Its virtues are (like the logistic) its simplicity, and its capacity to shed light on the factors that determine the outcome of a competitive interaction.

Within the logistic equation: $\dfrac{dN}{dt} = rN\dfrac{(K - N)}{K}$

the particular term that models intraspecific competition is $(K - N)/K$. Within this term, the larger N is (the bigger the population), the greater is the strength of intraspecific competition. The basis of the Lotka–Volterra model is the replacement of this term by one that models both intra- and interspecific competition. In the model, we call the population size of the first species N_1, and that of a second species N_2. Their carrying capacities and intrinsic rates of increase are K_1, K_2, r_1, and r_2.

By analogy with the logistic, we expect the *total* competitive effect on, say, species 1 (intra- *and* interspecific) to be greater the larger the values of N_1 and N_2; but we cannot just add them together, since the competitive effects of the two species on species 1 are unlikely to be the same. Suppose, though, that ten individuals of species 2 have, between them, only the same competitive effect on species 1 as does a single individual of species 1. The total competitive effect on species 1 (intra- and interspecific) will then be equivalent to the effect of $(N_1 + N_2 * 1/10)$ species 1 individuals. The constant (1/10 in the present case) is called a *competition coefficient* and is denoted by α_{12} (alpha one two). Thus multiplying N_2 by α_{12} converts it to a number of N_1 equivalents, and adding N_1 and $\alpha_{12}N_2$ together *does* give us the total competitive effect on species 1. (Note that $\alpha_{12} < 1$ means that individuals of species 2 have less inhibitory effect on individuals of species 1 than

individuals of species 1 have on others of their own species, and so on.)

The equation for species 1 can now be written:

$$\frac{dN_1}{dt} = r_1 N_1 \frac{(K_1 - [N_1 + \alpha_{12}N_2])}{K_1}$$

and for species 2 (with its own competition coefficient, converting species 1 individuals into species 2 equivalents):

$$\frac{dN_2}{dt} = r_2 N_2 \frac{(K_2 - [N_2 + \alpha_{21}N_1])}{K_2}.$$

These two equations constitute the Lotka–Volterra model.

The best way to appreciate its properties is to ask the question, "Under what circumstances does each species increase or decrease in abundance?" In order to answer, it is necessary to construct diagrams in which all possible combinations of N_1 and N_2 can be displayed. This has been done in Figure 6.7. Certain combinations (certain regions in Figure 6.7) give rise to increases in species 1 and/or species 2, whereas

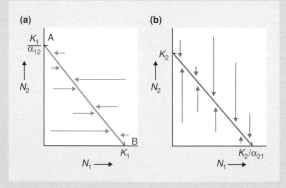

Figure 6.7
The zero isoclines generated by the Lotka–Volterra competition equations. (a) The N_1 zero isocline: species 1 increases below and to the left of it, and decreases above and to the right of it. (b) The equivalent N_2 isocline.

Box 6.1 (cont'd)

other combinations give rise to decreases. It follows inevitably that there must also therefore be a so-called *zero isocline* for each species: that is, a line with combinations leading to increase on one side of it and combinations leading to decrease on the other, but along which there is neither increase nor decrease.

We can map out the regions of increase and decrease in the figure for species 1 if we can draw its zero isocline, and we can do this by using the fact that *on* the zero isocline, $dN_1/dt = 0$ (the rate of change of species 1 abundance is zero, by definition). Rearranging the equation, this gives us, as the zero isocline for species 1:

$$N_1 = K_1 - \alpha_{21}N_2.$$

Below and to the left of this, species 1 increases in abundance (arrows in the figure, representing this increase, point from left to right, since N_1 is on the horizontal axis). It increases because numbers of both species are relatively low, and species 1 is thus subjected to only weak competition. Above and to the right of the line, however, numbers are high, competition is strong, and species 1 decreases in abundance (arrows from right to left). Based on an equivalent derivation, the figure also shows the species 2 zero isocline, with arrows, like the N_2 axis, running vertically.

In order to determine the outcome of competition in this model, it is necessary to determine, at each point on a figure, the behavior of the joint species 1–species 2 population, as indicated by the pair of arrows. There are, in fact, four different ways in which the two zero isoclines can be arranged relative to one another, and these can be distinguished by the intercepts of the zero isoclines (Figure 6.8). The outcome of competition will be different in each case.

Looking at the intercepts in Figure 6.8a, for instance,

$$\frac{K_1}{\alpha_{12}} > K_2 \quad \text{and} \quad K_1 > \frac{K_2}{\alpha_{21}}.$$

Rearranging these slightly gives us:

$K_1 > K_2\alpha_{12}$ and $K_1\alpha_{21} > K_2$.

The first inequality ($K_1 > K_2\alpha_{12}$) indicates that the inhibitory intraspecific effects that species 1 can exert on itself (denoted by K_1) are greater than the interspecific effects that species 2 can exert on species 1 ($K_2\alpha_{12}$). This means that species 2 is a weak interspecific competitor. The second inequality, however, indicates that species 1 can exert more of an effect on species 2 than species 2 can on itself. Species 1 is thus a *strong* interspecific competitor; and as the arrows in Figure 6.8a show, species 1 drives the weak species 2 to extinction and attains its own carrying capacity. The situation is reversed in Figure 6.8b. Hence Figure 6.8a and b describe cases in which the environment is such that one species invariably outcompetes the other, because the first is a strong interspecific competitor and the other weak.

In Figure 6.8c, by contrast:

$K_1 > K_2\alpha_{12}$ and $K_2 > K_1\alpha_{21}$.

In this case, both species have less competitive effect on the other species than those other species have on themselves; in this sense, both are weak competitors. This would happen, for example, if there were niche differentiation between the species—each competed mostly "within" its own niche. The outcome, as Figure 6.8c shows, is that all arrows point towards a stable, equilibrium combination of the two species, which all joint populations therefore tend to approach: that is, the outcome of this type of competition is the stable coexistence of the competitors. Indeed, it is only this type of competition (both species having more effect on themselves than on the other species) that does lead to the stable coexistence of competitors.

Finally, in Figure 6.8d:

$K_2\alpha_{12} > K_1$ and $K_1\alpha_{21} > K_2$.

Thus individuals of both species have a greater competitive effect on individuals of the other species than those other species do on themselves. This will

Box 6.1 (cont'd)

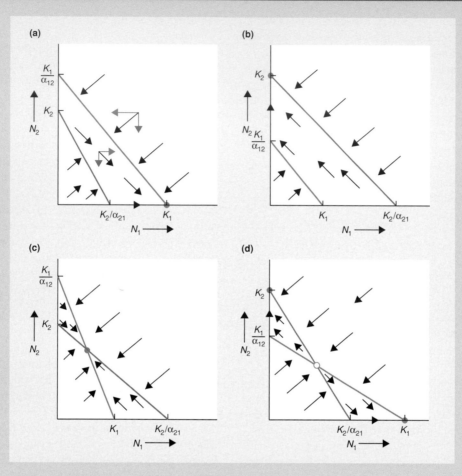

Figure 6.8
The outcomes of competition generated by the Lotka–Volterra competition equations for the four possible arrangements of the N_1 and N_2 zero isoclines. Black arrows refer to joint populations, and are derived as indicated in (a). The solid circles show stable equilibrium points. The open circle in (d) is an unstable equilibrium point. For further discussion, see text.

occur, for instance, when each species is more aggressive toward individuals of the other species than toward individuals of its own species. The directions of the arrows are rather more complicated in this case, but eventually they always lead to one or other of two alternative stable points. At the first, species 1 reaches its carrying capacity with species 2 extinct; at the second, species 2 reaches its carrying capacity with species 1 extinct. In other words, both species are capable of driving the other species to extinction, but which actually does so cannot be predicted with certainty. It depends on which species has the upper hand in terms of densities, either because they start with a higher density or because density fluctuations in some other way give them that advantage. Whichever species has this upper hand, capitalizes on that and drives the other species to extinction.

is left imprinted on the niches, the behavior, or the morphology of their present-day descendants. This particular question is taken up in Section 6.3.

Finally, the Competitive Exclusion Principle, as stated above, includes the word "stable". That is, in the habitats envisaged in the Principle, conditions and the supply of resources remain more or less constant—if species compete, then that competition runs its course, either until one of the species is eliminated or until the species settle into a pattern of coexistence within their realized niches. Sometimes this is a realistic view of a habitat, especially in laboratory or other controlled environments where the experimenter holds conditions and the supply of resources constant. However, most environments are not stable for long periods of time. How does the outcome of competition change when environmental heterogeneity in space and time are taken into consideration? This is the subject of the next section.

6.2.8 Environmental heterogeneity

As explained in previous chapters, spatial and temporal variations in environments are the norm rather than the exception. Environments are usually a patchwork of favorable and unfavorable habitats; patches are often only available temporarily; and patches often appear at unpredictable times and in unpredictable places. Under such variable conditions, competition may only rarely "run its course", and the outcome cannot be predicted simply by application of the Competitive Exclusion Principle. A species that is a "weak" competitor in a constant environment might, for example, be good at colonizing open gaps created in a habitat by fire, or a storm, or the hoofprint of a cow in the mud—or may be good at growing rapidly in such gaps immediately after they are colonized. It may then coexist with a strong competitor, as long as new gaps occur frequently enough. Thus, a realistic view of interspecific competition must acknowledge that it often proceeds not in isolation, but under the influence of, and within the constraints of, a patchy, impermanent or unpredictable world.

The following examples illustrate just two of the many ways in which environmental heterogeneity insures that the Competitive Exclusion Principle is very far from being the whole story when it comes to determining the outcome of an interaction between competing species.

The first concerns the coexistence of the sea palm *Postelsia palmaeformis* (a brown alga) and the mussel *Mytilus californianus* on the coast of Washington, USA (Paine, 1979) (Figure 6.9). *Postelsia* is an annual plant that must re-establish itself each year in order to persist at a site. It does so by attaching to the bare rock, usually in gaps in the mussel bed created by wave action. However, the mussels themselves slowly encroach on these gaps, gradually filling them and precluding colonization by *Postelsia*. In other words, in a stable environment, the mussels would outcompete and exclude *Postelsia*. But their environment is not stable—gaps are frequently being created. It turns out that these species coexist only at sites in which there is a relatively high average rate of gap formation (at least 7 percent of surface area per year), and in which this rate is approximately the same each year. Where the average rate is lower, or where it varies considerably from year to year, there is (either regularly or occasionally) a lack of bare rock for colonization by *Postelsia*. At the sites of coexistence, on the other hand, although *Postelsia* is eventually excluded from each gap, these are created with

competition may only rarely "run its course"

mussels, sea palms, and the frequency of gap formation

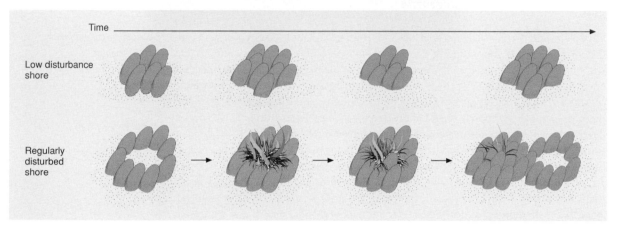

Figure 6.9
On shores in which gaps are not created, mussels are able to exclude the brown alga *Postelsia* (photo above [© Gerald and Buff Corsi, Visuals Unlimited]); but where gaps are created regularly enough the two species coexist, even though *Postelsia* is eventually excluded by the mussels from each gap.

sufficient frequency and regularity for there to be coexistence in the site as a whole. In short, there is coexistence of competitors—but not as a result of niche differentiation.

Many environments, by their very nature, are not simply variable but "ephemeral" (short lived). Among the more obvious examples are decaying corpses (carrion), dung, rotting fruit, and temporary ponds. But note too that a leaf or an annual plant can be seen as an ephemeral patch, especially if it is palatable to its consumer for only a limited period. Often these ephemeral patches have an unpredictable life span: a piece of fruit and its attendant insects, for instance, may be eaten at any time by a bird. In these cases, it is easy to imagine the coexistence of two species: a superior competitor and an inferior competitor that reproduces earlier than its competitive superior.

coexistence of the good and the fast

One example concerns two species of pulmonate snail, *Physa gyrina* and *Lymnaea elodes*, living in ponds in northeastern Indiana, USA (Brown, 1982) (Figure 6.10). Artificially altering the density of one or other species in the field showed that

Figure 6.10
Lymnaea (L) is the better competitor; but only *Physa* (P) reproduces at all in ponds that dry up quickly in the summer.

the fecundity of *Physa* was significantly reduced by interspecific competition from *Lymnaea*, but the effect was not reciprocated. *Lymnaea* was clearly the superior competitor when competition continued throughout the summer. Yet *Physa* reproduced earlier and at a smaller size than *Lymnaea*, and in the many ponds that dried up by early July it was often the only species to have produced drought-resistant eggs in time. The species therefore coexisted in the area as a whole, in spite of *Physa*'s apparent inferiority, and as a result not of niche differentiation, but of the differing effects on the species of a world composed of ephemeral patches.

These studies, and others like them, go a long way toward explaining the co-occurrence of species that in constant environments would probably exclude one another. The environment is rarely unvarying enough for competitive exclusion to run its course. Rather, competitive balances are often shifting, with one species and then another being favored by daily, seasonal, and longer term fluctuations in conditions and resources. Or, when new patches are created or old ones destroyed, competition may simply, for a time, be unimportant.

6.3   Evolutionary Effects of Interspecific Competition

Putting to one side the fact that environmental heterogeneity insures that the forces of interspecific competition are often much less profound than they would otherwise be, it is nonetheless the case that the potential of interspecific competition to adversely affect individuals is considerable, and that this potential is often realized. We have seen in Chapter 2 that natural selection in the past will have favored those individuals that,

evolutionary avoidance of competition

by their behavior, physiology, or morphology, have avoided adverse effects acting on other individuals in the same population. The adverse effects of extreme cold, for example, may have favored individuals with an enzyme capable of functioning effectively at low temperatures. Or, in the present context, the adverse effects of interspecific competition may have favored those individuals that, by their behavior, physiology, or morphology, avoided those competitive effects.

We can, therefore, expect species to have evolved characteristics that insure that they compete less, or not at all, with members of other species. How will this look to us at the present time? Coexisting species, with an apparent potential to compete, will exhibit differences in behavior, physiology, or morphology that insure that they compete little or not at all. Connell has called this line of reasoning to account for differences between coexisting species "invoking the ghost of competition past". Yet the pattern it predicts is precisely the same pattern supposed by the Competitive Exclusion Principle to be a prerequisite for the coexistence of species that *do* still compete. Coexisting present-day competitors, and coexisting species that have evolved an avoidance of competition, can look, at least superficially, the same.

UNANSWERED QUESTION: the
difficulty of distinguishing
ecological and evolutionary
effects

The question of how important either past *or* present competition are, actually, as forces structuring natural communities will be addressed in the last section of this chapter (Section 6.5). For now, we examine some examples of what interspecific competition *can* do as an evolutionary force. Note, however, that by invoking something that cannot be observed directly (evolution), it may be impossible to *prove* an evolutionary effect of interspecific competition, in the strict sense of "proof" that can be applied to mathematical theorems or carefully controlled experiments in the laboratory. Nonetheless, examples will be considered where an evolutionary (rather than an ecological) effect of interspecific competition is the most reasonable explanation for what is observed.

6.3.1   Character displacement and release in the Indian mongoose

In western parts of its range, the small Indian mongoose (*Herpestes javanicus*) coexists with one or two slightly larger species in the same genus (*H. edwardsii* and *H. smithii*), but these species are absent in the eastern part of its range (Figure 6.11). Simberloff et al. (2000) examined size variation in the upper canine tooth, the animal's principal prey-killing organ (note that female mongooses are smaller than males). In the east where it occurs alone (area VII in Figure 6.11), both males and females have larger canines than in the western areas (areas III, V, VI) where it coexists with the larger species (Figure 6.12). This is consistent with the view that where similar but larger predators are present, the prey-catching apparatus of *H. javanicus* has been selected for reduced size. This is likely to reduce the strength of competition with other species in the genus because smaller predators tend to take smaller prey than larger predators. Where *H. javanicus* occurs in isolation, its canine teeth are much larger.

the prey-killing apparatus of the
small Indian mongoose is smaller
where larger competitors are present

It is of particular interest that the small Indian mongoose was introduced about a century ago to many islands outside its native range (often as part of a naive attempt to control introduced rodents). In these places, the larger competitor mongoose species are absent. Within 100–200 generations the small Indian mongoose has increased in size (Figure 6.12), so that the sizes of island individuals are now intermediate between

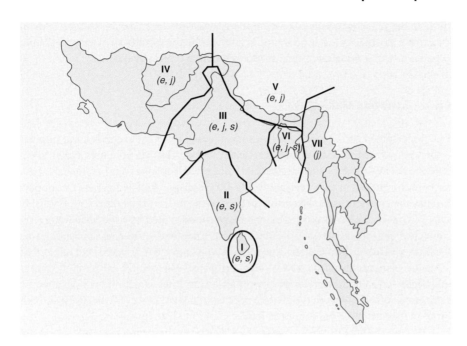

Figure 6.11
Native geographical ranges of *Herpestes javanicus* (*j*), *H. edwardsii* (*e*), and *H. smithii* (*s*). (From Simberloff et al., 2000.)

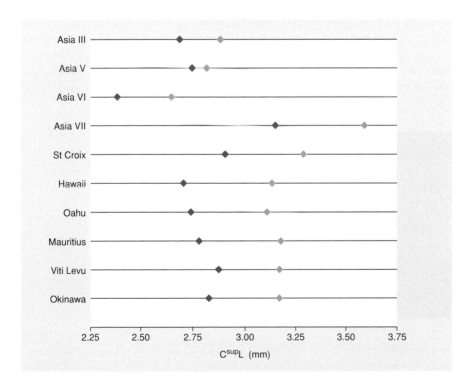

Figure 6.12
Maximum diameter (mm) of the upper canine (C^{sup}L) for *Herpestes javanicus* in its native (data only for areas III, V, VI, and VII from Figure 6.11) and introduced ranges. Symbols in blue represent mean female size and in orange mean male size. (From Simberloff et al., 2000.)

those in the region of origin (where they coexisted with other species and were small) and those in the east where they occur alone. On the islands they show variation consistent with "ecological release" from competition with larger species.

6.3.2   Canadian sticklebacks

The three-spined stickleback (*Gasterosteus aculeatus*) is a marine species, but populations were "left behind" in several freshwater lakes in British Columbia either when land in the area uplifted after deglaciation (about 12,500 years ago) or after the subsequent rise and fall of sea levels (about 11,000 years ago). Following these two opportunities for invasion, some lakes now support two species while others support only one. Wherever there are two species, one is always associated with the open water (it is "limnetic") and the other with the lake bottom ("benthic"). The first concentrates its feeding on plankton and has correspondingly long (and closely spaced) gill rakers that seive the plankton from the flow of ingested water. The second, with much shorter gill rakers, concentrates on larger prey that it consumes largely from vegetation or sediments. Wherever there is only one species in a lake, however, this exploits both kinds of food resource and is morphologically intermediate (Figure 6.13).

It is most unlikely to be by chance that whenever there have been two successful invasions of a lake the species just happen to be so different, whereas whenever there has been one invasion, the species is morphologically and behaviorally intermediate. There are, however, two plausible explanations for these patterns—though both are "historical" and both involve interspecific competition. First, it may be that a second successful invasion was only possible when the lake resident from the first invasion was either a limnetic or benthic specialist, whereas if the resident was a generalist there was insufficient empty niche space for a second species to invade and avoid being competitively excluded.

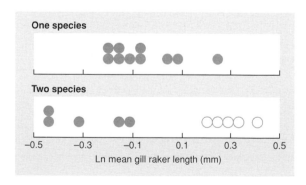

Figure 6.13

Character displacement in three-spined sticklebacks (*Gasterosteus aculeatus*) (© Glenn M. Oliver, Visuals Unlimited). In small lakes in coastal British Columbia supporting two stickleback species, the gill rakers of the benthic species (filled circle, below) are significantly shorter than those of the limnetic species (open circle, below), while those of species of sticklebacks that occupy comparable lakes alone (above) are intermediate in length. Lengths of gill rakers have been adjusted to take account of species differences in size overall. (After Schluter & McPhail, 1993.)

On the other hand, why should the first invaders have evolved into generalists in some lakes and specialists in other, similar lakes? A more likely alternative, therefore, is that following a second invasion, small initial differences between species became exaggerated over evolutionary time in response to interspecific competition, until now there is clear separation both behaviorally and morphologically (called "character displacement"). Moreover, because the differentiation here, and in the case of the small Indian mongoose, is associated with character displacement (morphological change), it is particularly suggestive of an *evolutionary* rather than an ecological response to competition. Another strong candidate for the evolutionary effects of interspecific competition, especially because of its association with character displacement, is provided by "Darwin's" finches of the genus *Geospiza* living on the Galapagos islands, discussed in Section 2.4.2.

character displacement in sticklebacks—and Indian mongooses—and Darwin's finches

6.4 Interspecific Competition and Community Structure

Interspecific competition, then, has the potential to either keep apart (Section 6.2) or drive apart (Section 6.3) the niches of coexisting competitors. How can these forces express themselves when it comes to the role of interspecific competition in molding the shape of whole ecological communities—who lives where and with whom?

6.4.1 Limiting resources and the regulation of diversity in phytoplankton communities

We begin by returning to the question of coexistence of competing phytoplankton species. In Section 6.2.5, we saw how two diatom species could coexist in the laboratory on two shared limiting resources—silicate and phosphate (Tilman, 1982). In fact, Tilman's resource competition theory predicts that the diversity of coexisting species should be proportional to the number of resources in a system that are at physiological limiting levels: the more limiting resources, the more coexisting competitors. Interlandi and Kilham (2001) tested this hypothesis directly in three lakes in the Yellowstone region of Wyoming, USA using an index ("Simpson's index") of the species diversity of phytoplankton there (diatoms and other species). If one species exists on its own, the index equals 1; in a group of species where biomass is strongly dominated by a single species, the index will be close to 1; when two species exist at equal biomass, the index is 2; and so on. According to resource competition theory, this index should therefore increase in direct proportion to the number of resources limiting growth. The spatial and temporal patterns in phytoplankton diversity in the three lakes for 1996 and 1997 are shown in Figure 6.14.

The principal limiting resources for phytoplankton growth are nitrogen, phosphorus, silicon, and light. These parameters were measured at the same depths and times that phytoplankton were sampled, and it was noted where and when any of the *potential* limiting factors *actually* occurred at levels below threshold limits for growth. Consistent with resource competition theory, species diversity increased as the number of resources at physiologically limiting levels increased (Figure 6.15).

as predicted, highest phytoplankton diversity occurred where many resources were limiting

Figure 6.14
Variation in phytoplankton species diversity (Simpson's index) with depth in 2 years in three large lakes in the Yellowstone region. Color indicates depth–time variation in a total of 712 discrete samples—red denotes high species diversity, blue denotes low species diversity. (From Interlandi & Kilham, 2001, with permission.)

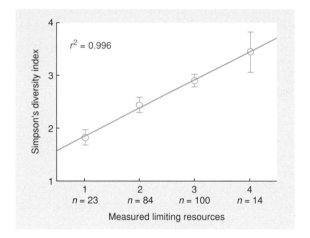

Figure 6.15
Phytoplankton diversity (Simpson's index; mean ± standard error) associated with samples with different numbers of measured limiting resources. It was possible to perform this analysis on 221 samples from those displayed in Figure 6.14—the number of samples (*n*) in each limiting resource class is shown. (From Interlandi & Kilham, 2001.)

These results suggest that even in the highly dynamic environments of lakes, where equilibrium conditions are rare, resource competition plays a role in continuously structuring the phytoplankton community. It is heartening that the results of experiments performed in the artificial world of the laboratory (Section 6.2.5) are echoed here in the much more complex natural environment.

6.4.2 Bumble bees in Colorado

Another study of niche differentiation and coexistence examined a number of species of bumble bee in Colorado (two of which were discussed in Section 6.2.4) where differences in resource utilization appeared to be actively maintained by current competition. Visits were made every 8 days during the summer to 17 sites along an altitudinal gradient (2,860 to 3,697 m), and on each visit the numbers of each of seven common bumble bee species visiting the flowers of various plant species were recorded. The bumble bee species fell into four groups in terms of both their proboscis length and the corolla length of the plants they visited preferentially (Figure 6.16). The long-proboscis bumble bees (*Bombus appositus* and *B. kirbyellus*) clearly favored plants with long corollas, particularly *Delphinium barbeyi*. The short-proboscis species (*B. sylvicolla*, *B. bifarius*, and *B. frigidus*) fed most frequently on various flower species in the family Compositae and on *Epilobium angustifolium*, all

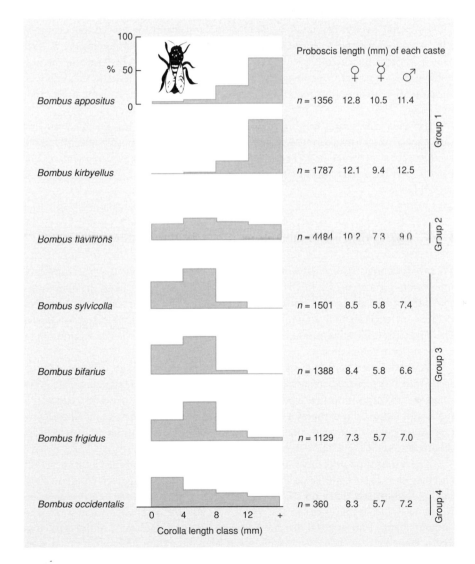

Figure 6.16
Percentage of all observations of each bumble bee species on plants in four classes of flower size. The total number of observations (*n*) is indicated in each case. Also shown for each species are the mean proboscis lengths of each caste (♀, queen; ☿, sterile female worker; ♂, male). The seven species can be divided into three groups according to proboscis length. *Bombus occidentalis* has been placed into a fourth group because of its unique mandible structure. (After Pyke, 1982.)

Proboscis length (mm) of each caste

	♀	☿	♂
Bombus appositus	n = 1356	12.8	10.5	11.4
Bombus kirbyellus	n = 1787	12.1	9.4	12.5
Bombus flavifrons	n = 4484	10.2	7.3	9.0
Bombus sylvicolla	n = 1501	8.5	5.8	7.4
Bombus bifarius	n = 1388	8.4	5.8	6.6
Bombus frigidus	n = 1129	7.3	5.7	7.0
Bombus occidentalis	n = 360	8.3	5.7	7.2

Group 1 (appositus, kirbyellus); Group 2 (flavifrons); Group 3 (sylvicolla, bifarius, frigidus); Group 4 (occidentalis)

Corolla length class (mm)

of which possess quite short corollas. The medium-proboscis species, B. *flavifrons,*
fed over the entire range of corolla lengths. Finally, another short-proboscis species,
B. *occidentalis,* fed as expected on plants with short corollas, but was also able to obtain
nectar from long corollas by using its large, powerful mandibles to bite through the
bases of the corollas and "rob" them of their nectar. For this reason it was placed in a
group of its own.

There was a clear tendency, in any single locality on the altitudinal gradient, for
the bumble bee community to be dominated by one long-proboscis species (*B. appositus* at low, *B. kirbyellus* at high altitude), one medium-proboscis species (*B. flavifrons*),
and one short-proboscis species (*B. bifarius* at low, *B. frigidus* at intermediate, and B.
sylvicola at high altitude). In addition, the nectar-robbing *B. occidentalis* was present at
sites where its preferred and exclusive nectar source was available (the plant *Ipomopsis
aggregata*). Thus, the bumble bees demonstrate niche complementarity: that is, niche
differentiation involves several niche dimensions, and coexisting species that occupy a
similar position along one dimension tend to differ along another dimension. B. *occidentalis* differs from the other six species in terms of food plant species and feeding
method (nectar robbing). Amongst the others, species differ from each other either in
the corolla length (and thus proboscis length), or in the altitude (i.e., habitat) favored,
or in both.

This study illustrates another important point: insofar as interspecific competition plays a role in structuring communities, it tends to do so not by affecting some
random sample of the members of that community, nor by affecting every member, but
by acting within guilds—groups of species exploiting the same class of resource in a
similar fashion.

*niche complementarity in nectar-
feeding bumble bees*

6.4.3   Aggregation alters competitive interactions in plant communities

We saw earlier how spatial and temporal variation can alter the outcome of competitive
interactions (Section 6.2.8). In this context, Stoll and Prati (2001) tested the hypothesis
that intraspecific aggregation can promote coexistence and thus maintain high species
richness in experimental communities of four annual terrestrial plants—*Capsella
bursa-pastoris, Cardamine hirsuta, Poa annua* and *Stellaria media. Stellaria* is known to
be the superior competitor.

Replicate three- and four-species mixtures were sown at high density, and the
seeds were either placed completely at random, or seeds of each species were aggregated in subplots within the experimental areas. Intraspecific aggregation decreased
the performance of the superior *Stellaria* in the mixtures, whereas in all but one
case aggregation improved the performance of the three inferior competitors
(Figure 6.17).

The results support the hypothesis that competitively inferior species grow better,
and may even coexist with superiors, when species are locally aggregated rather than
randomly dispersed. They do so, simply, because each species then competes more
within itself than with other species (which was also the case when there was niche
differentiation). Bear in mind that most species are aggregated most of the time. It is
likely, therefore, that spatial arrangement can be an important determinant of species
coexistence, especially in plant communities.

*intraspecific aggregation improves
the performance of competitively
inferior plants*

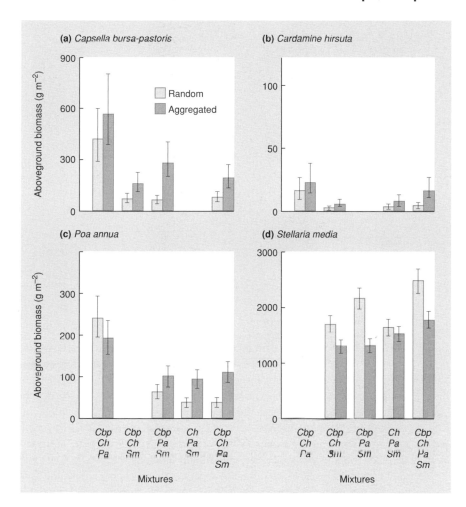

Figure 6.17
The effect of intraspecific aggregation on aboveground biomass (mean ± standard error) of four plant species (a–d) grown for 6 weeks in three- and four-species mixtures (four replicates of each). The normally competitively superior *Stellaria media* (*Sm*) did consistently less well when seeds were aggregated than when they were placed at random. In contrast, the three competitively inferior species almost always performed better when seeds had been aggregated. Note the different scales on the vertical axes. *Cbp, Capsella bursa-pastoris; Ch, Cardamine hirsuta; Pa, Poa annua.* (From Stoll & Prati, 2001.)

6.4.4   Niche differentiation amongst animals and amongst plants

In spite of the many examples where there is no direct connection between interspecific competition and niche differentiation, there is no doubt that niche differentiation is often the basis for the coexistence of species within natural communities.

There are a number of ways in which niches can be differentiated. One is resource partitioning or differential resource utilization. This can be observed when species living in precisely the same habitat nevertheless utilize different resources. Since the majority of resources for animals are individuals of other species (of which there are millions of types) or parts of individuals, there is no difficulty, in principle, in imagining how competing animals might partition resources amongst themselves.

Plants, on the other hand, all have very similar requirements for the same few potentially limited resources, and there is much less apparent scope for resource partitioning. Moreover, while mobile animals may compete for food, but still be free

less scope for resource partitioning by plants than by animals

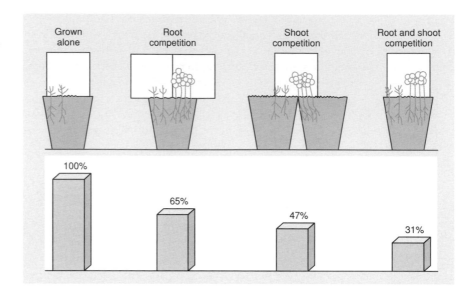

from competition for space, water, and other needs, for plants and other sessile organisms, competition for one resource often affects the ability of an organism to exploit another resource. If one plant species invades the canopy of another and deprives it of light, for example, the suppressed species will suffer directly from the reduction in light energy that it obtains, but this will also reduce its rate of root growth, and it will therefore be less able to exploit the supply of water and nutrients in the soil. This in turn will reduce its rate of shoot and leaf growth. Thus, when plant species compete, repercussions flow backwards and forwards between roots and shoots.

roots and shoots in plant competition

A number of workers have attempted to separate the effects of canopy and root competition by an experimental design in which two species are grown (1) alone, (2) together, (3) in the same soil but with their canopies separated, and (4) in separate soil with their canopies intermingling. One example is a study of subterranean clover and skeleton weed (Figure 6.18). The clover was not significantly affected under any circumstances (an example of one-sided competition), but the skeleton weed was affected when the roots intermingled (reduced to 65 percent of the control value of dry weight) and when the canopies intermingled (47 percent of the control), and when both intermingled the effect was multiplicative, dry weight being reduced to 31 percent of the control compared to the 30.6 percent (65×47 percent) that might have been expected. Clearly, both were important.

niche differentiation through spatial or temporal patterns in resources and/or conditions

In many cases, the resources used by ecologically similar species are separated spatially. Differential resource utilization will then express itself as either a microhabitat differentiation between the species (different species of fish, say, feeding at different depths) or even a difference in geographical distribution. Alternatively, the availability of the different resources may be separated in time; that is, different resources may become available at different times of the day or in different seasons. Differential resource utilization may then express itself as a temporal separation between the species.

The other major way in which niches can be differentiated is on the basis of conditions. Two species may use precisely the same resources; but if their ability to do so is influenced by environmental conditions (as it is bound to be), and if they respond differently to those conditions, then each may be competitively superior in different environments. This too can express itself as either a microhabitat differentiation, or a difference in geographical distribution, or a temporal separation, depending on whether the appropriate conditions vary on a small spatial scale, a large spatial scale, or over time. Of course, it is not always easy to distinguish between conditions and resources, especially with plants (Chapter 3). Niches may then be differentiated on the basis of a factor (such as water) which is both a resource and a condition.

6.5 How Significant is Interspecific Competition in Practice?

Competitors may exclude one another, or they may coexist if there is ecologically significant differentiation of their realized niches (Section 6.2). On the other hand, interspecific competition may exert neither of these effects if environmental heterogeneity prevents the process from running its course (Section 6.2.8). Evolution may drive the niches of competitors apart until they coexist but no longer compete (Section 6.3). All these forces may express themselves at the level of the ecological community (Section 6.4). Interspecific competition sometimes makes a high profile appearance by having a direct impact on human activity (Box 6.2). In this sense, competition can certainly be of practical significance.

Box 6.2   *Topical ECOncerns*

Competition in Action

When exotic plant species are introduced to a new environment, by accident or on purpose, they sometimes prove to be exceedingly good competitors and many native species suffer harmful consequences as a result. Some of them have even more far-reaching consequences for native ecosystems. This newspaper article by Beth Daley, published in the *Contra Costa Times* on June 27, 2001, concerns grasses that have invaded the Mojave Desert in the southern United States. Not only are the invaders outcompeting native wildflowers, they have also dramatically changed the fire regime.

INVADER GRASSES ENDANGER DESERT BY SPREADING FIRE—The newcomers crowd out native plants and provide fuel for once-rare

flames to damage the delicate ecosystem. Charred creosote bushes dot a mesa in the Mojave Desert, the ruins of what was likely the first fire in the area in more than 1,000 years.

Though deserts are hot and dry, they aren't normally much of a fire hazard because the vegetation is so sparse there isn't much to burn or any way for blazes to spread.

But, underneath these blackened creosote branches, the cause of the fire seven years ago has already grown back: flammable grasses fill the empty spaces between the native bushes, creating a fuse for the fire to spread again.

Tens of thousands of acres in the Mojave and other southwestern deserts have burned in the last decade, fueled by the red brome, cheat grass

Box 6.2 (cont'd)

and Sahara mustard, tiny grasses and plants that grow back faster than any native species and shouldn't be there in the first place.

. . . The grasses brought to America from Eurasia more than a century ago have no natural enemies, and little can stop their spread across empty desert pavement. And, once an area is cleared of native vegetation by one or repeated fires, the grasses grow in even thicker, sometimes outcompeting native wildflowers and shrubs.

. . . "These grasses could change the entire makeup of the Mojave Desert in short order," said William Schlesinger of Duke University, who has studied the Mojave Desert for more than 25 years. When he began his research in the 1970s, the grasses were in the Mojave, but there still were vast areas left untouched. Now, he said, the grasses are virtually everywhere and soon will be in concentrations large enough to fuel massive fires. "This is not an easy problem to solve," he said.

. . . Despite the harsh conditions, a rainbow of wildflowers blooms regularly in the desert, sometimes carpeting the ground with blossoms after a rainstorm. Zebra-tailed lizards, rattlesnakes, desert tortoises and kangaroo rats are able to get by for long periods without water and bear up under the sun. But the

innocuous-looking grasses threaten all these species by choking out wildflowers and killing off shelter and food that they rely on.

. . . Esque [of the US Geological Survey] has roped off 12 experimental sites, six of which he burned in 1999 to see how quickly invasive species re-establish themselves. But the result only showed the unpredictability of the desert: the first year, the invasive red brome took hold, but this year, native wildflowers came back in force.

. . . Esque said "It's not black and white with what is going on. We don't know if we are looking at coexistence or competition."

(All content © 2001 *Contra Costa Times* (Walnut Creek, CA) and may not be republished without permission.)

Some people have suggested bringing sheep into the desert to graze the invading grasses. Do you think this is a sensible idea? What further information would help you make a decision?

The US Geological Survey scientist found that red brome grass appeared to be outcompeting native flowers one year but not the next. Suggest some factors that may have changed the competitive outcome.

In a broader sense, however, the significance of interspecific competition rests not on a limited number of high profile effects, but on an answer to the question "How widespread are the ecological and evolutionary consequences of interspecific competition in practice?" The difficulty of answering this question can be illustrated by a study of the coexistence of five closely related woodland bird species: Lack's tits (Box 6.3).

The alternative explanations of patterns revealed by studies of niche differentiation, such as that in Box 6.3, raise a number of important questions, but we limit ourselves here to just two.

1   The first, dealt with in Section 6.5.1, is "How prevalent is current competition in natural communities?" To demonstrate current competition requires experimental field manipulations, in which one species is removed from or added to the community and the responses of the other species are monitored. It is important to answer this question, because where current competition is demonstrable, neither the ghost of competition past nor spatial and temporal variation are likely to have a crucial role.

Box 6.3   *Historical Landmarks*

Lack's Tits—Alternative Explanations for "Niche Differentiation"

In the 1960s, in an English broad-leaved woodland called Marley Wood, David Lack carried out a landmark study of five bird species: the blue tit (*Parus caeruleus*), the great tit (*P. major*), the marsh tit (*P. palustris*), the willow tit (*P. montanus*), and the coal tit (*P. ater*). Four of these species weigh between 9.3 and 11.4 g on average (the great tit weighs 20.0 g); all have short beaks and hunt for food chiefly on leaves and twigs, but at times on the ground; all eat insects throughout the year, and also seeds in winter; and all nest in holes, normally in trees. Nevertheless, despite their similarities (the overlaps between their niches), all five species bred where Lack studied them, and the blue, great, and marsh tits were common. In short, they coexisted but seemed very similar in where they lived, where they bred, and what they ate.

The closer Lack examined these species, however, the stronger became his conclusion that they were separated, at most times of the year, by precisely where within the trees they fed, by the size of their insect prey, and by the hardness of the seeds they took. He found, too, that this separation was associated with (often minor) differences in overall size, and in the size and the shape of the birds' beaks. Despite their similarities, these coexisting species exploited slightly different resources in slightly different ways.

Did this, though, have anything to do with competition? Lack thought so. He believed that the species coexisted as a result of evolutionary responses to interspecific competition (Connell's ghost of competition past). But there is a major problem—the absence of any direct proof. We cannot go back in time to check whether the tit species ever competed more than they do now, and it was not part of Lack's study to determine the extent of present-day competition. In fact, with such observational studies of niche differentiation, there are at least five plausible, alternative interpretations of the data.

1 The species are competitors now, and coexist as a result of the observed niche differentiation.

2 The species competed in the past but do not do so now—but their niche differentiation, which allows their coexistence, evolved in response to that past competition.

3 Competition in the past eliminated a number of other species, leaving behind only those that *were* different in their utilization of the habitat. Thus, we can still see the hand of the ghost of competition past—but acting as an ecological force (eliminating

The blue tit (left) and the great tit (right) (© Arthur Morris, Visuals Unlimited).

Box 6.3 (cont'd)

species) rather than an evolutionary one (changing them).

4 The species have, in the course of their evolution, responded to natural selection in different but entirely independent ways. They are distinct species, and they have distinctive niches. But they do not compete now, nor have they ever competed; they simply happen to be different. The coexistence of the tits, and the structure of the community of

which they are part, has nothing to do with competition.

5 Finally, although the niches of the species are distinct, they would not be different enough to allow all five to coexist if the environment was unvarying and competition ran its course. But it does not run its course, and the species therefore owe their coexistence to their responses, overall, to a patchy and ever-changing world.

And if current competition *is* prevalent, then interspecific competition is likely to be an important structuring force in nature. However, even if current competition is not prevalent, past competition, and therefore competition generally, may still have played a significant role in structuring communities.

2 The second problem, dealt with in Section 6.5.2, is to distinguish between interspecific competition (past *or* present) and either "mere chance" (the fourth alternative, Box 6.3) or spatial and temporal variation (the fifth). The many studies in which experimental field manipulations have *not* been possible can be examined to determine whether observed patterns provide strong evidence for a role for competition, or are open to alternative interpretations, as in Lack's study (Box 6.3).

6.5.1 The prevalence of current competition

surveys of published studies of competition indicate that current competition is widespread

Two surveys of field experiments on interspecific competition were published in 1983. Schoener examined the results of all the experiments he could find—164 studies in all. He found that approximately equal numbers of studies had dealt with terrestrial plants, terrestrial animals, and marine organisms, but that studies of freshwater organisms amounted to only about half the number in the other groups. Amongst the terrestrial studies, however, he found that most were concerned with temperate regions and mainland populations, and that there were relatively few dealing with phytophagous (plant-eating) insects. Any conclusions were therefore bound to be subject to the limitations imposed by what ecologists had chosen to look at. Nevertheless, Schoener (1983) found that approximately 90 percent of the studies had demonstrated the existence of interspecific competition, and that the figures were 89, 91 and 94 percent for terrestrial, freshwater, and marine organisms, respectively. Moreover, if he looked at single species or small groups of species (of which there were 390) rather than at whole studies which may have dealt with several groups of species, he found that 76 percent showed effects of competition at least sometimes, and 57 percent showed effects in all the conditions under which they were examined. Once again, terrestrial, freshwater, and marine organisms gave very similar figures.

Connell's review (1983) was less extensive than Schoener's: 72 studies, dealing with a total of 215 species and 527 different experiments. Interspecific competition was demonstrated in most of the studies, more than half of the species, and approximately 40 percent of the experiments. In contrast to Schoener, Connell found that interspecific competition was more prevalent in marine than in terrestrial organisms, and also that it was more prevalent in large than in small organisms.

Taken together, Schoener's and Connell's reviews certainly seem to indicate that active, current interspecific competition is widespread. Its percentage occurrence amongst species is admittedly lower than its percentage occurrence amongst whole studies, but this is to be expected, since, for example, if four species were arranged along a single niche dimension and all adjacent species competed with each other, this would still be only three out of six (or 50%) of all possible pairwise interactions.

Connell also found, however, that in studies of just one pair of species, interspecific competition was almost always apparent, whereas with more species the prevalence dropped markedly (from more than 90 percent to less than 50 percent). This can be explained to some extent by the argument outlined above, but it may also indicate biases in the particular pairs of species studied, and in the studies that are actually reported (or accepted by journal editors). It is highly likely that many pairs of species are chosen for study because they are "interesting" (because competition between them is suspected) and if none is found this is simply not reported. Judging the prevalence of competition from such studies is rather like judging the prevalence of debauched clergymen from the "gutter press". This is a real problem, only partially alleviated in studies on larger groups of species when a number of "negatives" can be conscientiously reported alongside one or a few "positives". Thus the results of surveys, such as those by Schoener and Connell, exaggerate, *to an unknown extent*, the frequency of competition.

UNANSWERED QUESTION: but these surveys exaggerate to an unknown extent the true frequency of competition

As previously noted, phytophagous insects were poorly represented in Schoener's data, but reviews of this group alone have tended to suggest either that it is relatively rare in this group overall (Strong et al., 1984) or rare in at least certain types of phytophagous insects, for example "leaf-biters" (Denno et al., 1995). On a more general level, it has been suggested that herbivores *as a whole* are seldom food limited, and are therefore not likely to compete for common resources (Hairston et al., 1960; Slobodkin et al., 1967). The bases for this suggestion are the observations that green plants are normally abundant and largely intact, they are rarely devastated, and most herbivores are scarce most of the time. Schoener found the proportion of herbivores exhibiting interspecific competition to be significantly lower than the proportions of plants, carnivores, or detritivores.

Taken overall, therefore, current interspecific competition has been reported in studies on a wide range of organisms and in some groups its incidence may be particularly obvious, for example amongst sessile organisms in crowded situations. However, in other groups of organisms, interspecific competition may have little or no influence. It appears to be relatively rare among herbivores generally, and particularly rare amongst some types of phytophagous insect.

6.5.2   Competition or mere chance?

When Lack's study of woodland birds is considered with the benefit of hindsight, it is easy to point to a dangerous tendency to interpret "mere differences" between the niches of coexisting species as confirming the importance of interspecific competition, when there may be other, equally plausible explanations. On the other hand, the theory of interspecific competition does more than predict "differences". It predicts not simply that the niches of competing species differ, but that they differ *more* than would be expected from chance alone.

neutral models   A more rigorous investigation of the role of interspecific competition, therefore, should address itself to the question: does the observed pattern, even if it appears to implicate competition, differ significantly from the sort of pattern that could arise in the community even in the absence of any interactions between species? This question was the driving force behind analyses that sought to compare real communities with so-called *neutral models*. These are hypothetical models of actual communities that retain certain of the characteristics of their real counterparts, but reassemble or reconstruct some of the community components in a way that specifically excludes the consequences of interspecific competition. In fact, the neutral model analyses are attempts to follow a much more general approach to scientific investigation, namely the construction and testing of *null hypotheses*. The idea is that the data are rearranged into a form (the neutral model or null hypothesis) representing what the data *would* look like in the *absence* of interspecific competition. Then, if the actual data show a significant statistical difference from the null hypothesis, the null hypothesis is rejected and the action of interspecific competition is strongly inferred.

The application of null hypotheses to community structure—that is, the reconstruction of natural communities with interspecific competition removed—has not been achieved to the satisfaction of all ecologists. But a brief examination of a study of lizard communities will at least make it possible to understand the potential and rationale of the neutral model approach (Box 6.4).

A study similar to that in Box 6.4 concerned spatial and temporal niche partitioning in grassland ant communities in Oklahoma (Albrecht & Gotelli, 2001). In this case, there was little evidence of niche partitioning on a seasonal basis. However, on a smaller spatial scale, at individual bating stations there was significantly less spatial niche overlap than expected by chance. This kind of pattern—sometimes a role for competition is confirmed, sometimes not—has been the general conclusion from the

---

Box 6.4   *Quantitative Aspects*

Neutral Models of Lizard Communities

Lawlor (1980) investigated differential resource utilization in ten North American lizard communities, consisting of four to nine species. For each community, there were estimates of the amounts of each of 20 food categories consumed by each species. This pattern of resource use allowed the calculation, for each pair of species in a community, of an index of

resource use overlap, which varied between 0 (no overlap) and 1 (complete overlap). Each community was then characterized by a single value: the mean resource overlap for all pairs of species present.

A number of "neutral models" of these communities were then created. They were of four types. The first type, for example, retained the

Box 6.4 (cont'd)

minimum amount of original community structure. Only the original number of species and the original number of resource categories were retained. Beyond that, species were allocated food preferences completely at random, such that there were far fewer species completely ignoring food in particular categories than in the real community. The niche breadth of each species was therefore increased. The fourth type, on the other hand, retained most of the original community structure: if a species ignored food in a particular category, then that was left unaffected, but among those categories where food was eaten, preferences were reassigned at random. These neutral models were then compared with their real counterparts in terms of their patterns of overlap

in resource use. If competition is a significant force in determining community structure, then the niches should be spaced out, and overlap in resource use in the real communities should be less—and statistically significantly less—than predicted by the neutral models.

The results (see figure below) were that in all communities, and for all four neutral models, the model mean overlap was higher than that observed for the real community, and that in almost all cases this was statistically significant. For these lizard communities, therefore, the observed low overlaps in resource use suggest that niches are segregated, and that interspecific competition plays an important role in community structure.

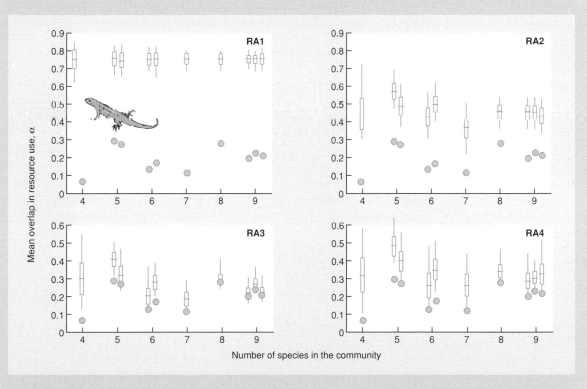

The mean indices of resource use overlap for each of ten North American lizard communities are shown as solid circles. These can be compared, in each case, with the mean (horizontal line), standard deviation (vertical rectangle), and range (vertical line) of mean overlap values for the corresponding set of 100 randomly constructed communities. The analysis was performed using four different types of reorganization algorithms (RA1 to RA4). (After Lawlor, 1980.)

A desert lizard of the southwestern United States.

neutral model approach. What then should be our verdict on it? Perhaps most fundamentally, its aim is undoubtedly worthy. It concentrates the minds of investigators, stopping them from jumping to conclusions too readily; it is important to guard against the temptation to see competition in a community simply because we are looking for it. On the other hand, the approach can never take the place of a detailed understanding of the field ecology of the species in question, or of manipulative experiments designed to reveal competition by increasing or reducing species abundances. It, like so many other approaches, can only be part of the community ecologist's armoury.

Summary

Ecological effects of interspecific competition

The essence of interspecific competition is that individuals of one species suffer a reduction in fecundity, survivorship, or growth as a result of exploitation of resources or interference by individuals of another species.

Species are often excluded by interspecific competition from locations at which they could exist perfectly well in the absence of interspecific competition.

With exploitation competition, the more successful competitor is the one that more effectively exploits shared resources. Two species exploiting two resources can compete but still coexist when each species holds one of the resources at a level that is too low for effective exploitation by the other species.

A fundamental niche is the combination of conditions and resources that allow a species to exist when considered in isolation from any other species; whereas its realized niche is the combination of conditions and resources that allow it to exist in the presence of other species that might be harmful to its existence—especially interspecific competitors.

The Competitive Exclusion Principle states that if two competing species coexist in a stable environment, then they do so as a result of differentiation of their realized niches. If, however, there is no such differentiation, or if it is precluded by the habitat, then one competing species will eliminate or exclude the other. However, whenever we see coexisting species with different niches it is *not* reasonable to jump to the conclusion that this is the principle in operation.

The only true test for whether competition occurs between species is to manipulate the abundance of each competitor and observe the response of its counterparts.

Environments are usually a patchwork of favorable and unfavorable habitats; patches are often only available temporarily; and patches often appear at unpredictable times and in unpredictable places. Under such variable conditions, competition may only rarely "run its course".

Evolutionary effects of interspecific competition

Although species may not be competing now, their ancestors may have competed in the past. We can expect species to have evolved characteristics that insure that they compete less, or not at all, with members of other species. Coexisting present-day competitors, and coexisting species that have evolved an avoidance of competition, can look, at least superficially, the same.

By invoking something that cannot be observed directly—"the ghost of competition past"—it is impossible to *prove* an evolutionary effect of interspecific competition. However, careful observational studies have sometimes revealed patterns that are difficult to explain in any other way.

Interspecific competition and community structure

Our perception of the role of interspecific competition in molding community structure depends on the scale at which it is examined. On a relatively small scale, we tend to see the co-occurrence of only a few species with complementary ecological niches. But if the scale is broadened, the community appears to contain more species, but these occur within a patchwork, with each patch supporting only a few.

Interspecific competition tends to structure communities by acting within guilds—groups of species exploiting the same class of resource in a similar fashion.

Niche complementarity can be discerned in some communities, where coexisting species that occupy a similar position along one niche dimension tend to differ along another dimension.

How significant is interspecific competition in practice?

Surveys of published studies of competition indicate that current competition is widespread but these exaggerate to an unknown extent the true frequency of competition.

The theory of interspecific competition predicts that the niches of competing species should be arranged regularly rather than randomly in niche space. Neutral models have been developed to determine what the community pattern *would* look like in the *absence* of interspecific competition. Real communities are sometimes structured in a way that makes an influence of competition difficult to deny.

Review Questions

1 Some experiments concerning interspecific competition have monitored both the population densities of the species involved *and* their impact on resources. Why is it helpful to do both?

2 Interspecific competition may be a result of exploitation of resources *or* of direct interference. Give an example of each and compare their consequences for the species involved.

3 Define fundamental niche and realized niche. How do these concepts help us to understand the effects of competitors?

4 With the help of one plant and one animal example, explain how two species may coexist by holding different resources at levels that are too low for effective exploitation by the other species.

5 ▲ Define the Competitive Exclusion Principle. When we see coexisting species with different niches is it reasonable to conclude that this is the principle in action?

6 Explain how environmental heterogeneity may permit an apparently "weak" competitor to coexist with a species that might be expected to exclude it.

7 ▲ What is the "ghost of competition past?" Why is it impossible to prove an evolutionary effect of interspecific competition?

8 Provide one example each of niche differentiation involving physiological, morphological, and behavioral properties of coexisting species. How may these differences have arisen?

9 Define "niche complementarity" and, with the help of an example, explain how it may help to account for the coexistence of many species in a community.

10 ▲ Discuss the pros and cons of the "neutral model" approach to evaluating the effects of competition on community composition.

▲ = Challenge Question.

Chapter 7

Organisms as Habitats

It is impossible to understand the ecology of any species without understanding its relationship with the habitat in which it lives, but for many, indeed most, species, this habitat is itself another organism: the host. This chapter examines the intimate interactions between hosts and the organisms that live on and within them.

Chapter Contents

7.1 Introduction

7.2 Hosts as habitats

7.3 Hosts as responsive habitats

7.4 Distribution and regulation of parasites and mutualists within hosts and their populations

7.5 Life cycles and dispersal

7.6 Coevolution

Key Concepts

In this chapter you will

- appreciate that the habitats of most organisms are other organisms, and understand the differences between commensalism, symbiosis, parasitism, and mutualism

- perceive that parasites are very diverse and often highly specialized, whereas mutualists are not so diverse or specialized but nevertheless dominate the productivity of many habitats

- understand that parasites, mutualists, and commensals may inhabit the surface or the body cavity or live within the tissues or cells of their hosts

- recognize that parasites and mutualists may "manipulate" the growth, form, or behavior of their hosts

- see that hosts may make defensive responses to the presence of parasites and mutualists

- appreciate that the life cycles of parasites are often complex and dominated by sex and dispersal, whereas the lives of mutualists are usually much simpler

- understand that parasites play a crucial and often subtle role in the evolution of their hosts, and vice versa

7.1   Introduction

symbiosis

More than half the world's species live in or on the bodies of other organisms, where they find conditions and sometimes also resources that support their growth. Almost every organism is the habitat of others—even bacteria are the habitat of their own viruses, bacteriophages. An intimate association between individuals of different species, in which one lives on or in the other, is known as a *symbiosis*.

A *parasite* is an organism that lives its life intimately associated with one or a very few individuals of another species, its host, deriving resources from its host and doing its host harm, but not predictably killing its host in the short term. More than 60 years ago, Charles Elton and his colleagues (see Section 1.1), in the kind of survey that would not be published in a scientific journal today, listed the organisms living on and in wood mice in a small woodland near Oxford, England. They found 47 different species, mainly parasites (including a species of spirochete, though the survey did not include other kinds of bacteria) (Table 7.1).

parasites are the major contributors to the biodiversity of nature . . .

We have now come to expect species-rich communities of parasites, not only on mammals like the wood mice, but also on fish, amphibians, reptiles, and birds, and if we survey the types of organism that live on plants in nature, or in cultivation, we find that almost every species of tree, shrub, and herb carries a community of small parasitic animals (mites, aphids, insect larvae) as well as parasitic fungi and bacteria (Figure 7.1). Parasites contribute overwhelmingly to the diversity of plant and animal communities in nature and to the diversity of life on Earth. A useful distinction between microparasites and macroparasites is drawn in Box 7.1.

Table 7.1
Species found on or in the bodies of wood mice (*Apodemus sylvaticus*) in a woodland near Oxford, England. (From Elton, 1940.)

On skin: 1 tick larva
On fur: At least 12 species of mite
    1 tick (adult)
    1 beetle
    11 fleas
    1 louse
On anus and genital organs: 1 mite
Under skin of limbs: 1 mite
In liver: 1 tapeworm larva
In stomach: 1 roundworm
In small intestine: 3 roundworms
    2 tapeworms
    3 flatworms
    2 flagellates
    1 ciliate
    1 amoeba
    2 coccidians
In cecum: 1 ciliate
In kidney: 1 spirochaete (no attempt was made to identify other bacteria or viruses)
In blood: 1 trypanosome

Box 7.1   *Historical Landmarks*

Microparasites and Macroparasites

Microparasites are small, are often extremely numerous, and they multiply directly within their host. Many are intracellular. It is always difficult, and usually impossible, to count the number of microparasites in a host, and so the number of infected hosts is usually studied, rather than the number of parasites. For example, a study of epidemics of measles will involve counting the number of cases of the disease, rather than the number of particles of the measles virus. Probably the most obvious microparasites are the bacteria and viruses that infect animals and plants (e.g., the yellow net viruses of beet and tomato and the bacterial crown gall disease). The other major group of microparasites affecting animals are the protozoa (e.g., the *Plasmodium* species that cause malaria). In plant hosts some of the simpler fungi behave as microparasites.

Macroparasites grow within their host but do not multiply there. They produce infective stages that are released to infect new hosts. The macroparasites of animals mostly live on the body or in the body cavities (e.g., the gut), rather than within the host cells. In plants they are generally intercellular. It is usually possible to count or at least estimate the numbers of macroparasites in or on a host (e.g., worms in an intestine or lesions on a leaf), and the numbers of parasites as well as the numbers of infected hosts can both be studied by the epidemiologist. Parasitic worms are major macroparasites of animals (e.g., tapeworms, flukes, and nematodes). In addition, there are lice, fleas, ticks, and mites and some fungi that attack animals. Plant macroparasites include the higher fungi that give rise to powdery mildews, downy mildews, rusts, and smuts, as well as the gall-forming and mining insects, and such flowering plants as the dodders and broomrapes that are themselves parasitic on other higher plants.

A less useful and certainly less clear distinction is sometimes drawn between parasites and *pathogens*. The term pathogen is more often applied to microparasites than to macroparasites and is usually reserved for parasites that give rise in their host to a "disease" with describable and presumably harmful "symptoms".

At least as striking as the diversity of parasites they support, though, are the populations of "mutualistic" fungi that live tightly integrated into the tissues of the great majority of plants. A *mutualism* is an association between species (usually a pair of species) that benefits both. A *mycorrhiza*, for example, is a mutualistic association between a fungus and the root of a higher plant (it is, in effect, the mycorrhiza, not the root, that extracts nutrients from the soil), and mycorrhizae are particularly strongly developed in trees and grasses. In addition, in a number of plants there are mutualistic bacteria and actinomycetes living in specialized nodules on the roots, capable of fixing atmospheric nitrogen and supplying it to the plant host. Other familiar mutualisms include those between flowering plants and insect pollinators. There are far fewer species of mutualists than of parasites, but mutualistic associations like mycorrhizae dominate the production of biomass in many habitats.

... but mutualists dominate productivity

We distinguish parasites and mutualists according to whether they benefit or harm the host. But words like *benefit* and *harm* must be used to mean something more than "what I think I would like or not like if I were that organism". Here, we use *harm* to mean increasing the death rate, reducing the birth rate, decreasing the growth rate, or

a parasite is not always a parasite

Figure 7.1
Higher plants (a) and animals such as humans (b) serve as specialized niches occupied by arthropods, worms, and protozoa. The two diagrams illustrate a little of this diversity but exclude the nematode worms that parasitize plants. Bacteria and viruses are omitted but comparable diagrams could be drawn for them—especially the diverse species that are parasitic on humans. Several of the worm species on humans migrate to different parts of the body during their life cycle. For example, *Schistosoma* occurs in mesenteric veins, but eggs accumulate in and even damage the liver and lungs. Fungi are shown in brown, worms in blue, arthropods in green, and protozoa in yellow.

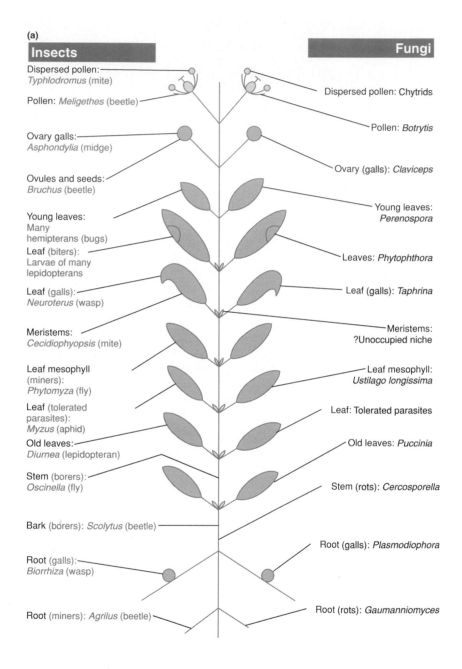

reducing the carrying capacity, and *benefit* to mean the converse of these. In practice, though, an organism that sometimes harms its host need not always do so. An activity that is harmful in one environment may be tolerated in another. Ticks draw blood from their hosts and might be thought to be obviously parasites. But individuals of the Australian sleepy lizard (*Tiliqua rugosa*) carry populations of ticks that seem not to harm them. An analysis of the relationship between the load of ticks on a lizard and its

Figure 7.1 *(cont'd)*

(b)

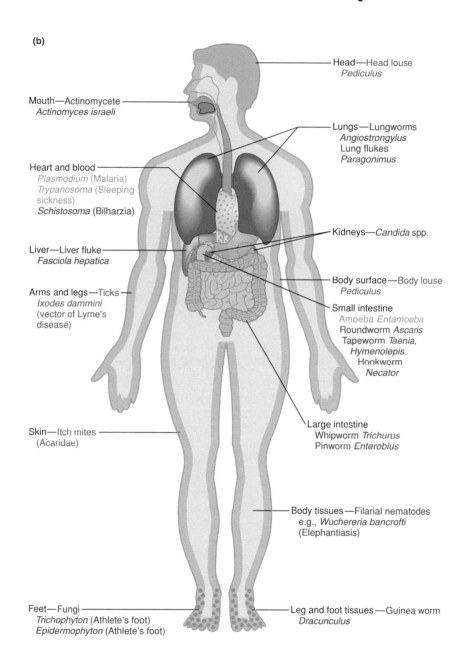

Head—Head louse *Pediculus*

Mouth—Actinomycete *Actinomyces israeli*

Lungs—Lungworms *Angiostrongylus* Lung flukes *Paragonimus*

Heart and blood *Plasmodium* (Malaria) *Trypanosoma* (Sleeping sickness) *Schistosoma* (Bilharzia)

Kidneys—*Candida* spp.

Liver—Liver fluke *Fasciola hepatica*

Body surface—Body louse *Pediculus*

Arms and legs—Ticks *Ixodes dammini* (vector of Lyme's disease)

Small intestine Amoeba *Entamoeba* Roundworm *Ascaris* Tapeworm *Taenia, Hymenolepis* Hookworm *Necator*

Large intestine Whipworm *Trichurus* Pinworm *Enterobius*

Skin—Itch mites (Acaridae)

Body tissues—Filarial nematodes e.g., *Wuchereria bancrofti* (Elephantiasis)

Feet—Fungi *Trichophyton* (Athlete's foot) *Epidermophyton* (Athlete's foot)

Leg and foot tissues—Guinea worm *Dracunculus*

longevity either showed that there was no correlation or that lizards with more ticks seemed to have lived longer. Perhaps these lizard ticks should be classified as *commensals* or "guests" (neither harmful nor beneficial to the host), bearing in mind that, like guests under other circumstances, they may be unwelcome when the hosts are ill or distressed! It is important not to forget these commensals in a preoccupation with the more dramatic interactions of mutualists or parasites.

a host is a habitat that can respond,
repair, remember, and evolve

In the remainder of this chapter we examine hosts as mosaics of microhabitats for other species. We consider ways in which the diversity of life cycles, growth form, dispersal, and specificity provides the match between colonists and the organism that they colonize. We also show how, in comparison with other kinds of habitats and environments, a living organism is a unique type of habitat. It can respond to being colonized, can repair some of the damage that may be caused, may have a "memory" of past experience, and can evolve. In Chapter 8 we consider the effects of parasites and other natural enemies on the dynamics of host populations.

7.2   Hosts as Habitats

7.2.1   Life on the surface of another organism

organisms may provide physical
support for other species

Solid surfaces provide potential support for living organisms and may allow them to gain access to conditions and resources that are otherwise unavailable. Trees and shrubs in particular provide habitats for the many species of birds, bats, and climbing and scrambling animals that are absent from treeless environments. Trees are particularly good examples of what have been called "ecological engineers". By their very presence, they create, modify, or maintain habitats for others (Figure 7.2). They provide sites for nesting and roosting and a degree of safety from predators on the ground. Lichens and mosses develop on tree trunks, and climbing plants such as ivy, vines, and figs, though they root in the ground, use tree trunks as support to extend their foliage up into a forest canopy. In warm and humid climates, *epiphytic* plants (plants living on plants) including ferns, orchids, and bromeliads, root in the branches and crevices of the upper canopy. They draw their water and mineral nutrient resources from the rain water that leaches down through the branches and even the feces of animals in the canopy. Water that collects in hollows on the trunks and in the cups formed by the leaf bases of bromeliads supports further microcosms of midge larvae and even the larvae of tree frogs. These are all commensals: they draw no resources from the host tree and do it no direct harm. But the community of organisms that lives on trees and shrubs also includes hemiparasites, such as the mistletoes, which penetrate the tissues of the host tree and depend on it for water and mineral nutrients, and true parasitic flowering plants, such as dodders, which depend on the host for photosynthetic products as well (see Section 7.2.3).

In aquatic communities, the solid surfaces of larger organisms are even more important contributors to biodiversity. Seaweeds and kelps normally grow only where they can be anchored on rocks, but their fronds are colonized in turn by filamentous algae, by tube-forming worms (*Spirorbis*), and by modular animals such as hydroids and bryozoans (Figure 7.3). These depend on the seaweeds and kelps for anchorage and access to resources in the moving waters of the sea. In fresh waters, the stems and leaves of aquatic flowering plants are exploited as habitats in the same way. Commensals have received far less study than parasites and mutualists, though many of them have ways of life that are quite as specialized and fascinating and their contribution to the diversity of communities may be quite as great.

life on roots and leaves: the
rhizosphere and phyllosphere

Plant roots draw water and mineral nutrients from the soil and so modify the local environments on or near their surface. They also shed root cap cells and release organic materials into the soil as the root hairs and outer cortical cells die and decompose. The

(a)

(b)

(c)

Figure 7.2
A variety of flowering plants use the bodies of other flowering plants as their habitats. (a) Spanish moss (*Tillandsia usnoides*) is an epiphytic flowering plant draped over trees in humid environments and with a characteristic lichen-like colour. (b) Mistletoe (*Phoraradendron* spp.) on sycamore. This is a hemiparasite, penetrating the host tree and extracting mineral nutrients and water from within the host tree, but capable of independent photosynthesis. (c) Dodder (*Cuscuta* spp.). This is a true parasite that repeatedly penetrates the host plant on which it climbs. It lacks chlorophyll and depends on its host for photosynthesis as well as for supplies of mineral nutrients and water. (© John D. Cunningham, Visuals Unlimited.)

volume of soil adjacent to and influenced by plant roots is called the *rhizosphere* and it carries an abundant microbial flora, different from that found in the main body of the soil. The rhizosphere flora differs from one host species to another (Westover et al., 1997), and there is evidence that it may sometimes protect the host plant against soil-borne parasites.

There is also a distinct *phyllosphere* of species that live on leaf surfaces. Aphids, for example, are *ectoparasites* of plants—they live outside the main body of the plant but tap its phloem with specialized mouthparts and subsist entirely on the extracted sap. It is easy to forget that such behavior is truly parasitism. Very similarly, fungal parasites on leaf surfaces penetrate the tissues of the host with their hyphae, while the main bodies of the fungi are carried on the surface of host leaves, where it is easy to disperse their spores. Examples are species of powdery mildews (e.g., *Erysiphe* spp., which cause devastating diseases of grapes) and downy mildews (e.g., *Phytophthora infestans*, which caused the Irish potato famine of the 1840s).

Fleas and lice are ectoparasites on the bodies of the larger animals. Most of these are (like the parasitic fungi on leaf surfaces) very tightly specialized to a very few hosts. The more intimate their association with a particular host individual, the more they

mildews, fleas, and lice: the ectoparasites

(a)

(b)

(c)

Figure 7.3
(a) A frond of the brown seaweed *Fucus serratus* bearing the polychaete worm *Spirorbis* (the spiral shells), the zigzag colonies of the hydroid *Obelia geniculata*, and (the brown area in the middle) the bryozoan *Bowerbankia*. (b) A frond of the brown seaweed *Ascophyllum nodosum* bearing tussocks of the red seaweed *Polysiphonia* (which is found almost uniquely living on *Ascophyllum*) and colonies of the hydroid *Obelia dichotoma*. (c) A frond of the brown seaweed *Ascophyllum nodosum* bearing colonies of *Clava multicornis*. (Courtesy of R.N. Hughes.)

tend to be restricted to a particular species of host. Thus most species of bird lice, which spend their entire lives on their host, exploit only one species of host, whereas louse flies, which move actively from one host individual to another, can use several species of host (Table 7.2).

The bird lice have chewing mouthparts and feed on the feathers and scales of the skin. Their crawling and gnawing irritate the birds, which may become restless and not feed or digest their food properly. The lice are therefore an economic problem in commercial poultry production. Lice that feed on the feathers of rock doves (*Columba livia*) reduce the protective insulation given by the feathers; infected birds spend more time

(d)

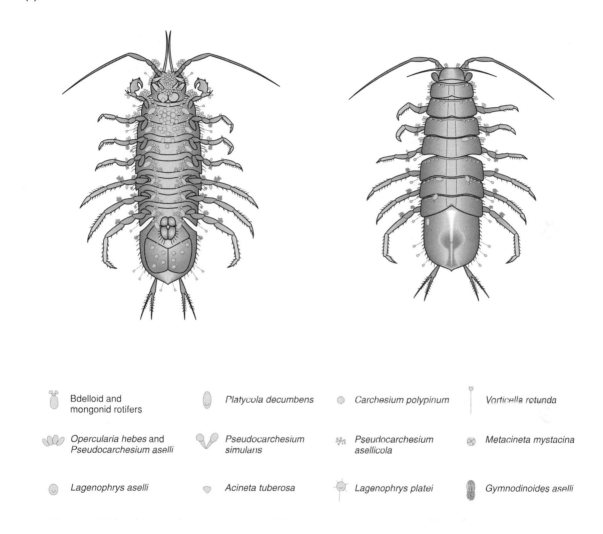

	Bdelloid and mongonid rotifers		Platycola decumbens		Carchesium polypinum		Vorticella rotunda
	Opercularia hebes and Pseudocarchesium aselli		Pseudocarchesium simulans		Pseudocarchesium asellicola		Metacineta mystacina
	Lagenophrys aselli		Acineta tuberosa		Lagenophrys platei		Gymnodinoides aselli

Figure 7.3 (cont'd)

(d) The water louse *Asellus aquaticus*, like most aquatic organisms, provides the habitat for a variety of other species that use their host as a vantage point from which to filter food from the water that flows past them. The figure shows *Asellus* from beneath (left) and in dorsal view (right) with 12 of the species that live on it shown in their characteristic (often very specialized) habitats. Apart from the rotifers, the species are all protozoa. (After Cook et al., 1998.)

preening and less on courtship and obtain fewer mates than less heavily parasitized males. Indeed there is now much evidence that the parasitized males of many species (especially birds) fail to attract mates—an important way in which parasites can harm their hosts because it translates so directly into effects on male fitness.

Table 7.2
Specialization in ectoparasites that feed on birds and mammals. (After Price, 1980.)

| | | | Percentage of species restricted to: | | |
		Number of species	1 host	2 or 3 hosts	More than 3 hosts
Philopteridae	Bird lice (spend whole life on one host)	122	87	11	2
Streblidae	Blood-sucking flies (parasitize bats)	135	56	35	9
Oestridae	Botflies (females fly between hosts)	53	49	26	25
Hystrichopsyllidae	Fleas (jump between hosts)	172	37	29	34
Hippoboscidae	Louse flies (are highly mobile)	46	17	24	59

extinction of a host species forces extinction of its specialist parasites

Because parasites are highly specialized to their hosts, the extinction of a host often causes the extinction of its parasites as well. The passenger pigeon became extinct in 1914. It was the host for two species of chewing lice, *Columbicola extinctus* and *Campanulotes defectus*. They became extinct at the same time. The hosts of parasites are their habitats: the extinction of a host species is the loss of a unique habitat.

7.2.2   Inhabitants of the body cavity

surface-dwelling organisms may disperse easily—but there are costs

Organisms that specialize in life on the surface of their hosts are usually well placed for dispersal (via spores, eggs, and so on) and the colonization ("infection") of other hosts. However, there are costs to a life spent on host surfaces, such as greater exposure to drought, heat, and cold and usually less intimate contact with the host as a resource. Surface dwellers are dangerously exposed to predators and parasites; for example, aphids on a leaf surface are searched for and attacked by parasitoid wasps and eaten by birds. Organisms that can invade the body cavities of an organism or even its tissues or cells are better placed to extract resources, are better hidden from all but the most specialized enemies, better insulated from the outside environment, and on warm-blooded animal hosts are maintained in a constant thermal environment.

life in the gut

Many species of parasites, commensals, and mutualists exploit the insulation and protection offered by cavities within the body of a host species. The alimentary canal from mouth to anus is a resource-rich habitat. It is largely protected from enemies and drought and contains a continuous flow of food that has been garnered, and partly processed, by the host. Many parasites live in the alimentary canal of animals. Gut inhabitants like tapeworms and roundworms, however, do not attack or consume any part of their host. Rather, they live in the food that the host is digesting, and can be thought of as competing with the host for these digestive products as they travel from mouth to anus. But they can certainly cause disease in their host. Some, indeed, cause further damage by releasing toxins: classic examples are *Vibrio cholerae*, the bacterium that causes cholera, and *Amoeba dysenterica*, the protist that causes amoebic dysentery. Note, too, that the diarrhea that these parasites induce favors their dispersal and epidemic spread.

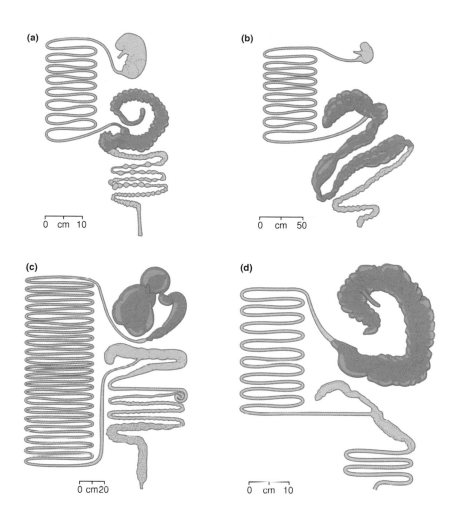

(a)

0 cm 10

(b)

0 cm 50

(c)

0 cm 20

(d)

0 cm 10

Figure 7.4
The digestive systems of herbivores are commonly modified to provide fermentation chambers inhabited by a rich fauna and flora of microbes. The figure shows the digestive tracts of four different herbivorous mammals with the fermentation chambers highlighted. (a) The rabbit has an expanded cecum. (b) The zebra has fermentation chambers in both the cecum and colon. (c) The sheep has fermentation in an enlarged and four-chambered stomach. (d) The kangaroo has a long fermentation chamber in the first part of the stomach. (After Stevens, 1988.)

The digestive tract is, of course, not just one environment but a continuum of changing conditions of pH, enzymic character, and aeration along its length (Figure 7.4), and the inhabitants are specialized to, and even migrate to find, local conditions within it. When nematode worms, *Nippostrongylus brasiliensis*, which infect rats, were experimentally transplanted into the anterior and posterior small intestine from their normal position between them in the jejunum, they migrated back to their original habitat.

Organisms that live in the guts of others gain protection and insulation, but dispersal from one host to another is much more difficult than for those that live on the surface. Many species of gut parasite solve the problem through complex life histories. The life cycle of the human tapeworm *Taenia saginata*, for example, involves the passage of parasite eggs in human feces to livestock, such as cattle, in contaminated food. The parasite then develops in the cattle and may be passed on to humans who eat meat from the cattle without cooking it adequately. An even more complex life cycle is illustrated in Figure 7.5.

Figure 7.5
The parasitic cestode worm *Diphyllobothrium latum* infects humans who eat infected fish and have not cooked it sufficiently to destroy the parasite. Eggs of the parasite pass in human feces into water, where they hatch and start to develop before they are eaten by copepods (small crustaceans) within which they continue their development. The copepods are eaten in turn by fish, within which they develop to an adult worm and may be eaten by humans (or dogs) in which they produce eggs and start the life cycle again. Notice that the release of eggs into the feces is the only point in the life cycle when the parasite multiplies. The parasite consumes most of the vitamin B_{12} in the host's diet causing severe anemia.

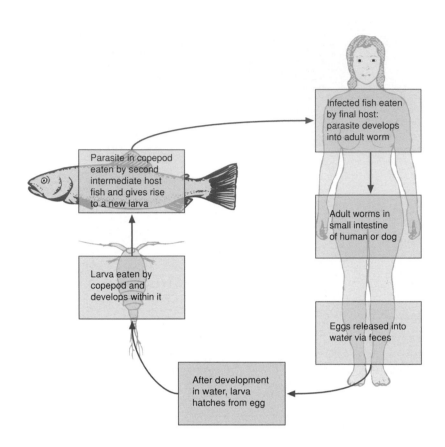

the floras and faunas of the digestive tract are crucial for the digestion of cellulose

Immensely diverse communities of mutualistic microorganisms inhabit the alimentary canal of herbivorous animals, playing a crucial role in the digestion of a diet of plants. Ruminants (deer, cattle, and antelopes) have four-chambered "stomachs," and chewed food passes from the first chamber to the second and is then regurgitated and chewed again (rumination) before it passes through to the final chambers and to the small intestine (Figure 7.4). The chambers of the ruminant's stomach provide a fermentation chamber with regulated pH and temperature, in which huge populations of bacteria and protozoa are stirred. In a sheep, the volume of the rumen is 4.7 liters, and 20 percent comprises the bodies of microorganisms. The inhabitants of the rumen are mainly unique to this peculiar habitat and are often anaerobic: many are killed by exposure to oxygen. The community there is comparable to that of a tropical forest in its diversity and complexity.

The primary importance of this microbial community to the host is that it digests cellulose. Cellulose is the most abundant and bulky of all plant products but is virtually indigestible by animals. Apart from a few insects and protozoa, animals cannot produce cellulases of their own. The anaerobic microbial community ferments cellulose to organic acids that are then absorbed and metabolized by the host. The stomach of

Figure 7.6
Gut flora of the termite (© Michael Abbey Photo).

ruminants is particularly well studied, but virtually all herbivorous mammals and some lizards and birds rely on a microbial flora to digest cellulose, and specialized fermentation chambers seem to have appeared repeatedly in the evolution of vertebrates (Figure 7.4). Some cellulose fermentation occurs before food reaches the stomach in kangaroos, camels, and ruminants, but in horses, rabbits, rodents, elephants, and some birds, it is concentrated in the colon and/or the cecum. Moreover, many species, including elephants and rabbits, practice *refection*—they consume their own feces and hence pass material twice through the alimentary canal. This doubles the period for which food is exposed to microbial digestion and, perhaps more importantly, doubles the period in which the symbiotic mutualists synthesize vitamins that are required by the host.

Many invertebrates also carry a symbiotic microflora that digests cellulose. The alimentary canal of termites, for example, is the habitat of a particularly diverse flora and fauna (Figure 7.6). Some species there can even fix atmospheric nitrogen—a valuable property in an associate of a host that spends its life eating something as non-nutritious as wood.

Plants offer a peculiar range of body cavities that become inhabited by microorganisms. Among the most remarkable are the hollow thorns developed by species of *Acacia* in tropical and subtropical America and used as nesting sites by ants. Probably the most studied of these are the hollow thorns of the bull's horn acacia *Acacia cornigera*, inhabited by the specialized ant *Pseudomyrmex ferruginea* (Figure 7.7). The *Acacia* species produce attractive nectar and also protein-rich *beltian bodies* on their leaves that the ants eat. The ants, for their part, prune off invading shoots from competing neighboring plants and also protect their host from herbivores. *Pseudomyrmex* is just one genus in a subfamily of ants that contains about 230 species. At least 37 of these are specialized inhabitants of plant cavities, and plant species in 20 genera and 14 families provide these peculiar habitats.

ants inhabit the hollow thorns of acacias

Figure 7.7
Structures of the bull's horn acacia (*Acacia cornigera*) that attract its ant mutualist. (a) Protein-rich beltian bodies at the tips of the leaflets (© Michael Fogden, Oxford Scientific Films). (b) Hollow thorns used by the ants as nesting sites (© C.P. Hickman, Visuals Unlimited).

(a)

(b)

7.2.3 Inhabitants of host tissues and cells

the mutualism of fungi and plant roots: mycorrhiza

Most forest trees carry a dense growth of fungi on their root surface. The fungus forms a sheath around the root and encloses its tip. The combined structure of root and fungus has its own morphological characteristics and is called a *mycorrhiza*. Filaments of the fungus spread out into the surrounding soil and also penetrate between, though not usually into, the cells of the host root. The fungus obtains energy-rich carbon resources from the host, while phosphate and probably other nutrients pass from the fungal filaments in the soil to the tree via the fungal sheath.

mycorrhizal plants form most of the biomass of terrestrial environments

A different type of mycorrhiza is found in most herbaceous plants. In these the fungus spreads from cell to cell of the host root and also out into the surrounding soil. The fungus is tolerated by the host cells and forms highly branched intrusions into them, making intimate intracellular contact between the fungus and host. Almost all trees, grasses, and perennial herbs are mycorrhizal. Together they account for by far the greatest part of the biomass of terrestrial environments, although, compared to parasites, relatively few species of fungus are involved.

sit-and-wait parasites of higher plants

We have already met organisms that penetrate host cells, but where the greater part of their body is on the host surface. For many parasites and mutualists, however, active life is spent entirely within host cells. For these, there is a problem of how to disperse from one host to another, and a wide variety of growth forms and life cycles have evolved which solve this problem. A common life cycle among species of parasite that live within the cells of plants involves dormant, often very long-lived, spores that germinate in the presence of a host. They release motile single-celled *zoospores*, which penetrate and infect the cells of the host. The parasite then multiplies within the host cells, kills them, and releases a new generation of dormant spores. This very simple life cycle is especially common among parasites that persist in the soil and infect plant roots or tubers (club root disease of crucifers [*Plasmodiophora brassicae*] and wart disease [*Synchitrium endobioticum*] of potatoes are examples). Long persistent spores from an infected plant remain dormant, "sitting and waiting" for a new host to disperse to them, rather than dispersing themselves in search of a new host. Agriculture then introduces a whole new element into the spread of these parasites, when infected host plants and tubers are carried by people from place to place (often with spores in the attached soil).

There is a specialized group of flowering plants that find their habitat on other plants, form intimate cellular contact with them, and behave as parasites. Each species has a very limited number of acceptable hosts. The dodders, species of *Cuscuta*, are *holoparasites*: they cannot photosynthesize and are entirely dependent on a host for energy, minerals, and water. When a dodder seed germinates, the seedling "searches" for a suitable host to colonize by rotating its shoot tip in a circle and then growing horizontally for a while before becoming erect and rotating again. This continues until either the seedling runs out of stored resources and dies or it contacts one of a very restricted range of acceptable hosts. It penetrates the host tissues with specialized pegs that make intimate contact within the host tissues. As it branches and spreads over the body of the host plant, the parasite penetrates it repeatedly, soon losing all contact with the ground.

Hemiparasites, on the other hand, contain chlorophyll and photosynthesize, but they use other flowering plants as their habitat and obtain water and minerals through intimate connections with the host roots or shoots. Mistletoes are hemiparasites specialized for growth on the branches of trees: their seeds are usually dispersed there by birds. A great variety of mistletoes is found in tropical forests: at least 11 genera and 27 species are found on 126 host species in Sri Lanka alone. Some species of mistletoe (e.g., *Dendrophthoe falcata*) can develop as epidemics in forests and seriously damage forest productivity. *Dendrophthoe*, though, has its own "hyperparasitic" mistletoe, *Viscum capitellatum* (hyperparasite means "parasite of a parasite") which apparently helps to keep epidemics under control.

Many of the parasites of animals that have their habitat within the cells of the host have no dormant persistent stage but depend on the movements of their hosts to transmit them. Transmission may be "direct", as, for example, when a host infected with a human respiratory pathogen moves close to an uninfected host and passes the infection on via droplets (coughs and sneezes). Alternatively, the parasite may spend different phases of its life cycle in two different species of host, one of which plays a particularly crucial role in transmission and persistence. The malarial parasite, *Plasmodium*, for example, alternates between humans and mosquitoes and multiplies in both, but the mosquito is said to be the "vector" of the parasite since it seeks out humans (rather than the other way round) (Figure 7.8).

Many important species of symbiotic mutualists also inhabit and multiply in host cells. Of particular significance are the nitrogen-fixing bacteria of the genus *Rhizobium*, which infect the roots of legumes such as clover, beans, and peas. They stimulate the formation of nodules, which form a specialized habitat in which the bacteria receive resources from the host plant and fix atmospheric nitrogen. They

flowering plants parasitic on others

directly and indirectly transmitted animal parasites

nitrogen-fixing microbes within the cells of host plants

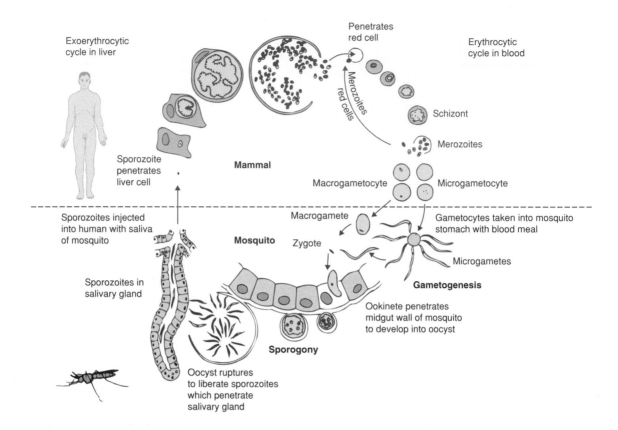

Figure 7.8

The parasite that causes malaria (species of protozoa of the genus *Plasmodium*) lives within the cells of primates and of mosquitoes. It has no life outside its two hosts and must alternate from one to the other. There is no dormant stage in the life cycle. In humans the parasite lives and multiplies in red blood cells, and, as it multiplies, it continually invades new blood cells, causing recurrent bouts of fever in the host. Mosquitoes of the genus *Anopheles* draw blood from the human and the parasite begins a new intracellular life in the cells of the insect gut and later in the cells of its salivary glands. When the mosquito next draws blood from a human host it infects it with the parasite, which is carried in the insect saliva. The parasite now enters the liver cells of the human and multiplies again. It then escapes from the liver cells and invades the red blood cells. The parasite forms its gametes in the human, but sexual fusion occurs in the mosquito.

contribute this fixed nitrogen to the host plant, which then becomes independent of supplies of nitrate or ammonium from the soil. Their life cycle is illustrated in Figure 7.9.

In some ways the plant behaves as if it is capturing the bacteria and harnessing them to provide its own nitrogen nutrition. The plant even appears to stimulate *Rhizobium* in the rhizosphere by releasing special root exudates. The bacteria multiply to form a colony on a root hair, and this colony stimulates active cell division within the root, leading to the formation of a nodule. Once inside a root hair, the bacteria form a tube or "infection thread" which branches and carries the infection to a number of cells which continue to divide and form much of the body of the nodule. In the infected cells, the bacteria multiply further, and when the cells are tightly packed,

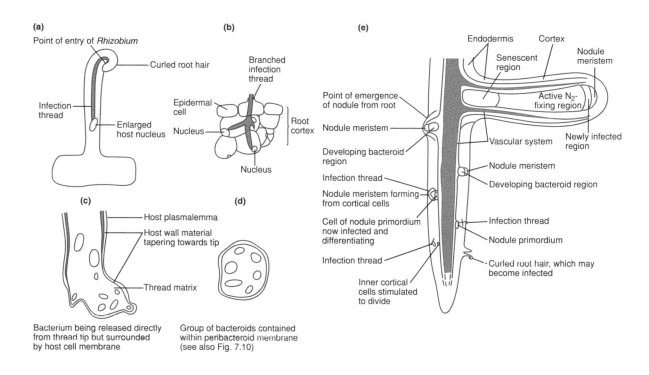

Figure 7.9
A typical life cycle of *Rhizobium*, an intracellular nitrogen-fixing mutualist of legumes such as clover, beans, and peas. (a–c) Infection of a host root hair by *Rhizobium*. The hair becomes curled before entry. Subsequently an infection thread carries infection from cell to cell. (d) A mature infected host cell contains tightly packed bacteria ("bacteroids") which have ceased to multiply. (e) The sequence of development of nodules on a host root from infection behind the root tip to the emergence of mature nodules on older parts of the root. (After Sprent & Sprent, 1990.)

the bacteria change their shape and become branched *bacteroids* (Figure 7.10), and multiplication stops.

Rhizobial bacteria are widely distributed in the soil and grow and multiply in the absence of host plants. But in this free-living condition they do not fix atmospheric nitrogen because this is an anaerobic process. The host provides a paradoxical environment in its nodules: it has a rapid diffusive flow of oxygen (allowing the rhizobia to respire aerobically) but extremely low oxygen concentrations (allowing them to fix nitrogen with an effectively anaerobic nitrogenase enzyme).

Enormous research effort has been put into disentangling the details of the evolution and the morphological, anatomical, physiological, and genetic characteristics of the interaction between nitrogen-fixing organisms and the environment within their hosts, but unanswered questions remain at all these levels. There are many more species of nitrogen-fixing symbionts than previously recognized, and different host species provide quite different environments in their nodules. One important ecological and evolutionary question is why higher plants have never evolved the ability to fix nitrogen for themselves and how it has happened that only a few species have found a solution by providing habitats and resources for nitrogen-fixing microbes. The vast

great research opportunities and unanswered questions remain

Figure 7.10
Transmission electron micrograph of nitrogen-fixing bacteria (*Rhizobium*) within a cell of a root nodule of soybean (*Glycine max*). The bacteria are grouped in clusters, each cluster enclosed in a peribacterial membrane. The dark bodies at the edge of the cell are host cell inclusions within the cell cytoplasm. (Courtesy of Janet Sprent.)

majority of higher plants have to depend on competing among themselves for nitrates or ammonium salts in the soil.

unicellular algae inhabit specialized cells in many aquatic invertebrates

Most unicellular green algae are free-living organisms in aquatic or very humid environments. But almost all corals, jellyfish, and sea anemones, and many mollusks in shallow tropical waters, contain intracellular algae as symbionts—usually flagellates of the genus *Symbiodinium*. In fresh water, too, the sponges, hydras, and some mollusks are hosts to intracellular algae, although here they are usually species of *Chlorella*. As much as 90 percent of the carbon fixed in photosynthesis by symbiotic marine intracellular algae is released to the hosts that provide them with their habitat. Bivalve mollusks that carry algal symbionts capture them directly from free-living populations, and flatworms seem to acquire their algae in the same way. However, many corals, anemones, and jellyfish that grow by budding, transmit their symbionts from parent to offspring as well as capturing some from free-living forms.

intracellular bacteria inhabit specialized cells in some insects

Intracellular symbionts are usually confined to special cells in the host. An extreme case of this occurs in the many insects that provide a habitat for bacterial symbionts that appear to synthesize B vitamins that the hosts can neither make for themselves nor find in their diet. Aphids carry bacteria of this type (belonging to the genus *Buchnera*) in special cells called *mycetocytes*. The bacteria are transmitted directly from parent to offspring and colonize the mycetocytes in the aphid embryo. If the aphids are freed from their *Buchnera* infection by antibiotic treatment, they cease to reproduce. In effect, *Buchnera* bacteria have become incorporated into the normal life and body of the aphids—neither can live without the other.

7.3 Hosts as Responsive Habitats

7.3.1 Changes in the growth and form of hosts caused by their inhabitants

All organisms change the environment in which they live. Often they make it less habitable for themselves (they pollute it!—see Chapter 13), though they may transform it into a condition more suitable for others. This generalization is as true for organisms that inhabit living habitats as it is for those that have independent lives.

The most dramatic change that an organism can cause to its living habitat is to kill it. For specialized parasites a dead host is no longer habitable. However, some "parasites" kill their host and then continue to grow and reproduce on the dead body. These are *facultative parasites* (that is, organisms that are not obliged always to be parasitic), but by starting to decompose their food resource before their host is dead they have an advantage over all those that have to wait for death before they can begin to colonize the body. The blowfly, *Lucilia*, is a facultative parasite that commonly kills its host habitat. It may start its life as an egg laid on the body of a mammal such as a sheep. The larvae feed on dung close to the skin or on a wound, and then spread into living tissue, often killing the sheep or making it more likely to die of other causes. They then continue to feed, mature, pupate, and hatch to give further generations, which continue to consume the dead body.

Gray mold (*Botrytis cinerea*) and damping-off (species of *Pythium*) are unspecialized fungal parasites of higher plants. Like blowflies on a sheep, they attack and kill their host and then grow and reproduce on the dead tissues. Facultative parasites may use a dead host as a food base from which to invade and kill others.

Most parasites, however, are obligatorily parasitic. If their host dies, they cease to grow on it. If they persist on the dead host, it is only as an infective dormant residue (e.g., the long-lived spores of *Anthrax* bacilli). Commonly, therefore, though not always, the success of a parasitic way of life depends on keeping the host alive, but twisting its growth, form, and behavior in ways that enhance the fitness of the parasite. In response, the host usually reacts to limit the activity of organisms that parasitize it.

A common host plant reaction to the presence of a parasite is to isolate it with a scab. Potato tubers react to invasion by the parasite *Actinomyces scabies* by forming a layer of cork that limits it to the tuber surface. Many plants react to leaf parasites by shedding the infected leaves. Some species of parasites and mutualists, however, take control of the local growth of the host and force it to create habitable microenvironments in which the growth of the parasites is supported but confined. The parasitic bacterium *Agrobacterium tumefaciens* induces the formation of galls on plant tissues, especially of woody plants (crown gall disease), and the bacteria then multiply within the enlarged tissues. As in the initiation of nodules on leguminous plants by *Rhizobium* (see Section 7.2.3), the host cells divide in anticipation of the entry of the parasite, and gall tissue can be recovered that lacks the parasite but has now been set in its new pattern of morphogenesis: it continues to produce gall tissue—even in tissue culture. The parasite has induced a genetic transformation of the host cells.

Insect galls, caused by agromyzid flies and cecidomyid and cynipid wasps, are some of the most remarkable examples of this takeover of the host's development to provide a habitat for a parasite (Figure 7.11). In the formation of a typical oak gall, the insect lays eggs in the plant tissues, which then stimulate it to grow in a quite alien fashion. The presence, just for a short time, of the parasite's egg appears to be sufficient to start the host into developing a specialized gall. A great many parasites induce galls on species of oak (*Quercus* spp.) and each induces its own specific form of gall. Many of the gall-forming insects have a sexual generation in the spring and an asexual generation in the summer and fall, and even these induce quite different-shaped galls on quite different parts of the plant. The galls provide a supply of food for the parasitic insect larvae and give them some protection from parasites of their own.

Figure 7.11
Galls formed by wasps of the genus *Andricus* on oaks (*Quercus petraea, Q. robur, Q. pubescens,* or *Q. cerris*). Each figure (a–s) shows a section through a gall induced by a different species of *Andricus*. The dark shaded areas are the gall tissue and the central unshaded areas are the cavities containing the insect larva. (From Stone & Cook, 1998.) The photographs below show the morphological changes that are induced in oak trees by gall wasps to create the habitats for their larvae; they are very precisely controlled by specific interactions of the genes from both the oak and the wasp. (t) Brightly colored "oak apples" are produced by a cynipid wasp on the brances of the oak *Quercus engelmanii* (© D. Cavagnaro, Visuals Unlimited). (u) Flat mushroom-like outgrowths are developed from the underside of oak leaves by the gall wasp *Cecidomyia paculum* (© John D. Cunningham, Visuals Unlimited).

7.3.2   Changes in host behavior caused by their inhabitants

Organisms that use a host as a habitat may manipulate its behavior to their advantage. The human threadworm *Enterobius vermicularis* lives mainly in the cecum, but at nights

Figure 7.12
The effect of infection by the malaria parasite (*Plasmodium falciparum*) on the probability that the mosquito (*Anopheles gambiae*) will have more than one meal. The bars show the percentage of multiple meals in each size group. The pink bars represent the uninfected and the blue bars the infected mosquitoes. (From Koella et al., 1998.)

the adult females migrate to the anus and lay eggs on the skin that surrounds it. This causes the host to itch and scratch: as a result the eggs contaminate the hands and transmit easily to reinfect through the mouth; infection spreads rapidly in families. Plant parasites such as late blight of potatoes cause the leaves to curl in ways that expose fungal spores to dispersal. *Anopheles* mosquitoes infected with the malarial parasite *Plasmodium* take more blood and take it from more individuals than uninfected mosquitoes—thus increasing the chances of parasite transmission. In a field sample of infected mosquitoes, 22 percent contained blood from at least two people compared with only 10 percent of uninfected mosquitoes (Figure 7.12). There can be few more effective ways in which *Plasmodium* could change its host's behavior to its own reproductive ends!

The gordian worm (*Gordius dimorphus*) infects the aquatic larvae of insects such as mayflies, caddis flies, stoneflies, and alderflies. The parasite remains in the aquatic host until it metamorphoses to an adult and leaves the water. The adult insect may then be eaten by large carnivorous insects such as the praying mantis or the New Zealand weta, and the parasite develops further at the expense of their internal organs. The parasite grows until it occupies the entire abdomen, coiled up tightly like a watch spring. The infected host now develops a desperate craving for water and jumps in when it finds it. The parasite emerges immediately as a long black worm, and the life cycle starts again with the infection of aquatic insect larvae. The craving for water that the parasite induces in its terrestrial host is so extreme that mantises rescued from the water will quickly make their way back to it and throw themselves in again.

Many of the organisms that use hosts as habitats use the host's sexual activity to hitch a ride from host to host—most obviously, perhaps, the bacteria that cause venereal diseases of humans (gonorrhea, syphilis). Similarly, though, many viruses of trees are transmitted "sexually" from tree to tree by pollen, and in several species of plant (e.g., the loose smuts of cereals, *Ustilago tritici*, and anther smut of clover, *Botrytis anthophila*) the fungal parasite replaces the pollen or seeds of the host with its own spores.

microbial inhabitants may twist host behavior to their own advantage

7.3.3   Immune and other defensive responses of the host

When an organism reacts to the presence of another in its body it is recognizing and distinguishing between what is "self" and "not self". In invertebrates, populations of phagocytic cells are responsible for most of a host's response to invaders, even to inanimate particles. Phagocytes may engulf and digest small alien bodies and encapsulate and isolate larger ones. In vertebrates there is also a phagocytic response to material that is not self, but their armory of defense also includes a much more elaborate process: the immune response (Figure 7.13).

the immune responses of vertebrates allow hosts to eliminate populations of bacterial and viral parasites

The ability of vertebrates to recognize and reject tissue grafts from unrelated members of their own species is just one example of the immune response in action (Figure 7.13). In terms of the ecology of parasites, an immune response has two vital features. First, it may enable a host to eliminate a parasite and recover from disease. Second, it can give a once-infected host (= habitat) a "memory" that changes its behavior if the parasite (= colonist) strikes again; the host has become immune to reinfection. In mammals, the transmission of immunoglobulins to the offspring in the mother's milk can sometimes even extend protection to the early life of the next generation. The response of vertebrates to infection therefore protects the individual host and increases its potential for survival to reproduction. The response of invertebrates gives poorer protection to the individuals and, when populations recover after disease, this depends more on the high reproductive potential of the survivors than on the recovery of those that had been infected.

Figure 7.13
A vertebrate host is a reactive habitat in which the immune response of the host converts it from being habitable into being uninhabitable. This sort of reaction distinguishes living organisms as habitats from the sorts of environment more normally considered by ecologists. The mechanisms mediating resistance to infection can be divided into "natural" or "nonspecific" (left) and "adaptive" (right), each composed of both cellular elements (lower half) and humoral elements (i.e., free in the serum or body fluids) (upper half). The adaptive response begins when the immune system is stimulated by an antigen that is taken up and processed by a macrophage (MAC). The antigen is a part of the parasite, such as a surface molecule. The processed antigen is presented to T and B lymphocytes. T lymphocytes respond by stimulating various clones of cells, some of which are cytotoxic (NK, natural killer cells), as others stimulate B lymphocytes to produce antibodies. The parasite that bears the antigen can now be attacked in a variety of ways. PMN, polymorphonuclear neutrophil. (After Playfair, 1996.)

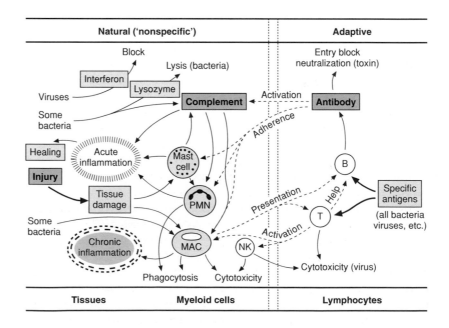

For many viral and bacterial infections of higher animals, the colonization of the host is a brief and transient episode. The parasites multiply within the host and elicit a strong immunological response. Assuming the host survives, the parasite population is eliminated and the host acquires lasting, often lifelong, immunity to reinfection. By contrast, the immune responses elicited by many of the worms and protozoan parasites, and indeed, for example, by herpesviruses, tend to be relatively ineffective. The infections themselves, therefore, tend to be persistent, and hosts are subject to continual reinfection. For the dog or cat lover this means that treatment for fleas and intestinal worms must continually be repeated through the life of the pet, unlike immunization to some viral infections, which may be effective for life.

The modular structure of plants, the presence of cell walls, and the absence of a true circulating system (such as blood or lymph) make any form of immunological response an inefficient protection. There is no migratory population of phagocytes in plants that can be mobilized to deal with invaders. Many plants, however, respond even to the earliest stages of infection by "hypersensitive" reactions. The infected cells die, and they and the immediately surrounding cells produce *phytoalexins*— antimicrobial compounds that accumulate to inhibitory levels in the locality of infection. There is also increasing evidence of signaling mechanisms from one part of a plant to another. These may allow one infected part to induce the formation of defenses in the rest of the plant body, and there is some evidence that an infected plant (for example, by aphids) may use gaseous signals to induce defenses in other members of the population. This is an exciting area of biology with many unanswered questions.

transient and persistent infections

plants have no immune system that distinguishes self from not self

7.4   Distribution and Regulation of Parasites and Mutualists within Hosts and their Populations

The distribution of parasites within populations of hosts is rarely random. It is usually the case that many hosts harbor few or no parasites, and a few hosts harbor many—the distributions are usually aggregated or clumped (Figure 7.14). Some of the aggregation results when microparasites (see Box 7.1) multiply within randomly infected hosts and quickly generate large differences between the zero populations in uninfected hosts and the dense populations in infected hosts. However, when dispersal is weak or slow, local aggregations of infected hosts can occur simply because most transmission occurs between immediate neighbors rather than randomly within a population. Aggregations also arise if individual hosts vary in their susceptibility to infection (whether as a result of genetic, behavioral, or environmental factors).

parasites are usually aggregated

A widely used statistic to describe the distribution of parasites on a population of hosts is the proportion of the host population that is infected. This is termed the *prevalence* of the parasite. It is a statistic with obvious ecological meaning—it is the proportion of habitable patches in an environment that are actually inhabited. The number of parasites in or on a particular host is referred to as the *intensity* of infection. This is comparable to the local density of birds on an island or insects in a patch of vegetation. The *mean intensity* of infection is then the mean number of parasites per host in a population, including those hosts that are not infected, just as the mean

prevalence, intensity, and mean intensity of infection

Figure 7.14
Examples of aggregated distributions of parasite numbers per host. (a) Crayfish, *Orconectes rusticus*, infected with the flatworm *Paragonimus kellicotti*. The distribution is significantly different from Poisson or random ($X^2 = 723$; $P < 0.001$) but conforms well with a "negative binomial" that is good at describing aggregated distributions: $X^2 = 12$; $P \approx 0.4$. (After Stromberg et al., 1978; Shaw & Dobson, 1995.) (b) Distribution of *Onchocerca volvulus* worms, that cause onchocerciasis or "river blindness" in human Yanomami communities in southern Venezuela. Again the distributions, plotted as cumulative frequencies (green lines), conform well to a negative binomial distribution (blue lines), whether the typical intensity of infection is low (hypoendemic), moderate (mesoendemic), or high (hyperendemic). (After Vivas-Martinez et al., 2000. Reprinted with the permission of Cambridge University Press.)

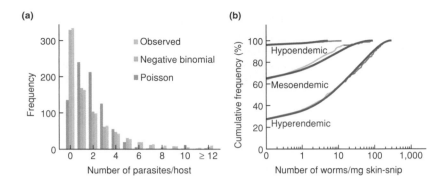

populations of organisms that inhabit a host are regulated

density of fish in a series of ponds includes those ponds that contain no fish. In locally aggregated populations, however, the mean density of parasites (mean intensity of infection) may have rather little useful meaning. In a population of humans in which only one person is infected with anthrax, the mean intensity of *Bacillus anthracis* is a particularly useless piece of information.

Almost all parasites and symbiotic mutualists are smaller than their hosts and able to grow much faster. Yet their growth is constrained within the hosts, and this implies that there are density-dependent processes (crowding effects—see Chapters 3 and 5) at work. It is technically extremely difficult to study the population dynamics of microparasites that multiply within the host cells (imagine how one might try to follow the growth of a population of measles virus particles in the cells of a child). However, it is quite practicable to study the effects of population size on the behavior of some macroparasites (see Box 7.1) and even to control levels of infection experimentally. The results of experiments and observations like those shown in Figures 7.15 and 8.19, for example (and see Figure 3.22), are remarkably similar to the density-dependent relationships found to result from intraspecific competition generally (Chapters 3 and 5).

The number of mutualistic algal cells per host cell in corals and lichens, and of rhizobial bacteria in the cells of legume nodules, must also all be regulated. The algal symbionts of the coral *Stylophora pistillata* can increase 100 times faster than their hosts, but in practice they do not outgrow them. The ways in which such symbiont populations are regulated are obscure, but it is probable that the growth rate of the algae is limited by mineral nutrients and/or carbon—just as in more traditional habitats, density-dependent regulation often involves the effects of resource shortage on birth and death rates.

7.5   Life Cycles and Dispersal

The most complicated of all life cycles are found among the parasites; some of these have already been illustrated. Most parasites are extremely specialized and are able to

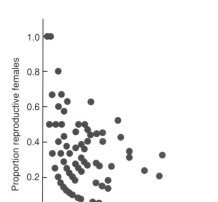

Figure 7.15
Density-dependent responses of macroparasites within their hosts. As the number of adult nematodes, *Anguillicola crassus*, observed at Albertcanal (Belgium) in the swimbladder of eels, *Anguilla anguilla*, increased, the proportion of those that were in reproductive condition decreased. (After Ashworth & Kennedy, 1999.) (See also Figure 8.19.)

infect, grow, and reproduce on one or a very few, usually closely related, species of host. Almost all organisms are totally resistant to almost all parasites! It is therefore all the more remarkable that the life cycles of a number of important parasites require them to spend alternate periods of their lives on two different and usually quite unrelated hosts. Where there is such an alternation, the parasite usually has its sexual phase on one host and most of its multiplication on the other. This means that most genetic recombinations and heritable variations are generated on one host and most of the natural selection of its results occurs on the other.

Complex life cycles such as that of the tapeworm in Figure 7.5 imply a quite extraordinary degree of habitat specialization of the parasite. These specialists, in a lifetime that extends from parent zygote to offspring zygote, must pass in a defined progression from one defined host to another. Similarly, many of the rust fungi (Uredinales) have life cycles that involve spending phases on two quite unrelated species of host plant (e.g., black stem rust, Figure 7.16). The closest parallels in traditional habitats are the life cycles of amphibians, which must pass from an aquatic habitat for the larvae to a terrestrial habitat for the adults, or of butterflies and moths, which must pass from the larval habitat of a leafy host to the flowers frequented by the adults.

Two relationships are often confused when considering complex life cycles. We refer to *alternate hosts* when there is an obligatory alternation of the parasite between one host and another, as in the tapeworm and black stem rust examples. In contrast, we use the phrase *alternative hosts* to indicate that a particular phase of a parasite's life cycle may occur on different host species. *Berberis* and *Mahonia* are examples of alternative hosts in the life of black stem rust, but *Berberis/Mahonia* and wheat are alternate hosts (Figure 7.16). Similarly, at least 150 species of mammal, including humans, can serve as alternative hosts for the trypanosome *Trypanosoma cruzi*, which causes Chagas' disease in the human host, and the parasite itself multiplies in the gut of 50–60 alternative species of hemipteran insects. But the mammal and the bug are alternate hosts: the parasite must pass from mammal to bug to mammal to bug and so on. Note that

alternate and alternative hosts

Figure 7.16
The life cycle of black stem rust of wheat (*Puccinia graminis tritici*). The parasite infects woody shrubs of the genus *Berberis* or of the closely related *Mahonia*. The fungus spreads through the cells and tissues of the host, and, after a sexual process has occurred, the fungus forms pustules, which break out from the lower surface of the leaves and release spores (aeciospores). These are quite unable to infect (colonize) *Berberis* or *Mahonia* and can infect only wheat (*Puccinia graminis*). On the wheat host the fungus produces a new type of spore called a urediospore that can only infect further wheat plants. Cycles of infection and reinfection of wheat plants can repeat indefinitely. Epidemics develop in crops of wheat if climatic conditions are suitable. At the end of the host's growing season a different type of spore is produced (a teliospore), which cannot infect anything but is thick-walled and can remain alive but dormant. Eventually the teliospores germinate, and, at last, sexual fusion occurs, followed almost immediately by meiosis and the formation of yet a further type of spore (the basidiospore), which can infect only *Berberis* or *Mahonia*.

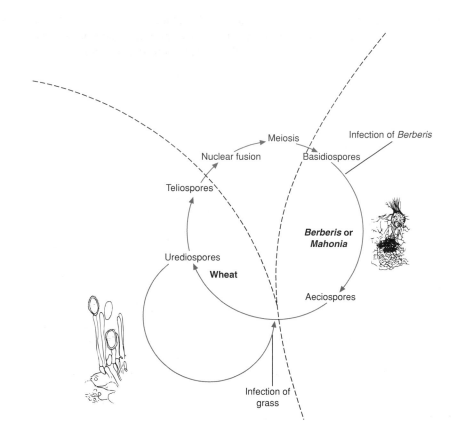

alternate hosts and vectors

alternative hosts are usually closely related (often members of the same genus), whereas alternate hosts are never phylogenetically close.

A second relationship that can be misunderstood is that between alternate hosts and vectors. Many parasites are spread from one host to another by a vector organism, but the parasite need not draw resources from the vector. A vector disperses a parasite, but often simply as a contaminant, for example on its mouthparts. This is the most common route of the passage of plant viruses from their habitat in one plant to another. Mosaic virus of corn is transmitted by a vector, the plant hopper *Peregrinus maidis*, and the vector of maize bushy stunt virus is a leaf hopper (*Dalbulus* spp.). Aphids are particularly effective vectors of plant viruses. They insert their feeding stylet through the tissues of the host plant into the living cells of the plant's phloem transport system, and this places any contaminant virus or bacterium in a position optimal for its spread.

Among animals, probably the most noteworthy groups of vectors are flies, especially mosquitoes. These transmit a number of important human diseases, including malaria and yellow fever. The adult fly takes several blood meals in its short life and transmits infection when consecutive meals are taken from an infectious and then an uninfected, susceptible host (see also Figure 7.12). Ticks are also important vectors. They take only one blood meal in each of their life stages and therefore harbor infections for long periods, often more than a year, while they metamorphose from one stage to the next. Many species of tick are capable of feeding from several alternative host species, and they have recently gained increasing notoriety as the bearers of infections, such as Lyme disease, that normally circulate in wildlife species, because ticks can transmit the disease to humans.

7.6   Coevolution

Parasites and mutualists act as selective forces on the evolution of the plants and animals that they use as hosts. Between hosts and mutualists there is no conflict—hosts gain in fitness if they allow or encourage invasion by mutualists, which also then benefit—but this is clearly not the case for parasites. Hosts benefit if they evolve resistance or tolerance to parasites and pathogens, but parasites gain fitness by penetrating the defenses of their hosts. One result is to lead parasites deeper and deeper into the rut of extreme specialization: as we have already noted, all species of plant and animal are totally resistant to all but a very few species of parasite. The relationship between hosts and the parasites that use them as habitats involves as precise a match between organism and environment as occurs anywhere in nature.

Indeed, the specialization may go further than that between species. Within species of obligate parasites it is common to find a high degree of genetic variation in the virulence of parasites and/or in the resistance or immunity of hosts. Every few years, a new strain of the influenza virus evolves of sufficient virulence and novelty to generate a widespread epidemic and mortality in human populations that had been relatively resistant to previously circulating strains. No strain has been more devastating than the worldwide epidemic (*pandemic*) of Spanish flu that followed World War I in 1918–19 and killed 20 million people—many more than died in the war itself. Human diseases can also provide examples of variation in host resistance. When the American Indians of the Canadian Plains were forcibly settled onto reservations in the 1880s, their death rate due to tuberculosis (TB) initially exploded but then gradually declined (Figure 7.17). Environmental factors (inadequate diet, overcrowding, spiritual demoralization) undoubtedly played some part in this, but variation in resistance is also likely to have been significant. The mortality rate among the Indians was often 20 times that of the surrounding white population living in similar conditions but who had been previously exposed to TB. Some Indian families had a particularly low mortality rate in the 1880s epidemic, and many of the survivors in the 1930s were descendants of those families (Ferguson, 1933; Dobson & Carper, 1996).

parasite and host variation: influenza and TB

In some cases, especially in plant diseases—flax and flax rust, powdery mildew of barley, stem rust of wheat—for every gene controlling resistance in the host, there is a corresponding gene in the parasite controlling pathogenicity (a precise gene-for-gene relationship). Although this relatively simple system is far from being a general, invariable rule, it provides a useful framework within which to think about the perpetual struggle of plants in nature to keep one step ahead of their pathogens.

gene-for-gene relationships and plant breeding

A most dramatic example of the ecological and evolutionary interaction of a parasite and its host involves the rabbit and the myxoma virus, which causes the disease myxomatosis. The disease originated in the South American jungle rabbit *Sylvilagus brasiliensis*, in which it is a mild disease that only rarely kills the host. It is usually fatal, however, when it infects the European rabbit *Oryctolagus cuniculus*. In one of the greatest

myxomatosis: coevolution in action

Figure 7.17
The mortality rate due to tuberculosis in
three generations of Canadian Plains
American Indians after their forced
settlement onto reservations. (After
Ferguson, 1933; Dobson & Carper, 1996.)

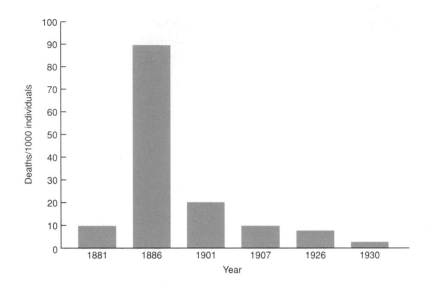

Figure 7.17
The mortality rate due to tuberculosis in three generations of Canadian Plains American Indians after their forced settlement onto reservations. (After Ferguson, 1933; Dobson & Carper, 1996.)

examples of the biological control of a pest, the myxoma virus was introduced into Australia in the 1950s to control the European rabbit, which had become a pest of grazing lands. The disease spread rapidly in 1950–1, and rabbit populations were greatly reduced—by more than 90 percent in some places. A little later, the virus was introduced to England and France, where it also resulted in huge reductions in the rabbit populations. The evolutionary changes that then occurred in Australia were followed in detail by Fenner and his associates, who had the brilliant foresight to establish baseline genetic strains of both rabbits and virus. They used these to measure subsequent changes in the virulence of the virus and the resistance of the host as they evolved in the field.

When the disease was first introduced to Australia it killed more than 99 percent of infected rabbits. This *case mortality* rate fell to 90 percent within a year and then declined further (Fenner & Ratcliffe, 1965). The virulence of the virus was graded according to the survival time and case mortality rate of control rabbits. The original, highly virulent virus (1950–1) was grade I, which killed >99 percent of infected laboratory rabbits. Already by 1952 most of the virus isolates from the field were grade III and IV (Figure 7.18). At the same time the rabbit population in the field was increasing in resistance. When injected with a standard grade III strain of the virus, field samples of rabbits in 1950–1 had a case mortality of nearly 90 percent, which had declined to less than 30 percent only 8 years later (Marshall & Douglas, 1961).

parasites do not evolve to preserve their hosts

This evolution of resistance in the European rabbit is easy to understand. The case of the virus, however, is more subtle. The contrast between the virulence of the myxoma virus in the European rabbit and its lack of virulence in the American host with which it had coevolved, combined with the attenuation of its virulence in Australia and Europe after its introduction, fit a commonly held view: namely, that parasites evolve toward becoming benign to their hosts in order to prevent the parasite eliminating its host and thus eliminating its habitat. This view is quite wrong. The parasites favored by natural

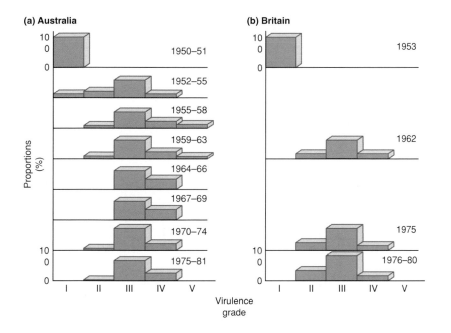

Figure 7.18
(a) The percentages in which various virulence grades (I–V) of myxoma virus have been found in wild populations of rabbits in Australia at different times from 1950 to 1981. Grade I is the most virulent and grade V the least virulent. (b) Similar data for wild populations of rabbits in Britain from 1953 to 1980. (After May & Anderson, 1983; from data collected by Fenner, 1983.)

selection are those with the greatest fitness (broadly, the greatest reproductive rate). Sometimes this is achieved through a decline in virulence, but sometimes it is not.

In the myxoma virus, an initial decline in virulence was indeed favored—but further declines were not. The myxoma virus is blood-borne and is transmitted from host to host by blood-feeding insect vectors. In Australia in the first 20 years after its introduction, the main vectors were mosquitoes (especially *Anopheles annulipes*), which feed only on live hosts. The problem with grade I and II viruses is that they kill the host so quickly that there is only a very short time in which the mosquito can transmit them. Effective transmission may be possible at very high host densities, but as soon as densities decline, it is not. Hence, there was selection against grades I and II and in favor of less virulent grades, giving rise to longer periods of host infectiousness. At the other end of the virulence scale, however, the mosquitoes are unlikely to transmit grade V of the virus because it produces very few infective particles in the host skin that could contaminate the vectors' mouthparts. The situation was complicated in the late 1960s when an alternative vector of the disease, the rabbit flea *Spilopsyllus cuniculi* (the main vector in England), was introduced to Australia. There is some evidence that more virulent strains of the virus may be favored when the flea is the main vector.

Overall, however, there has been selection in the rabbit–myxomatosis system not for decreased virulence but for *increased transmissibility* (and hence increased fitness)—which happens in this system to be maximized at intermediate grades of virulence. Many parasites of insects rely on killing their host for effective transmission. In these, very high virulence is favored. In yet other cases natural selection acting on parasites has clearly favored very low virulence: for example, the human herpes simplex virus may do very little tangible harm to its host but effectively gives it lifelong infectiousness. These differences reflect differences in the underlying host–parasite ecologies,

Box 7.2   *Historical Landmarks*

Some Contrasts between Parasites and Mutualists

In comparison with species of parasites, there are remarkably few species of symbiotic mutualists, because they are much less specialized to particular hosts. The greatest diversity is found in the lichens (each lichen is a symbiotic association between a fungus and an alga), but although there are at least 13,500 species of lichen, their algal symbionts are represented by only 30–40 species of green algae and 12 or so species of blue–green algae. More than half the known species of lichen, in fact, carry just one genus of algal symbiont, *Trebouxia*. Similarly, algal symbionts of marine invertebrates are drawn almost entirely from just one genus of dinoflagellate algae (*Symbiodinium*) and one genus of cyanobacteria (*Prochloron*), which inhabits sea squirts. Just three genera of rhizobial bacteria account for all the nitrogen-fixing symbionts of legumes.

Symbiotic mutualists have rather simple life cycles and, unlike parasites, none of them requires an alternation between hosts: there are no alternate hosts but lots of alternative hosts.

The lives of parasites are dominated by dispersal. Their continued existence depends typically on transmission and infection from one unique type of host to another of the same sort or one that is equally unique. But symbiotic mutualists lack any structures or behavior specialized for dispersal from host to host and vectors are not involved. Legumes, for example, acquire their rhizobia from free-living forms in the soil. Mollusks and some other marine invertebrates acquire their algal symbionts from free-living cells in the sea. When animal host species are modular and grow by budding or branching (corals, anemones, and so on) algal symbionts pass directly to the tissues of the buds and branches. The intracellular symbiotic bacteria of aphids, *Buchnera*, are vertically transmitted through the mother to her eggs. They have become so completely assimilated into the life of their hosts that they appear to have lost any vestige of a free-living or dispersive stage or even any independent life of their own.

A remarkable feature of the lives of mutualists living in intimate association is the almost complete suppression of sex. This contrasts with the lives of parasites, and, since sexual reproduction generates so many new variants, the lifeblood of evolution, this presumably reflects the continual evolutionary battle between parasites and their hosts. By contrast, there is little or no conflict between mutualists and their hosts.

but what the examples have in common is that there has been evolution toward increased parasite fitness.

hosts as patches—but special patches

In this chapter we have focused on organisms as the habitats of both parasites and mutualists. Some of the contrasts between them are summarized in Box 7.2. A susceptible host is a habitable patch (or an inhabitable "island") in the environment. The dynamics of populations of parasites and pathogens depend on the colonization of these patches and on the rate at which they become locally extinct. These in turn depend on the distance apart and longevity of the patches and the dispersibility of the parasite. We discuss the broader ecological issues of patch and island population behavior and metapopulation dynamics in Chapters 9 and 10. But we need to remember that parasites represent a special case: motile animals are motile habitats, moving patches, migrating islands. They pair and mate, they may have social behavior, and they can run away. This adds a delightful subtlety to thinking about hosts as habitats.

Summary

Symbiosis, parasitism, and mutualism

An intimate association between individuals of different species, in which one lives on or in the other, is known as a *symbiosis*. A *parasite* is an organism that lives its life intimately associated with one or a very few individuals of another species, its host, deriving resources from its host and doing its host harm, but not predictably killing its host in the short term. A *mutualism* is an association between species (usually a pair of species) that benefits both. *Commensals* are neither harmful nor beneficial to their host. Parasites contribute overwhelmingly to the diversity of life on Earth, but mutualistic associations often dominate the production of biomass.

Life on the surface of another organism

The bodies of many organisms provide anchorage for others and give them access to conditions and resources that are otherwise unavailable. These ecological engineers create, modify, maintain, or create habitats for others.

Plant roots shed root cap cells and release organic materials to the adjacent soil. This *rhizosphere* carries its own abundant microbial flora. There is also often a distinct *phyllosphere* of species on leaf surfaces.

Fleas and lice are examples of *ectoparasites* on the bodies of larger animals. Organisms that live on the surface of their hosts are usually well placed for the colonization of other hosts, but there are costs in exposure to ambient hazards and usually less intimate contact with the host.

Inhabitants of the body cavity

The alimentary canal is a resource-rich, protected, and regulated habitat. Many parasites live there, competing with the host for its digestive products. Most have complex life histories that carry infection from

the feces of the host and disperse it widely among others in the host population. Immensely diverse communities of mutualistic microorganisms inhabit the alimentary canal of herbivorous animals, playing a most crucial role in the digestion of cellulose.

Inhabitants of host tissues and cells

The combined structure of root and fungus is called a *mycorrhiza*. The fungus obtains energy-rich carbon resources from the host, and phosphate and other nutrients pass from the fungus to the plant.

Many parasites that live within the cells of plants have dormant, long-lived spores that germinate in the presence of a host. Many of the parasites of animals that have their habitat within the cells of the host have no dormant persistent stage but depend absolutely on alternating phases of their life cycle between two different species of host.

Nitrogen-fixing mutualistic bacteria of the genus *Rhizobium* infect the roots of leguminous plants (though there are other nitrogen-fixing mutualisms). They contribute fixed nitrogen to the host plant, which is then independent of supplies of nitrogen from the soil. Much of the carbon fixed in photosynthesis by marine intracellular algal symbionts is released to hosts that provide their habitat.

Some "parasites" kill their host and then continue to grow and reproduce on the dead body, pre-empting those that have to wait for an organism to die. Most parasites, however, are obligatorily parasitic. If their host dies, they cease to grow on it. Other parasites (and mutualists) take control of the local growth of their hosts and force them to create habitable microenvironments. Still others manipulate host behavior.

In invertebrates, phagocytic cells are responsible for most of a host's response to invaders. In vertebrates there is also a much more elaborate process: the immune response. This may enable a host to eliminate a parasite and recover from disease and can also make a once-infected host immune to reinfection. There are no migratory phagocytes in plants, but many respond to infection by producing phytoalexins that are strictly localized and nonspecific in their action.

Parasites are usually aggregated within populations of their hosts. The *prevalence* of a parasite is the proportion of the host population that is infected. The number of parasites in or on a particular host is the *intensity* of infection. The *mean intensity* of infection is the mean number of parasites per host in a population. Almost all populations of parasites and symbiotic mutualists are constrained within their hosts by density-dependent processes (crowding effects).

The life cycles of many parasites require them to spend alternate periods of their lives on two different and usually quite unrelated "alternate" hosts. Many parasites are spread from one host to another by a vector organism, but the parasite need not draw resources from the vector.

All species of plants and animals are totally resistant to all but a very few species of parasites. In some cases, especially for plant diseases, a *gene-for-gene* relationship has been suggested: for every gene controlling host resistance, there is a corresponding gene in the parasite controlling pathogenicity.

Myxomatosis illustrates coevolution of parasite and host, but also indicates that parasites do not evolve to preserve their hosts. Rather, the parasites favored by natural selection are those with the greatest fitness (broadly, the greatest reproductive rate). Sometimes this is achieved through a decline in virulence, but sometimes it is not.

Review Questions

1 Define and give examples of symbiosis, parasitism, mutualism, and commensalism.

2 Contrast the communities of parasites and mutualists inhabiting the vertebrate alimentary canal.

3 What are mycorrhizae and what is their significance?

4 ▲ Discuss the following propositions: "Most herbivores are not really herbivores but consumers of the by-products of the mutualists living in their gut," and "Most gut parasites are not really parasites but competitors with their hosts for food that the host has captured."

5 ▲ Provide a careful interpretation of Figure 7.19.

6 Account for the fact that the blowfly, *Lucilia*, often kills its host, whereas most parasites do not.

7 Give examples of parasites that change (1) the morphological characteristics and (2) the behavior of their host. Can these changes be interpreted as defenses by the host or opportunities gained by the parasite?

8 ▲ What do we really mean when we say that a parasite harms its host?

9 Compare and contrast the defensive responses to parasites of higher vertebrates, invertebrates, and plants.

10 Account for the decline in virulence of the myxomatosis virus in European rabbits after its initial introductions in Australia and Europe.

▲ = Challenge Question

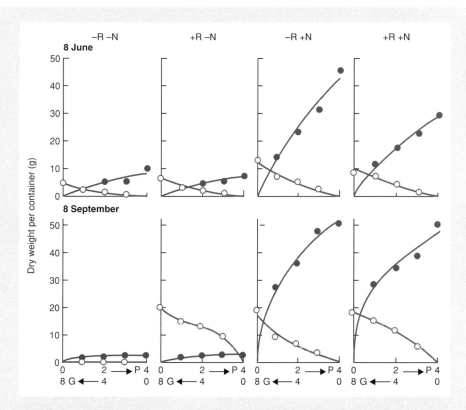

Figure 7.19
Soybeans (*Glycine max*, G — a legume that can form a mutualistic association with *Rhizobium*) and a grass (*Paspalum*, P) are grown alone and in mixtures with and without nitrogen fertilizer and with and without inoculation with nitrogen-fixing *Rhizobium*. The plants were grown in pots containing zero to four plants of the grass and zero to eight plants of *Glycine*. The horizontal scale on each figure shows the proportions of plants of the two species in each container. −R −N, no *Rhizobium*, no fertilizer; +R −N, inoculated with *Rhizobium* but no fertilizer; −R +N, no *Rhizobium* but nitrate fertilizer was applied; +R +N, inoculated with *Rhizobium* and nitrate fertilizer supplied. (After de Wit et al., 1966.)

Chapter 8

Predation, Grazing, and Disease

Every living organism is either a consumer of other living organisms, or is consumed by other living organisms, or—in the case of most animals—is both. We cannot hope to understand the structure and dynamics of ecological populations and communities until we understand the links between consumers and their prey.

Chapter Contents

8.1 Introduction

8.2 Prey fitness and abundance

8.3 The subtleties of predation

8.4 Predator behavior—foraging and transmission

8.5 Population dynamics of predation

8.6 Predation and community structure

Key Concepts

In this chapter you will

- distinguish the similarities and differences among "true predators", grazers, and parasites
- understand the subtleties of predation, including the capacity of prey to compensate
- appreciate the value of the optimal foraging approach for analyzing predator choices
- recognize the underlying tendency of populations of predators and prey to cycle and the "damping" effect of crowding and patchy distributions
- understand the consequences of predation for community composition

8.1   Introduction

Ask most people to name a predator and they are almost certain to say something like "lion", "tiger", or "grizzly bear"—something big, potentially ferocious, instantly lethal. However, from an ecological point of view, "predator" is a term that can be used to encompass a whole variety of organisms that obtain their resources by consuming other living organisms. They are not all large, aggressive, or instantly lethal—and they need not even be animals. By considering these consumers together, we can more readily understand the part each plays in determining the structure and dynamics of ecological systems.

predator: a term extending beyond the obvious examples

In its broadest sense, then, a predator may be defined as any organism which consumes all or part of another living organism (its "prey" or "host") thereby benefitting itself, but, under at least some circumstances, reducing the growth, fecundity, or survival of the prey. Notice that this definition extends beyond the likes of lions and tigers by including those that consume all or *part* of their prey, and those that merely *reduce* their prey's growth, fecundity, or survival. Within this broad definition, however, three main types of predator can be distinguished.

"true" predators, grazers, and parasites

1 "True" predators:
   - invariably kill their prey and do so more or less immediately after attacking them;
   - and consume several or many prey items in the course of their life.

   True predators therefore include lions, tigers, and grizzly bears, but also spiders, baleen whales that filter plankton from the sea, zooplanktonic animals within that community that consume the phytoplankton, birds, rodents, and ants that eat seeds (each one an individual organism), "cutworm" caterpillars that eat through the roots of seedlings and young plants and invariably kill them, carnivorous plants, and so on.

2 Grazers:
   - also attack several or many prey items in the course of their life;
   - but consume only part of each prey item;
   - and do not usually kill their prey, especially in the short term.

   Grazers therefore include cattle, sheep, and locusts, mobile, nibbling caterpillars that move from plant to plant, but also, for example, blood-sucking leeches that take a small, relatively insignificant blood meal from several vertebrate prey over the course of their life.

3 Parasites:
   - also consume only part of each prey item (usually called their "host");
   - and also do not usually kill their prey, especially in the short term;
   - but attack one or very few prey items in the course of their life, with which they therefore often form a relatively intimate association.

   Parasites therefore include some obvious examples: animal parasites and pathogens such as tapeworms and the tuberculosis bacterium, plant pathogens like tobacco mosaic virus, parasitic plants like mistletoes, and the tiny wasps that form "galls" on oak leaves. But aphids that extract sap from one or a very few plants with which they enter into an intimate association, and even caterpillars that spend their whole life on one host plant, are also, in effect, parasites. A useful distinction between microparasites and macroparasites was drawn in Box 7.1.

A parasitoid wasp, which uses its long ovipositor to insert its eggs into the larvae of other insects, where they develop by consuming their host.

On the other hand, these distinctions between "true" predators, grazers, and parasites, as with most categorizations of the living world, have been drawn in large part for convenience—certainly not because every organism fits neatly into one and only one category. We could, for example, have included a fourth class, the parasitoids: little known to nonbiologists, but extensively studied by ecologists (and immensely important in the biological control of insect pests—see Chapter 12). Parasitoids are flies and wasps whose larvae consume their insect-larva host from within, having been laid there as an egg by their mother. Parasitoids therefore straddle the "parasite" and "true predator" categories (only one host individual, which it always kills), fitting neatly into neither and confirming the impossibility of constructing clear boundaries.

There is, moreover, no truly satisfactory term to describe all the "animal consumers of living organisms" to be discussed in this chapter. Detritivores and plants are also "consumers" (of dead organisms, or of water, radiation, and so on); whilst the term "predator" inevitably tends to suggest a "true" predator even after we have defined it to encompass grazers and parasites too. But neither is it very satisfactory to be continually using the qualifier "true" when discussing conventional predators like big cats or ladybugs. Thus, throughout this chapter, "predator" will often be used as a shorthand term to encompass true predators, grazers, and parasites, when general points are being made; but it will also be used to refer to predators in the more conventional sense, when it is obvious that this is what is being done.

8.2   Prey Fitness and Abundance

The fundamental similarity between predators, grazers, and parasites is that each, in obtaining the resources it needs from its prey, reduces either the fecundity or the chances of survival of individual prey, and therefore may reduce prey abundance. The effect of true predators on the survival of individual prey hardly needs illustrating—they die. That the effects of grazers and parasites can be equally profound, if rather more subtle, is illustrated by the following two examples.

When the sand-dune willow, *Salix cordata*, was grazed by a flea beetle in two separate years—1990 and 1991—the reduction in the growth rate of the willow was marked in both years, but the consequences were rather different (Figure 8.1). Only in 1991 were the plants also subject to a severe shortage of water—thus it was only in 1991 that the reduced growth rate was translated into plant mortality: 80 percent of the plants died in the high grazing treatment, 40 percent died in the low grazing treatment, but none of the control, ungrazed plants died.

The pied flycatcher is a bird that migrates early each summer from tropical West Africa to Finland (and elsewhere in northern Europe) to breed. Males that arrive relatively early are particularly successful at finding mates. Late arrival therefore has a serious detrimental effect on the expected "fecundity" of a male: the number of offspring that it can expect to father. Significantly, the later arrivals are disproportionately infected with the blood parasite *Trypanosoma* (Figure 8.2). Infection with the parasite therefore has a profound effect on the reproductive output of individual birds.

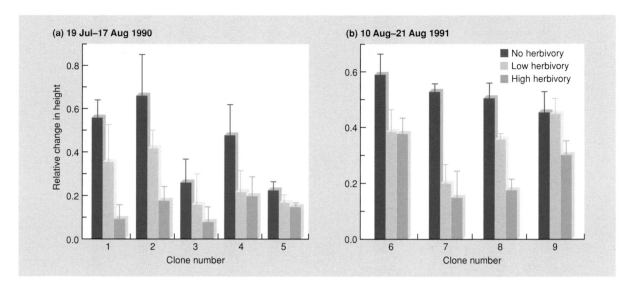

Figure 8.1
Relative growth rates (changes in height, with standard errors) of a number of different clones of the sand-dune willow, *Salix cordata*, (a) in 1990 and (b) in 1991, subjected either to no herbivory, low herbivory (four flea beetles per plant), or high herbivory (eight beetles per plant). (After Bach, 1994.)

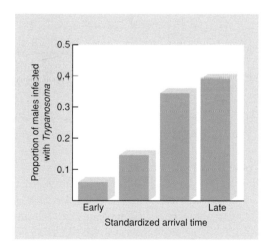

Figure 8.2
The proportion of males of male pied flycatchers (*Ficedula hypoleuca*) infected with *Trypanosoma* amongst groups of migrants arriving in Finland at different times. (After Rätti et al., 1993.)

It is not so straightforward, though, to demonstrate that reductions in the survival or fecundity of individual prey translate into reductions in prey abundance—we need to be able to compare prey populations in the presence and the absence of predators. As so often in ecology, we cannot rely simply on observation; we need experiments—either ones we set up ourselves, or natural experiments set up for us by nature.

For example, Figure 8.3 contrasts the dynamics of laboratory populations of an important pest, the Indian meal moth, with and without a parasitoid wasp, *Venturia*

predators can reduce prey abundance—but do not necessarily do so

Figure 8.3
Long-term population dynamics in
laboratory population cages of a host
(*Plodia interpunctella*), with and without its
parasitoid (*Venturia canescens*). (a) Host
and parasitoid, and (b) the host alone.
(After Begon et al., 1995.)

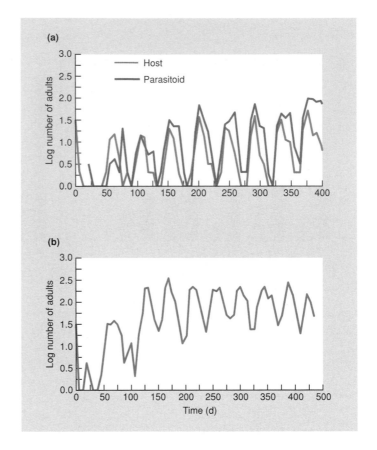

canescens. Ignoring the rather obvious regular fluctuations (cycles) in both moth and wasp, it is apparent that the wasp reduced moth abundance to less than one-tenth of what it would otherwise be (notice the logarithmic scale in the figure).

A particularly graphic example of the impact parasites can have is provided by the story of the invasion of Lake Moon Darra in North Queensland (Australia) by *Salvinia molesta*, a water fern that originated in Brazil. In 1978, the lake carried an infestation of 50,000 metric tons fresh weight of the fern. Amongst possible control agents collected from *Salvinia*'s native range in Brazil, the black long-snouted weevil (*Cyrtobagous* spp.) was known to graze only on *Salvinia*. On June 3, 1980, 1,500 adults were released at an inlet to the lake and a further release was made on January 20, 1981. By April 18, 1981, throughout the lake the *Salvinia* was dying—at this time there was an estimated total population of one billion beetles. By August 1981, less than 1 tonne of *Salvinia* remained. This may have been the most rapid success of any attempted biological control of one organism by the introduction of another. It was a "controlled" experiment in that other lakes in the region continued to bear large populations of *Salvinia*.

All sorts of predators can cause reductions in the abundance of their prey. We shall see as this chapter develops, however, that they do not *necessarily* do so.

8.3   The Subtleties of Predation

There is much to be gained by stressing the similarities between different types of predators. On the other hand, it would be wrong to make this an excuse for over-simplification (there *are* important differences between true predators, grazers, and parasites), or to give the impression that all acts of predation are simply a question of "prey dies—predator takes one step closer to the production of its next offspring".

8.3.1   Interactions with other factors

Grazers and parasites, in particular, often exert their harm not by killing their prey immediately like true predators, but by making the prey more vulnerable to some other form of mortality. One especially important reason for grazing herbivores having a more drastic effect than is initially apparent is the interaction between grazing and plant competition. This can be seen in the interaction between a chrysomelid beetle, *Gastrophysa viridula*, and two species of dock plant, *Rumex obtusifolius* and *R. crispus*. *R. crispus* is only moderately affected by competition from *R. obtusifolius* and is little affected by light beetle grazing. Indeed, the beetle feeds preferentially on *R. obtusifolius*. Light grazing and competition together, however, have a considerable impact on *R. crispus* (Figure 8.4).

Parasites, too, may disturb competitive relationships. One example, from a natural population, comes from a study of two *Anolis* lizards that live on the Caribbean island of St. Maarten. *A. gingivinus* is the stronger competitor and appears to exclude *A. wattsi* from most of the island. However, the malarial parasite *Plasmodium azurophilum* very commonly affects *A. gingivinus* but rarely affects *A. wattsi*. Wherever the parasite infects *A. gingivinus*, *A. wattsi* is present—but wherever the parasite is absent, even in very local communities, only *A. gingivinus* occurs (Schall, 1992).

grazers and parasites may make prey more vulnerable to other forms of mortality

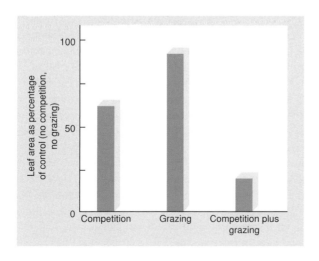

Figure 8.4
The effects of grazing by the beetle *Gastrophysa viridula* on the dock plant *Rumex crispus* are increased significantly when combined with competition from another dock, *R. obtusifolius*. Plant performance is measured as leaf area, relative to control plants (no grazing, no interspecific competition). (After Bentley & Whittaker, 1979.)

Infection may also make hosts more susceptible to predation. For example, post-mortem examination of red grouse (*Lagopus lagopus scoticus*) showed that birds killed by predators in the spring and summer carried significantly greater burdens of the gut nematode parasite *Trichostrongylus tenuis* than the birds that remained in the fall (Hudson et al., 1992).

8.3.2 Compensation and defense by individual prey

compensatory plant responses

The effects of parasites and grazers, however, are not always more profound than they first seem. They are often *less* profound because, for example, individual plants can compensate in a variety of ways for the effects of herbivory (Strauss & Agrawal, 1999). The removal of leaves from a plant may decrease the shading of other leaves and thereby increase their rate of photosynthesis. Or, in the immediate aftermath of an attack from a herbivore, many plants compensate by utilizing reserves stored in a variety of tissues and organs. Herbivory frequently alters the distribution of newly photosynthesized material within the plant, the general rule being apparently that a balanced root : shoot ratio is maintained. When shoots are defoliated, an increased fraction of net production is channeled to the shoots themselves; and when roots are destroyed, the switch is towards the roots. Often, there is compensatory regrowth of defoliated plants when buds that would otherwise remain dormant are stimulated to develop. There is also, commonly, a reduced subsequent death rate of surviving plant parts.

An example of compensation to herbivory is provided by field gentians (*Gentianella campestris*). When herbivory on this biennial plant is simulated by clipping to remove half its biomass (Figure 8.5a), subsequent production of fruits is increased (Figure 8.5b). The outcome depends on the timing of clipping—fruit production was much increased over controls if clipping occurred between July 12 and 20, but if clipping occurred later than this fruit production was less in clipped plants than in unclipped controls. The period when the plants show compensation coincides with the time when damage by herbivores normally occurs.

defensive plant responses

Apart from compensating for the attacks of grazers, plants may respond by initiating or increasing their production of defensive structures or chemicals. For example, a few weeks of grazing on the brown seaweed *Ascophyllum nodosum* by snails (*Littorina obtusata*) induces substantially increased concentrations of phlorotannins (Figure 8.6a), which reduce further snail grazing (Figure 8.6b). In this case, simple clipping of the plants did not have the same effect as the herbivore. Indeed, grazing by another herbivore, the isopod *Idotea granulosa*, failed to induce the chemical defense. The snails can stay and feed on the same plant for long time periods (the isopods are much more mobile), so that induced responses that take time to develop can still be effective in reducing damage.

In at least some cases—though it is not always easy to prove—the plants' responses do seem to be genuinely harmful to the herbivores. When larch trees were defoliated by the larch budmoth, *Zeiraphera diniana*, for example, the survival and adult fecundity of the moths were reduced throughout the succeeding 4–5 years as a combined result of delayed leaf production, tougher leaves, higher fiber and resin concentration, and lower nitrogen levels (Baltensweiler et al., 1977).

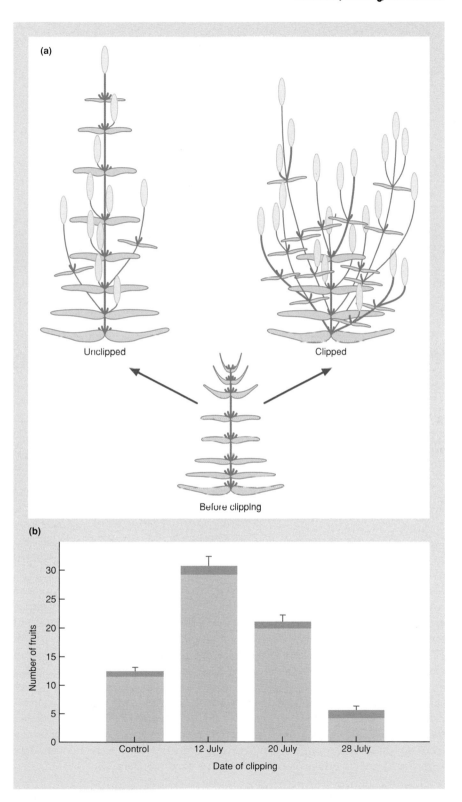

Figure 8.5
(a) The clipping of field gentians to simulate herbivory causes changes in architecture and the number of flowers produced.
(b) Production of mature (pink histograms) and immature fruits (purple histograms) of unclipped control plants and plants clipped on different occasions from July 12 to 28, 1992. Means and standard errors are shown and all means are significantly different from each other ($P < 0.05$). Plants clipped on July 12 and 20 developed significantly more fruits than unclipped controls. Plants clipped on July 28 developed significantly fewer fruits than controls. (From Lennartsson *et al.*, 1998.)

Figure 8.6
(a) Phlorotannin content of *Ascophyllum nodosum* plants after exposure to simulated herbivory (removing tissue with a hole punch) or grazing by real herbivores of two species. Means and standard errors are shown—only the snail *Littorina obtusata* had the effect of inducing increased concentrations of the defensive chemical in the seaweed. Different letters indicate that means are significantly different ($P < 0.05$).
(b) In a subsequent experiment, the snails were presented with algal shoots from the control and the snail-grazed treatments in (a)—the snails ate significantly less of plants with high phlorotannin content. (From Pavia & Toth, 2000.)

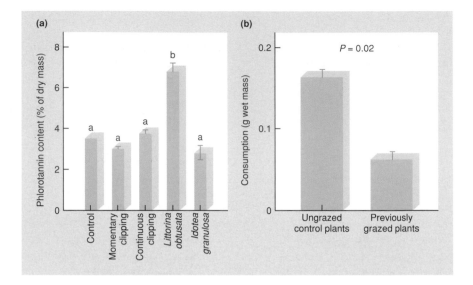

The defensive responses of hosts to attack by parasites were discussed in Chapter 7. The effectiveness of, for example, the immune response of vertebrates is perhaps most clearly demonstrated by those "immunocompromised" individuals in which the response is lacking or severely weakened. Many of the symptoms of people who have AIDS (acquired immunodeficiency syndrome caused by infection with human immunodeficiency viruses—HIV) actually result from infections with other pathogens that are otherwise extremely rare; this provides persuasive testimony to the threat that parasites pose to an undefended host.

8.3.3 From individual prey to prey populations

In spite of these various qualifications, however, the general rule is undoubtedly that predators are harmful to individual prey. But the effects of predation on a population of prey are not always so predictable. In the first place, there may be compensatory changes in the growth, survival, or reproduction of the surviving prey: they may experience reduced competition for a limiting resource, or produce more offspring, or other predators may take fewer of the prey. In other words, whilst predation is bad for the prey that get eaten, it may be good for those that do not.

compensatory reactions amongst surviving prey . . .

The impact of predation is most commonly limited by compensatory reactions amongst the survivors as a result of reduced intraspecific competition. For example, in an experiment in which large numbers of woodpigeons (*Columba palumbus*) were shot, the *overall* level of winter mortality was not increased, and stopping the shooting led to no increase in pigeon abundance (Murton et al., 1974). This was because the number of surviving pigeons was determined ultimately not by shooting but by food availability, and so when shooting reduced density, there were compensatory reductions in intraspecific competition and in natural mortality, as well as density-dependent immigration of birds moving in to take advantage of the unexploited food.

Predation may also have a negligible impact on prey abundance if an increased loss of prey to predators at one stage of the prey's life simply leads to a decreased loss to predators at some other stage. If, for example, recruitment to a population of adult plants is not limited by the number of seeds produced, then insects that reduce seed production are unlikely to have an important effect on plant population dynamics. The point is illustrated by a study of the shrub, *Haplopappus venetus*, which decreased steadily in abundance along a 60 km gradient inland from the coast in California (Louda, 1982, 1983). The level of insect damage to the developing flowers and seeds was high. Experimental exclusion of flower and seed predators, therefore, caused an 11 percent decrease in the abortion of flower heads, a 19 percent increase in pollination success, and a 104 percent increase in the number of developing seeds escaping damage. This led to an increase in the number of seedlings established. But subsequently this was followed by a much greater loss of seedlings, probably to vertebrate herbivores, especially inland. As a consequence, the original abundance gradient was re-established in spite of the short-term importance of the seed predators.

Compensation, however, is by no means always perfect. Figure 8.7, for example, shows the results of an experiment in which Douglas fir seeds were sown both in open plots and in plots screened from vertebrate seed eaters. The immediate effect of this was an enormous reduction in the loss of seeds to these rodents and birds (though the screens were not totally effective). There were, however, compensatory increases in the losses to fungal infection of seeds prior to germination, of seeds during the germination period, and of seedlings and young plants in the year following germination. Nonetheless, in spite of this compensation, the overall effect of screening was to more than double the number of seedlings still surviving 1 year after germination.

... but compensation is often imperfect

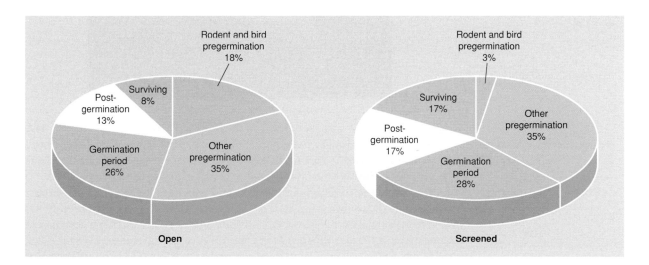

Figure 8.7
When Douglas fir seeds are protected from vertebrate predation by screens, the lowered mortality is compensated for (but not *fully* compensated for) by increased mortality from other sources. (After Lawrence & Redoske, 1962.)

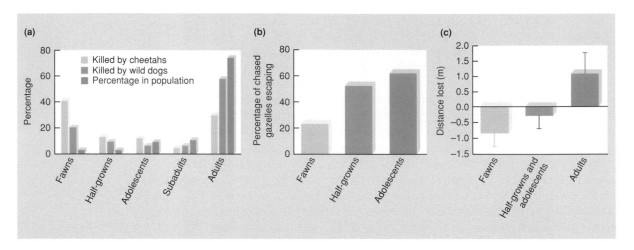

Figure 8.8
(a) The proportions of different age classes (determined by tooth wear) of Thomson's gazelles in cheetah and wild dog kills is quite different from their proportions in the population as a whole. (b) Age influences the probability for Thomson's gazelles of escaping when chased by cheetahs. (c) When prey (Thomson's gazelles) "zigzag" to escape chasing cheetahs, prey age influences the mean distance lost by the cheetahs. (After FitzGibbon & Fanshawe, 1989; FitzGibbon, 1990.)

Thomson's gazelle.

predators often attack the weakest and most vulnerable

Predators may also have little impact on prey populations as a whole because of the particular individuals they attack. Many large carnivores, for example, concentrate their attacks on the old (and infirm), on the young (and naive), or on the sick. Thus, a study in the Serengeti found that cheetahs and wild dogs killed a disproportionate number from the younger age classes of Thomson's gazelles (Figure 8.8a) because: (1) these young animals were easier to catch (Figure 8.8b); (2) they had lower stamina and running speeds; (3) they were less good at outmaneuvering the predators (Figure 8.8c); and (4) they may even have failed to recognize the predators. The effects of predation on the prey population will therefore have been less than would otherwise have been the case, because these young gazelles will have been making no present reproductive contribution to the population, and many would have died anyway, from other causes, before they were able to do so. Similarly, the weevil *Phyllobius argentatus* has been observed to feed mainly on the lower, shaded leaves towards the center of a beech tree, and it therefore has little impact on a tree's overall productivity (Nielsen & Ejlerson, 1977).

It is apparent, then, that the effects of a predator on an individual prey are crucially dependent on the response of the prey; and the effects on prey populations are equally dependent on which prey are attacked and on the responses of other prey individuals and other natural enemies of the prey. The effect of a predator may be more drastic than it appears, or less drastic. It is only rarely what it seems.

8.4 Predator Behavior—Foraging and Transmission

So far, we have been looking, in effect, at what happens *after* a predator finds its prey. Now, we take a step back and examine how contact is established in the first place. This is crucially important, because this pattern of contact is critical in determining the predator's "consumption rate", which goes a long way to determining its own level of benefit and the harm it does to the prey, which determines, in turn, the impact on the dynamics of predator and prey populations, and so on.

True predators and grazers typically "forage." Many move around within their habitat in search of their prey, and their pattern of contact is therefore itself determined by the predators' behavior—and sometimes by the evasive behavior of the prey (Figure 8.9a). This foraging behavior is discussed below. Other predators, web-spinning spiders for instance, "sit and wait" for their prey, though almost always in a location they (or sometimes their parent) has selected (Figure 8.9b). The carnivorous, net-spinning aquatic larvae of the caddis fly *Plectrocnemia conspersa*, for example, when fed on fly larvae in a laboratory stream, simply abandon unprofitable areas more rapidly than they abandon profitable ones (Figure 8.10a). Furthermore, the likelihood that a larva spins a net in the first place depends on whether it happens to encounter a food item (which it can consume even without a net). Larvae that have fed begin net building immediately, whereas unfed larvae continue wandering and are more likely to leave a patch. Overall, therefore, in the natural stream environment, there are most predators "sitting and waiting" (with their nets) where there are most prey (Figure 8.10b).

With parasites and pathogens, on the other hand, we usually talk about transmission rather than foraging. This may be direct transmission between infectious and uninfected hosts when they come into contact with one another (Figure 8.9c), or free-living stages of the parasite may be released from infected hosts, so that it is the pattern of contact between these and uninfected hosts that is important (Figure 8.9d). The simplest assumption we can make for directly transmitted parasites—and one that often is made when attempting to understand their dynamics (discussed in Section 8.5)—is that transmission depends on infectious and uninfected hosts "bumping into one another". In other words, the overall rate of parasite transmission depends both on the density of uninfected, susceptible hosts (since these represent the size of the "target") and on the density of infectious hosts (since this represents the risk of the target being "hit") (Figure 8.9c).

sit-and-wait predators

parasite transmission

Figure 8.9
The different types of foraging and transmission. (a) Active predators seeking (possibly active) prey. (b) Sit-and-wait predators waiting for active prey to come to them. (c) Direct parasite transmission—infectious and uninfected hosts "bumping into each other". (d) Transmission between free-living stages of a parasite shed by a host and new, uninfected hosts.

Figure 8.10
(a) On arrival in a patch, fifth-instar *Plectrocnemia conspersa* larvae that encounter and eat a chironomid prey item at the beginning of the experiment ("fed") quickly cease wandering and commence net building. Predators that fail to encounter a prey item ("unfed") exhibit much more widespread movement during the first 30 min of the experiment, and are significantly more likely to move out of the patch. (b) Directly density-dependent aggregative response of fifth-instar larvae in a natural environment expressed as mean number of predators against the combined biomass of chironomid and stonefly prey per 0.0625 m^2 sample of streambed (*n* = 40). (After Hildrew & Townsend, 1980; Townsend & Hildrew, 1980.)

8.4.1   Foraging behavior

There are many questions we might ask about the behavior of a foraging predator. Where, within the habitat available to it, does it concentrate its foraging? How long does it tend to remain in one location before moving on to another? And so on. Ecologists address all such questions from two points of view. The first is from the viewpoint of the *consequences* of the behavior for the dynamics of predator and prey populations. We turn to this in Section 8.5.

The second is the viewpoint of "behavioral ecology" or "optimal foraging". The aim is to seek to understand why particular patterns of foraging behavior have been favored by natural selection. Most readers will be familiar with the general approach as applied, for example, to the anatomy of the bird's wing—we may seek to understand why a particular surface area, a particular combination of bone strength and weight, a particular arrangement of feathers, have been favored by natural selection for the effectiveness they bring to the bird's powers of flight. Quite rightly, in so doing, it would not occur to anybody to suggest that this implies even a basic understanding of aerodynamics theory on the bird's part—only that those birds with the most effective wings have been favored in the past by natural selection and have passed their effectiveness on to their offspring. Likewise, in applying this approach to foraging behavior, there is no question of suggesting "conscious decision-making" on the predator's part.

the evolutionary, optimal foraging approach

What, though, is the appropriate measure of "effectiveness" in foraging behavior —the equivalent of flying ability as a criterion for a successful bird's wing? Usually, the *net* rate of energy intake has been used—that is, the amount of energy obtained per unit time, *after* account has been taken of the energy expended by the predator in carrying out its foraging. The term *optimal* foraging is used, then, because in seeking to understand why particular patterns of foraging behavior have been favored by natural selection, it is assumed that they will have been so favored because they give rise to the highest net rate of energy intake. For many consumers, however, the efficient gathering of energy may be less critical than some other dietary constituent (e.g., nitrogen), or it may be of prime importance for the forager to consume a mixed and balanced diet. The predictions of optimal foraging theory do not apply to all the foraging decisions of every predator.

A range of the aspects of foraging behavior to which the optimal foraging approach has been applied is illustrated in Figure 8.11. These are elaborated on briefly here, before the whole approach is demonstrated by examining just one of them in detail.

applying the optimal foraging approach to a range of foraging behaviors

- Where, within the habitat available to it, does a predator concentrate its foraging (Figure 8.11a)? Does it concentrate where the long-term expectation of net energy intake is highest *or* where the risk of extended periods of low intake is lowest?

- Does the location chosen by a predator reflect just the expected energy intake? Or does there appear to be some balancing of this against the risk of being preyed upon by its own predators (Figure 8.11b)? We can note in passing, that a recent study has shown individual mice (*Peromyscus leucopus*) continue to harvest food patches until they are depleted to low levels when cover from predators is available, whereas they quit patches where they are more vulnerable while there is still plenty of food available (Morris & Davidson, 2000).

- How long does a predator tend to remain in one location—one patch, say, of a patchy environment—before moving on to another (Figure 8.11c)? Does it remain for extended periods and hence avoid unproductive trips from one patch to another? Or does it leave patches early, before the resources there are depleted? Morris and

Figure 8.11
The types of foraging "decisions" considered by optimal foraging theory. (a) Choosing between habitats. (b) The conflict between increasing input and avoiding predation. (c) Patch stay time decisions. (d) The "ideal free" decision—the conflict between patch quality and competitor density. (e) Optimal diets—to include or not to include an item in the diet (when something better might be "round the corner").

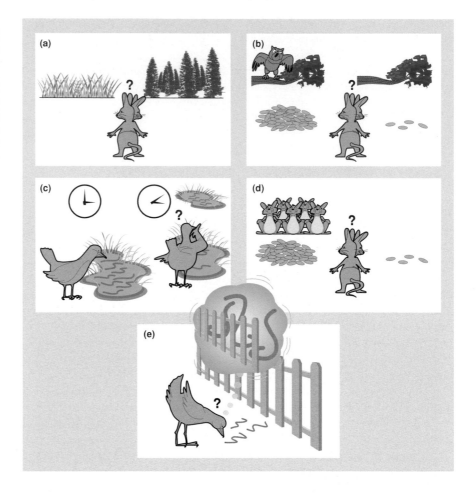

Davidson (2000) also reported that *Peromyscus leucopus* abandoned foraging patches while still harvesting at high rates if the habitat was richer in food as a whole.

■ What are the effects of other, competing predators foraging in the same habitat (Figure 8.11d)? The expected net energy intake from a location is now presumably a reflection of both its intrinsic productivity and the number of competing foragers. What is the expected distribution of the predators as a whole over the various habitat patches?

The remaining "question", in Figure 8.11e, and the one to which we now turn in Box 8.1 for a fuller illustration of the optimal foraging approach, is that of "diet width". No predator can possibly be capable of consuming all types of prey. Simple design constraints prevent shrews from eating owls (even though shrews are carnivores) and prevent humming-birds from eating seeds. Even within their constraints, however, most animals consume a narrower range of food types than they are morphologically capable of consuming.

Box 8.1   *Quantitative Aspects*

Optimal Diet Width

Diet width is the range of food types consumed by a predator. In order to derive widely applicable predictions about when diets are likely to be broad or narrow, we need to strip down the act of foraging to its bare essentials. So, we can say that to obtain food, any predator must expend time and energy, first in searching for its prey, and then in handling it (i.e., pursuing, subduing, and consuming it). While searching, a predator is likely to encounter a wide variety of food items. Diet width, therefore, depends on the responses of predators once they have encountered prey. Generalists, those with a broad diet, pursue a large proportion of the prey they encounter. Specialists, those with a narrow diet, continue searching except when they encounter prey of their specifically preferred type.

Generalists have the advantage of spending relatively little time searching—most of the items they find they pursue and, if successful, consume. But they suffer the disadvantage of including relative low profitability items in their diet. That is, generalists enjoy a net intake of energy much of the time—but their *rate* of intake is often relatively low. Specialists, on the other hand, have the advantage of only including high profitability items in their diet. But they suffer the disadvantage of spending a relatively large amount of their time searching for them. Thus, specialists spend relatively long periods with a net expenditure of energy—but when they do take in energy it is at a relatively high rate. Determining the predicted optimal foraging strategy for a particular predator amounts to determining how these pros and cons should be balanced so as to maximize the *overall* net rate of energy intake, while searching for *and* handling prey (MacArthur & Pianka, 1966; Charnov, 1976).

We can start by taking it for granted that any predator will include the single most profitable type of prey in its diet: that is, the one for which the net

rate of energy intake is highest. But should it include the next most profitable type of item too? Or, when it comes across such an item, should it ignore it and carry on searching for the *most* profitable type? And if it does include the second most profitable type, what about the third? and the fourth? and so on.

Consider first this "second most profitable food type". When will it pay a predator to include an item of this type in its diet (in energetic terms)? The answer is when, having found the item, its expected rate of energy intake over the time spent handling it exceeds its expected rate of intake if, instead, it continued to search for, and then handled, an item of the *most* profitable type. (The *expected* times are simply the average times for items of a particular type.) Expressing this in symbols, we call the expected searching and handling times for the most profitable type s_1 and h_1, and its energy content E_1, and the expected handling time for the second most profitable type h_2, and its energy content E_2. Then it pays the predator to increase the width of its diet if E_2/h_2 (i.e., the rate of intake, energy per unit time, if it handles the second-best type) is greater than $E_1/(s_1 + h_1)$ (the rate of intake if instead it searches for the most profitable type).

Suppose now that it did pay the predator to expand its diet. What about the third most profitable type? We argue in the same way as before: it will pay a predator to include this in its diet if, when it has found it, its expected rate of intake over the time spent handling it, h_3, exceeds the expected rate if it searches for and handles *either* of the two most profitable types, both already included in its diet. Thus, if we call $\bar{s}$, $\bar{h}$, and $\bar{E}$ the searching and handling times and energy content for items already in the diet, it will pay the predator to expand its diet if E_3/h_3 exceeds $\bar{E}/(\bar{s} + \bar{h})$, or, more generally, if E_n/h_n exceeds $\bar{E}/(\bar{s} + \bar{h})$, where n refers generally to the "next" most profitable prey type (not already in the diet). The ecological implications of this rule are considered in the main text.

In summary, Box 8.1 suggests that a predator should continue to add increasingly less profitable items to its diet as long as this increases its overall rate of energy intake. This will serve to maximize its overall rate of energy intake.

predictions of the optimal diet model

This "optimal diet model", then, leads to a number of predictions.

1 Predators with handling times that are typically short compared to their search times should be generalists, because in the short time it takes them to handle a prey item that has already been found, they can barely begin to search for another prey item. (In terms of Box 8.1: E_n/h_n is large [h_n is small] for a wide range of prey types, whereas $\bar{E}/[\bar{s} + \bar{h}]$ is small [$\bar{s}$ is large] even for broad diets.) This prediction seems to be supported by the broad diets of many insectivorous birds feeding in trees and shrubs. Searching is always moderately time consuming, but handling the minute, stationary insects takes negligible time and is almost always successful. A bird, therefore, has something to gain and virtually nothing to lose by consuming an item once found, and overall profitability is maximized by a broad diet.

2 By contrast, predators with handling times that are long relative to their search times should be specialists: maximizing the rate of energy intake is achieved by including only the most profitable items in the diet. For instance, lions live more or less constantly in sight of their prey so that search time is negligible; handling time, on the other hand, and particularly pursuit time, can be long (and very energy consuming). Lions consequently specialize on those prey that can be pursued most profitably: the immature, the lame, and the old.

3 Other things being equal, a predator should have a broader diet in an unproductive environment (where prey items are relatively rare and search times generally are relatively large) than in a productive environment (where search times are generally smaller). This prediction is supported by the two examples shown in Figure 8.12: in experimental arenas, both bluegill sunfish (*Lepomis macrochirus*) and great tits (*Parus major*) had more specialized diets when prey density was higher. A related result has been reported from predators in their natural setting—brown and black bears (*Ursos arctos* and *U. americanus*) feeding on salmon in Bristol Bay in Alaska. When salmon availability was high, bears consumed less biomass per captured fish, targeting energy-rich fish (those that had not spawned) or energy-rich body parts (eggs in females, brain in males); in essence their diet became more specialized when prey were abundant (Gende et al., 2001).

4 The equation in Box 8.1 does not depend on the search time for the "next most profitable" item in the diet and hence does not depend on their abundance. In other words, predators should ignore insufficiently profitable food types irrespective of their abundance. Re-examining the examples in Figure 8.12, we can see that these both refer to cases in which the optimal diet model does indeed predict that the least profitable items should be ignored completely. The foraging behavior was very similar to this prediction—but in both cases the animals consistently took slightly more than expected of the less profitable food types. In fact, this sort of discrepancy has been uncovered repeatedly, and there are a number of reasons why it may occur, which can be summarized crudely by noting that the animals are not "all knowing". The optimal diet model, however, like all optimal foraging and indeed all evolutionary models, does not predict a perfect correspondence between observation and expectation. It predicts the sort of strategy that will be favored by natural selection,

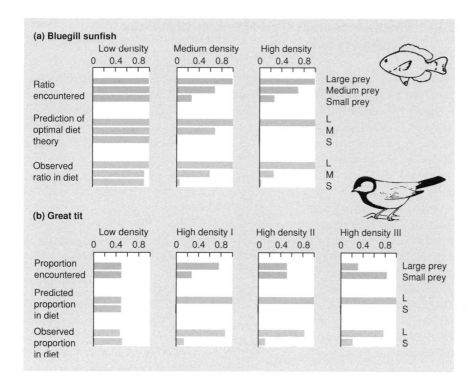

Figure 8.12
Two studies of optimal diet choice that show a clear but limited correspondence with the predictions of the optimal diet width model. Diets are more specialized at high prey densities; but more low profitability items are included than predicted by the theory. (a) Bluegill sunfish preying on different size classes of *Daphnia*: the histograms show the ratios of encounter rates with each size class at three different densities, together with the predicted and observed ratios in the diet. (After Werner & Hall, 1974.) (b) Great tits preying on large and small pieces of mealworm. The histograms in this case refer to the proportions of the two types of item taken. (After Krebs et al., 1977; Krebs, 1978.)

and says that the animals that come *closest* to this strategy will be *most* favored. From this point of view, the correspondence between data and theory in Figure 8.12 seems much more satisfactory.

Overall, then, we can see how an evolutionary, optimal foraging approach can help us make sense of predators' foraging behavior—how it makes predictions of what that behavior might be expected to be, and that these predictions may be supported by real examples.

8.5   Population Dynamics of Predation

What roles do predators play in driving the dynamics of their prey, or prey play in driving the dynamics of their predators? Are there common patterns of dynamics that emerge? The preceding sections should have made it plain that there are no simple answers to these questions. It depends on the detail of the behavior of individual predators and prey, on possible compensatory responses at individual and population levels, and so on. Rather than despair at the complexity of it all, however, we can build an understanding of these dynamics by starting simply and then adding additional features one by one to construct a more realistic picture.

UNANSWERED QUESTION: why do different predator–prey systems display different dynamics?

8.5.1   Underlying dynamics of predator–prey interactions: a tendency to cycle

building a picture from simple beginnings

We begin by consciously oversimplifying: ignoring everything but the predator and the prey, and asking what underlying tendency there might be in the dynamics of their interaction. It turns out (this section and Sections 8.5.2 and 8.5.3) that the underlying tendency is to exhibit coupled oscillations—cycles—in abundance. With this established, we can turn to the many other important factors—further predators, shortage of food for prey, environmental patchiness, etc.—that might modify or override this underlying tendency. Rather than explore each and every one of them, however, Sections 8.5.4 and 8.5.5 examine just two of the more important ones: the effects of crowding in either the predator or the prey, and some of the consequences of spatial patchiness. Between them, these two factors cannot, of course, tell the whole story; but they illustrate how the differences in predator–prey dynamics, from example to example, might be explained by the varying influences of the different factors with a potential impact on those dynamics.

Starting simply then, suppose there is a large population of prey (Figure 8.13). Predators presented with this population should do well: they should consume many prey and hence increase in abundance themselves. The large population of prey thus gives rise to a large population of predators. But this increasing population of predators increasingly takes its toll of the prey. The large population of predators therefore gives rise to a small population of prey. Now the predators are in trouble: large numbers of them and very little food. Their abundance declines. But this takes the pressure off the prey: the small population of predators gives rise to a large population of prey—and the populations are back to where they started. There is, in short, an underlying tendency for predators and their prey to undergo coupled oscillations in abundance—population cycles (Figure 8.13)—essentially because of the time delays in the response of predator abundance to that of the prey, and vice versa. (A "time delay" in response means, for example, that a high predator abundance reflects a high prey abundance *in the past*, but it *coincides* with declining prey abundance, and so on.) A simple mathematical model—the Lotka–Volterra model—conveying essentially the same message is described in Box 8.2.

8.5.2   Predator–prey cycles in practice

the "expectation" of cycles is only rarely fulfilled

This underlying tendency for predator–prey interactions to generate coupled oscillations in abundance could produce an "expectation" of such cycles in real populations. Remember, however, that there were many aspects of predator and prey ecology that had to be ignored in order to demonstrate this underlying tendency, and these can greatly modify expectations. It should come as no surprise, then, that there are rather few good examples of clear predator–prey cycles—albeit ones that have received a great deal of attention from ecologists. And even if a herbivore population, for example, does fluctuate in abundance, that may reflect its interaction with its food *or* with its predators—or it may simply be *tracking* fluctuations in the abundance of its food, which is exhibiting cycles for some quite different reason. Predator–prey interactions

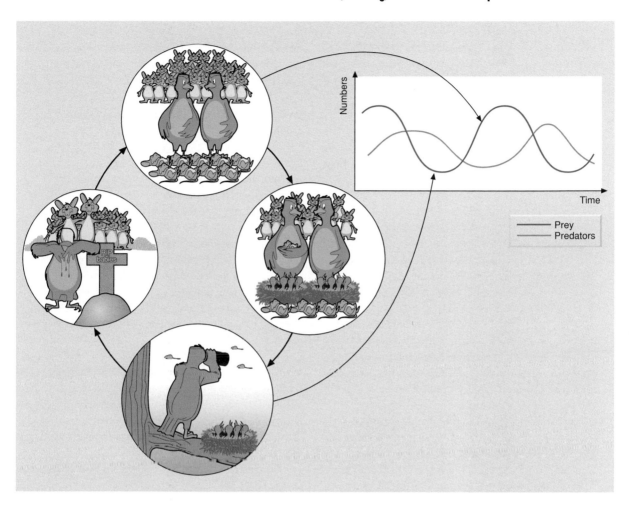

Figure 8.13
The underlying tendency for predators and prey to display coupled oscillations in abundance as a result of the time delays in their responses to each other's abundance.

can generate regular cycles in the abundance of both interacting populations, and they can reinforce such cycles if they exist for some other reason; but attributing a cause to regular cycles in nature is generally a difficult task. Nonetheless, in trying to make sense of predator–prey population dynamics, cycles—the underlying tendency—are a good place to start.

They do occur sometimes. It has been possible in several cases, for example, to generate coupled predator–prey oscillations, several generations in length, in the laboratory (one is shown later in Figure 8.22c). Amongst field populations, there are a number of examples in which regular cycles of prey and predator abundance can be discerned. Cycles in hare populations, in particular, have been discussed by ecologists since the 1920s, and were recognized by fur trappers more than 100 years earlier (Keith, 1983; Krebs et al., 1992). Most famous of all is the snowshoe hare, *Lepus*

plants, hares, and lynx in North America . . .

Box 8.2   *Quantitative Aspects*

The Lotka–Volterra Predator–Prey Model

Here, as in Boxes 5.4 and 6.1, one of the foundation-stone mathematical models of ecology is described and explained. The model is known (like the model of interspecific competition in Box 6.1) by the name of its originators: Lotka and Volterra (Volterra, 1926; Lotka, 1932). It has two components: P, the numbers present in a predator (or consumer) population, and N, the numbers or biomass present in a prey or plant population.

It is assumed that in the absence of consumers the prey population increases exponentially (Box 5.4):

$$dN/dt = rN.$$

But now we also need a term signifying that prey individuals are removed from the population by predators. They will do this at a rate that depends on the frequency of predator–prey encounters, which will increase with increasing numbers of predators (P) and prey (N). The exact number encountered and consumed, however, will also increase with the searching and attacking efficiency of the predator, denoted by a. The consumption rate of prey will thus be aPN, and overall:

$$dN/dt = rN - aPN. \tag{1}$$

Turning to predator numbers, in the absence of food these are assumed to decline exponentially through starvation:

$$dP/dt = -qP$$

where q is their mortality rate. But this is counteracted by predator birth, the rate of which is assumed to depend on (i) the rate at which food is consumed, aPN and (ii) the predator's efficiency, f, at turning this food into predator offspring. Overall:

$$dP/dt = faPn - qP. \tag{2}$$

Equations 1 and 2 constitute the Lotka–Volterra model.

The properties of this model can be investigated by finding zero isoclines (Box 6.1). There are separate predator and prey zero isoclines, both of which are drawn on a graph of prey density (x-axis) against predator density (y-axis) (Figure 8.14). The prey zero isocline joins combinations of predator and prey densities that lead to an unchanging prey population, $dN/dt = 0$, while the predator zero isocline joins combinations of predator and prey densities that lead to an unchanging predator population, $dP/dt = 0$.

In the case of the prey, we "solve" for $dN/dt = 0$ in equation (1), giving the equation of the isocline as:

$$P = r/a.$$

Thus, since r and a are constants, the prey zero isocline is a line for which P itself is a constant (Figure 8.14a): prey increase when predator abundance is low (P is less than r/a) but decrease when it is high (P is greater than r/a).

Similarly, for the predators, we solve for $dP/dt = 0$ in equation (2), giving the equation of the isocline as:

$$N = q/fa.$$

The predator zero isocline is therefore a line along which N is constant (Figure 8.14b): predators decrease when prey abundance is low (N is less than q/fa) but increase when it is high (N is greater than q/fa).

Putting the two isoclines (and two sets of arrows) together in Figure 8.15 shows the behavior of joint populations. The various combinations of increases and decreases, listed above, mean that the populations undergo "coupled oscillations" or "coupled cycles" in abundance; "coupled" in the sense that the rises and falls of the predators and prey are linked, with predator abundance tracking that of the prey (discussed biologically in the main text).

Box 8.2 (cont'd)

Figure 8.14

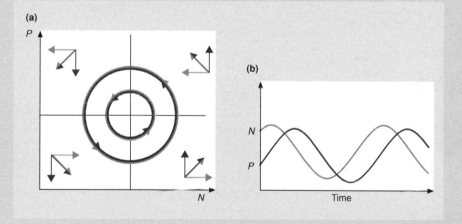

Figure 8.15

It is important to realize, however, that the model does not "predict" the exact patterns of abundance that it generates. This is because it exhibits *neutral stability*, which means that the populations follow precisely the same cycles indefinitely unless the environment changes and shifts them to new values, after which they follow new cycles indefinitely (Figure 8.16). But environments *are* continually changing, and populations would continually be "shifted to new values". Hence a population following the Lotka–Volterra model would not exhibit regular cycles but, because of repeated disturbance, fluctuate erratically. No sooner would it start one cycle than it would be diverted to a new one. Nonetheless, the model does capture the essential tendency for coupled cycles in predator–prey interactions.

Box 8.2 (cont'd)

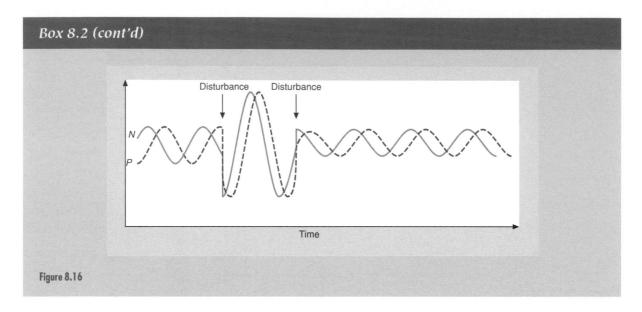

Figure 8.16

americanus, which in the boreal forests of North America follows a "10-year cycle" (although in reality this varies in length between 8 and 11 years; see Figure 8.17). The snowshoe hare is the dominant herbivore of the region, feeding on the terminal twigs of numerous shrubs and small trees. A number of predators, including the Canada lynx (*Lynx canadensis*), have associated cycles of similar length. The hare cycles often involve 10–30-fold changes in abundance, and 100-fold changes can occur in some habitats. They are made all the more spectacular by being virtually synchronous over a vast area from Alaska to Newfoundland.

The declines in hare abundance are accompanied by low birth rates, low juvenile survivorship, and low growth rates or even weight loss. All of these features can be induced experimentally by food shortages. Direct measurements, too, have suggested that there are often shortages of accessible, high-quality food during periods of peak hare abundance (Keith et al., 1984; Smith et al., 1988). Hence, the hares are forced to eat high-fiber (low-quality) food. Indeed, the quantity of relatively palatable, mature shoots does not then recover until after a time lag of another 2–3 years. This is likely to have some direct effect on their body condition, perhaps making them more susceptible to predation. But the hares also spend more time searching for food, exposing them more to predation. As a consequence, the predators concentrate, behaviorally, on hares, and also increase numerically, and so drive hare numbers down again.

. . . but how are the cycles generated?   Is this, then, a hare–plant cycle or predator–hare cycle? Experiments, as ever, have been instructive in suggesting an answer. Normally, with both plant and predator effects present, there are cycles. But if food is added and predators are excluded from experimental areas (neither effects are active), then hare numbers increase ten-fold and stay there—the cycles are lost. However, if either predators are excluded but no food is added (food effect alone), or food is added in the presence of predators (predator effect alone), the numbers of hares double but then drop again—the "cycle" is retained. Both

Figure 8.17
Coupled oscillations in the abundance of the snowshoe hare (*Lepus americanus*) and the Canada lynx (*Lynx canadensis*) as determined by the number of pelts lodged with the Hudson Bay Company. (After MacLulick, 1937.)

The Canada lynx and the snowshoe hare—a predator and prey that may show coupled oscillations.

the hare–plant and the predator–hare interactions have some propensity to cycle on their own—but in practice the cycle seems normally to be generated by the interaction between the two.

Interestingly, moreover, a rather sophisticated statistical analysis of the time series in Figure 8.17 suggests that whereas the cycles in lynx abundance reflect their interaction with the hares, the cycles in hare abundance reflect their interaction with both their predators (essentially the lynx) *and* their food (Stenseth et al., 1997). This warns us that even when we have a predator–prey pair both exhibiting cycles (the hare and lynx) we may still not be observing simple predator–prey oscillations. Apparent instances of predator–prey cycles sometimes make the news—see Box 8.3 for an example.

8.5.3   Disease dynamics and cycles

Cycles are also apparent in the dynamics of many parasites, especially microparasites (bacteria, viruses, etc.). To understand the dynamics of any parasite, the best starting point is its basic reproductive rate, R_p. For microparasites, R_p is the average number of new infected hosts that would arise from a single infectious host in a population of susceptible hosts. An infection will eventually die out for $R_p < 1$ (each present infection

basic reproductive rate and the transmission threshold

Box 8.3   *Topical ECOncerns*

A Cyclical Outbreak of a Forest Insect in the News

Large outbreaks of forest tent caterpillars occur about every 10 years, and each lasts for 2–4 years. During these outbreaks, massive damage is done to the foliage of forest trees over large tracks of land. This article appeared in the *Telegraph Herald* (Dubuque, Iowa) on June 11, 2001.

CATERPILLARS MAKING A MEAL OUT OF NORTHERN FORESTS

Forest tent caterpillars have munched their way through much of northern Wisconsin, eating aspen, sugar maple, birch and oak from Tomahawk to southern Canada.

The insects move across roads in waves that make the pavement seem to crawl and hang from trees in large clumps. . . . "One lady from Eagle River said they were on her house and on her driveway and on her sidewalk, and she was ready to move back to Oak Creek," said Jim Bishop, public affairs manager for the Department of Natural Resource's northern region.

Shane Weber, a DNR forest entomologist from Spooner, said the caterpillars on sidewalks, driveways and highways are a good sign. "Whenever they start these mass overland moves, suddenly moving in waves across the ground, it means that they're starving, looking for another source of food," he said.

In Superior, customers have inundated Dan's Feed Bin [general store], looking for ways to rid their yards and homes of the insects. Employee Amy Connor said some customers held their telephones up to the window so

Connor could hear the worms falling like hail. "It's terribly gross," she said.

The caterpillars have eaten most of the leaves in the Upper Peninsula, said Jeff Forslund, of Hartland, who drove to Ramsey, Michigan. "My grandfather has about 500 acres of aspen, and there isn't a leaf left," Forslund said.

Most of the trees will survive and the caterpillars should start spinning cocoons by mid-June, the DNR said. Forest entomologist Dave Hall said he expects the outbreak to peak this year. "I can't imagine it getting much worse," he said. The last infestation of the native forest tent caterpillars in Wisconsin was in the late 1980s and early 1990s . . . During the last tent caterpillar outbreak, several serious traffic collisions in Canada were blamed on slick roads from squashed tent caterpillars.

About 4 million of the fuzzy crawlers can be found per acre at the peak of the cyclical infestation, the DNR said.

(All content © 2001 *Telegraph Herald* (Dubuque, IA) and may not be republished without permission.)

From what you have learnt about population cycles in this chapter, suggest an ecological scenario to account for the periodic outbreaks of these caterpillars.

Do you believe the comment attributed to a Department of Natural Resources employee that mass movements of the caterpillars is a good sign? How would you determine whether this behavior heralds an end to the peak phase of the cycle?

leads to less than one infection in the future), but an infection will spread for $R_p > 1$. There is therefore a "transmission threshold" when $R_p = 1$, which must be crossed if a disease is to spread. A derivation of R_p for microparasites with direct transmission (see Figure 8.9c) is given in Box 8.4.

Box 8.4   *Quantitative Aspects*

Transmission Threshold for Microparasites

Putting it simply, for microparasites with direct transmission, the basic reproductive rate, R_p, measures the average number of new infections arising from a single infected individual in a population of susceptible hosts. It increases with the average period of time over which an infected host remains infectious, L, since a long infectious period means plenty of opportunity to transmit to new hosts; it increases with the number of susceptible individuals in the host population, S, because more susceptible hosts offer more opportunities ("targets") for transmission of the parasite; and it increases with the transmission rate of the infection, β, because this itself increases first with the infectiousness of the parasite—the probability that contact leads to transmission—but also with the likelihood of infectious and susceptible hosts coming into contact as a reflection of

the pattern of host behavior (Anderson, 1982). Thus, overall:

$$R_p = S \cdot \beta L.$$

We know that $R_p = 1$ is a transmission threshold, in that below this the infection will die out but above it the infection will spread. But this in turn allows us to define a critical *threshold population size* S_T: the number of susceptibles that give rise to $R_p = 1$. At that threshold, making $R_p = 1$ in the equation means:

$$S_T = 1/\beta L.$$

In populations with fewer susceptibles than this, the infection will die out ($R_p < 1$), but with more than this, the infection will spread ($R_p > 1$).

Box 8.4 provides us with a crucial insight into disease dynamics—for each directly transmitted microparasite there is a critical "threshold population size" that needs to be exceeded for a parasite population to be able to sustain itself. For example, measles has been calculated to have a threshold population size of around 300,000 individuals and is unlikely to have been of great importance until quite recently in human biology. However, it has generated major epidemics in the growing cities of the industrialized world in the 18th and 19th centuries, and in the growing concentrations of population in the developing world in the 20th century. Current estimates suggest that around 900,000 deaths occur each year from measles infection in the developing world (Walsh, 1983).

Moreover, the immunity induced by many bacterial and viral infections, combined with death from the infection, reduces the number of susceptibles in a population, reduces R_p, and therefore tends to lead to a decline in the incidence of the disease itself. In due course, though, there will be an influx of new susceptibles into the population (as a result of new births or perhaps immigration), an increase in R_p, an increase in incidence, and so on. There is thus a marked tendency with such diseases to generate a sequence from "high incidence", to "few susceptibles", to "low incidence", to "many susceptibles", to "high incidence", etc.—just like any other predator–prey cycle. This undoubtedly underlies the observed cyclic incidence of many human diseases (especially prior to modern immunization programs), with the differing lengths of cycle reflecting the differing characteristics of the diseases: measles with peaks every

threshold population sizes and
microparasite cycles

Figure 8.18
(a) Reported cases of measles in England
and Wales from 1948 to 1968, prior to the
introduction of mass vaccination.
(b) Reported cases of pertussis (whooping
cough) in England and Wales from 1948 to
1982. Mass vaccination was introduced in
1956. (After Anderson & May, 1991.)

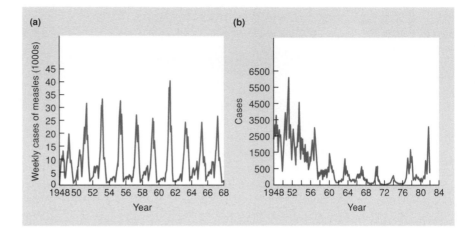

1 or 2 years, pertussis (whooping cough) every 3–4 years, diphtheria every 4–6 years, and so on (Figure 8.18).

8.5.4 Crowding

*mutual interference amongst
predators reduces the predation rate*

One of the most fundamental features that has been ignored so far is the fact that no predator lives in isolation: all are affected by other predators. The most obvious effects are competitive; many predators experience exploitation competition for limited amounts of food when their density is high or the amount of food is small, and this results in a reduction in the consumption rate per individual as predator density increases (Chapter 3). However, even when food is not limited, the consumption rate per individual can be reduced by increases in predator density by a number of processes known collectively as "mutual interference". For example, many predators interact behaviorally with other members of their population, leaving less time for feeding and therefore depressing the overall feeding rate. Humming-birds actively and aggressively defend rich sources of nectar; badgers patrol and visit the "latrines" around the boundaries between their territories and those of their neighbors; and females of certain parasitoid wasps will threaten and, if need be, fiercely drive away an intruding female from their own area of tree trunk. Alternatively, an increase in consumer density may lead to an increased rate of emigration, or of consumers stealing food from one another (as do many gulls) or the prey themselves may respond to the presence of consumers and become less available for capture.

In all such cases, the underlying pattern is the same: the consumption rate per individual predator declines with increasing predator density. This reduction is likely to have an adverse effect on the fecundity, growth, and mortality of individual predators, which intensifies as predator density increases. The predator population is thus subject to density-dependent regulation (Chapters 3 and 5).

*a similar density-dependent effect in
parasites*

With parasites, too, it is to be expected that individuals will often interfere with each other's activities, and that there will be intraspecific competition between para-

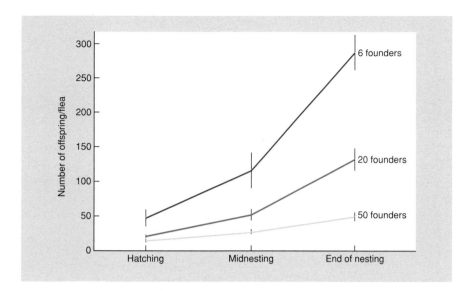

Figure 8.19
Influence of the size of the founder flea population in blue tits' nests on the number of offspring per founder flea (mean ± standard error) at three stages of the bird's nesting period. (From Tripet & Richner, 1999.)

sites and density dependence in their growth and in their birth and/or death rates. In experimental infestations of nests of blue tits (*Parus caeruleus*) with the bird flea *Ceratophyllus gallinae*, for example, the total number of offspring per flea declined with increasing size of the founder adult flea population (Figure 8.19). Competition between flea larvae, rather than food limitation among adults, is the main process underlying this density-dependent pattern in flea reproductive success (Tripet & Richner, 1999).

Moreover, it is, of course, not only the predators that may be subject to the effects of crowding. Prey, too, are likely to suffer reductions in growth, birth, and survival rates as their abundance increases and their individual intake of resources declines.

The effect of either predator or prey crowding on their dynamics is, in a general sense, fairly easy to predict. Prey crowding prevents their abundance from reaching as high a level as it would otherwise do, which means in turn that predator abundance is also unlikely to reach the same peaks. Predator crowding, similarly, prevents predator abundance from rising so high, but also tends to prevent them from reducing prey abundance as much as they would otherwise do. Overall, therefore, crowding is likely to have a damping effect on any predator–prey cycles, reducing their amplitude or removing them altogether; not just because crowding chops off the peaks and troughs, but also because each peak in a cycle tends itself to generate the next trough (e.g., high prey abundance → high predator abundance → *low* prey abundance), so that the lowering of peaks in itself tends to raise troughs.

crowding tends to dampen or eliminate predator–prey cycles

There are certainly examples that appear to confirm the stabilizing effects of crowding in predator–prey interactions. For instance, there are two groups of primarily herbivorous rodents that are widespread in the Arctic: the microtine rodents (lemmings and voles) and the ground squirrels. The microtines are renowned for their dramatic, cyclic fluctuations in abundance, but the ground squirrels have populations that remain remarkably constant from year to year. Significantly, the ground squirrels are strongly self-limited by their aggressive territorial defense of burrows used for breeding and hibernating, and it is to this that their stability has been convincingly attributed (Batzli, 1983).

8.5.5   Predators and prey in patches

The second feature that was ignored initially but will be examined here is the fact that many populations of predators and prey exist not as a single, homogeneous mass, but as a "metapopulation"—an overall population divided, by the patchiness of the environment, into a series of subpopulations, each with its own internal dynamics, but linked to other subpopulations by movement (dispersal) between patches (a topic developed further in Section 9.3).

dispersal and asynchrony dampen cycles

It is possible to get a good idea of the general effect of this spatial structure on predator–prey dynamics by considering the simplest imaginable metapopulation: one consisting of just two subpopulations. If the patches are displaying the same dynamics, and dispersal is the same in both directions, then the dynamics are unaffected: patchiness and dispersal have no effect in their own right (Figure 8.20a). Differences between the patches, however, either in the dynamics within subpopulations or in the dispersal between them, tend, in themselves, to stabilize the interaction: to dampen any cycles that might exist (Ives, 1992; Murdoch et al., 1992; Holt & Hassell, 1993) (Figure 8.20b). The reason is that any difference leads to asynchrony in the fluctuations in the patches. Inevitably, therefore, a population at the peak of its cycle tends to lose more by dispersal than it gains, a population at a trough tends to gain more than it loses, and so on. Dispersal and asynchrony together, therefore—and some degree of asynchrony is likely to be the general rule—tend to dampen fluctuations in predator–prey dynamics.

a computer simulation of a metapopulation leads to a range of persistent dynamics

Computer simulations allow this "theoretical" consideration of metapopulation predator–prey dynamics to be taken a stage further (Figure 8.21). In these, the environment consists of a patchwork of squares. In each generation, two processes occur in sequence. First, a fraction of predators and a fraction of prey disperse from each square to the eight neighboring squares. At the same time, predators and prey from the eight neighboring squares disperse into the first square. The second phase then consists of one generation of standard predator–prey dynamics. Simulations are started with random prey and predator populations in a single patch, with all other patches empty.

Within individual squares, if they existed in isolation, the dynamics in these simulations would be unstable, ever-increasing fluctuations. But within the patchwork of squares as a whole, highly persistent patterns can readily be generated (Figure 8.21). The general message is similar to the results that we have already seen: that stability can be generated by dispersal in metapopulations in which different patches are fluctuating asynchronously. The explicitly spatial aspects of this model have, however, quite literally, added another dimension to the results. Depending on the dispersal fractions and the prey reproductive rate, a number of quite different spatial structures can be generated (although they tend to blur into one another) (Figure 8.21a–c). The model, therefore, makes the point very graphically that persistence at the level of a whole

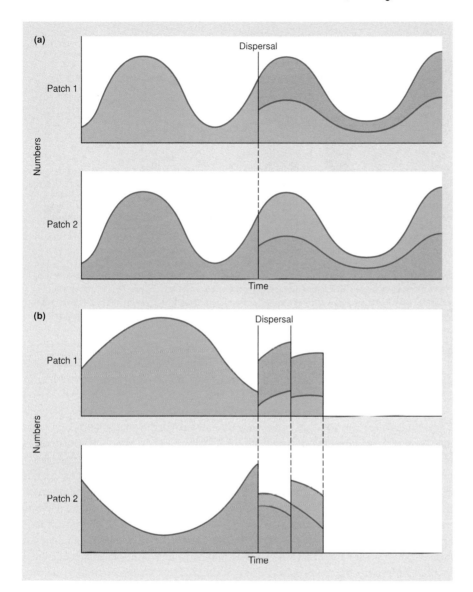

Figure 8.20
Patch separation and dispersal alone, in the absence of asynchrony (a), has no effect on dynamics, but with asynchrony (b), tends to dampen cycles. At each dispersal event, half the individuals leave each patch and join the other patch. In (a) there is an equal exchange and hence the overall dynamics are unaltered. In (b), however, the patch near its peak loses more and donates more than the patch near its trough. Hence, peaks tend to be not so high, nor troughs so low—the fluctuations are "dampened".

population does not necessarily imply either uniformity across the population or stability in individual parts of it.

Is it possible, though, to see the stabilizing influence of this type of metapopulation structure in practice? One famous example is experimental work on a laboratory system in which a predatory mite *Typhlodromus occidentalis* fed on a herbivorous mite *Eotetranychus sexmaculatus*, which fed on oranges interspersed amongst rubber balls in a tray. In the absence of its predator, *Eotetranychus* maintained a fluctuating but persistent population (Figure 8.22a); but if *Typhlodromus* was added during the early stages of prey population growth, it rapidly increased its own population size, consumed all

stabilizing metapopulation effects in Huffaker's mites . . .

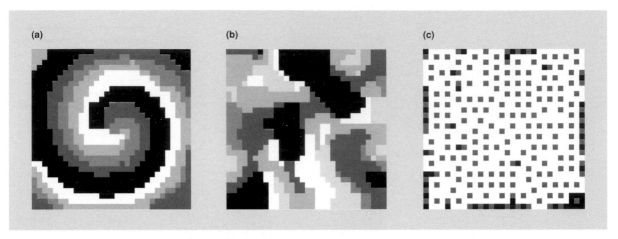

Figure 8.21
Instantaneous maps of population density for simulations of a computer model of dispersal between patches in a metapopulation, where the local dynamics are unstable predator–prey cycles. Different levels of shading represent different densities of prey and predators. Black squares represent empty patches; dark shades becoming paler represent patches with increasing prey densities; light shades to white represent patches with prey and increasing predator densities. (a) Spirals, (b) spatial chaos, (c) a "crystalline lattice", at different rates of predator and prey dispersal. (The maps are single frames from simulations over many generations, which took about 0.2 s per generation on a 33 MHz 80486 personal computer.) (After Comins et al., 1992.)

of its prey and then became extinct itself (Figure 8.22b): the underlying predator–prey dynamics were unstable.

The interaction was altered, however, when the habitat was made more "patchy". The oranges were spread further apart and partially isolated from each other by placing a complex arrangement of petroleum jelly barriers in the tray, which the mites could not cross. The dispersal of *Eotetranychus* was facilitated, however, by inserting a number of upright sticks from which they could launch themselves on silken strands carried by air currents. Dispersal between patches was therefore much easier for prey than it was for predators. In a patch occupied by both, the predators consumed all the prey and then either became extinct themselves or dispersed (with a low rate of success) to a new patch. In patches occupied by prey alone, there was rapid, unhampered growth accompanied by successful dispersal to new patches. And in a patch occupied by predators alone, there was usually death of the predators before their food arrived. Predators and prey were therefore ultimately doomed to extinction in each patch—that is, the patch dynamics were unstable. But overall, at any one time, there was a mosaic of unoccupied patches, prey–predator patches heading for extinction, and thriving prey patches; and this mosaic was capable of maintaining persistent populations of both predators and prey (Figure 8.22c). Amezcua and Holyoak (2000) have described a similar result, where provision of extinction-invulnerable, prey-only refuges promoted the persistence of an interaction between a protozoan predator and its protozoan prey.

... and in starfish and mussels   A similar example, from a natural population, is provided by work off the coast of southern California on the predation by starfish of clumps of mussels (Murdoch & Stewart-Oaten, 1975). Clumps that are heavily preyed upon are liable to be dislodged by heavy seas so that the mussels die; the starfish are continually driving

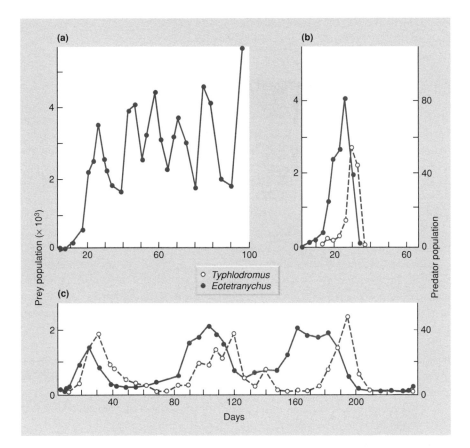

Figure 8.22
Predator–prey interactions between the mite *Eotetranychus sexmaculatus* and its predator, the mite *Typhlodromus occidentalis*. (a) Population fluctuations of *Eotetranychus* without its predator. (b) A single oscillation of the predator and prey in a simple system. (c) Sustained oscillations in a more complex system. (After Huffaker, 1958.)

patches of their mussel prey to extinction. The mussels, however, have planktonic larvae which are continually colonizing new locations and initiating new clumps, whereas the starfish disperse much less readily. They aggregate at the larger clumps, but there is a time lag before they leave an area when the food is gone. Thus, patches of mussels are continually becoming extinct, but other clumps are growing prior to the arrival of the starfish. As with the mites, the combination of patchiness, the aggregation of predators in particular patches, and a lack of synchrony between the behavior of different patches appears capable of stabilizing the dynamics of a predator–prey interaction.

In a similar vein, it is significant that populations of the "cyclic" snowshoe hare (Section 8.5.2) never show cyclic behavior in habitats that are a mosaic of habitable and uninhabitable areas. In mountainous regions, and in areas fragmented by the incursions of agriculture, the snowshoe hare maintains relatively stable, non-cyclic populations (Keith, 1983).

A metapopulation structure, then, like crowding, can have an important influence on predator–prey dynamics. More generally, however, the message from this section is that predator–prey dynamics can take a wide variety of forms, but there are good grounds for believing that we can make sense of this variety through seeing it as a

an explanation for the variety of predator–prey dynamics begins to emerge reflection of the way in which the different aspects of predator–prey interactions combine to play out variations on an underlying theme.

8.6   Predation and Community Structure

What roles can predation play when we broaden our perspective from populations to whole ecological communities? Before this question is addressed, it is worth reflecting that predation, in many of its effects, is just one of the forces acting on communities that can be described as a "disturbance". The result of a predator opening up a gap in a community for colonization by other organisms, for example, is often essentially indistinguishable from that of battering by waves on a rocky shore or a hurricane in a forest.

It turns out that many of the effects of predation (and other disturbances) on community stucture are the result of its interaction with the processes of competitive exclusion (taking up a theme introduced in Section 6.2.8). In an undisturbed world, the most competitive species (the ones that are most efficient at converting limited resources into descendants) might be expected to drive less competitive species to extinction. However, this assumes first that the organisms are actually competing, which in turn implies that resources are limiting. Yet there are many situations where predation may hold down the densities of populations, so that resources are not limiting and individuals do not compete for them. When predation promotes the coexistence of species amongst which there would otherwise be competitive exclusion (because the densities of some or all of the species are reduced to levels at which competition is relatively unimportant) this is known as "predator-mediated coexistence".

For example, in a study of nine Scandinavian islands, pigmy owls (*Glaucidium passerinum*) occurred on only four of the islands, and the pattern of occurrence of three species of tit had a surprising relationship with this distribution. The five islands without the predatory owl were home to only one species, the coal tit (*Parus ater*). However, in the presence of the owl, the coal tit was always joined by two larger tit species, the willow tit (*P. montanus*) and the crested tit (*P. cristatus*). Kullberg and Ekman (2000) argued that the coal tit is superior in exploitation competition for food; but the two larger species have an advantage in interference competition for foraging sites that are safer from predators, i.e., they are less affected than the coal tit by predation from the owl. It seems that the owl may be responsible for predator-mediated coexistence, by reducing the competitive dominance enjoyed by coat tits in its absence.

In another example, grazing by local zebu cows and oxen in natural pasture in the Ethiopian highlands was manipulated to provide a no-grazing control and four grazing intensity treatments (several replicates of each) in two sites. Figure 8.23 shows how the mean number of plant species varied in the sites in October, the period when plant productivity was at its highest (Mwendera et al., 1997). Significantly more species occurred at intermediate levels of grazing than where there was no grazing or heavier grazing ($P < 0.05$).

In the ungrazed plots, several highly competitive plant species, including the grass *Bothriochloa insculpta*, accounted for 75–90 percent of ground cover. At intermediate

predation as an interruptor of competitive exclusion: predator-mediated coexistence

an intriguing pattern in the distribution of owls and tits on Scandinavian islands

grazing by cattle can promote the coexistence of plants

most species occur at intermediate levels of predation?

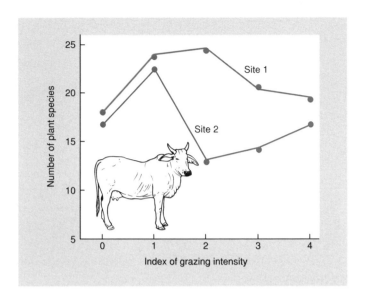

Figure 8.23
Mean species richness of pasture vegetation in plots subjected to different levels of cattle grazing in two sites in the Ethiopian highlands in October. 0, no grazing; 1, light grazing; 2, moderate grazing; 3, heavy grazing; 4, very heavy grazing (estimated according to cattle stocking rates). (After Mwendera et al., 1997.)

levels of grazing, however, the cattle apparently kept the aggressive, competitively dominant grasses in check and allowed a greater number of plant species to persist. But at very high intensities of grazing, species numbers were reduced as the cattle were forced to turn from heavily grazed, preferred plant species to less preferred species, driving some to extinction (Figure 8.23). Where grazing pressure was particularly intense, grazing-tolerant species such as *Cynodon dactylon* became dominant. Thus, overall the number of species was greatest at intermediate levels of predation.

It can be deduced from these cattle data that, as a generalization, *selective* predation should favor an increase in species numbers in a community as long as the preferred prey are competitively dominant, although species numbers may also be low at very high predation pressures. Along the rocky shores of New England, the most abundant and important herbivore in mid and low intertidal zones is the periwinkle snail *Littorina littorea*. The snail will feed on a wide range of algal species but is relatively selective: it shows a strong preference for small, tender species and in particular for the green alga *Enteromorpha intestinalis*. The least-preferred foods are much tougher (e.g., the perennial red alga *Chondrus crispus* and brown algae). They are either never eaten by the periwinkles or are eaten only if no other food has been available for some time.

Is *Enteromorpha*, the periwinkles' preferred food, a competitive dominant in their absence? In a normal *Chondrus* pool, periwinkles feed on microscopic plants and the young stages of many ephemeral algae that settle on *Chondrus* (including *Enteromorpha*). However, if periwinkles are artificially removed from a *Chondrus* pool, *Enteromorpha* and several other algae settle, grow, and become abundant. *Enteromorpha* achieves competitive dominance, and *Chondrus* individuals become bleached and then disappear. The other way round, the addition of periwinkles to *Enteromorpha* pools leads, in a year, to a decline in percentage cover of the green alga from almost 100 percent to less than 5 percent, as *Chondrus* colonizes and eventually comes to

selective predation on a rocky shore

Figure 8.24
The effect of *Littorina littorea* density on
species richness (a) in tide pools and (b) on
emergent substrata. (After Lubchenco,
1978.)

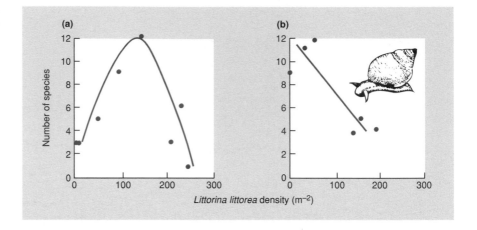

dominate. It seems clear that the periwinkles are responsible for the dominance of *Chondrus* in *Chondrus* pools.

The natural composition of tide pools in the rocky intertidal region varies from almost pure stands of *Enteromorpha* to almost pure stands of *Chondrus*. Is grazing by the periwinkle responsible? A survey suggests that it is (Figure 8.24a). When periwinkles were absent or rare, *Enteromorpha* appeared to competitively exclude other species and the number of algal species was low. At very high densities of periwinkles, however, all palatable algal species were consumed to extinction and were prevented from reappearing, leaving almost pure stands of the tough *Chondrus*. As with the cattle, therefore, it was when periwinkles were present in intermediate densities that the abundance of *Enteromorpha* and other ephemeral algal species was reduced, competitive exclusion was prevented, and many species, both ephemeral and perennial, coexisted.

Why then do some pools contain periwinkles while others do not? Predation is again the answer. The periwinkle colonizes pools while it is in an immature, planktonic stage. Although planktonic periwinkles are just as likely to settle in *Enteromorpha* pools as *Chondrus* pools, the crab *Carcinus maenas*, which can shelter in the *Enteromorpha* canopy, feeds on the young periwinkles and prevents them from establishing a new population. The final thread in this tangled web of predator–prey interactions is the effect of gulls which prey on crabs where the dense green algal canopy is absent. Thus there is no bar to continuing periwinkle recruitment in *Chondrus* pools.

selection of inferiors

The picture is quite different when the preferred prey species is competitively inferior to other prey. Here, increased predation pressure should simply reduce the number of prey species in the community. This can also be illustrated on the rocky shores of New England, where the competitive dominance of the most abundant tide pool plants is actually reversed when the plant species interact on emergent substrata rather than in tide pools. Here it is perennial brown and red algae that predominate, at least in sites where periwinkles are rare or absent, while a number of ephemeral algal species manage to maintain only a precarious foothold. Any increase in the grazing pressure, however, decreases the algal diversity, as the preferred, ephemeral species are consumed totally and prevented from re-establishing themselves (Figure 8.24b).

Overall, then, predation can have an important role in developing our understanding of the structure of ecological communities, not least in reminding us that the patterns we saw in Chapter 6 when we were focusing on interspecific competition may never get a chance to express themselves because communities in the real world rarely proceed smoothly to an equilibrium state.

Summary

Predation, true predators, grazers, and parasites

A predator may be defined as any organism that consumes all or part of another living organism (its "prey" or "host") thereby benefitting itself, but, under at least some circumstances, reducing the growth, fecundity, or survival of the prey.

"True" predators invariably kill their prey and do so more or less immediately after attacking them, and consume several or many prey items in the course of their life. Grazers also attack several or many prey items in the course of their life, but consume only part of each prey item and do not usually kill their prey. Parasites also consume only part of each host, and also do not usually kill their host, especially in the short term, but attack one or very few hosts in the course of their life, with which they therefore often form a relatively intimate association.

The subtleties of predation

Grazers and parasites, in particular, often exert their harm not by killing their prey immediately like true predators, but by making the prey more vulnerable to some other form of mortality.

The effects of grazers and parasites on the organisms they attack are often *less* profound than they first seem because individual plants can compensate for the effects of herbivory and hosts may have defensive responses to attack by parasites.

The effects of predation on a population of prey are complex to predict because the surviving prey may experience reduced competition for a limiting resource, or produce more offspring, or other predators may take fewer of the prey.

Predator behavior

True predators and grazers typically "forage", moving around within their habitat in search of their prey. Other predators "sit and wait" for their prey, though almost always in a selected location. With parasites and pathogens, there may be direct transmission between infectious and uninfected hosts, or contact between free-living stages of the parasite and uninfected hosts may be important.

Optimal foraging theory aims to understand why particular patterns of foraging behavior have been favored by natural selection (because they give rise to the highest net rate of energy intake).

Generalist predators spend relatively little time searching but include relatively low profitability items in their diet. Specialists only include high profitability items in their diet but spend a relatively large amount of their time searching for them.

Population dynamics of predation

There is an underlying tendency for predators and prey to exhibit cycles in abundance, and cycles are observed in some predator–prey and host–parasite interactions. However, there are many important factors that can modify or override the tendency to cycle.

Crowding of either predator or prey is likely to have a damping effect on any predator–prey cycles.

Many populations of predators and prey exist as a "metapopulation". In theory, and in practice, asynchrony in population dynamics in different patches and the process of dispersal tend to dampen any underlying population cycles.

Predation and community structure

There are many situations where predation may hold down the densities of populations, so that resources are not limiting and individuals do not compete for them. When predation promotes the coexistence of species amongst which there would otherwise be competitive exclusion (because the densities of some or all of the species are reduced to levels at which competition is relatively unimportant) this is known as "predator-mediated coexistence".

The effects of predation generally on a group of competing species depend on which species suffers most. If it is a subordinate species, then this may be driven to extinction and the total number of species in the community will decline. If it is the competitive dominants that suffer most, however, the results of heavy predation will usually be to free space and resources for other species, and species numbers may then increase.

It is not unusual for the number of species in a community to be greatest at intermediate levels of predation.

Review Questions

1 With the aid of examples, explain the feeding characteristics of true predators, grazers, parasites, and parasitoids.

2 ▲ True predators, grazers, and parasites can alter the outcome of competitive interactions that involve their "prey" populations: discuss this assertion using one example for each category.

3 Discuss the various ways that plants may "compensate" for the effects of herbivory.

4 Predation is "bad" for the prey that get eaten. Explain why it may be good for those that do not get eaten.

5 ▲ Discuss the pros and cons, in energetic terms, of (1) being a generalist as opposed to a specialist predator, and (2) being a sit-and-wait predator as opposed to an active forager.

6 In simple terms, explain why there is an underlying tendency for populations of predators and prey to cycle.

7 ▲ You have data that shows cycles in nature among interacting populations of a true predator, a grazer, and a plant. Describe an experimental protocol to determine whether this is a grazer–plant cycle or a predator–grazer cycle.

8 Define mutual interference and give examples for true predators and parasites. Explain how mutual interference may dampen inherent population cycles.

9 Discuss the evidence presented in this chapter that suggests environmental patchiness has an important influence on predator–prey population dynamics.

10 With the help of an example, explain why most prey species may be found in communities subject to an intermediate intensity of predation.

▲ = Challenge Question.

Chapter 9

Population Processes— the Big Picture

In previous chapters, we generally dealt with individual species or pairs of species in isolation, as ecologists often do. Ultimately, however, we must recognize that every population exists within a web of interactions with myriad other populations, across several trophic levels. Each population must be viewed in the context of the whole community, and we need to understand that populations occur in patchy and inconstant environments in which disturbance and local extinction may be common.

Chapter Contents

9.1 Introduction
9.2 Multiple determinants of the dynamics of populations
9.3 Dispersal, patches, and metapopulation dynamics
9.4 Temporal patterns in community composition
9.5 Food webs

Key Concepts

In this chapter you will

- appreciate the variety of interacting abiotic and biotic factors that account for the dynamics of populations
- distinguish between the determination and the regulation of population abundance
- understand how patchiness and dispersal between patches influence the dynamics of both populations and communities
- recognize the influence of disturbance on community patterns and understand the nature of community succession
- appreciate the importance of direct and indirect effects and distinguish between bottom-up and top-down control of food webs
- understand the relationship between the structure and stability of food webs

Figure 9.1
Community matrix illustrating how each
species may interact with several others in
competitive interactions (among plant
species 1, 2, and 3; or between grazers 4
and 5; or between predators 6 and 7) and
predator–prey interactions (such as
between 6 and 4, or 5 and 2).

9.1   Introduction

Single-species populations have been the focus for many of the questions posed in previous chapters. In attempting to answer the most fundamental ecological question of all—what determines a species' abundance and distribution—we have chosen to ask separately about the role of conditions and resources, of migration, of competition (both intra- and interspecific), of mutualism, and of predation and parasitism. In reality, the dynamics of any population reflect a combination of these effects, though their relative importance varies from case to case. Now, though, we need to view the population in the context of the whole community, since each exists within a web of predator–prey and competitive interactions (Figure 9.1), and each responds differently to the prevailing abiotic conditions.

In Section 9.2 we consider how abiotic and biotic factors combine to determine the dynamics of species populations. Then, in Section 9.3, we revisit one of the major themes of this book—the importance of patchiness and dispersal between patches in ecological dynamics—and discuss especially the importance of the concept of the metapopulation. Disturbances, such as forest fires and the storm battering of seashores, also play an important role in the dynamics of many populations and the composition of most communities. After each disturbance, there is a pattern of re-establishment of species that is played out against a background of changing conditions, resources, and population interactions. We deal with temporal patterns in community composition, including community succession, in Section 9.4. Finally, in Section 9.5 we broaden our view further to examine food webs, like the one illustrated in Figure 9.1, with usually at least three trophic levels (plant–herbivore–predator), emphasizing the importance not only of direct but also of indirect effects that a species may have on others on the same trophic level or on levels below or above it.

9.2   Multiple Determinants of the Dynamics of Populations

fluctuations in abundance are caused by a wide variety of biotic and abiotic factors

Why are some species rare and others common? Why does a species occur at low population densities in some places and at high densities in others? What factors cause fluctuations in a species' abundance? These are crucial questions when we wish to conserve rare species, or control pests, or manage natural, living resources, or when we wish simply to understand the patterns and dynamics of the natural world. To provide complete answers for even a single species in a single location, we need to know the physicochemical conditions, the level of resources available, the organism's life cycle, and the influence of competitors, predators, parasites, and so on—and how all these factors influence abundance through effects on birth, death, dispersal, and migration. We now bring these factors together and consider how we might discover which actually matter in particular examples.

The raw material for the study of abundance is usually some estimate of the numbers of individuals in a population. However, a record of numbers alone can hide vital information. Picture three human populations, shown to contain identical numbers of individuals. One is an old people's residential area, the second is a population of young children, and the third is a population of mixed age and sex. In the absence of information beyond mere numbers, it would not be clear that the first population was doomed to extinction (unless maintained by immigration), the second would grow fast but only after a delay, and the third would continue to grow steadily. The most satisfactory studies, therefore, estimate not only the numbers of individuals (and their parts, in the case of modular organisms) but also those of different ages, sexes, and sizes.

what total numbers can and cannot tell us

The data that accumulate from estimates of abundance may be used to establish correlations with external factors like weather. Correlations may be used to suggest causal relationships, but they cannot prove them. For example, low populations of a pest may regularly follow cold winters. Such a correlation might suggest that cold winters kill the overwintering stage of the pest, but it might equally mean that cold winters favor its enemies or deplete part of its food resources. Correlational data may also indicate a relationship between the size of a population and its growth rate—for instance, that it grew fastest when numbers were low and declined when numbers were high. But what was the causal mechanism behind this correlation? It may be that when the population is large many individuals starve to death, or fail to reproduce, or become aggressive and drive out the weaker members. It is only by observing what is happening to the individuals that we can discover why a population changes in size.

what correlations can and cannot tell us

9.2.1 Fluctuation or stability?

Some populations appear to change very little in size. One population study that covers an extended time span—though it is not necessarily the most scientific!—has examined the swifts (*Micropus apus*) in the village of Selborne in southern England. In one of the earliest published works on ecology, Gilbert White, who lived in the village, wrote in 1778:

many populations are very stable . . .

> I am now confirmed in the opinion that we have every year the same number of pairs invariably. . . . The number that I constantly find are eight pairs, about half of which reside in the church, and the rest in some of the lowest and meanest thatched cottages.

More than 200 years later, Lawton and May (1984) visited the village and, not surprisingly, found major changes. Swifts were unlikely to have nested in the church for 50 years, and the thatched cottages had either disappeared or their roofs had been covered with wire. Yet the number of breeding pairs of swifts regularly to be found in

Figure 9.2
Changes in population size of creeping buttercup (*Ranunculus repens*) at site C (see text). Open circles, cumulative gains from seed germination and clonal growth; closed circles, cumulative losses; open squares, net population size. (After Sarukhán & Harper, 1973.)

the village was then 12. In view of the many changes that have taken place in the intervening centuries, this number is remarkably close to the eight pairs so consistently found by White.

... but stability need not mean "nothing changes"

However, the stability of a population may conceal complex underlying dynamics. In a study of creeping buttercup (*Ranunculus repens*) in an old permanent pasture in North Wales, detailed maps of the distribution of plants and seedlings in three separate sites within the pasture allowed the fate of every individual to be followed (an approach that is rarely possible for mobile animals). Overall, over the 2 years of the study the number of buttercups in the pasture (as estimated from the three sites) fell slightly—from 650 to 518. However, at site A the population halved over the 2 years, at site B it scarcely changed, and at site C it increased by 50 percent (Figure 9.2). Moreover, 1,054 new individuals were gained by the population and 1,189 individuals were lost from it. This illustrates clearly that different patches in the same "population" may show quite different dynamics, and that even when there is relatively little overall change in a population, very rapid fluxes of births and deaths may occur.

9.2.2   Theories of species abundance

Is the move from eight to 12 pairs of swifts over 200 years, or from 650 to 518 buttercups over 2 years, an indication of consistency or of change? Is the similarity between eight and 12 of more interest—or the difference between them? Some investigators have emphasized the apparent constancy of populations; others have emphasized the fluctuations.

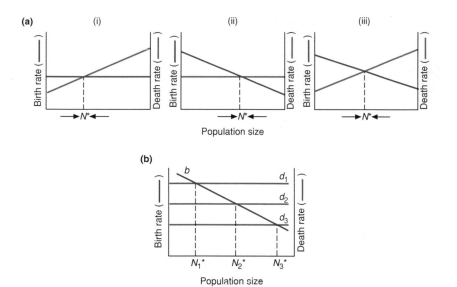

Figure 9.3
(a) Population regulation with (i) density-independent birth and density-dependent death; (ii) density-dependent birth and density-independent death; and (iii) density-dependent birth and death. Population size increases when the birth rate exceeds the death rate and decreases when the death rate exceeds the birth rate. N^* is therefore a stable equilibrium population size. The actual value of the equilibrium population size is seen to depend on both the magnitude of the density-independent rate and the magnitude and slope of any density-dependent process. (b) Population regulation with density-dependent birth, b, and density-independent death, d. Death rates are determined by physical conditions which differ in three sites (death rates d_1, d_2, and d_3). Equilibrium population size varies as a result (N_1^*, N_2^*, N_3^*).

Those who have emphasized constancy argue that we need to look for stabilizing forces within populations to explain why the populations do not exhibit unfettered increase or a decline to extinction (generally density-dependent forces, for instance competition between crowded individuals for limited resources). Those who have emphasized fluctuations often look to external factors, weather, or disturbance, to explain the changes. Disagreements between the two camps (often more like warring factions) dominated much of ecology in the middle third of the 20th century. By considering some of these arguments, it will be easier to appreciate the details of the modern consensus.

First, however, it is important to understand clearly the difference between questions about the ways in which abundance is *determined* and questions about the way in which abundance is *regulated*. *Regulation* is the tendency of a population to decrease in size when it is above a particular level, but to increase in size when it is below that level. In other words, regulation of a population can, by definition, occur only as a result of one or more density-dependent processes (Chapters 3 and 5) that act on rates of birth and/or death and/or movement (Figure 9.3a). Various potentially density-dependent processes have been discussed in earlier chapters on competition, predation, and parasitism. We must look at regulation, therefore, to understand how it is that a population tends to remain within defined upper and lower limits.

On the other hand, the precise abundance of individuals will be *determined* by the combined effects of all the factors and all the processes that affect a population, whether they are dependent or independent of density (Figure 9.3b). We must look at the determination of abundance, therefore, to understand how it is that a particular population exhibits a particular abundance at a particular time, and not some other abundance.

the distinction between determination and regulation of abundance

the view that abundance is most
influenced by the effects of crowding

The first point of view traces its roots back to the Australian A.J. Nicholson, a theoretical and laboratory animal ecologist (e.g., Nicholson, 1954), who believed that density-dependent, biotic interactions play the main role in determining population size, holding populations in a state of balance in their environments. Nicholson recognized, of course, that "factors which are uninfluenced by density may produce profound effects upon density" (see Figure 9.3b), but he considered that density dependence "is merely relaxed from time to time and subsequently resumed, and it remains the influence which adjusts population densities in relation to environmental favourability".

the view that abundance is most
influenced by the length of periods of
population growth

The other point of view can be traced back to two other Australian ecologists, Andrewartha and Birch (1954), whose research was concerned mainly with the control of insect pests in the wild. It is likely, therefore, that their views were conditioned by the needs of being able to predict abundance, and predict, especially, the timing and intensity of pest outbreaks. They believed that the most important factor limiting the numbers of organisms in natural populations was the shortage of time when the rate of increase in the population was positive. In other words, populations could be viewed as passing through a repeated sequence of setbacks and recovery—a view that can certainly be applied to many insect pests that are sensitive to unfavorable environmental conditions but are able to bounce back rapidly. They also rejected any subdivision of the environment into density-dependent and density-independent "factors", preferring instead to see populations as sitting at the center of an ecological web, where the essence was that various factors and processes interacted in their effects on the population.

resolving the contrasting views

With the benefit of hindsight, it seems clear that the first camp was preoccupied with what regulates population size and the second with what determines population size—and both are perfectly valid interests. Disagreement seems to have arisen because of some feeling within the first camp that whatever regulates *also* determines; and some feeling in the second camp that the determination of abundance is, for practical purposes, all that really matters. It is indisputable, however, that no population can be absolutely free of regulation—long-term unrestrained population growth is unknown, and unrestrained declines to extinction are rare. Furthermore, any suggestion that density-dependent processes are rare or generally of only minor importance would be wrong. A very large number of studies have been made of various kinds of animals, especially of insects. Density dependence has by no means always been detected but it is commonly seen when studies are continued for many generations. For instance, density dependence was detected in 80 percent or more of studies of insects that lasted more than 10 years (Hassell et al., 1989; Woiwod & Hanski, 1992).

On the other hand, in the kind of study that Andrewartha and Birch focused on, weather was typically the major determinant of abundance and other factors were of relatively minor importance. For instance, in one famous classic study of a pest, the apple thrips, weather accounted for 78 percent of the variation in thrips numbers (Davidson & Andrewartha, 1948)—for predicting thrips abundance, information on the weather is of paramount importance. Hence, it is clearly not necessarily the case that whatever regulates the size of a population also determines its size for most of the time. And it would also be wrong to give regulation or density dependence some kind of pre-eminence. It may be occurring only infrequently or intermittently. And even when regulation is occurring, it may be drawing abundance toward a level that is itself changing in response to changing levels of resources. It is likely that no natural population is ever truly at equilibrium. Rather, it seems reasonable to expect to find

Figure 9.4
Idealized diagrams of population dynamics: (a) dynamics dominated by phases of population growth after disasters; (b) dynamics dominated by limitations on environmental carrying capacity—carrying capacity high; (c) same as (b) but carrying capacity low; (d) dynamics within a habitable site dominated by population decay after more or less sudden episodes of colonization or recruitment.

some populations in nature that are almost always recovering from the last disaster (Figure 9.4a), others that are usually limited by an abundant resource (Figure 9.4b) or by a scarce resource (Figure 9.4c), and others that are usually in decline after sudden episodes of colonization (Figure 9.4d).

9.2.3   Key-factor analysis

We can distinguish clearly between what regulates and what determines the abundance of a population, and see how regulation and determination relate to one another, by examining an approach known as *key-factor analysis*. It has been applied to many insects and some other animals and plants and is based on calculating what are known as *k*-values for each phase of the life cycle. In fact, key-factor analysis is poorly named, since it identifies key *phases* (rather than key factors) in the life of a study organism (those most important in determining abundance). Details are described in Box 9.1, but the approach can be understood simply by appreciating that the *k*-values measure the amount of mortality: the higher the *k*-value, the greater the mortality (*k* stands for "killing power").

Box 9.1 *Quantitative Aspects*

Determining *k*-Values for Key-Factor Analysis

Table 9.1 sets out a typical set of life table data, collected by Harcourt (1971) for the Colorado potato beetle, *Leptinotarsa decemlineata*, in Canada. The first column lists the various phases of the life cycle. *Spring adults* emerge from hibernation around the middle of June, when potato plants are breaking through the ground. Within 3 or 4 days egg laying begins, and it continues for about 1 month. The eggs are laid in

clusters (approximately 34 eggs) on the lower leaf surface, and the larvae crawl to the top of the plant, where they feed throughout their development, passing through four stages. When mature, they drop to the ground and form pupal cells in the soil. *Summer adults* emerge in early August, feed, and then re-enter the soil at the beginning of September to hibernate and become the next season's spring adults.

Table 9.1
Life table data for the Canadian Colorado potato beetle.

Age interval	Numbers per 96 potato hills	Numbers dying	Mortality factor	Factor $\log_{10} N$	k-value	
Eggs	11,799	2,531	Not deposited	4.072	0.105	(k_{1a})
	9,268	445	Infertile	3.967	0.021	(k_{1b})
	8,823	408	Rainfall	3.946	0.021	(k_{1c})
	8,415	1,147	Cannibalism	3.925	0.064	(k_{1d})
	7,268	376	Predators	3.861	0.023	(k_{1e})
Early larvae	6,892	0	Rainfall	3.838	0	(k_2)
Late larvae	6,892	3,722	Starvation	3.838	0.337	(k_3)
Pupal cells	3,170	16	Parasitism	3.501	0.002	(k_4)
Summer adults	3,154	−126	Sex (52% ♀)	3.499	−0.017	(k_5)
♀ × 2	3,280	3,264	Emigration	3.516	2.312	(k_6)
Hibernating adults	16	2	Frost	1.204	0.058	(k_7)
Spring adults	14			1.146		
					2.926	(k_{total})

Box 9.1 (cont'd)

The next column lists the estimated numbers (per 96 potato hills) at the start of each phase, and the third column then lists the numbers dying in each phase, before the start of the next. This is followed, in the fourth column, by what were believed to be the main causes of deaths in each stage of the life cycle. The fifth and sixth columns then show how the *k*-values are calculated. In the fifth column, the logarithms of the numbers at the start of each phase are listed. The *k*-values in the sixth column are then simply the differences between successive values in column 5. Thus, each value refers to deaths in one of the phases, and, similarly to column 3, the total of the column refers to the total death throughout the life cycle. Moreover, each *k*-value measures the rate or intensity of mortality in its own phase, whereas this is not true for the values in column 3—there, values tend to be higher earlier in the life cycle simply because there are more individuals "available" to die. These useful characteristics of *k*-values are put to use in *key-factor analysis*.

An adult Colorado potato beetle (*Leptinotarsa decemlineata*) taking off from its host plant. Emigration by summer adults represents the key phase in the population dynamics of potato beetles.

For a key-factor analysis to be carried out, data are compiled in the form of a life table (Chapter 5), such as that for a Canadian population of the Colorado potato beetle (*Leptinotarsa decemlineata*) in Box 9.1. The sampling program in that case provided estimates of the population at seven stages: eggs, early larvae, late larvae, pupae, summer adults, hibernating adults, and spring adults. One further category was included, "females × 2", to take account of any unequal sex ratios among the summer adults.

The first question we can ask is how much of the total "mortality" tends to occur in each of the phases? (Mortality is in inverted commas because it refers to all losses from the population.) The question can be answered by calculating the mean *k*-values for each phase, in this case determined over ten seasons (that is, from ten tables like the one in Box 9.1). These are presented in Table 9.2. Thus, here, most loss occurred amongst summer adults—in fact, mostly through emigration rather than mortality as such. There was also substantial loss of older larvae (starvation), of hibernating adults (frost-induced mortality), of young larvae (rainfall), and of eggs (cannibalization and "not being laid").

It is usually more valuable, however, not to ask that first question: "How much of the total mortality occurs in each of the phases?", but to ask a second: "What is the relative importance of these phases as determinants of year-to-year *fluctuations* in mortality, and hence of year-to-year fluctuations in abundance?" This is rather different. For instance, a phase might repeatedly witness a significant toll being taken from a

the Colorado potato beetle

when does most mortality occur?

the phases that determine abundance . . .

Table 9.2
Summary of the life table analysis for Canadian Colorado beetle populations—see Box 9.1. (Afer Harcourt, 1971.)

		Mean	Coefficient of regression on k_{total}
Eggs not deposited	k_{1a}	0.095	−0.020
Eggs infertile	k_{1b}	0.026	−0.005
Rainfall on eggs	k_{1c}	0.006	0.000
Eggs cannibalized	k_{1d}	0.090	−0.002
Eggs predation	k_{1c}	0.036	−0.011
Larvae 1 (rainfall)	k_2	0.091	0.010
Larvae 2 (starvation)	k_3	0.185	0.136
Pupae (parasitism)	k_4	0.033	−0.029
Unequal sex ratio	k_5	−0.012	0.004
Emigration	k_6	1.543	0.906
Frost	k_7	0.170	0.010
	k_{total}	2.263	

population (a high mean k-value), but if that toll is always roughly the same, it will play little part in determining the particular rate of mortality (and thus the particular population size) in any particular year. In other words, this second question is much more concerned with discovering what *determines* particular abundances at particular times, and it can be addressed in the following way.

Mortality during a phase that is important in determining population change—referred to as a *key phase*—will vary in line with total mortality in terms of both size and direction. It is a key phase in the sense that when mortality during it is high, total mortality tends to be high and the population declines—whereas, when phase mortality is low, total mortality tends to be low and the population tends to remain large, and so on. By contrast, a phase with a k-value that varies quite randomly with respect to total k will, by definition, have little influence on changes in mortality and hence will have little influence on population size. We need, therefore, to measure the relationship between phase mortality and total mortality and this is achieved by the *regression coefficient* of the former on the latter. The largest regression coefficient will be associated with the key phase causing population change, whereas phase mortality that varies at random with total mortality will generate a regression coefficient close to zero.

In the present example (Table 9.2), the summer adults, with a regression coefficient of 0.906, are the key phase. Other phases (with the possible exception of older larvae) have a negligible effect on the changes in generation mortality.

... and the factors that regulate abundance

What, though, about the possible role of these phases in the *regulation* of the Colorado beetle population? In other words, which, if any, act in a density-dependent way? This can be answered most easily by plotting k-values for each phase against the numbers present at the start of the phase. For density dependence, the k-value should be highest (that is, mortality greatest) when density is highest. For the beetle population, two phases are notable in this respect: for both summer adults (the key phase) and older larvae there is evidence that losses are density dependent (Figure 9.5), and thus there is a possible role of these losses in regulating the size of the beetle population. In

Figure 9.5
(a) Density-dependent emigration by Colorado beetle "summer" adults (slope = 2.65). (b) Density-dependent starvation of larvae (slope = 0.37). (After Harcourt, 1971.)

Table 9.3
Key-factor (or key phase) analysis for wood frog populations from three areas in the United States: Maryland (two ponds, 1977–82), Virginia (seven ponds, 1976–82), and Michigan (one pond, 1980–93). In each area, the phase with the highest mean k-value, the key phase, and any phase showing density dependence are highlighted in bold. (After Berven, 1995.)

Age Interval	Mean k-value	Coefficient of regression on k_{total}	Coefficient of regression on log (population size)
Maryland			
Larval period	1.94	**0.85**	**Pond 1: 1.03 ($P = 0.04$)**
			Pond 2: 0.39 ($P = 0.50$)
Juvenile: up to 1 year	0.49	0.05	0.12 ($P = 0.50$)
Adult: 1–3 years	**2.35**	0.10	0.11 ($P = 0.46$)
Total	4.78		
Virginia			
Larval period	**2.35**	**0.73**	0.58 ($P = 0.09$)
Juvenile: up to 1 year	1.10	0.05	−0.20 ($P = 0.46$)
Adult: 1–3 years	1.14	0.22	**0.26 ($P = 0.05$)**
Total	4.59		
Michigan			
Larval period	1.12	**1.40**	1.18 ($P = 0.33$)
Juvenile: up to 1 year	0.64	1.02	0.01 ($P = 0.96$)
Adult: 1–3 years	**3.45**	−1.42	**0.18 ($P = 0.005$)**
Total	5.21		

this case, therefore, the phases with the largest role in determining abundance are also those that seem likely to play the largest part in regulating abundance. But, as we see next, this is by no means a general rule.

Key-factor analysis has been applied to a great many insect populations, but to far fewer vertebrate or plant populations. Examples of these, though, are shown in Table 9.3 and Figure 9.6.

two further examples of key-factor analysis

Figure 9.6
Key-factor analysis of the sand dune annual plant *Androsace septentrionalis*. A graph of total generation mortality (k_{total}) and of various k-factors is presented. The values of the regression coefficients of each individual k-value on k_{total} are given in brackets. The largest regression coefficient signifies the key phase and is shown as a colored line. Alongside is shown the one k-value that varies in a density-dependent manner. (After Symonides, 1979; analysis in Silvertown, 1982.)

In populations of the wood frog (*Rana sylvatica*) in three regions of the United States (Table 9.3), the larval period was the key phase determining abundance in each region (second data column), largely as a result of year-to-year variations in rainfall during the larval period. In low rainfall years, the ponds could dry out, reducing larval survival to catastrophic levels, sometimes as a result of a bacterial infection. Such mortality, however, was inconsistently related to the size of the larval population (one pond in Maryland and only approaching significance in Virginia—third data column) and hence played an inconsistent part in regulating the sizes of the populations. Rather, in two of the regions it was during the adult phase that mortality was clearly density dependent and hence regulatory (apparently as a result of competition for food), and indeed, in two of the regions, mortality was also most intense in the adult phase (first data column).

The key phase determining abundance in a Polish population of the sand dune annual plant *Androsace septentrionalis* (Figure 9.6) was found to be the seeds in the soil. Once again, however, mortality there did not operate in a density-dependent manner, whereas mortality of seedlings, which were not the key phase, was found to be density

Box 9.2   *Topical EConcerns*

Acorns, Mice, Ticks, Deer, and Human Disease—Complex Population Interactions

Ecologists have been trying to uncover the complex interactions among acorn production, populations of mice and deer, parasitic ticks, and ultimately a bacterial pathogen carried by the ticks that can affect people. It is clear that a thorough understanding of the abiotic factors that determine the size of the acorn crop and of the various population interactions can enable scientists to predict years when the risk of human disease is high. This is the topic of the following newspaper article in the *Contra Costa Times* on Friday, February 13, 1998, by Paul Recer:

More Acorns May Mean a Rise in Lyme Disease
A big acorn crop last fall could mean a major outbreak of Lyme disease next year, according to a study that linked acorns, mice and deer to the number of ticks that carry the Lyme disease parasite.

Based on the study, researchers at the Institute of Ecosystem Studies in Millbrook, New York, say that 1999 may see a dramatic upswing in the number of Lyme disease cases among people who visit the oak forests of the Northeast.

"We had a bumper crop of acorns this year, so in 1999, two years after the event, we should also have a bumper year for Lyme disease," said Clive G. Jones, a researcher at the Institute of Ecosystem Studies; "1999 should be a year of high risk for Lyme disease."

Lyme disease is caused by a bacterium carried by ticks. The ticks normally live on mice and deer, but they can bite humans. Lyme disease first causes a mild rash, but left untreated can damage the heart and nervous system and cause a type of arthritis.

Jones, along with researchers at the University of Connecticut, Storrs, and Oregon State University, Corvallis, found that the number of mice, the number of ticks, the deer population and even the number of gypsy moths are linked directly to the production of acorns in the oak forest.

Jones said that in years following a big acorn crop, the number of tick larvae is eight times greater than in years following a poor acorn crop. Additionally, he said, there are about 40 percent more ticks on each mouse.

The researchers tested the effect of acorns by manipulating the population of mice and the availability of acorns in forest plots along the Hudson River. Jones said the work, extended over several seasons, proved the theory that mice and tick populations rise and fall based on the availability of acorns.

(All content © 1998 *Contra Costa Times* and may not be republished without permission. Send comments or questions to newslib@infi.net.) All archives are stored on a SAVE™ newspaper library system from MediaStream Inc., a Knight-Ridder Inc. company. (Photo © Robert Calantine, Visuals Unlimited.)

How could a key-factor analysis be used to pinpoint the phases of importance in determining risk of human disease?

dependent. Seedlings that emerge first in the season stand a much greater chance of surviving; this suggests that competition for resources may be intense (and density dependent).

Overall, therefore, key-factor analysis (its rather misleading name apart) is useful in identifying important phases in the life cycles of study organisms, and useful too in distinguishing the variety of ways in which phases may be important: in contributing significantly to the overall sum of mortality; in contributing significantly to variations in mortality, and hence in *determining* abundance; and in contributing significantly to the *regulation* of abundance by virtue of the density dependence of the mortality. Box 9.2 presents an account of a topical problem, an understanding of which could benefit from key-factor analysis.

9.3 Dispersal, Patches, and Metapopulation Dynamics

dispersal is ignored at the ecologist's peril

In many studies of abundance, the assumption has been made that the major events all occur within the study area, and that immigrants and emigrants can be safely ignored. But migration can be a vital factor in determining and/or regulating abundance. We have already seen that emigration was the predominant reason for the loss of summer adults of the Colorado potato beetle, which was both the key phase in determining population fluctuations and one in which loss was strongly density dependent. We have also discussed previously a detailed study of the effect of seed dispersal on the population dynamics of a Canadian annual plant (*Cakile edentula*—see Section 5.4.1).

populations may show a patchwork of dynamics

Most populations are fragmented and patchy and may include patches that have quite different dynamics. This was clear in the study of buttercups in a pasture, summarized in Figure 9.2. On a different scale, the aphids on one leaf may be crowded, but on the next leaf new colonists may be in the early phases of unimpeded population growth, while on other, older leaves, the aphids may be becoming extinct as the leaf dies. Yet other leaves may still be unoccupied: "habitable sites" that dispersers have not yet reached. Indeed, patchiness is the essence of many ecological processes. The patchy distribution of prey interacts profoundly with the searching behavior of predators (Chapter 8). A population of pathogens persists if there are patches (i.e., hosts) that are successfully occupied (infected), and remain occupied long enough to become infective, and are close enough to allow dispersal to (infection of) uncolonized hosts (Chapter 7).

habitable sites and dispersal distance

The processes determining the abundance of patchily distributed organisms can be linked through the two concepts of "habitable site" and "dispersal distance." Then (as originally suggested by Gadgil, 1971) a population may be small because:

1 There are few patches (habitable sites) that provide the conditions and resources that it requires.
2 The habitable sites support few individuals.
3 The habitable sites remain habitable for only a short time.
4 The dispersal distance between habitable sites is great relative to the dispersibility of the species.

Habitable sites, though, may not always be easy to identify in practice, since they remain uninhabited simply because individuals fail to disperse into them. If we are to discover the limitations that the number of habitable sites places on abundance, it is necessary to identify habitable sites that are not inhabited. One method involves identifying characteristics of habitat patches to which a species is restricted and then determining the distribution and abundance of similar, unoccupied patches. The water vole (*Arvicola terrestris*) lives in river banks, and in a survey of 39 sections of river bank in northern England, ten contained breeding colonies of voles (core sites), 15 were visited by voles but they did not breed there (peripheral sites), and 14 were apparently never used or visited. The core sites were carefully characterized, and on this basis a further 12 unoccupied or peripheral sites that should have been suitable for breeding voles (i.e., habitable sites) were identified. Overall, about 30 percent of habitable sites were in practice uninhabited by voles, because they were too isolated to be colonized or in some cases suffered high levels of predation by mink (Lawton & Woodroffe, 1991).

It is also possible to identify habitable sites for a number of rare butterflies, because their larvae feed on only one or a few patchily distributed plant species. Thus, by identifying habitable sites whether or not they were inhabited, Thomas et al. (1992) found that the silver-studded blue butterfly *Plebejus argus* was able to colonize virtually all habitable sites less than 1 km from existing populations, but those further away (beyond the dispersal powers of the butterfly) remained uninhabited. Indeed, the habitability of some of these isolated sites was established when the butterfly was successfully introduced there (Thomas & Harrison, 1992). This, after all, is the crucial test of whether an uninhabited "habitable" site is really habitable or not.

A radical change in the way ecologists think about populations has involved combining patchiness, dispersal, and within-patch dynamics in the concept of a *metapopulation*, the origins of which are described in Box 9.3. A population can be described as a metapopulation if it can be seen to comprise a collection of subpopulations, each one of which has a realistic chance of going extinct. The essence is a change of focus: less emphasis given to the birth, death, and movement processes going on within a single population, assumed to be more or less homogeneous; but much more emphasis given to the birth (= colonization) and death (= extinction) of subpopulations within the metapopulaion as a whole, and the movement between subpopulations. From this perspective, it becomes apparent that a metapopulation may persist, stably, as a result of the balance between extinctions and recolonizations, even though none of the local subpopulations is stable in its own right. An example of this is shown in Figure 9.7, where within a persistent, highly fragmented metapopulation of the Glanville fritillary butterfly (*Melitaea cinxia*) in Finland, even the largest subpopulations had a high probability of declining to extinction within 2 years.

metapopulations

Even simple models of metapopulations can generate stable, equilibrium populations in which the rates of colonization and extinction are balanced (Box 9.3). But in practice, the observable dynamics of a metapopulation may have more to do with "transient" behavior, far from equilibrium. For example, the silver-spotted skipper butterfly (*Hesperia comma*) declined steadily in Great Britain from a widespread distribution over most calcareous (limestone) hills in 1900, to 46 or fewer refuge localities in ten regions by the early 1960s (Thomas & Jones, 1993). The probable reasons were

transient dynamics may be as important as equilibria

Box 9.3   *Historical Landmarks*

The Genesis of Metapopulation Theory

A classic book, *The Theory of Island Biogeography*, written by MacArthur and Wilson and published in 1967, was an important catalyst in radically changing ecological theory. They showed how the distribution of species on islands could be interpreted as a balance between the opposing forces of extinctions and colonizations (Chapter 10) and focused attention especially on situations in which those species were all available for repeated colonization of individual islands from a common source—the mainland. They developed their ideas in the context of the floras and faunas of real (i.e., oceanic) islands, but their thinking has been rapidly assimilated into much wider contexts with the realization that patches everywhere have many of the properties of true islands—ponds as islands of water in a sea of land, trees as islands in a sea of grass, and so on.

At about the same time as MacArthur and Wilson's book was published, a simple model of "metapopulation dynamics" was proposed by Levins (1969). The concept of *metapopulation* was introduced to refer to a subdivided and patchy population in which the population dynamics operate at two levels:

1 The dynamics of individuals within patches (determined by the usual demographic forces of birth, death, and movement).
2 The dynamics of the occupied patches (or "subpopulations") themselves, within the overall metapopulation (determined by the rates of

colonization of empty patches and of extinction within occupied patches).

Although both this and MacArthur and Wilson's theory embraced the idea of patchiness, and both focused on colonization and extinction rather than the details of local dynamics, MacArthur and Wilson's theory was based on a vision of mainlands as rich sources of colonists for whole archipelagos of islands, whereas in a metapopulation there is a collection of patches but no such dominating mainland.

Levins introduced the variable $p(t)$, the fraction of habitat patches occupied at time t. Note that the use of this single variable carries the profound notion that not all habitable patches are always inhabited. The rate of change in $p(t)$ depends on the rate of local extinction of patches and the rate of colonization of empty patches. It is not necessary to go into the details of Levin's model; suffice to say that as long as the intrinsic rate of colonization exceeds the intrinsic rate of extinction within patches, the total metapopulation will reach a stable, equilibrium fraction of occupied patches, even though none of the local populations is stable in its own right.

Perhaps because of the powerful influence on ecology of MacArthur and Wilson's theory, the whole idea of metapopulations was largely neglected during the 20 years after Levins's initial work. The 1990s, however, saw a great flowering of interest, both in underlying theory and in populations in nature that might conform to the metapopulation concept (Hanski, 1994).

changes in land use—increased plowing of grasslands, reduced stocking with grazing animals, and the virtual elimination of rabbits by myxomatosis with its consequent profound vegetational changes. Throughout this nonequilbrium period, rates of local extinction generally exceeded those of recolonization. In the 1970s and 1980s, however, reintroduction of livestock and recovery of the rabbits led to increased grazing, and suitable habitats increased again. Recolonization exceeded local extinction—

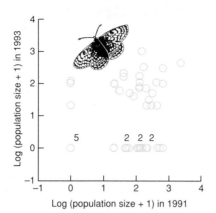

Figure 9.7
Comparison of the subpopulation sizes in June 1991 (adults) and August 1993 (larvae) of the Glanville fritillary butterfly (*Melitaea cinxia*) on Åland Island in Finland. Multiple data points are indicated by numbers. Many 1991 populations, including many of the largest, had become extinct by 1993. (After Hanski et al., 1995.)

but the spread of the skipper remained slow, especially into localities isolated from the 1960s refuges. Even in southeast England, where the density of refuges was greatest, it is predicted that the abundance of the butterfly will increase only slowly —and remain far from equilibrium—for at least 100 years. Thus, it is suggested that around a century of "transient" decline in the dynamics of the metapopulation is to be followed by another century of transient increase—except that the environment will no doubt alter again before that transient phase conveys the metapopulation to equilibrium.

In reality, moreover, there is likely to be a continuum of types of metapopulation: from collections of nearly identical local populations, all equally prone to extinction, to metapopulations in which there is great inequality between local populations, some of which are effectively stable in their own right. This contrast is illustrated in Figure 9.8 for the silver-studded blue butterfly (*Plejebus argus*) in North Wales.

a continuum of metapopulation types

9.4   Temporal Patterns in Community Composition

9.4.1   Founder-controlled and dominance-controlled communities

The metapopulation concept is important when population dynamics are considered in the context of environmental patchiness, but when communities are the focus of attention we usually refer to the *patch dynamics* concept of community organization. The concepts are very closely related. Both accept that populations and communities are usually open systems—with dispersal between patches and differential dynamics within patches—and that a combination of patchiness and movement between patches can give rise to dynamics quite different from those that would be observed if there was just one, homogeneous patch.

Disturbances that open up gaps (patches) are common in all kinds of community. In forests, they may be caused by high winds, lightning, earthquakes, elephants,

disturbances and the patch dynamics concept of community organization

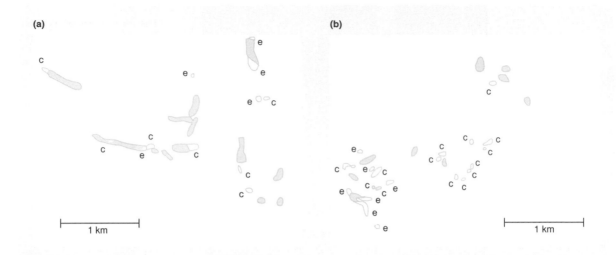

Figure 9.8
Two metapopulations of the silver-studded blue butterfly (*Plejebus argus*) in North Wales. Filled outlines, present in both 1983 and 1990 ("persistent"); open outlines, not present at both times; e, present only in 1983 (presumed extinction); c, present only in 1990 (presumed colonization). (a) In a limestone habitat, where there was a large number of persistent (often larger) local populations among smaller, much more ephemeral local populations (extinctions and colonizations). (b) In a heathland habitat, where the proportion of smaller and ephemeral populations was much greater. (After Thomas & Harrison, 1992.)

lumberjacks, or simply the death of a tree through disease or old age. Agents of disturbance in grassland include frost, burrowing animals, and the teeth, feet, or dung of grazers. On rocky shores or coral reefs, gaps in algal or sessile animal communities may be formed as a result of severe wave action during hurricanes, tidal waves, battering by logs or moored boats, fins of careless scuba divers, or the action of predators.

From the viewpoint of the effects of gap creation and reoccupancy, two fundamentally different kinds of community organization can be recognized (Yodzis, 1986). Situations in which all species are good colonists and essentially equal competitors are described as *founder controlled*, whereas those in which some species are strongly competitively superior can be described as *dominance controlled*. The dynamics of these two situations are quite different, and we deal with them in turn.

founder-controlled communities— competitive lotteries

In founder-controlled communities, species are approximately equivalent in their ability to invade gaps, are equally tolerant of the abiotic conditions, and can hold the gaps against all comers during their lifetime. Hence, the probability of competitive exclusion in the community as a whole may be much reduced where gaps appear continually and randomly. This can be referred to as a "competitive lottery". Figure 9.9 illustrates how the occupancy of a series of gaps is likely to change through time. On each occasion that an organism dies (or is killed) the gap is reopened for invasion. All conceivable replacements are possible, and species richness can be expected to be maintained at a high level in the system as a whole.

Some tropical reef communities of fish may conform to this model (Sale & Douglas, 1984). They are extremely rich in species. For example, the number of species of fish on the Great Barrier Reef off the east coast of Australia ranges from

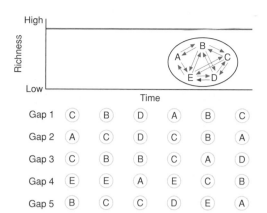

Figure 9.9
Founder-controlled community dynamics. There are five patches (arranged vertically, below) that change in their occupancy over time. Each of species A to E is equally likely to fill a gap, regardless of the identity of its previous occupant (this is illustrated in the inset). The detailed sequence of occupancy is therefore unpredictable. Species richness remains high and relatively constant. (Compare with Figure 9.10.)

900 in the south to 1,500 in the north, and more than 50 resident species may be recorded on a single patch of reef 3 m in diameter. Only a proportion of this species richness is likely to be attributable to resource partitioning of food and space—indeed the diets of many of the coexisting species are very similar. In this community, vacant living space seems to be a crucial limiting factor, and it is generated unpredictably in space and time when a resident dies or is killed. The lifestyles of the species match this state of affairs. They breed often, sometimes year-round, and produce numerous clutches of dispersive eggs or larvae. It can be argued that the species compete in a lottery for living space in which larvae are the tickets, and the first arrival at the vacant space wins the site, matures quickly, and holds the space for its lifetime.

In dominance-controlled communities, by contrast, some species are competitively superior to others, and an initial colonizer of a patch cannot necessarily maintain its presence there. Dispersal between patches, or growth of an individual within a patch, will bring about a reshuffle, and species may be competitively excluded locally. In these cases, disturbances that open up gaps lead to reasonably predictable species sequences because different species have different strategies for exploiting resources—early species are good colonizers and fast growers, whereas later species can tolerate lower resource levels and grow to maturity in the presence of early species, eventually outcompeting them. Such sequences are examples of community *successions*. The effect of the disturbance is to knock the community back to an earlier stage of succession (Figure 9.10). The open space is colonized by one or more of a group of opportunistic, early successional species (p_1, p_2, etc., in Figure 9.10). As time passes, more species invade, often those with poorer powers of dispersal. These eventually reach maturity, dominating midsuccession (m_1, m_2, etc.), and many or all of the pioneer species are driven to extinction. Later still, the community regains a *climax* stage when the most efficient competitors (c_1, c_2, etc.) oust their neighbors. In this sequence, if it runs its full course, the number of species first increases (because of colonization) then decreases (because of competition).

One study that has provided support for this pattern was carried out on the rocky shores of southern California (Sousa, 1979a, 1979b). In an intertidal algal community associated with boulders of various sizes, wave action disturbs small boulders more often than large ones. Hence, the successional sequence that follows the clearing of a

The Great Barrier Reef (© Dave Fleetham, Visuals Unlimited).

dominance-controlled communities and community succession

illustrated by rocky shore data

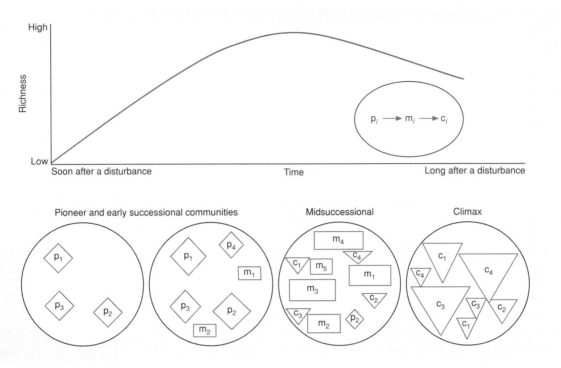

Figure 9.10
Hypothetical succession in a gap—an example of dominance control. The occupancy of gaps is reasonably predictable. Richness begins at a low level as a few pioneer (p_i) species arrive; reaches a maximum in midsuccession when a mixture of pioneer, midsuccessional (m_i) and climax (c_i) species occur together; and drops again as competitive exclusion by climax species takes place. (Compare with Figure 9.9.)

space (but in the absence of further disturbance) could be established by studying both large boulders that had been experimentally cleared and concrete blocks that had been implanted. Within the first month the surfaces were colonized by a mat of an ephemeral green alga, *Ulva* spp. Then, in the fall and winter of the first year, several species of perennial red alga became established, including *Gelidium coulteri*, *Gigartina leptorhynchos*, *Rhodoglossum affiine*, and *Gigartina canaliculata*. However, the last of these gradually came to dominate the community, holding 60–90 percent of the surface area after 2–3 years. Thus, as described in Figure 9.10, the number of species on a boulder ("in a patch") increased during early stages of succession through a process of colonization but declined again subsequently because of competitive exclusion by *G. canaliculata*. The same succession occurred on small boulders that had been made artificially stable.

9.4.2 Community succession

the importance of phasing   Some disturbances are synchronized over extensive areas. A forest fire may destroy a huge tract of a climax community. The whole area then proceeds through a more or less synchronous succession, with diversity increasing through the early colonization phase and falling again through competitive exclusion as the climax is approached. Other disturbances are much smaller and produce a patchwork of habitats. If these disturbances are out of phase with one another, the resulting community comprises a mosaic of patches at different stages of succession. This is true of the Californian shore, described previously, where different boulders are disturbed at different times. For a more detailed description of a succession, however, we turn next to a much more synchronized disturbance.

Successions on old fields have been studied primarily in the eastern United States, where many farms were abandoned by farmers who moved west after the frontier was opened up in the 19th century. Indeed, in many places, a series of sites are available that have been abandoned for different, recorded periods of time. Most of the pre-colonial mixed conifer–hardwood forest had been destroyed, but regeneration was swift after the "disturbance" caused by farmers came to an end. The early pioneers of the American West left behind exposed land that was colonized by pioneers of a very different kind!

community succession on fields abandoned by farmers

The typical sequence of dominant vegetation is:

annual weeds → herbaceous perennials → shrubs → early successional trees → late successional trees

The pioneer species are those that can establish themselves quickly in the disturbed habitat of a recently cultivated field, either by rapid dispersal into the site or from propagules that are already present. Perhaps the most common annual of old field succession in the eastern United States is *Ambrosia artemisiifolia*, which has seeds that survive for many years in the soil and germinate when disturbance brings them to the surface, where they experience unfiltered light, reduced carbon dioxide concentration, and fluctuating temperatures—all conditions that induce germination in many annuals. However, summer annuals such as *Ambrosia* spp. are often quickly eclipsed in importance by winter annuals that have small seeds with little or no dormancy but disperse over long distances. Their seeds germinate soon after they land, usually in late summer or fall. Next spring they have a head start over summer annuals and thus they pre-empt the resources of light, water, space, and nutrients.

Early successional plants have a fugitive lifestyle. Their continued survival depends on dispersal to other disturbed sites. They cannot persist in competition with later species, and thus they must grow and consume the available resources rapidly. High growth and photosynthetic rates are crucial properties of the fugitive. Those of later successional plants are much lower (Table 9.4).

In contrast to the pioneer annuals, seeds of later successional plants can germinate in the shade—for example, beneath a forest canopy. They can continue to grow at these low light intensites, too—quite slowly but faster than the species they replace (Figure 9.11).

The trees that take part in the later stages of an old field succession can themselves be grouped into early and late successional classes. Many early successional trees have multilayered foliage. Leaves extend deep into the canopy, where they still receive enough light to add more to the plant (in photosynthate) than they take from it. Species such as the eastern red cedar (*Juniperus virginiana*) are thus able to use the abundant light that is available early in the tree stage of the succession. In contrast, sugar maple

early and late successional species have different properties

Table 9.4
Some representative photosynthetic rates (mg CO_2 dm^{-2} h^{-1}) of plants in a successional sequence. Late successional trees are arranged according to their relative successional position. (Afer Bazzaz, 1979.)

Plant	Rate	Plant	Rate
Summer annuals		Early successional trees	
Abutilon theophrasti	24	Diospyros virginiana	17
Amaranthus retroflexus	26	Juniperus virginiana	10
Ambrosia artemisiifolia	35	Populus deltoides	26
Ambrosia trifida	28	Sassafras albidum	11
Chenopodium album	18	Ulmus alata	15
Polygonum pensylvanicum	18		
Setaria faberii	38	Late successional trees	
		Liriodendron tulipifera	18
Winter annuals		Quercus velutina	12
Capsella bursa-pastoris	22	Fraxinus americana	9
Erigeron annuus	22	Quercus alba	4
Erigeron canadensis	20	Quercus rubra	7
Lactuca scariola	20	Aesculus glabra	8
		Fagus grandifolia	7
Herbaceous perennials		Acer saccharum	6
Aster pilosus	20		

(*Acer saccharum*) and American beech (*Fagus grandifolia*) are monolayered species. They possess a single layer of leaves in a shell around the tree and are more efficient in the crowded canopy of late succession. If a monolayered and a multilayered tree simultaneously colonize an open site, the multilayered tree will usually grow faster and dominate until it is crowded by its neighbors, when the more slowly growing monolayered tree emerges to dominance.

The early colonists among trees usually have efficient seed dispersal; this in itself makes them likely to be early on the scene. They are usually precocious reproducers and are soon ready to leave descendants in new sites elsewhere. The late colonists are those with larger seeds, poorer dispersal, and long juvenile phases. The contrast is between the lifestyles of the "quickly come, quickly gone" and "what I have, I hold".

experiments unravel the multiple causes of an old field succession

A particularly detailed study of old field succession has been performed at the Cedar Creek Natural History Area in Minnesota on well-drained and nutrient-poor soil (see Section 1.3.2, where this study is discussed in detail). One consequence of the poor growing conditions is that species replacement is slow: annual plants persist as dominants for up to 40 years after field abandonment and woody plants (mostly vines and shrubs) still provide only 13 percent of ground cover in 60-year-old fields. In other respects, however, the pattern of species replacement is similar to what has been described elsewhere (see Figure 1.11).

animals are often affected by, but may also affect, plant successions

The fact that plants dominate most of the structure and succession of communities does not mean that the animals always follow the communities that plants dictate. This will often be the case, of course, because the plants provide the starting point for all food webs and determine much of the character of the physical environment in which

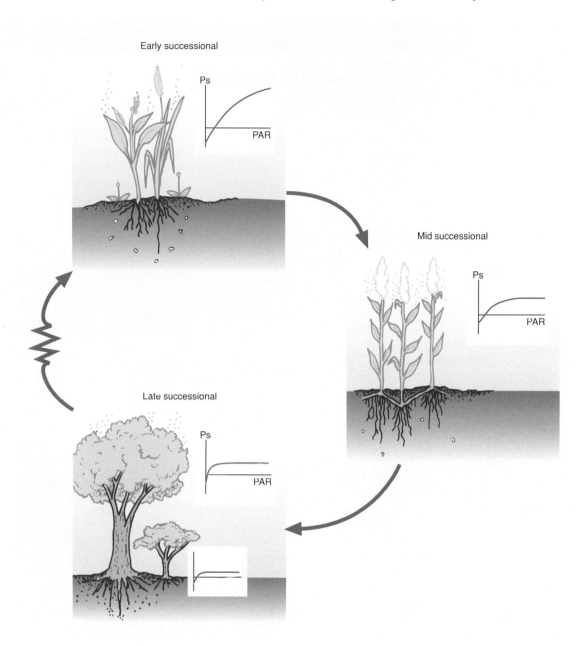

Figure 9.11
Idealized light saturation curves (photosynthetic rate, Ps, plotted against the quantity of photosynthetically active radiation, PAR) for early, mid, and late successional plants.
(After Bazzaz, 1996.)

Box 9.4   *Topical ECOncerns*

Conservation Sometimes Requires
Manipulation of a Succession

Some endangered animal species are associated with particular stages of a succession. Their conservation then depends on a full understanding of the successional sequence, and intervention may be required to maintain their habitat at an appropriate successional stage.

An intriguing example is provided by a giant New Zealand insect, the weta *Deinacrida mahoenuiensis* (Orthoptera, Anostostomatidae). This species, which is believed to have been formerly widespread in forest habitat, was discovered in the 1970s in an isolated patch of gorse (*Ulex europaeus*). Ironically, in New Zealand gorse is an introduced weed that farmers spend much time and effort attempting to control. Its dense, prickly sward provides a refuge for the giant weta against other introduced pests, particularly rats but also hedgehogs, stoats, and possums, which could readily capture weta in their original forest home. Mammalian predation is believed to be responsible for weta extinction elsewhere.

New Zealand's Department of Conservation purchased this important patch of gorse from the landowner, who insisted that his cattle should still be permitted to overwinter in the reserve. Conservationists were unhappy about this, but the cattle subsequently proved to be part of the weta's salvation. By opening up paths through the gorse, cattle provided entry for feral goats that browse the gorse, producing a dense hedge-like sward and preventing the habitat from succeeding to a stage inappropriate to the wetas.

This story involves a single endangered endemic insect together with a whole suite of introduced pests

(gorse, rats, goats, etc.) and introduced domestic animals (cattle). Before the arrival of people in New Zealand, the island's only land mammals were bats, and New Zealand's endemic fauna has proved to be extraordinarily vulnerable to the mammals that arrived with people. However, by maintaining gorse succession at an early stage, the grazing goats provide a habitat in which the weta can escape the attentions of rats and other predators.

Because of its economic cost to farmers, ecologists have been trying to find an appropriate biological control agent for gorse, ideally one that would eradicate it. How would you weigh up the needs of a rare insect against the economic losses associated with gorse on farms?

A giant weta on a gorse branch. (Courtesy of Greg Sherley, Department of Conservation, Wellington, New Zealand.)

animals live. But it is also sometimes the animals that determine the nature of the plant community, for example, through heavy grazing or trampling (Box 9.4). More often, though, animals are passive followers of successions among the plants. This is the case for bird species in an old field succession (see Figure 4.10).

Do successions reach a climax? In fact, it is generally very difficult to identify a stable "climax" community in the field. Usually we can do no more than point out that the rate of change of succession slows to the point where any change is imperceptible to us. The succession of seaweeds on overturned boulders (Sousa, 1979a, 1979b) is unusual because it may reach a climax in only a few years. Old field successions, on the other hand, might take 100–300 years to reach a climax, but in that time the probabilities of fire or severe hurricanes, which occur every 70 or so years in New England, are so high that a process of succession may never go to completion. Bearing in mind that forest communities in northern temperate regions, and probably also in the tropics, are still recovering from the last glaciation, it is questionable whether the idealized climax vegetation is often reached in nature.

concept of the climax

Finally, we again return to issues of scale that feature in almost every chapter of this book. A forest, or a rangeland, that appears to have reached a stable community structure when studied on a scale of hectares will always be a mosaic of miniature successions. Every time a tree falls or a grass tussock dies an opening is created in which a new succession starts. The patch dynamics pattern of many communities is a result of the dynamic processes of deaths, replacements, and microsuccessions that the broad view may conceal.

9.5 Food Webs

No predator–prey, parasite–host, or grazer–plant pair exists in isolation. Each is part of a complex web of interactions with other predators, parasites, food sources, and competitors within its community. Ultimately, it is these food webs that ecologists wish to understand. However, it has been useful to isolate groups of competitors as we did in Chapter 6, and mutualists, symbionts, and parasite–host pairs as in Chapter 7, and predator–prey pairs as in Chapter 8, simply because we have little or no hope of understanding the whole unless we have some understanding of the component parts. Toward the end of Chapter 8 (see Section 8.6), our field of view was expanded to include the effects of predators on groups of competitors and to show, for example, the importance of predator-mediated coexistence.

We now take this approach a stage further to focus on systems with at least three trophic levels (plant–herbivore–predator), and consider not only direct but also indirect effects that a species may have on others on the same or on other trophic levels. The effects of a predator on both individuals and the populations of its herbivorous prey, for example, are direct and relatively straightforward. But these effects may also be felt by any plant population on which the herbivore feeds, or by other predators and parasites of the herbivore, or by other consumers of the plant, or competitors of the herbivore and the plant, or by the myriad species linked even more remotely in the food web.

food webs —shifting the focus to systems with at least three trophic levels

9.5.1 Indirect and direct effects

The deliberate removal of a species from a community can be a powerful tool in unraveling the workings of a food web. We might expect such a removal to lead to an increase in the abundance of a competitor, or, if the species removed is a predator, to an increase in the abundance of its prey. Sometimes, however, when a species is removed, a competitor may actually decrease in abundance, and the removal of a predator can lead to a decrease in a prey species population. Such unexpected effects arise when direct effects are less important than effects that occur through indirect pathways. Thus, removal of a species might increase the density of one competitor, which in turn causes another competitor to decline. Or the removal of a predator might increase the abundance of a prey species that is competitively superior to another, leading to a decrease in the density of the latter.

direct and indirect effects of
shorebirds on limpet populations

In one study, for example, predation by birds was experimentally manipulated over a 2-year period in an intertidal community on the northwest coast of the United States to determine the consequences for three limpet species and their algal food. Glaucous-winged gulls (*Larus glaucescens*) and oystercatchers (*Haematopus bachmani*) were excluded by means of wire cages from large areas (each 10 m²) in which limpets were common. It became evident that excluding the birds increased the overall abundance of one of the limpet species, *Lottia digitalis*, as might have been expected, but a second limpet species (*L. strigatella*) became rarer, and the third, *L. pelta*, which was the one most frequently consumed by the birds, did not vary in abundance. The reasons are complex and go well beyond the direct effects of birds eating limpets (Figure 9.12).

Lottia digitalis, a light-colored limpet, tends to occur on light-colored goose barnacles (*Pollicipes polymerus*), where it is camouflaged, whereas the dark *L. pelta* occurs primarily on dark Californian mussels (*Mytilus californianus*). Predation by birds normally reduces the area covered by goose barnacles, and so excluding the birds increased goose barnacle abundance and also increased the abundance of *L. digitalis* (Figure 9.12). Increasing barnacle abundance through bird exclusion also led to a decrease in the area covered by mussels, because they were now subject to more intense competition from the barnacles (Figure 9.12). This, one imagines, might have led to a decrease in the abundance of *L. pelta* living predominantly on those mussels. However, the third limpet species, *L. strigatella*, is competitively inferior to the others, and the increase in abundance of *L. digitalis* when birds were excluded therefore led to a decrease in the abundance of *L. strigatella*, which in turn released pressure on *L. pelta* such that overall its abundance remained effectively unchanged (Figure 9.12).

The effects of bird predation also cascade down to the plant trophic level, because by consuming limpets, the birds normally reduce the grazing pressure of the limpets on fleshy algae, and by consuming goose barnacles, the birds normally free up space for algal colonization. Hence, when the birds were excluded, algal cover decreased (Figure 9.12). It is apparent that short-term studies of pairwise interactions would have failed to reveal the rich array of direct and indirect interactions in this food web.

keystones in food web architecture

Some species are more intimately and tightly woven into the fabric of the food web than others. A species whose removal would produce a significant effect (extinction or a large change in density) in at least one other species may be thought of as a strong interactor. Removal of some strong interactors would lead to significant changes spreading throughout the food web—we refer to these as *keystone species*. In building construction, a keystone is the wedge-shaped block at the highest point of an arch that locks the other pieces together. Removal of the keystone species, just like removal of the keystone in an arch, leads to collapse of the structure. More precisely, it leads to extinction or large changes in abundance of several species, producing a community

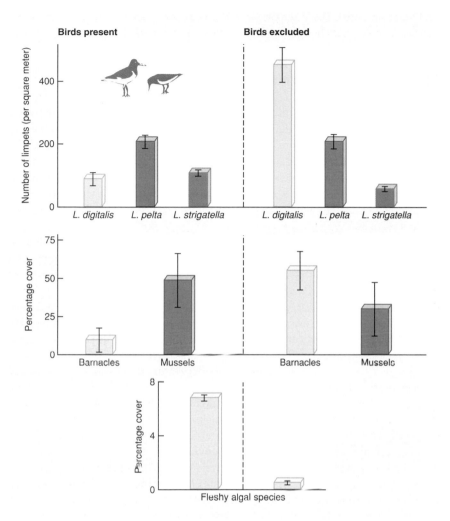

Figure 9.12
When birds are excluded from the intertidal community, barnacles increase in abundance at the expense of mussels, and three limpet species show marked changes in density, reflecting changes in the availability of cryptic habitat and competitive interactions as well as the easing of direct predation. Algal cover is much reduced in the absence of effects of birds on intertidal animals (means ± standard errors are shown). (After Wootton, 1992.)

with a very different species composition and, to our eyes, an obviously different physical appearance.

Although the term was originally applied only to predators, it is now widely accepted that *keystone species* can occur at any trophic level. For example, lesser snow geese (*Chen caerulescens caerulescens*) are herbivores that breed in large colonies in coastal marshes along the west coast of Hudson Bay in Canada. At their nesting sites in the spring, before growth of aboveground foliage begins, adult geese grub for roots and rhizomes of plants in dry areas and eat the swollen bases of shoots of sedges in wet areas. Their activity creates bare areas (1–5 m²) of peat and sediment. Few pioneer plant species are able to recolonize these patches, and recovery is very slow. Furthermore, in areas of intense summer grazing, "lawns" of *Carex* and *Puccinellia* spp. become established. Here, therefore, high densities of grazing geese are essential to maintain the species composition of the vegetation and its aboveground production

(Kerbes et al., 1990). The lesser snow goose is a keystone species—the whole structure and composition of these communities are drastically altered by its presence.

9.5.2   Top-down or bottom-up control of food webs?

One of the most fundamental questions about food webs is whether they are controlled from the top down or from the bottom up. *Top-down control* refers to situations in which the structure (abundance, species number) of lower trophic levels depends on the effects of consumers from higher trophic levels—"predators" controlling "prey". *Bottom-up control* refers to a dependence of community structure on factors, such as nutrient concentration and prey availability, that influence a trophic level from below; in this case, populations within a trophic level are affected predominantly by competition not predation.

dependence on the number of trophic levels

It is helpful to dissect the question of whether predators or resources will dominate the dynamics of populations by considering first a hypothetical community with just one (plant) trophic level and then successively adding extra trophic levels, one at a time.

It is not easy to envisage real communities in nature that consist of only one trophic level, but if one existed, predation would be absent by definition, control would be bottom up, and competition would be the predominant population interaction (Figure 9.13).

Two-trophic-level systems can be identified if we allow ourselves to isolate limited but significant components of real systems. For example, grazing by the giant tortoise on the most remote island on Earth, Aldabra, crops the turf down to less than 5 mm over extensive areas (Strong, 1992). When the tortoises are excluded by fences, many species of trees, shrubs, and grasses, normally excluded by grazing, grow and dominate the plant community. No predators of the giant tortoises, or their eggs or young, have arrived on the island. In this two-trophic-level system, then, there is clear top-down control of the bottom trophic level, where the effects of predation predominate, and bottom-up control of the herbivores (Figure 9.13).

The saline Great Salt Lake of Utah provides a case of a three-trophic-level system, when the usual zooplankton–phytoplankton system is augmented by a third trophic level (a predatory insect, *Trichocorixa verticalis*) in unusually wet years when salinity declines. Normally, the zooplankton keep phytoplankton biomass at a low level. But

Figure 9.13
Diagrammatic representation of communities with one, two, three, or four trophic levels, illustrating for each trophic level whether control is predicted to be bottom up (B) or top down (T), and whether population dynamics are determined primarily by competition (comp) or predation (pred).

	1	2	3	4
Secondary carnivore				B/comp
Primary carnivore			B/comp	T/pred
Herbivore		B/comp	T/pred	B/comp
Plant	B/comp	T/pred	B/comp	T/pred

Figure 9.14
Variation in the pelagic ecosystem of the
Great Salt Lake during three periods that
differed in salinity. Decreased salinity in
1985–6 allowed invasion of the predator
Trichocorixa, which reduced the density of
the grazer *Artemia*, which allowed the
density of phytoplankton to rise (more
chlorophyll and lowered water
transparency). (After Wurtsbaugh, 1992.)

when salinity declined from above 100 to 50 g l^{-1} in 1985, *Trichocorixa* invaded and
reduced zooplankton biomass from 720 to 2 mg m^{-3}, leading to a 20-fold increase in
phytoplankton concentration (Figure 9.14). In this three-trophic-level community at
least, the plants are subject to bottom-up control, having been released from heavy
grazing pressure by the effects of carnivores on the herbivores. The herbivores are thus
subject to top-down and the carnivores to bottom-up control (Figure 9.13).

The top-down view was first introduced in a famous paper that proposed that "the
world is green" (Hairston et al., 1960). It argued that green plant biomass accumulates
because carnivores keep herbivores in check. The Great Salt Lake study is one of
several that tend to support this argument. An alternative view, however, which has
been described as "the world is prickly and tastes bad" (Pimm, 1991), emphasizes that
many plants have evolved physical and chemical defenses that make life difficult for
herbivores (see Section 3.4.2). Thus, the world may be green—where it is green—
because plants are wholly or partly inedible and not because herbivores are kept in

UNANSWERED QUESTION: how
green is our world?

UNANSWERED QUESTION: is the
world green or prickly and bad
tasting?

Figure 9.15
Two examples of food webs with four trophic levels. Arrows indicate the strength of predation or grazing (rather than the flow of energy). (a) The absence of omnivory (feeding at more than one trophic level) in this North American stream community means it functions, as predicted in Figure 9.13, as a four-trophic-level system. (After Power, 1990.) (b) On the other hand, this web from a terrestrial Bahamanian community functions as a three-trophic-level web. This is because of the strong direct effects of omnivorous top predators on herbivores and their less influential effects on intermediate predators. (After Spiller & Schoener, 1994.)

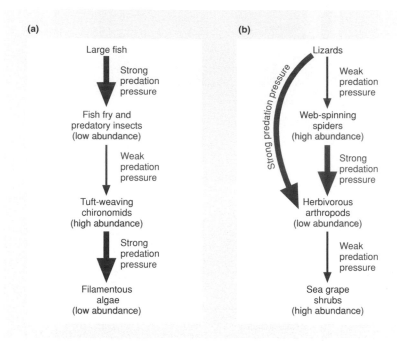

check by their predators. Which of these views has most support, and under what circumstances, remains to be seen.

If the pattern developing thus far is continued, we expect that in a four-trophic-level system, the plants and the primary carnivores will be limited top down, whereas the herbivores and secondary carnivores will be limited bottom up (Figure 9.13). This is precisely what was found in a study of the food web in Eel River, northern California. Large fish reduced the abundance of fish fry and invertebrate predators, allowing their prey, tuft-weaving midge larvae, to attain high densities and to exert intense grazing pressure on filamentous algae, whose biomass was thus kept low (Figure 9.15a).

However, in a four-trophic-level terrestrial community in the Bahamas, consisting of sea grape shrubs, which were fed upon by herbivorous arthropods, and then web spiders (primary carnivores) and lizards (secondary or "top" carnivores), the results of experimental manipulations indicated a strong effect of the lizards on the herbivores but a weaker effect of the lizards on the spiders. Consequently, the net effect of the top predators on the plants was positive and there was less leaf damage in the presence of lizards. In essence, this four-trophic-level community functions as if it has only three levels (compare Figure 9.15a and b).

Overall, therefore, it is apparent, first, that although the food webs of the real world may be disassembled into series of simple predator–prey links for our convenience of study, these isolated pairs can never tell the whole story. It is also apparent, however, that not even the first line of that story could be written unless it were based on a solid foundation of understanding predator–prey dynamics.

What Do We Mean by "Community Stability"?

Among several, there are two important qualifications that can be made when we come to decide what we mean by stability. The first is the distinction between the resilience of a community and its resistance. A *resilient community* is one that returns rapidly to something like its former structure after that structure has been altered. A *resistant community* is one that undergoes relatively little change in its structure in the face of a disturbance.

The second distinction is between *fragile* and *robust stability*. A community has only *fragile stability* if it remains essentially unchanged in the face of a small disturbance but alters utterly when subjected to a larger disturbance, whereas one that stays roughly the same in the face of much larger disturbances is said to have stability that is *dynamically robust*.

To illustrate these distinctions by analogy, consider the following:

- a pool or billiard ball balanced carefully on the end of a cue,

- the same ball resting on the table,
- the ball sitting snugly in its pocket.

The ball on the cue is stable in the narrow sense that it will stay there forever as long as it is not disturbed—but its stability is fragile, and both its resistance and its resilience are low: the slightest touch will send the ball to the ground, far from its former state (low resistance), and it has not the slightest tendency to return to its former position (low resilience).

The same ball resting on the table has a similar resilience: it has no tendency to return to exactly its former state (assuming the table is level), but its resistance is far higher: pushing or hitting it moves it relatively little. And its stability is also relatively robust—it remains "a ball on the table" in the face of all sorts and all strengths of assault with the cue.

The ball in the pocket, finally, is not only resistant but resilient too—it moves little and then returns—and its stability is highly robust: it will remain where it is in the face of almost everything other than a hand that carefully plucks it away.

9.5.3   Community stability and food web structure

Of all the imaginable food webs in nature, are there particular types that we tend to observe repeatedly? Do actual (as opposed to imaginable) food webs have particular properties? Are some food web structures more stable than others? (We discuss what *stable* means in Box 9.5.) These are important practical questions. We require answers if we are to determine whether some communities are more fragile (and more in need of conservation) than others; or whether there are certain "natural" structures that we should aim for when we construct communities ourselves; or whether communities that have been restored are likely to stay "restored". Progress in answering these questions, though, depends critically on the quality of data that are gathered from natural communities. This is not always as good as it needs to be. There are many unresolved questions.

The aspect of structure that has received most attention from the point of view of stability has been community complexity. For a long time, the conventional wisdom, arrived at largely through "logical" argument, was that increased complexity within a

a long-standing belief that complexity leads to stability . . .

community leads to increased stability (MacArthur, 1955; Elton, 1958). For example, it was argued that in more complex communities, with more species and more interactions, there were more possible pathways by which energy passed through the community. Hence, if there was a perturbation to the community (a change in the density of one of the species), this would affect only a small proportion of the energy pathways and would have relatively little effect on the densities of other species: the complex community would be resistant to change (Box 9.5).

. . . that is not supported by mathematical models

However, as analyses of mathematical models of food webs became more sophisticated, the conventional wisdom by no means always received support (reviewed by May, 1981). Briefly, these model food webs were characterized by (1) the number of species they contained, (2) the *connectance* of the web (the fraction of all possible pairs of species that interact directly with one another—as competitors, mutualists, or predators and prey), and (3) the average interaction strength between pairs of species. Most models came to similar conclusions: increases in the number of species, increases in connectance, and increases in average interaction strength—each representing an increase in complexity—all tended to decrease resilience, that is the tendency of the community to return to its former state following a disturbance (Box 9.5). Thus these models suggest that complexity leads to *instability*, and certainly indicate that there is no necessary, unavoidable connection linking stability to complexity.

complexity and stability in practice

What is the evidence from real communities? A number of studies have sought to build on the mathematical models by examining the relationships among number of species, connectance, and interaction strength. The argument runs as follows. The only communities we can observe are those that are stable enough to exist. Hence, those with more species can only be sufficiently stable if there are compensatory decreases in connectance and/or interaction strength. But data on interaction strengths for whole communities are unavailable. Hence, we can assume, for simplicity, that average interaction strength is constant. Thus, communities with more species will only retain stability if there is an associated reduction in average connectance.

a prediction supported by some studies but not others

In fact, the first analysis of a compilation of published food webs (Briand, 1983) found, as predicted, that connectance decreased with species number (Figure 9.16a). This compilation suffered from a very serious drawback, however—the data on which it was based were not collected for the purpose of quantitative study of food web properties. In particular, the accuracy of identification varied substantially from web to web, and even in the same web components were sometimes grouped at the level of kingdom (e.g., "plants"), sometimes as a family (e.g., Diptera), and sometimes as a species (polar bear) (see review by Hall & Raffaelli, 1993). More recent studies, then, in which food webs were more rigorously documented, indicate that connectance may decrease with species number (as predicted) (Figure 9.16b), or may be independent of species number (Figure 9.16c), or may even increase with species number (Figure 9.16d). Thus, the stability argument does not receive consistent support from food web analyses either.

a complex community that is less stable when disturbed . . .

The relationship between complexity and stability has also been investigated by manipulating two sets of plant communities. In the first, the perturbation consisted of adding plant nutrients to the soil; in the second it involved introducing African buffalo to graze the plants. In both, the effect was monitored in both species-rich and species-poor communities, and in both the perturbation significantly reduced the diversity of the species-rich but not the species-poor community (Table 9.5)—suggesting

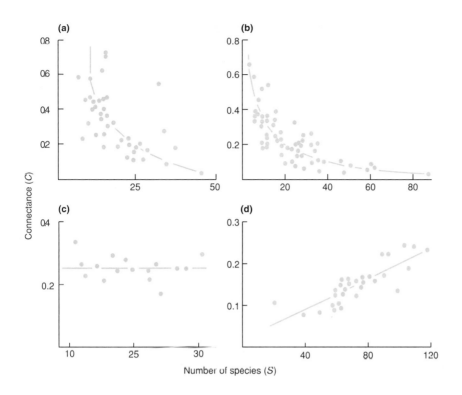

Figure 9.16
The relationships between connectance and species richness. (a) For a compilation from the literature of 40 food webs from terrestrial, freshwater, and marine environments. (After Briand, 1983.) (b) For a compilation of 95 insect-dominated webs from various habitats. (After Schoenly et al., 1991.) (c) For seasonal versions of a food web for a large pond in northern England, varying in species richness from 12 to 32. (After Warren, 1989.) (d) For food webs from swamps and streams in Costa Rica and Venezuela. (Winemiller, 1990; after Hall & Raffaelli, 1993.)

that more complex communities are *less* likely to resist change in the face of a perturbation.

On the other hand, in a "natural experiment" involving a set of eight varied grassland communities, those that were more diverse in species were more stable in community composition in the face of a severe summer drought in Yellowstone National Park, USA (Figure 9.17). Experimental studies too, therefore, provide conflicting results in relation to the predicted complexity/stability relationship.

... and another that is more stable

The conflicting results among both models and experiments, and the variety of types of stability that they have used, suggest, at least, that no single relationship between complexity and stability will be appropriate in all communities. It would be wrong to replace one sweeping generalization with another. Even if complexity and instability were connected in models, moreover, this would not necessarily mean that we should expect to see an association between complexity and instability in real communities. Unstable communities will fail to persist when they experience environmental conditions that reveal their instability. But the range and predictability of environmental conditions will vary from place to place. In a stable and predictable environment, a community that is dynamically fragile may nevertheless persist. But in a variable and unpredictable environment, only a community that is dynamically robust will be able to persist. Hence, we might expect to see (1) complex and fragile communities in stable and predictable environments, with simple and robust communities in variable and unpredictable environments; and (2) approximately the same

environmental predictability linked to community fragility?

Table 9.5
The influence of (a) nutrient addition on species richness and diversity (Shannon's index, H; see Box 10.1) in two fields and (b) grazing by African buffalo on species diversity in two areas of vegetation. (After McNaughton, 1977.)

	Control plots	Experimental plots	Statistical significance
(a) *Nutrient addition*			
Species richness per 0.5 m² plot			
Species-poor plot	20.8	22.5	n.s.
Species-rich plot	31.0	30.8	n.s.
Equitability			
Species-poor plot	0.660	0.615	n.s.
Species-rich plot	0.793	0.740	$P < 0.05$
Diversity			
Species-poor plot	2.001	1.915	n.s.
Species-rich plot	2.722	2.532	$P < 0.05$
(b) *Grazing*			
Species diversity			
Species-poor plot	1.069	1.357	n.s.
Species-rich plot	1.783	1.302	$P < 0.005$

n.s., not statistically significant.

Figure 9.17
Relationship between resistance in grassland community composition and species diversity (Shannon's index, H) for a number of grassland areas in Yellowstone National Park. The resistance measure (R) is designed to be inversely related to the cumulative differences in species abundances at sites between 1988 (a year of severe drought) and 1989 (a year of normal rainfall). Thus a high value for R indicates that relative abundances changed little in the face of the drought, whereas a low value means they changed considerably. (After Frank & McNaughton, 1991.)

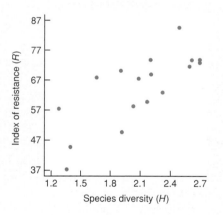

observed stability (in terms of population fluctuations, and so forth) in all communities, since this will depend on the inherent stability of the community combined with the variability of the environment.

This line of argument, moreover, carries a further, very important implication for the likely effects of unnatural perturbations caused by humans on communities. We might expect these to have their most profound effects on the dynamically fragile, complex communities of stable environments, which are relatively unaccustomed to perturbations, and least effect on the simple, robust communities of variable environments, which have previously been subjected to repeated (albeit natural) perturbations.

Figure 9.18
In New Zealand streams, less disturbed sites support more "complex" communities (more species and greater connectance between species). (a) Web size (number of species) and (b) connectance or the average number of feeding links per animal species (number of prey species in diet) increases with the intensity of flow-related disturbances to the streambed. (After Townsend et al., 1998.)

One study addressing the question of food web complexity and stability and environmental stability investigated ten small streams in New Zealand that differ in the intensity and frequency of flow-related disturbances to their beds (Figure 9.18). Food webs in the more disturbed streams are characterized by smaller numbers of species and fewer links between species. This conforms to the prediction that simpler food webs should be found in more variable environments because they possess a structure that confers stability. These results are therefore intriguing, but inevitably fall short of establishing a general rule. There is a pressing requirement for more rigorous descriptions of food webs involving a standard methodology, and for analyses from related sets of habitats that nonetheless differ in key, crucial features. Many more empirical and experimental studies are needed before our understanding of food webs approaches maturity.

UNANSWERED QUESTION: so far more questions than answers have arisen in the important field of food web theory

Summary

Multiple determinants of the dynamics of populations

To understand the factors responsible for the population dynamics of even a single species in a single location, it is necessary to have a knowledge of physicochemical conditions, available resources, the organism's life cycle, and the influence of competitors, predators, and parasites on rates of birth, death, immigration, and emigration.

There are contrasting theories to explain the abundance of populations. At one extreme, researchers emphasize the apparent stability of populations and point to the importance of forces that stabilize them (density-dependent factors). At the other extreme, those who place more emphasis on density fluctuations may look at external (often density-independent) factors to explain the changes. Key-factor analysis is a technique that can be applied to life table studies to throw light both on determination and on regulation of abundance.

Dispersal, patches, and metapopulation dynamics

Movement can be a vital factor in determining and/or regulating abundance. A radical change in the way ecologists think about populations has involved combining ideas about patchiness, dispersal, and population dynamics within patches in *metapopulation theory*.

Temporal patterns in community composition

Disturbances that open up gaps (patches) are common in all kinds of community. *Founder-controlled* communities are those in which all species are approximately equivalent in their ability to invade gaps and are equal competitors that can hold the gaps against all comers during their lifetime. On the other

hand, *dominance-controlled* communities are those in which some species are competitively superior to others so that an initial colonizer of a patch cannot necessarily maintain its presence there.

The phenomenon of dominance control is responsible for many examples of community succession. Successions on old fields, for example, typically involve a vegetation sequence in which pioneer species are those that can establish themselves quickly in the disturbed habitat of a recently cultivated field, to be replaced later by species that are slower to arrive and establish but eventually outcompete the pioneers. It can be very difficult to identify when a succession reaches a stable climax community, since this may take centuries to achieve and in the meantime further disturbances are likely to occur.

Food webs

No predator–prey, parasite–host, or grazer–plant pair exists in isolation. Each is part of a complex food web involving other predators, parasites, food sources, and competitors within the various trophic levels of a community.

The effect of one species (say, a predator) on another (its herbivorous prey) may be *direct* and straightforward. But *indirect* effects may also be felt by any of the myriad species linked more remotely in the food web.

Some species are more tightly woven into the food web than others. A species whose removal would produce a significant effect (extinction or a large change in density) in at least one other species may be thought of as a strong interactor. Removal of some strong interactors would lead to significant changes that would spread throughout the food web. We refer to these as *keystone species*.

Top-down control of a food web occurs in situations in which the structure (abundance, species number) of lower trophic levels depends on the effects of consumers from higher trophic levels. *Bottom-up control*, on the other hand, occurs in a dependence of community structure on factors, such as nutrient concentration and prey availability, that influence a trophic level from below. The relative importance of these forces varies according to the trophic level under investigation and the number of trophic levels present.

The old conventional wisdom in ecology was that increased food web complexity leads to increased stability (though care is needed in deciding what is meant by stability). This view has been undermined by the analysis of mathematical models that have often concluded that complexity leads to instability. The evidence from real communities is conflicting, but many of the data gathered so far are unreliable. More rigorous studies will be needed before ecologists fully understand the causes and consequences of food web patterns.

Review Questions

1 ▲ Construct a flow diagram (boxes and arrows) with a named population at its center to illustrate the wide range of abiotic and biotic factors that influence its pattern of abundance.

2 Population census data can be used to establish correlations between abundance and external factors such as weather. Why can such correlations not be used to prove a causal relationship that explains the dynamics of the population?

3 ▲ Consider the information given in Section 9.2.1 about the apparent constancy in numbers of swifts in the village of Selborne. Now imagine first that you are H.G. Andrewartha and then A.J. Nicholson and construct an argument that you think each would have made to account for the population dynamics of swifts.

4 Distinguish between the determination and regulation of population abundance.

5 What is meant by a "metapopulation" and how does it differ from a simple "population"?

6 Define founder control and dominance control as they apply to community organization. Why are communities in each category so different?

7 What factors are responsible for changes in species composition during an old field succession?

8 ▲ Draw up a food web of, say, six or seven species with which you are familiar and which spans at least three trophic levels. Take each species in turn and suggest the kind of community organization that would be necessary for this to be a "keystone species".

9 What are meant by bottom-up and top-down control? How is the importance of each likely to vary with the number of trophic levels in a community?

10 Discuss what is understood about the relationship between the complexity and stability of food webs.

▲ = Challenge Question

Chapter 10

Patterns in Species Richness

Recognizing and conserving the world's biological resources are becoming increasingly important. To conserve biodiversity we must understand why species richness varies widely across the face of the Earth. Why do some communities contain more species than others? Are there patterns or gradients in this biodiversity? If so, what are the reasons for these patterns?

Chapter Contents

10.1 Introduction

10.2 A simple model of species richness

10.3 Spatially varying factors that influence species richness

10.4 Temporally varying factors that influence species richness

10.5 Gradients of species richness

10.6 Patterns in taxon richness in the fossil record

10.7 Appraisal of patterns in species richness

Key Concepts

In this chapter you will

- understand the meanings of *species richness*, *diversity indices*, and *rank–abundance diagrams*

- appreciate that species richness is limited by available resources, the average portion of the resources used by each species (niche breadth), and the degree of overlap in resource use

- recognize that species richness may be highest at intermediate levels of both productivity and predation intensity but tends to increase with spatial heterogeneity

- appreciate the importance of habitat area and remoteness in determining richness, especially with reference to the *equilibrium theory of island biogeography*

- understand richness gradients with latitude, altitude, and depth, and during community succession, and the difficulties of explaining them

- appreciate how theories of species richness can also be applied to the fossil record

10.1   Introduction

Why the number of species varies from place to place, and from time to time, are questions that present themselves not only to ecologists but to anybody who observes and ponders the natural world. They are interesting questions in their own right—but they are also questions of practical importance. The number of species in a community is a crucial aspect of that community's biodiversity. The exact meaning of *biodiversity* is discussed in Chapter 14, but for now it is clear that if we wish to conserve or restore biodiversity, then we must understand how species numbers are determined and how it comes about that they vary. We will see that there are plausible answers to the questions we ask, but these answers are by no means conclusive. Yet this is not so much a disappointment as a challenge to ecologists of the future. Much of the fascination of ecology lies in the fact that many of the problems are blatant, whereas the solutions are not. We will see that a full understanding of patterns in species richness must draw on our knowledge of all the areas of ecology discussed so far in this book.

The number of species in a community is referred to as its species richness. Counting or listing the species present in a community may sound a straightforward procedure, but in practice it is often surprisingly difficult, partly because of taxonomic problems, but also because only a subsample of the organisms in an area can usually be counted. The number of species recorded then depends on the number of samples that have been taken, or on the volume of the habitat that has been explored. The most common species are likely to be represented in the first few samples, and as more samples are taken, rarer species will be added to the list. At what point does one cease to take further samples? Ideally, the investigator should continue to sample until the number of species reaches a plateau. At the very least, the species richness of different communities should be compared only if they are based on the same sample sizes (in terms of area of habitat explored, time devoted to sampling, or, best of all, number of individuals included in the samples). *determining species richness*

An important aspect of the structure of a community is completely ignored, though, when its composition is described simply in terms of the number of species present— namely, that some species are rare and others common. Intuitively, a community of ten species with equal numbers in each seems more diverse than another, again consisting of ten species, with 91 percent of the individuals belonging to the most common species and just 1 percent in each of the other nine. Yet, each community has the same species richness. *Diversity indices* are designed to combine both species richness and the evenness or equitability of the distribution of individuals among those species (Box 10.1). Moreover, attempts to describe a complex community structure by one single attribute, such as richness, or even diversity, can still be criticized because so much valuable information is lost. A more complete picture of the distribution of species abundance in a community is therefore sometimes provided in a *rank–abundance diagram* (Box 10.1). *diversity indices and rank–abundance diagrams*

Nonetheless, for many purposes, the simplest measure, species richness, suffices. In the following sections, therefore, we examine the relationships between species richness and a variety of factors that may, in theory, influence richness in ecological communities. It will become clear that it is often extremely difficult to come up with unambiguous predictions and clean tests of hypotheses when dealing with something as complex as a community.

Box 10.1   *Quantitative Aspects*

Diversity Indices and
Rank–Abundance Diagrams

The measure of the character of a community that is most commonly used to take into account both species richness and the relative abundances of those species is known as the *Shannon* or the *Shannon–Weaver diversity index* (denoted by *H*). This is calculated by determining, for each species, the proportion of individuals or biomass (P_i for the *i*th species) that that species contributes to the total in the sample. Then, if *S* is the total number of species in the community (i.e., the richness):

diversity, $H = -\Sigma P_i \ln P_i$,

where the Σ ("summation") sign indicates that the product ($P_i \ln P_i$) is calculated for each of the *S* species in turn and these products are then summed. As required, the value of the index depends on both the species richness and the evenness (equitability) with which individuals are distributed among the species. Thus, for a given richness, *H* increases with equitability, and for a given equitability, *H* increases with richness.

An example of an analysis using diversity indices is provided by the unusually long-term study that commenced in 1856 in an area of pasture at Rothamsted in England. Experimental plots received a fertilizer treatment once every year, and control plots did not. Figure 10.1 shows how the species diversity (*H*) of the grass species changed between 1856 and 1949. While the unfertilized area remained essentially unchanged, the fertilized area showed a progressive decline. This is discussed in Section 10.3.1.

Rank–abundance diagrams, on the other hand, make use of the full array of P_i values by plotting P_i

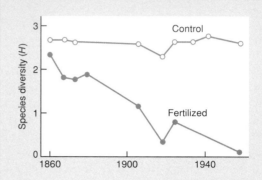

Figure 10.1
Species diversity (*H*) of a control plot and a fertilized plot in the Rothamstead Parkgrass experiment. (After Tilman, 1982.)

against "rank"; i.e., the most abundant species takes rank 1, the second most abundant rank 2, and so on, until the array is completed by the rarest species of all. Rank–abundance diagrams, like indices of richness and diversity, should be viewed as abstractions of the highly complex structure of communities that may nonetheless sometimes be useful when making comparisons. For example, the steeper the slope of a rank–abundance diagram, the greater the dominance of common species over rare species in the community (a steep slope means a sharp drop in relative abundance, P_i, for a given drop in rank). Thus, in the case of the Rothamsted experiment, Figure 10.2 shows how the dominance of commoner species steadily increased (steeper slope) while species richness decreased over time.

Box 10.1 (cont'd)

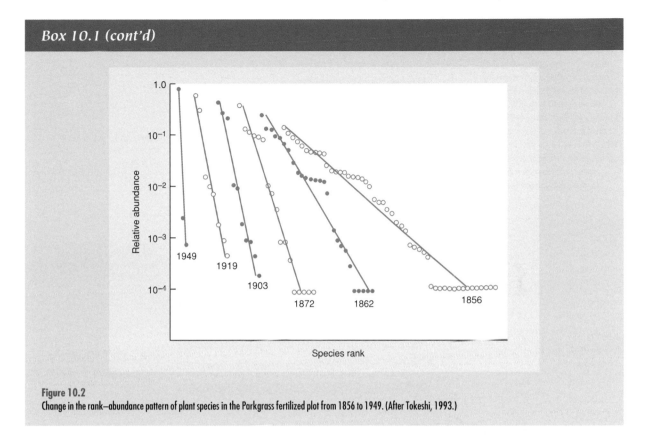

Figure 10.2
Change in the rank–abundance pattern of plant species in the Parkgrass fertilized plot from 1856 to 1949. (After Tokeshi, 1993.)

10.2   A Simple Model of Species Richness

To try to understand the determinants of species richness, it will be useful to begin with a simple model (Figure 10.3). Assume, for simplicity, that the resources available to a community can be depicted as a one-dimensional continuum, R units long. Each species uses only a portion of this resource continuum, and these portions define the *niche breadths* (n) of the various species: the average niche breadth within the community is $\bar{n}$. Some of these niches overlap, and the overlap between adjacent species can be measured by a value o. The average niche overlap within the community is then $\bar{o}$. With this simple background, it is possible to consider why some communities should contain more species than others.

First, for given values of $\bar{n}$ and $\bar{o}$, a community will contain more species the larger the value of R, i.e., the greater the range of resources (Figure 10.3a).

Second, for a given range of resources, more species will be accommodated if $\bar{n}$ is smaller, i.e., if the species are more specialized in their use of resources (Figure 10.3b).

Alternatively, if species overlap to a greater extent in their use of resources (greater $\bar{o}$), then more may coexist along the same resource continuum (Figure 10.3c).

Figure 10.3
A simple model of species richness. Each species utilizes a portion n of the available resources (R), overlapping with adjacent species by an amount o. More species may occur in one community than in another (a) because a greater range of resources is present (larger R), (b) because each species is more specialized (smaller average n), (c) because each species overlaps more with its neighbors (larger average o), or (d) because the resource dimension is more fully exploited. (After MacArthur, 1972.)

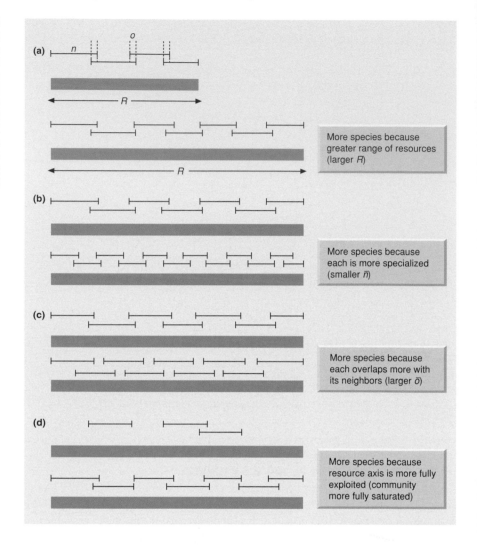

competition and predation may influence species richness

Finally, a community will contain more species the more fully saturated it is; conversely, it will contain fewer species when more of the resource continuum is unexploited (Figure 10.3d).

We can now consider the relationship between this model and two important kinds of species interactions described in previous chapters—interspecific competition and predation. If a community is dominated by interspecific competition (Chapter 6), the resources are likely to be fully exploited. Species richness will then depend on the range of available resources, the extent to which species are specialists, and the permitted extent of niche overlap (Figure 10.3a–c). We will examine a range of influences on each of these three.

Predation, on the other hand, is capable of exerting contrasting effects (Chapter 8). First, we know that predators can exclude certain prey species; in the absence of these species the community may then be less than fully saturated, in the sense that some available resources may be unexploited (Figure 10.3d). In this way, predation may reduce species richness. Second though, predation may tend to keep species below their carrying capacities for much of the time, reducing the intensity and importance of direct interspecific competition for resources. This may then permit much more niche overlap and a *greater* richness of species than in a community dominated by competition (Figure 10.3c).

The next two sections examine a variety of determinants of species richness. To organize these, Section 10.3 focuses on factors that usually vary from place to place (though they can vary from time to time): productivity, predation intensity, spatial heterogeneity, and "harshness". Section 10.4 then focuses on factors that usually vary from time to time (though they can vary from place to place): climatic variation, disturbance, and evolutionary age.

10.3 Spatially Varying Factors that Influence Species Richness

10.3.1 Productivity and resource richness

For plants, the productivity of the environment can depend on whichever nutrient or condition is most limiting to growth (dealt with in detail in Chapter 11). Broadly speaking, the productivity of the environment for animals follows the same trends as for plants, both as a result of the changes in resource levels at the base of the food chain, and as a result of the changes in critical conditions such as temperature.

If higher productivity is correlated with a wider *range* of available resources, then this is likely to lead to an increase in species richness (Figure 10.3a). However, a more productive environment may have a higher rate of supply of resources but not a greater variety of resources. This might lead to more individuals per species rather than more species. Alternatively again, it is possible, even if the overall variety of resources is unaffected, that rare resources in an unproductive environment may become abundant enough in a productive environment for extra species to be added, because more specialized species can be accommodated (Figure 10.3b).

In general, then, we might expect species richness to increase with productivity— a contention that is supported by an analysis of the species richness of trees in North America in relation to climatic variables. Two measures of available environmental energy were used. The first, *potential* evapotranspiration (PET), is the amount of water that would evaporate or be transpired from a saturated surface, and can be taken as a crude measure of available energy. But the *actual* amount of water evaporated depends also on how much precipitation there has been, and thus how much water is available (surfaces may not be saturated). This is measured by the actual evapotranspiration (AET), which reflects the joint availability of energy *and* water. It can be regarded as a good correlate of the energy that can actually be captured by plants, and is therefore a good index of their productivity. Tree species richness was most strongly correlated with AET (Figure 10.4).

When this work was extended to four vertebrate groups, species richness was found to be correlated to some extent with tree species richness itself. However, the best correlations were consistently with PET (Figure 10.5). Why should animal species richness be positively correlated with crude atmospheric energy? The answer is not

increased productivity might be expected to lead to increased richness . . .

. . . and often does

UNANSWERED QUESTION: why should there be more animal species in regions that receive more energy?

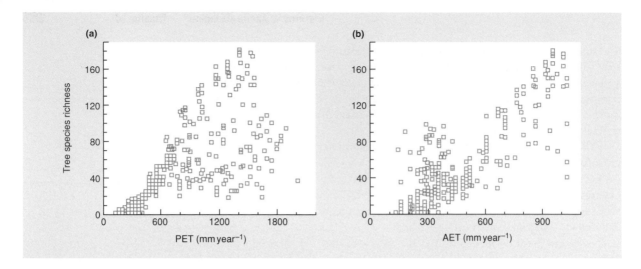

Figure 10.4
Species richness of trees in North America north of the Mexican border (in which the continent has been divided into 336 quadrats following lines of latitude and longitude) in relation to (a) potential evapotranspiration (PET), and (b) actual evapotranspiration (AET). The relationship with AET is strongest. (After Currie & Paquin, 1987; Currie, 1991.)

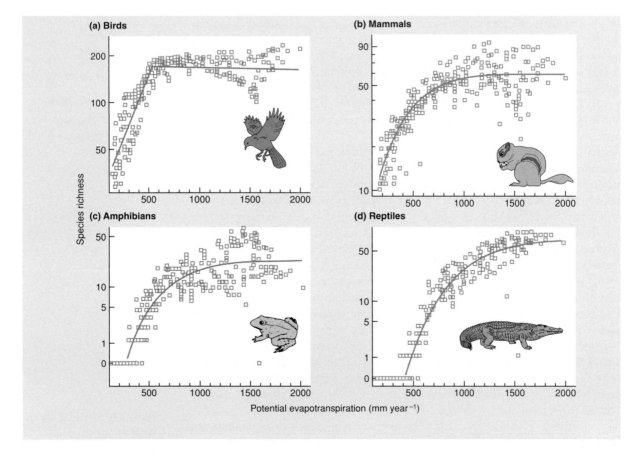

Figure 10.5
Species richness of (a) birds, (b) mammals, (c) amphibians, and (d) reptiles in North America in relation to potential evapotranspiration. (After Currie, 1991.)

known with any certainty, but it may be because for an ectotherm, such as a reptile, extra atmospheric warmth would enhance the intake and utilization of food resources; while for an endotherm, such as a bird, the extra warmth would mean less expenditure of resources in maintaining body temperature and more available for growth and reproduction. In both cases, then, this could lead to faster individual and population growth and thus to larger populations. Warmer environments might therefore allow species with narrower niches to persist and such environments may therefore support more species in total (Turner et al., 1996) (Figure 10.3b).

Sometimes there seems to be a direct relationship between animal species richness and plant productivity. Thus, there are strong positive correlations between species richness and precipitation for both seed-eating ants and seed-eating rodents in the southwestern deserts of the United States (Figure 10.6a). In such arid regions, it is well established that mean annual precipitation is closely related to plant productivity and thus to the amount of seed resources available. It is particularly noteworthy that in the species-rich sites, the communities contain more species of very large ants (which consume large seeds) and more species of very small ants (which take small seeds) (Davidson, 1977). It seems that either the range of sizes of seeds is greater in the more productive environments (Figure 10.3a) or the abundance of seeds becomes sufficient to support extra consumer species with narrower niches (Figure 10.3b).

On the other hand, an increase in diversity with productivity is by no means universal, as shown for example, by the unique Parkgrass experiment which started in 1856 at Rothamsted in England (see Box 10.1). An 8-acre (3.25 ha) pasture was divided into 20 plots, two serving as controls and the others receiving a fertilizer treatment once a year. While the unfertilized areas remained essentially unchanged, the fertilized areas showed a progressive decline in species richness (and diversity).

other evidence shows richness declining with productivity . . .

Such declines have long been recognized. Rosenzweig (1971) referred to them as illustrating "the paradox of enrichment". One possible resolution of the paradox is that high productivity leads to high rates of population growth, bringing about the extinction of some of the species present because of a speedy conclusion to any potential competitive exclusion (see Section 6.2.7). At lower productivity, the environment is more likely to have changed before competitive exclusion is achieved. An association between high productivity and low species richness has been found in several other studies of plant communities (reviewed by Tilman, 1986). It can be seen, for example, when an increased input of plant resources like nitrates and phosphates into lakes, rivers, estuaries, and coastal marine regions (cultural eutrophication) consistently leads to a decrease in species richness of phytoplankton (despite an increase in their productivity).

. . . and further evidence suggests a "humped" relationship

It is perhaps not surprising, then, that several studies have demonstrated both an increase and a decrease in richness with increasing productivity—that is, that species richness may be highest at intermediate levels of productivity. Species richness declines at the lowest productivities because of a shortage of resources, but also declines at the highest productivities where competitive exclusions speed rapidly to their conclusion. For instance, there are humped curves when the number of Malaysian rain forest woody species is plotted against an index of phosphorus and potassium concentration, expressing the resource richness of the soil (Figure 10.6b), and when the species richness of desert rodents is plotted against precipitation (and

Figure 10.6
(a) Patterns of species richness of seed-eating rodents (triangles) and ants (circles) inhabiting sandy soils in a geographic gradient of precipitation and productivity. (After Brown & Davidson, 1977.) (b) Species richness of woody species in several Malaysian rain forests plotted against an index of phosphorus (P) and potassium (K) concentration. (After Tilman, 1982.) (c) Species richness of desert rodents in Israel plotted against rainfall. (After Abramsky & Rosenzweig, 1983.) (d) Percentage of published studies on plants and animals showing various patterns of relationship between species richness and productivity. (After Mittelbach et al., 2001.)

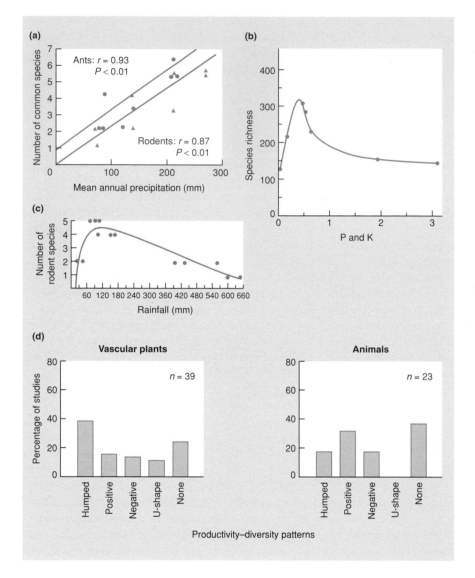

thus productivity) along a gradient in Israel (Figure 10.6c). Indeed, an analysis of a wide range of such studies found that when communities differing in productivity but of the same general type (e.g., tallgrass prairie) were compared (Figure 10.6d) a positive relationship was the most common finding in animal studies (with fair numbers of humped and negative relationships), whereas with plants, humped relationships were most common, with smaller numbers of positives and negatives (and even some U-shaped curves—cause unknown!). Clearly, increased productivity can and does lead to increased or decreased species richness—or both.

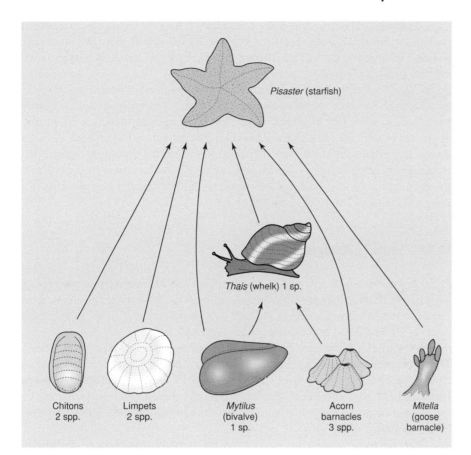

Figure 10.7
Paine's rocky shore community. (After Paine, 1966.)

10.3.2 Predation intensity

The possible effects of predation on the species richness of a community were examined in Chapter 8: predation may increase richness by allowing otherwise competitively inferior species to coexist with their superiors (*predator-mediated coexistence*); but intense predation may reduce richness by driving prey species (whether or not they are strong competitors) to extinction. Overall, therefore, there may also be a humped relationship between predation intensity and species richness in a community, with greatest richness at intermediate intensities, such as that shown by the effects of cattle grazing (illustrated in Figure 8.23, Section 8.6).

A further example of predator-mediated coexistence is provided by a study that established the concept in the first place: the work of Paine (1966) on the influence of a top carnivore on community structure on a rocky shore (Figure 10.7). The starfish *Pisaster ochraceus* preys on sessile filter-feeding barnacles and mussels, and also on browsing limpets and chitons and a small carnivorous whelk. These species, together with a sponge and four macroscopic algae (seaweeds), form a typical community on rocky shores of the Pacific coast of North America. Paine removed all starfish from a stretch of shoreline about 8 m long and 2 m deep and continued to exclude them

predator-mediated coexistence by starfish on a rocky shore

for several years. The structure of the community in nearby control areas remained unchanged during the study, but the removal of *Pisaster* had dramatic consequences. Within a few months, the barnacle *Balanus glandula* settled successfully. Later mussels (*Mytilus californicus*) crowded it out, and eventually the site became dominated by these. All but one of the species of alga disappeared, apparently through lack of space, and the browsers tended to move away, partly because space was limited and partly because there was a lack of suitable food. The main influence of the starfish *Pisaster* appears to be to make space available for competitively subordinate species. It cuts a swathe free of barnacles and, most importantly, free of the dominant mussels that would otherwise outcompete other invertebrates and algae for space. Overall, there is usually predator (starfish) mediated coexistence, but the removal of starfish led to a reduction in number of species from 15 to eight.

10.3.3   Spatial heterogeneity

Environments that are more spatially heterogeneous can be expected to accommodate extra species because they provide a greater variety of microhabitats, a greater range of microclimates, more types of places to hide from predators, and so on. In effect, the extent of the resource spectrum is increased (Figure 10.3a).

richness and the heterogeneity of the abiotic environment

In some cases, it has been possible to relate species richness to the spatial heterogeneity of the abiotic environment. For instance, a study of plant species growing in 51 plots alongside the Hood River, Canada, revealed a positive relationship between species richness and an index of spatial heterogeneity (based, among other things, on the number of categories of substrate, slope, drainage regimes, and soil pH present) (Figure 10.8a).

animal richness and plant spatial heterogeneity

Most studies of spatial heterogeneity, though, have related the species richness of animals to the structural diversity of the plants in their environment (Figure 10.8b, c), occasionally as a result of experimental manipulation of the plants, as with the spiders in Figure 10.8b, but more commonly through comparisons of different natural communities (Figure 10.8c). However, whether spatial heterogeneity arises intrinsically from the abiotic environment or is provided by other biological components of the community, it is capable of promoting an increase in species richness.

10.3.4   Environmental harshness

Environments dominated by an extreme abiotic factor—often called harsh environments—are more difficult to recognize than might be immediately apparent. An anthropocentric view might describe as extreme both very cold and very hot habitats, unusually alkaline lakes, and grossly polluted rivers. However, species have evolved and live in all such environments, and what is very cold and extreme for us must seem benign and unremarkable to a penguin.

We might try to get around the problem of defining environmental harshness by "letting the organisms decide". An environment may be classified as *extreme* if organisms, by their failure to live there, show it to be so. But if the claim is to be

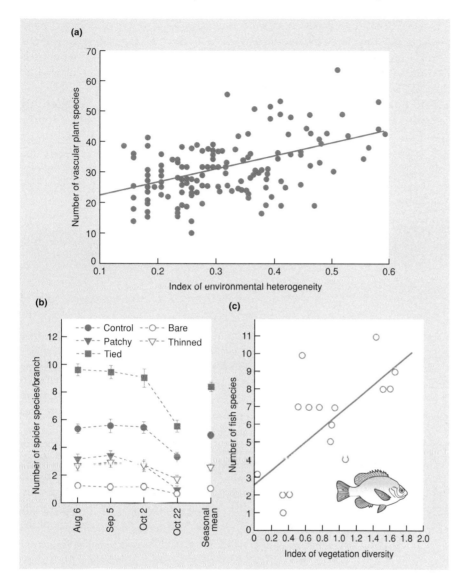

Figure 10.8
(a) Relationship between the number of plants per 300 m² plot beside the Hood River, Northwest Territories, Canada, and an index (ranging from 0 to 1) of spatial heterogeneity in abiotic factors associated with topography and soil. (After Gould & Walker, 1997.) (b) In an experimental study, the number of spider species living on Douglas fir branches increases with their structural diversity. Those "bare", "patchy", or "thinned" were less diverse than normal ("control") by virtue of having needles removed; those "tied" were more diverse because their twigs were entwined together. (After Halaj et al., 2000.) (c) Relationship between animal species richness and an index of structural diversity of vegetation for freshwater fish in 18 Wisconsin lakes. (After Tonn & Magnuson, 1982.)

made—as it often is—that species richness is lower in extreme environments, then this definition is circular, and is designed to prove the very claim we wish to test.

Perhaps the most reasonable definition of an extreme condition is one that requires, of any organism tolerating it, a morphological structure or biochemical mechanism that is not found in most related species and is costly, either in energetic terms or in terms of the compensatory changes in the biological processes of the organism that are needed to accommodate it. For example, plants living in highly acidic soils (low pH) may be affected directly through injury by hydrogen ions or indirectly via deficiencies in the availability and uptake of important resources such as phosphorus,

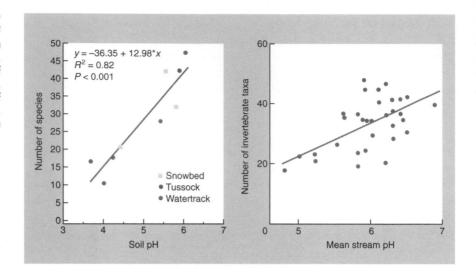

magnesium, and calcium. In addition, aluminum, manganese, and heavy metals may have their solubility increased to toxic levels, and mycorrhizal activity and nitrogen fixation may be impaired. Plants can only tolerate low pH if they have specific structures or mechanisms allowing them to avoid or counteract these effects.

Environments that experience low pHs can thus be considered harsh, and the mean number of plant species recorded per sampling unit in a study in the Alaskan Arctic tundra was indeed lowest in soils of low pH (Figure 10.9a—see also Figure 10.20c). Similarly, the species richness of benthic stream invertebrates in the Ashdown Forest (southern England) was markedly lower in the more acidic streams (Figure 10.9b). Further examples of extreme environments that are associated with low species richness include hot springs, caves, and highly saline water bodies such as the Dead Sea. The problem with these examples, however, is that they are also characterized by other features associated with low species richness such as low productivity and low spatial heterogeneity. In addition, many occupy small areas (caves, hot springs), or areas that are rare compared to other types of habitat (only a small proportion of the streams in southern England are acidic). Hence extreme environments can often be seen as small and isolated islands. We will see in Section 10.5.1 that these features, too, are usually associated with low species richness. Although it appears reasonable that intrinsically extreme environments should as a consequence support few species, this has proved an extremely difficult proposition to establish.

10.4 Temporally Varying Factors that Influence Species Richness

Temporal variation in conditions and resources may be predictable or unpredictable and operate on timescales from minutes through to centuries and millennia. All may influence species richness in profound ways.

UNANSWERED QUESTION: are harsh environments species-poor?

10.4.1 Climatic variation

The effects of climatic variation on species richness depend on whether the variation is predictable or unpredictable (measured on timescales that matter to the organisms involved). In a predictable, seasonally changing environment, different species may be suited to conditions at different times of the year. More species might therefore be expected to coexist in a seasonal environment than in a completely constant one (Figure 10.3a). Different annual plants in temperate regions, for instance, germinate, grow, flower, and produce seeds at different times during a seasonal cycle; while phytoplankton and zooplankton pass through a seasonal succession in large, temperate lakes with a variety of species dominating in turn as changing conditions and resources become suitable for each.

temporal niche differentiation in seasonal environments

On the other hand, there are opportunities for specialization in a nonseasonal environment that do not exist in a seasonal environment. For example, it would be difficult for a long-lived obligate fruit eater to exist in a seasonal environment when fruit is available for only a very limited portion of the year. But such specialization is found repeatedly in nonseasonal, tropical environments where fruit of one type or another is available continuously.

specialization in nonseasonal environments

Unpredictable climatic variation (climatic instability) could have a number of effects on species richness. On the one hand: (1) stable environments may be able to support specialized species that would be unlikely to persist where conditions or resources fluctuated dramatically (Figure 10.3b); (2) stable environments are more likely to be saturated with species (Figure 10.3d); and (3) theory suggests that a higher degree of niche overlap will be found in more stable environments (Figure 10.3c). All these processes could increase species richness. On the other hand, populations in a stable environment are more likely to reach their carrying capacities, the community is more likely to be dominated by competition, and species are therefore more likely to be excluded by competition ($\bar{o}$ is smaller, Figure 10.3c).

UNANSWERED QUESTION: does climatic instability increase or decrease richness?

Some studies have seemed to support the notion that species richness increases as climatic variation decreases. For example, there is a significant negative relationship between species richness and the range of monthly mean temperatures for birds, mammals, and gastropods that inhabit the west coast of North America (from Panama in the south to Alaska in the north) (Figure 10.10). However, this correlation does not prove causation, since there are many other things that change between Panama and Alaska. There is no established relationship between climatic instability and species richness.

10.4.2 Disturbance

Previously, in Section 9.4, the influence of disturbance on community structure was examined, and it was demonstrated that when a disturbance opens up a gap, and the community is *dominance controlled* (strong competitors can replace residents), there tends in a community succession to be an initial increase in richness as a result of colonization, but a subsequent decline in richness as a result of competitive exclusion.

If the frequency of disturbance is now superimposed on this picture, it seems likely that very frequent disturbances will keep most patches in the early stages of

intermediate disturbance hypothesis

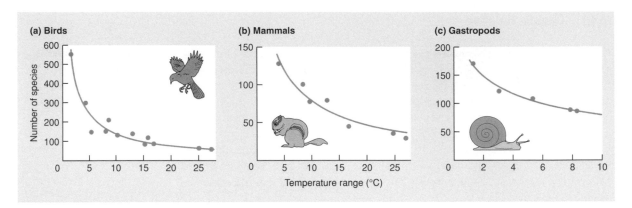

Figure 10.10
Relationships between species richness and the range of monthly mean temperatures at sites along the west coast of North America for (a) birds, (b) mammals, and (c) gastropods. (After MacArthur, 1975.)

succession (where there are few species) but also that very rare disturbances will allow most patches to become dominated by the best competitors (where there are also few species). This suggests an *intermediate disturbance hypothesis* (Figure 10.11), in which communities are expected to contain most species when the frequency of disturbance is neither too high nor too low (Connell, 1978). The intermediate disturbance hypothesis was originally proposed to account for patterns of richness in tropical rain forests and coral reefs. It has occupied a central place in the development of ecological theory because all communities are subject to disturbances that exhibit different frequencies and intensities.

supported by more rocky shore data . . . Among a number of studies that have provided support for this hypothesis, we turn first to the study of green and red algae on different-sized boulders on the rocky shores of southern California, discussed in Section 9.4 (Sousa, 1979a, 1979b). Small boulders had a monthly probability of movement of 42 percent, intermediate-sized boulders a probability of 9 percent, and large boulders a probability of only 0.1 percent. The successional sequence that occurs in the absence of disturbance was described in Section 9.4: first *Ulva*, the ephemeral green alga; then several species of perennial red alga; and then predominantly the competitive dominant, *Gigartina canaliculata*.

The percentage of bare space decreased from small to large boulders, and mean species richness was lowest on the regularly disturbed small boulders (Figure 10.12a). These were dominated most commonly by *Ulva* spp. The highest levels of species richness were consistently recorded on the intermediate boulder class. Most held mixtures of three to five abundant species from all successional stages. The largest boulders had a lower mean species richness than the intermediate class, though a *G. canaliculata* monoculture was achieved on only a few boulders.

. . . and from studies of small streams and plankton Disturbances in small streams often take the form of bed movements during periods of high discharge, and because of differences in flow regimes and in the substrata of streambeds, some stream communities are disturbed more frequently than others. This variation was assessed in 54 stream sites in the Taieri River in New Zealand. The pattern of richness of macroinvertebrate species conformed with the

Figure 10.11
The intermediate disturbance hypothesis. I, no disturbance; II, intermediate disturbance; III, repeated, frequent disturbance. See text for further details.

intermediate disturbance hypothesis (Figure 10.12b). Finally, in controlled field experiments, natural phytoplankton communities in Lake Plußsee (North Germany) were disturbed (at intervals between every 2 and 12 days) by disrupting the normal stratification in the water column with bubbles of compressed air. Again, both species richness and Shannon's diversity index were highest at intermediate frequencies of disturbance (Figure 10.12c).

10.4.3   Environmental age: evolutionary time

It has also often been suggested that communities that are "disturbed" only on very extended timescales may nonetheless lack species because they have yet to reach an ecological or evolutionary equilibrium. Thus communities may differ in species

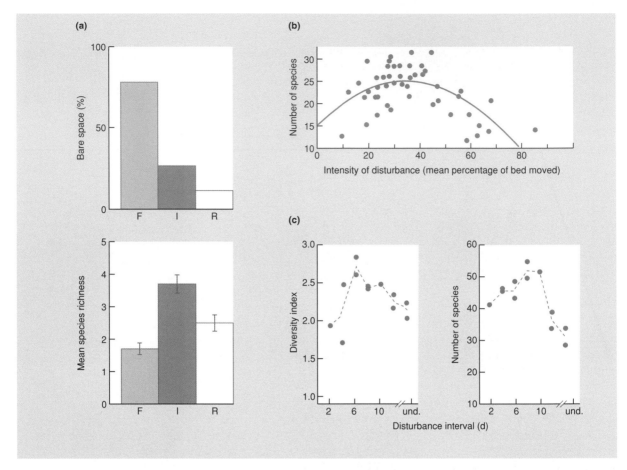

Figure 10.12
(a) Patterns in bare space and species number (± standard error) on boulders in each of three classes categorized according to the force (in newtons, N) required to move them. The three classes are frequently disturbed (F: requires less than 49 N to move), disturbed at an intermediate rate (I: requires 50–294 N to move), and rarely disturbed (R: requires more than 294 N to move). (After Sousa, 1979b.) (b) Relationship between insect species richness and intensity of disturbance: average percentage of the bed that moved (polynomial regression fitted, relationship significant at $P < 0.001$) assessed at 54 stream sites in the Taieri River, New Zealand. (After Townsend et al., 1997b.) (c) Both species richness (right) and diversity (left) of natural phytoplankton communities were highest at intermediate frequencies of disturbance in controlled field experiments in Lake Plußsee (North Germany). und., undisturbed. (After Flödder & Sommer, 1999.)

UNANSWERED QUESTUION: are temperate communities less saturated with species?

richness because some are closer to equilibrium and are therefore more saturated than others (Figure10.3d).

For example, many have argued that the tropics are richer in species than are more temperate regions, at least in part because the tropics have existed over long and uninterrupted periods of evolutionary time, whereas the temperate regions are still recovering from the Pleistocene glaciations. It seems, however, that the long-term stability of the tropics has in the past been greatly exaggerated by ecologists. Whereas the climatic and biotic zones of the temperate region moved toward the equator during the glaciations, the tropical forest appears to have contracted to a limited number of small refuges surrounded by grasslands. A simplistic contrast between the unchanging tropics and the disturbed and recovering temperate regions is therefore untenable. If the lower species richness of communities nearer the poles is to be attributed partly to their being well below an evolutionary equilibrium, then some complex (and unproven) argument must be constructed. (Perhaps movements of temperate zones to a different latitude caused many more extinctions than the contractions of the tropics into smaller areas at the same latitude.) Thus, although it seems likely that some communities are further below equilibrium than others, it is impossible at present to pinpoint those communities with certainty or even confidence.

10.5   Gradients of Species Richness

Sections 10.3 and 10.4 have demonstrated how difficult it is to formulate and test explanations for variations in species richness. It is easier to describe patterns, especially gradients, in species richness. These are discussed next. Explanations for these, too, however, are often very uncertain.

10.5.1   Habitat area and remoteness—island biogeography

It is well established that the number of species on islands decreases as island area decreases. Such a *species–area* relationship is shown in Figure 10.13a for plants on cays (small islands) off the northeast coast of Andros in the Bahamas.

species–area relationships on oceanic islands

"Islands", however, need not be islands of land in a sea of water. Lakes are islands in a "sea" of land, mountaintops are high altitude islands in a low altitude ocean, gaps in a forest canopy where a tree has fallen are islands in a sea of trees, and there can be islands of particular geological types, soil types, or vegetation types surrounded by dissimilar types of rock, soil, or vegetation. Species–area relationships can be equally apparent for these types of islands (Figure 10.13b–d).

habitat islands and areas of mainland

The relationship between species richness and habitat area is one of the most consistent of all ecological patterns. However, the pattern raises an important question: is the impoverishment of species on islands more than would be expected in comparably small areas of mainland? In other words, does the characteristic isolation of islands contribute to their impoverishment of species? These are important questions for an understanding of community structure since there are many oceanic islands, many lakes, many mountaintops, many woodlands surrounded by fields, many isolated trees, and so on.

"island effects" and community structure

Probably the most obvious reason why larger areas should contain more species is that larger areas typically encompass more different types of habitat. However, MacArthur and Wilson (1967) believed this explanation to be too simple. In their *equilibrium theory of island biogeography*, they argued that island size and isolation themselves played important roles: that the number of species on an island is determined by a balance between immigration and extinction; that this balance is dynamic, with species continually going extinct and being replaced (through immigration) by the same or by different species; and that immigration and extinction rates may vary with island size and isolation (Box 10.2).

MacArthur and Wilson's theory makes several predictions:

Figure 10.13
Species–area relationships: (a) for plants on cays off the northeast coast of Andros, Bahamas (after Morrison, 1997); (b) for birds inhabiting lakes in Florida (after Hoyer & Canfield, 1994); (c) for bats inhabiting different-sized caves in Mexico (after Brunet & Medellin, 2001); and (d) for fish living in Australian desert springs that have source pools of different sizes (after Kodric-Brown & Brown, 1993).

Box 10.2   *Historical Landmarks*

MacArthur and Wilson's Equilibrium Theory of Island Biogeography

Taking immigration first, imagine an island that as yet contains no species at all. The rate of immigration of *species* will be high, because any colonizing individual represents a species new to that island. However, as the number of resident species rises, the rate of immigration of new, unrepresented species diminishes. The immigration rate reaches zero when all species from the source pool (i.e., from the mainland or from other nearby islands) are present on the island in question (Figure 10.14a).

The immigration graph is drawn as a curve, because the immigration rate is likely to be particularly high when there are low numbers of residents and many of the species with the greatest powers of dispersal are yet to arrive. In fact, the curve should really be a blur rather than a single line, since the precise curve will depend on the exact sequence in which species arrive, and this will vary by chance. In this sense, the immigration curve can be thought of as the *most probable curve*.

Box 10.2 (cont'd)

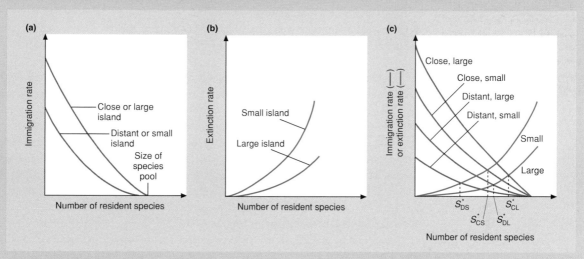

Figure 10.14

MacArthur and Wilson's (1976) equilibrium theory of island biogeography. (a) The rate of species immigration on to an island, plotted against the number of resident species on the island, for large and small islands and for close and distant islands. (b) The rate of species extinction on an island, plotted against the number of resident species on the island, for large and small islands. (c) The balance between immigration and extinction on small and large and on close and distant islands. In each case, S^* is the equilibrium species richness. C, close; D, distant; L, large; S, small.

The exact immigration curve will depend on the degree of remoteness of the island from its pool of potential colonizers (Figure 10.14a). The curve will always reach zero at the same point (when all members of the pool are resident), but it will generally have higher values on islands close to the source of immigration than on more remote islands, since colonizers have a greater chance of reaching an island the closer it is to the source. It is also likely that immigration rates will generally be higher on a large island than on a small island, since the larger island represents a larger target for the colonizers (Figure 10.14a).

The rate of species extinction on an island (Figure 10.14b) is bound to be zero when there are no species there, and it will generally be low when there are few species. However, as the number of resident species rises, the extinction rate is assumed by the theory to increase, probably at a more than proportionate rate. This is thought to occur because with more species, competitive exclusion becomes more likely, and the population size of each species is on average smaller, making it more vulnerable to chance extinction. Similar reasoning suggests that extinction rates should be higher on small than on large islands—population sizes will typically be smaller on small islands (Figure 10.14b). As with immigration, the extinction curves are best seen as "most probable" curves.

In order to see the net effect of immigration and extinction, their two curves can be superimposed (Figure 10.14c). The number of species where the curves cross (S^*) is a dynamic equilibrium and should be the characteristic species richness for the island in question. Below S^*, richness increases (immigration rate exceeds extinction rate); above S^*, richness decreases (extinction exceeds immigration). The theory, then, makes a number of predictions, described in the text.

1 The number of species on an island should eventually become roughly constant through time.

2 This should be a result of a continual *turnover* of species, with some becoming extinct and others immigrating.

3 Large islands should support more species than small islands.

4 Species number should decline with the increasing remoteness of an island.

partitioning variation between habitat diversity and area itself

On the other hand, a higher richness on larger islands would be expected simply as a consequence of larger islands having more habitat types. Does richness increase with area at a rate *greater* than could be accounted for by increases in habitat diversity alone? Some studies have attempted to partition species–area variation on islands into that which can be entirely accounted for in terms of habitat diversity, and that which remains and must be accounted for by island area in its own right. For beetles on the Canary Islands, the relationship between species richness and habitat diversity (as measured by plant species richness) is much stronger than that with island area, and this is particularly marked for the herbivorous beetles, presumably because of their particular food plant requirements (Figure 10.15a).

However, in a study of a variety of animal groups living on the Lesser Antilles islands in the West Indies, the variation in species richness from island to island was partitioned, statistically, into that attributable to island area alone, that attributable to habitat diversity alone, that attributable to correlated variation between area and habitat diversity (and hence not attributable to either alone), and that attributable to neither. For reptiles and amphibians (Figure 10.15b), like the beetles of the Canary Islands, habitat diversity was far more important than island area. But for bats, the reverse was the case; and for birds and butterflies, both area itself and habitat diversity had important parts to play. Overall, therefore, studies like this suggest a separate area effect (larger islands are larger targets for colonization; populations on larger islands have a lower risk of extinction) beyond a simple correlation between area and habitat diversity.

bird species richness on Pacific islands decreases with remoteness

An example of species impoverishment on more remote islands can be seen in Figure 10.16 for nonmarine, lowland birds on tropical islands in the southwest Pacific. With increasing distance from the large source island of Papua New Guinea, there is a decline in the number of species, expressed as a percentage of the number present on an island of similar area but close to Papua New Guinea.

species missing because of insufficient time for colonization

A more transient but nonetheless important reason for the species impoverishment of islands, especially remote islands, is the fact that many lack species that they could potentially support, simply because there has been insufficient time for the species to colonize. An example is the island of Surtsey, which emerged in 1963 as a result of a volcanic eruption (Fridriksson, 1975). The new island, 40 km southwest of Iceland, was reached by bacteria and fungi, some seabirds, a fly, and seeds of several beach plants within 6 months of the start of the eruption. Its first established vascular plant was recorded in 1965, and the first moss colony in 1967. By 1973, 13 species of vascular plant and more than 66 mosses had become established (Figure 10.17).

evolution rates on islands may be faster than colonization rates

Finally, it is important to reiterate that no aspect of ecology can be fully understood without reference to evolutionary processes (Chapter 2), and this is particularly true for an understanding of island communities. On isolated islands, the rate at which new species evolve may be comparable to or even faster than the rate at which they arrive as new colonists. Clearly, the communities of these islands will be incompletely

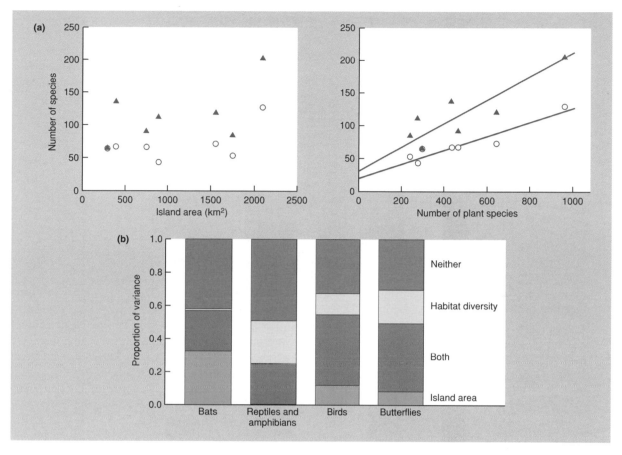

Figure 10.15
(a) The relationships between species richness of herbivorous (circles) and carnivorous (triangles) beetles of the Canary Islands and both island area and plant species richness. (After Becker, 1992.) (b) Proportion of variance, for four animal groups, in species richness among islands in the Lesser Antilles related uniquely to island area (green), uniquely to habitat diversity (yellow), to correlated variation between area and habitat diversity (purple), and unexplained by either (orange). (After Ricklefs & Lovette, 1999.)

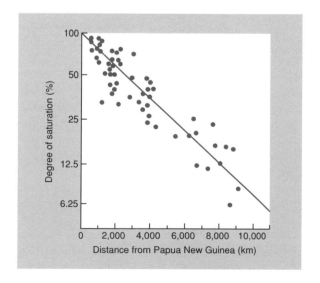

Figure 10.16
The number of resident, nonmarine, lowland bird species on islands more than 500 km from the larger source island of Papua New Guinea expressed as a proportion of the number of species on an island of equivalent area but close to Papua New Guinea and plotted as a function of island distance from Papua New Guinea. (After Diamond, 1972.)

Figure 10.17
The number of species of mosses and
vascular plants recorded on the new island
of Surtsey from 1965 to 1973. (After
Fridriksson, 1975.)

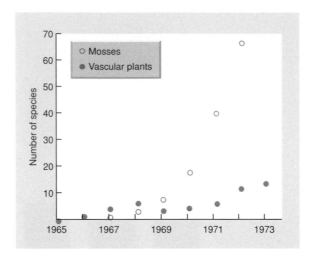

understood by reference only to ecological processes. Take the remarkable numbers of
Drosophila species (fruitflies) found on the remote volcanic islands of Hawaii. There are
probably about 1,500 *Drosophila* species worldwide but at least 500 of these are found
on the Hawaiian islands; they have evolved, almost entirely, on the islands themselves.
The communities of which they are a part are clearly much more strongly affected by
local evolution and speciation than by the processes of invasion and extinction.

10.5.2   Latitudinal gradients

One of the most widely recognized patterns in species richness is the increase that
occurs from the poles to the tropics. This can be seen in a wide variety of groups,
including marine invertebrates, butterflies, lizards, and trees (Figure 10.18). The pattern can be seen, moreover, in terrestrial, marine, and freshwater habitats.

A number of explanations have been put forward for the general latitudinal trend in
species richness, but not one of these is without problems. In the first place, the richness
of tropical communities has been attributed to a greater intensity of predation and to more
specialized predators. More intense predation could reduce the importance of competition, permitting greater niche overlap and promoting higher richness (Figure 10.3c),
but predation cannot readily be forwarded as the root cause of tropical richness, since
this begs the question of what gives rise to the richness of the predators themselves.

productivity as an explanation?   Second, increasing species richness may be related to an increase in productivity
as one moves from the poles to the equator. Certainly, on average, there are both more
heat and more light energy in more tropical regions and, as discussed in Section 10.3.1,
both of these have tended to be associated with greater species richness, though
increased productivity in at least some cases has been associated with reduced richness.

Moreover, light and heat are not the only determinants of plant productivity.
Tropical soils tend, on average, to have lower concentrations of plant nutrients than
temperate soils. The species-rich tropics might therefore be seen, in this sense, as
reflecting their *low* productivity. In fact, tropical soils are poor in nutrients because

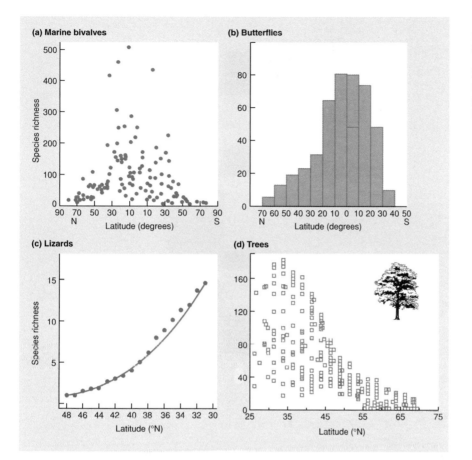

Figure 10.18
Latitudinal patterns in species richness: (a) marine bivalves (after Flessa & Jablonski, 1995); (b) swallowtail butterflies (after Sutton & Collins, 1991); (c) lizards in the United States (after Pianka, 1983); and (d) trees in North America (after Currie & Paquin, 1987).

most of the nutrients are locked up in the large tropical biomass. A productivity argument might therefore have to run as follows. The light, temperature, and water regimes of the tropics lead to high biomass communities but not necessarily to diverse communities. This, though, leads to nutrient-poor soils and perhaps a wide range of light regimes from the forest floor to canopy far above. These in turn lead to high plant species richness and thus to high animal species richness. There is certainly no *simple* "productivity explanation" for the latitudinal trend in richness.

Some ecologists have invoked the climate of low latitudes as a reason for their high species richness. Specifically, equatorial regions are generally less seasonal than temperate regions, and this may allow species to be more specialized (i.e., have narrower niches, Figure 10.3b). This suggestion has now received a number of tests. For example, a comparison of bird communities in temperate Illinois and tropical Panama showed that both tropical shrub and tropical forest habitats contained very many more breeding species than their temperate counterparts, and up to a half of the increase in richness consisted of specialist fruit eaters in the tropical habitats—a specialized way of life that could not be sustained year-round nearer the poles. In contrast, there are also more species of bark and ambrosia beetles in the tropics than in temperate regions, but in this case the tropical species are less specialized (Figure 10.19).

climatic variation as an explanation?

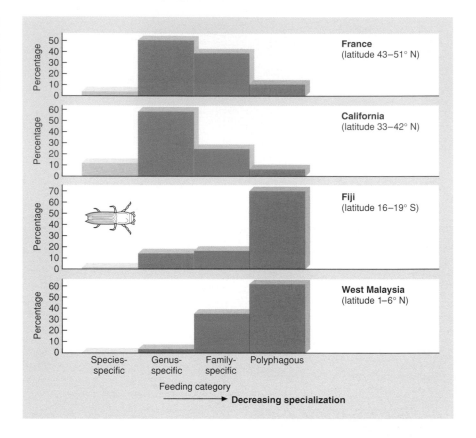

The greater evolutionary "age" of the tropics has also been proposed as a reason for their greater species richness, and another line of argument suggests that the repeated fragmentation and coalescence of tropical forest refugia promoted genetic differentiation and speciation, accounting for much of the high richness in tropical regions. These ideas, too, are plausible but very far from proven.

Overall, therefore, the latitudinal gradient lacks an unambiguous explanation. This is hardly surprising. The components of a possible explanation—trends with productivity, climatic stability, and so on—are themselves understood only in an incomplete and rudimentary way, and the latitudinal gradient intertwines these components with one another, and with other, often opposing forces: isolation, harshness, and so on.

10.5.3　Gradients with altitude and depth

In terrestrial environments, a decrease in species richness with altitude is a phenomenon almost as widespread as a decrease with latitude. Examples are given in Figure 10.20 for birds and mammals in the Himalayan mountains of Nepal (Figure 10.20a), and for plants in the Sierra Manantlán, Mexico (Figure 10.20b). On the other hand, the pattern is not universal, as is apparent from a study of plants in Spain and Portugal (Figure 10.20c.)

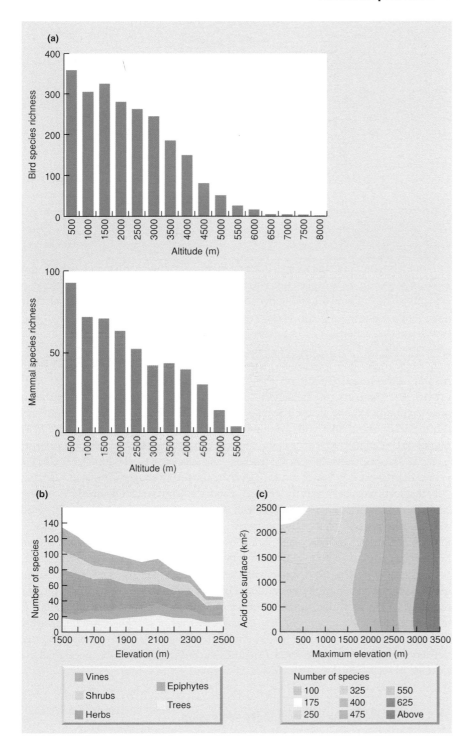

Figure 10.20
Relationship between species richness and altitude: (a) in the Nepalese Himalayas for breeding bird species and mammals (after Hunter & Yonzon, 1992); (b) for plants in the Sierra Manantlán, Mexico (after Vázquez & Givnish, 1998); and (c) for plants in Spain and Portugal (after Lobo et al., 2001). Note also, in this latter case, the decrease in species richness at low pH (increased acidity) at the lower altitudes.

Figure 10.21
Depth gradient in species richness of the
megabenthos (fish, decapods, holothurians,
and asteroids) in the ocean southwest of
Ireland. (After Angel, 1994.)

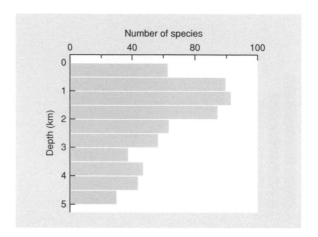

Figure 10.21
Depth gradient in species richness of the
megabenthos (fish, decapods, holothurians,
and asteroids) in the ocean southwest of
Ireland. (After Angel, 1994.)

At least some of the factors instrumental in the latitudinal trend in richness are also likely to be important as explanations for this altitudinal trend (though the problems in explaining the latitudinal trend apply equally to altitude). In addition, high altitude communities almost invariably occupy smaller areas than lowlands at equivalent latitudes, and they will usually be more isolated from similar communities than lowland sites. The effects of area and isolation are certain to contribute to the decrease in species richness with altitude. Exceptions, however, emphasize the need to be cautious in rushing towards generalizations. The study of plants in Spain and Portugal (Figure 10.20c) compared the richness in different 50 × 50 km squares and related it the maximum altitude in each square. In this case, it was the lowland squares (hottest and driest) that had the lowest productivity, and these were also most influenced by human alterations to the landscape that reduce species richness. Patterns (such as the one between species richness and altitude) must ultimately reflect a variety of processes, and these can vary with the context in which they occur.

In aquatic environments, the change in species richness with depth shows strong similarities to the terrestrial gradient with altitude. In larger lakes, the cold, dark, oxygen-poor abyssal depths contain fewer species than the shallow surface waters. Likewise, in marine habitats, plants are confined to the photic zone (where they can photosynthesize), which rarely extends below 30 m. In the open ocean, therefore, there is a rapid decrease in richness with depth, reversed only by the variety of often bizarre animals living on the ocean floor. Interestingly, however, in coastal regions the effect of depth on the species richness of benthic (bottom-dwelling) animals is to produce not a single gradient, but a peak of richness at about 1000 m, possibly reflecting higher environmental predictability there (Figure 10.21). At greater depths, beyond the continental slope, species richness declines again, probably because of the extreme paucity of food resources in abyssal regions.

10.5.4   Gradients during community succession

Section 9.4 described how, in community successions, if they run their full course, the number of species first increases (because of colonization) but eventually decreases

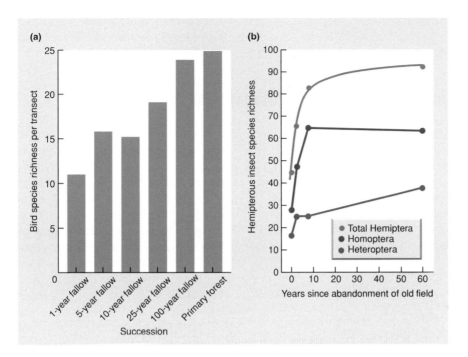

Figure 10.22
The increase in species richness during successions. (a) Birds following shifting cultivation in a tropical rain forest in northeast India. (After Shankar Raman et al., 1998.) (b) Hemipterous insects following an old field succession. (After Brown & Southwood, 1983.)

(because of competition). This is most firmly established for plants, but the few studies that have been carried out on animals in successions indicate, at least, a parallel increase in species richness in the early stages of succession. Figure 10.22 illustrates this for birds following shifting cultivation in a tropical rain forest in northeast India, and for insects associated with old field successions.

To a certain extent, the successional gradient is a necessary consequence of the gradual colonization of an area by species from surrounding communities that are at later successional stages, that is, later stages are more fully saturated with species (Figure 10.3d). However, this is a small part of the story, since succession involves a process of replacement of species and not just the mere addition of new ones.

Indeed, as with the other gradients in species richness, there is something of a cascade effect with succession: one process that increases richness kick-starts a second, which feeds into a third, and so on. The earliest species will be those that are the best colonizers and the best competitors for open space. They immediately provide resources (and introduce heterogeneity) that were not previously present. For example, the earliest plants generate resource depletion zones in the soil that inevitably increase the spatial heterogeneity of plant nutrients. The plants themselves provide a new variety of microhabitats, and for the animals that might feed on them they provide a much greater range of food resources (Figure 10.3a). The increase in herbivory and predation may then feed back to promote further increases in species richness (predator-mediated coexistence: Figure 10.3c), which provides further resources and more heterogeneity, and so on. In addition, temperature, humidity, and wind speed are much less variable (over time) within a forest than in an exposed early successional stage, and the enhanced constancy of the environment may provide a stability of conditions and resources that permits specialist species to build up populations and to

a cascade effect?

persist (Figure 10.3b). As with other gradients, the interaction of many factors makes it difficult to disentangle cause from effect. But with the successional gradient of richness, the tangled web of cause and effect appears to be of the essence.

10.6   Patterns in Taxon Richness in the Fossil Record

the Cambrian explosion— exploiter-mediated coexistence?

Finally, it is of interest to take the processes that are believed to be instrumental in generating present-day gradients in richness and apply them to trends occurring over much longer time spans. The imperfection of the fossil record has always been the greatest impediment to the palaeontological study of evolution. Nevertheless, some general patterns have emerged, and our knowledge of six important groups of organisms is summarized in Figure 10.23.

Until about 600 million years ago, the world was populated virtually only by bacteria and algae, but then almost all the phyla of marine invertebrates entered the

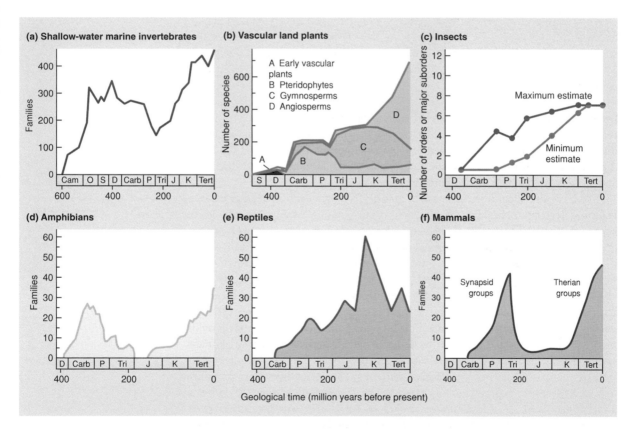

Figure 10.23

Curves showing patterns in taxon richness through the fossil record. (a) Families of shallow-water invertebrates. (After Valentine, 1970.) (b) Species of vascular land plants in four groups—early vascular plants, pteridophytes, gymnosperms, and angiosperms. (After Niklas et al., 1983.) (c) Major orders and suborders of insects. The minimum values are derived from definite fossil records; the maximum values include "possible" records. (From Strong et al., 1984.) (d–f) Vertebrate families of amphibians, reptiles, and mammals, respectively. (After Webb, 1987.) Key to geological periods: Cam, Cambrian; O, Ordovician; S, Silurian; D, Devonian; Carb, Carboniferous; P, Permian; Tri, Triassic; J, Jurassic; K, Cretaceous; Tert, Tertiary.

fossil record within the space of only a few million years (Figure 10.23a). Given that the introduction of a higher trophic level can increase richness at a lower level, it can be argued that the first single-celled herbivorous protist was probably instrumental in the Cambrian explosion in species richness. The opening up of space by cropping of the algal monoculture, coupled with the availability of recently evolved eukaryotic cells, may have caused the biggest burst of evolutionary diversification in Earth's history.

In contrast, the equally dramatic decline in the number of families of shallow-water invertebrates at the end of the Permian (Figure 10.23a) could have been a result of the coalescence of the Earth's continents to produce the single supercontinent of Pangaea; the joining of continents produced a marked reduction in the area occupied by shallow seas (which occur around the periphery of continents) and thus a marked decline in the area of habitat available to shallow-water invertebrates. Moreover, at this time the world was subject to a prolonged period of global cooling in which huge quantities of water were locked up in enlarged polar caps and glaciers, causing a widespread reduction of warm, shallow sea environments. Thus, a species–area relationship may be invoked to account for the reduction in richness of this fauna.

the Permian decline—a species–area relationship?

The analysis of the fossil remains of vascular land plants (Figure 10.23b) reveals four distinct evolutionary phases: (1) a Silurian–mid-Devonian proliferation of early vascular plants, (2) a subsequent late-Devonian–Carboniferous radiation of fern-like lineages, (3) the appearance of seed plants in the late Devonian and the adaptive radiation to a gymnosperm-dominated flora, and (4) the appearance and rise of flowering plants (angiosperms) in the Cretaceous and Tertiary. It seems that after initial invasion of the land, made possible by the appearance of roots, the diversification of each plant group coincided with a decline in species numbers of the previously dominant group. In two of the transitions (early plants to gymnosperms, and gymnosperms to angiosperms), this pattern may reflect the competitive displacement of older, less specialized taxa by newer and presumably more specialized taxa.

competitive displacement among the major plant groups?

The first undoubtedly phytophagous insects are known from the Carboniferous. Thereafter, modern orders appeared steadily (Figure 10.23c) with the Lepidoptera (butterflies and moths) arriving last on the scene, at the same time as the rise of the angiosperms. Reciprocal evolution and counterevolution between plants and herbivorous insects has almost certainly been, and still is, an important mechanism driving the increase in richness observed in both land plants and insects through their evolution.

Toward the end of the last ice age, the continents were much richer in large animals than they are today. For example, Australia was home to many genera of giant marsupials; North America had its mammoths, giant ground sloths, and more than 70 other genera of large mammals; and New Zealand and Madagascar were home to giant flightless birds, the moas (Dinorthidae) and elephant bird (*Aepyornis*), respectively. Over the past 30,000 years or so, a major loss of this biotic diversity has occurred over much of the globe. The extinctions particularly affected large terrestrial animals (Figure 10.24a); they were more pronounced in some parts of the world than others; and they occurred at different times in different places (Figure 10.24b). The extinctions mirror patterns of human migration. Thus, the arrival in Australia of ancestral aborigines occurred between 40,000 and 30,000 years ago; stone spear points became abundant throughout the United States about 11,500 years ago; and humans have been in both Madagascar and New Zealand for 1,000 years. It can be convincingly argued, therefore, that the arrival of efficient human hunters led to the rapid over-

extinctions of large animals in the Pleistocene—prehistoric overkill?

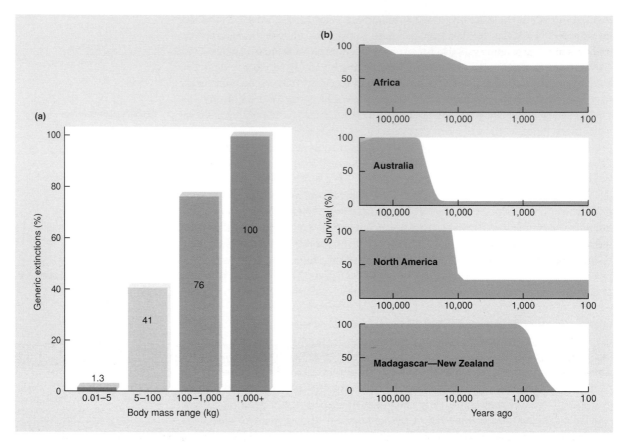

Figure 10.24
(a) The percentage of genera of large mammalian herbivores that have gone extinct in the last 130,000 years is strongly size dependent (data from North and South America, Europe, and Australia combined). (After Owen-Smith, 1987.) (b) Percentage survival of large animals on three continents and two large islands (New Zealand and Madagascar). (After Martin, 1984.)

exploitation of vulnerable and profitable large prey. Africa, where humans originated, shows much less evidence of loss, perhaps because coevolution of large animals alongside early humans provided ample time for them to develop effective defenses (Owen-Smith, 1987).

The Pleistocene extinctions heralded the modern age, in which the influence upon natural communities of human activities has been increasing dramatically.

10.7   Appraisal of Patterns in Species Richness

There are many generalizations that can be made about the species richness of communities. We have seen how richness may peak at an intermediate level of disturbance frequency and how richness declines with a reduction in island area or an increase in island remoteness. We find also that species richness decreases with increasing latitude, altitude, and (after an initial rise) depth in the ocean. In more productive environments, a wider range of resources may promote species richness, whereas just more of the same can lead to a reduction in richness. A positive correlation between species richness and temperature has also been documented. Species richness increases with an increase in spatial heterogeneity but may decrease with an increase in temporal heterogeneity (increased climatic variation). It increases, at least initially, during the course of succession and with the passage of evolutionary time. However, for many of these generalizations important exceptions can be found, and for most of them the current explanations are not entirely adequate.

It also needs to be recognized that global patterns of species richness have been disrupted in dramatic ways by human activities, such as land-use development, pollution, and the introduction of exotic species (Box 10.3).

Unraveling richness patterns is one of the most difficult and challenging areas of modern ecology. Clear, unambiguous predictions and tests of ideas are often very difficult to devise and will require great ingenuity from future generations of ecologists. Because of the increasing importance of recognizing and conserving the world's biodiversity, though, it is crucial that we come to understand thoroughly these patterns in species richness. We will assess the adverse effects of human activities, and how they may be remedied, in Chapters 12–14.

Box 10.3   *Topical ECOncerns*

The Flood of Exotic Species

Throughout the history of the world, species have invaded new geographic areas, as a result of chance colonizations (e.g., dispersed to remote areas by wind or to remote islands on floating debris; see Section 10.5.1) or during the slow northwards spread of forest trees in the centuries since the last ice age (see Section 4.3). However, human activities have increased this historical trickle to a flood, disrupting global patterns of species richness.

Some human-caused introductions are an accidental consequence of human transport. Other species have been introduced intentionally, perhaps to bring a pest under control (see Section 12.5), to produce a new agricultural product, or to provide new recreational opportunities. Many invaders become part of natural communities without obvious consequences. But some have been responsible for driving native species extinct or changing natural communities in significant ways (see Section 14.2.3).

The alien plants of the British Isles illustrate a number of general points about invaders. Species inhabiting areas where people live and work are more likely to be transported to new regions, where they will tend to be deposited in habitats like those where they originated. As a result, more alien species are found in disturbed habitats close to human transport centers (docks, railways, and cities) and fewer in remote mountain areas (Figure 10.25a). Moreover, more invaders to the British Isles are likely to arrive from nearby geographic locations (e.g., Europe) or from remote locations whose climate matches that of Britain (e.g., New Zealand) (Figure 10.25b). Note the small number of alien plants from tropical environments; these species usually lack the frost-hardiness required to survive the British winter.

Review the options available to governments to prevent (or reduce the likelihood) of invasions of undesirable alien species.

Box 10.3 *(cont'd)*

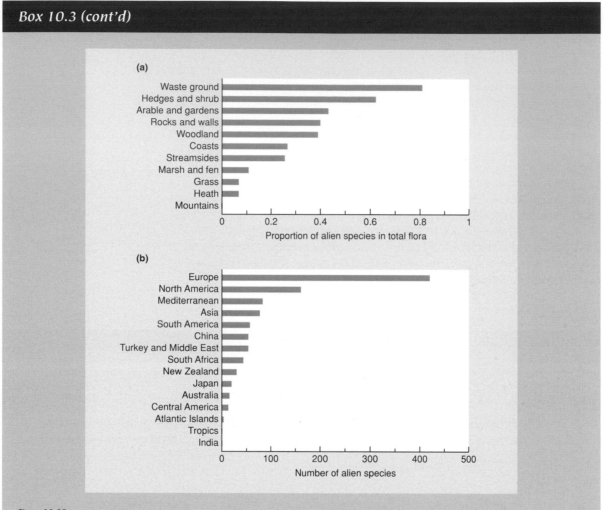

Figure 10.25
The alien flora of the British Isles (a) according to community type (note the large number of aliens in open, disturbed habitats close to human settlements) and (b) by geographical origin (reflecting proximity, trade, and climatic similarity). (After Godfray & Crawley, 1998.)

Summary

Richness and diversity

The number of species in a community is referred to as its *species richness*. Richness, though, ignores the fact that some species are rare and others common. *Diversity indices* are designed to combine species richness and the evenness of the distribution of individuals among those species. Attempts to describe a complex community structure by one single attribute, such as richness or diversity, can still be criticized because so much valuable information is lost. A more complete picture is therefore sometimes provided in a *rank–abundance diagram*.

A simple model of species richness

A simple model can help us understand the determinants of species richness. Within it, a community will contain more species the greater the range of resources, if the species are more specialized in their use of resources, if species overlap to a greater extent in their use of resources, or if the community is more fully saturated.

Productivity and resource richness

If higher productivity is correlated with a wider *range* of available resources, then this is likely to lead to an increase in species richness, but more of the same might lead to more individuals per species rather than more species. In general, though, species richness often increases with the richness of available resources and productivity, although in some cases the reverse has been observed—the paradox of enrichment—and others have found species richness to be highest at intermediate levels of productivity.

Predation intensity

Predation can exclude certain prey species and reduce richness or permit more niche overlap and a greater

richness (predator-mediated coexistence). Overall, therefore, there may be a humped relationship between predation intensity and species richness in a community, with greatest richness at intermediate intensities.

Spatial heterogeneity

Environments that are more spatially heterogeneous often accommodate extra species because they provide a greater variety of microhabitats, a greater range of microclimates, more types of places to hide from predators, and so on—the resource spectrum is increased.

Environmental harshness

Environments dominated by an extreme abiotic factor—often called harsh environments—are more difficult to recognize than might be immediately apparent. Some apparently harsh environments do support few species, but any overall association has proved extremely difficult to establish.

Climatic variation

In a predictable, seasonally changing environment, different species may be suited to conditions at different times of the year. More species might therefore be expected to coexist than in a completely constant environment. On the other hand, there are opportunities for specialization (e.g., obligate fruit eating) in a nonseasonal environment that do not exist in a seasonal environment. Unpredictable climatic variation (climatic instability) could decrease richness by denying species the chance to specialize or increase richness by preventing competitive exclusion. There is no established relationship between climatic instability and species richness.

Disturbance

The *intermediate disturbance hypothesis* suggests that very frequent disturbances keep most patches in the early stages of succession (where there are few species), but very rare disturbances allow most patches to become dominated by the best competitors (where there are also few species). Originally proposed to account for patterns of richness in tropical rain forests and coral reefs, the hypothesis has occupied a central place in the development of ecological theory.

Environmental age: evolutionary time

It has often been suggested that communities may differ in richness because some are closer to equilibrium and therefore more saturated than others, and that the tropics are rich in species in part because the tropics have existed over long and uninterrupted periods of evolutionary time. A simplistic contrast between the unchanging tropics and the disturbed and recovering temperate regions, however, is untenable.

Habitat area and remoteness— island biogeography

Islands need not be islands of land in a sea of water. Lakes are islands in a sea of land; mountaintops are high altitude islands in a low altitude ocean. The number of species on islands decreases as island area decreases, in part because larger areas typically encompass more different types of habitat. However, MacArthur and Wilson's *equilibrium theory of island biogeography* argues for a separate island effect based on a balance between immigration and extinction, and the theory has received much support. In addition, on isolated islands especially, the rate at which new species evolve may be comparable to or even faster than the rate at which they arrive as new colonists.

Gradients in species richness

Richness increases from the poles to the tropics. Predation, productivity, climatic variation, and the greater evolutionary age of the tropics have been put forward as explanations, but not one of these is without problems.

In terrestrial environments, richness often (but not always) decreases with altitude. Factors instrumental in the latitudinal trend are also likely to be important in this, but so are area and isolation. In aquatic environments, richness usually decreases with depth for similar reasons.

In successions, if they run their full course, richness first increases (because of colonization) but eventually decreases (because of competition). There may also be a *cascade effect*: one process that increases richness kick-starts a second, which feeds into a third, and so on.

Patterns in taxon richness in the fossil record

The Cambrian explosion of taxa may have been an example of exploiter-mediated coexistence. The Permian decline may have been the result of a species–area relationship when the Earth's continents coalesced into Pangaea. The changing pattern of plant taxa may reflect the competitive displacement of older, less specialized taxa by newer, more specialized ones. The extinctions of many large animals in the Pleistocene may reflect the hand of human predation and hold lessons for the present day.

Review Questions

1 Explain species richness, diversity index, and the rank–abundance diagram and compare what each measures.

2 What is the paradox of enrichment, and how can the paradox be resolved?

3 Explain, with examples, the contrasting effects that predation can have on species richness.

4 ▲ Researchers have reported a variety of hump-shaped patterns in species richness, with peaks of richness occurring at intermediate levels of productivity, predation pressure, disturbance, and depth in the ocean. Review the evidence and consider whether these patterns have any underlying mechanisms in common.

5 Why is it so difficult to identify harsh environments?

6 Explain the intermediate disturbance hypothesis.

7 Islands need not be islands of land in an ocean of water. Compile a list of other types of habitat islands over as wide a range of spatial scales as possible.

8 ▲ An experiment was carried out to try to separate the effects of habitat diversity and area on arthropod species richness on some small mangrove islands in the Bay of Florida. These consisted of pure stands of the mangrove species *Rhizophora mangle*, which support communities of insects, spiders, scorpions, and isopods. After a preliminary faunal survey, some islands were reduced in size by means of a power saw and brute force! Habitat diversity was not affected, but arthropod species richness on the three islands nonetheless diminished over a period of 2 years (Figure 10.26). A control island, the size of which was unchanged, showed a slight increase in

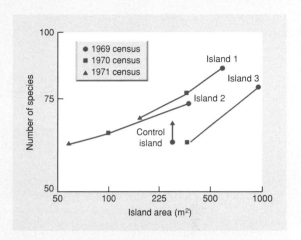

Figure 10.26
The effect on the number of arthropod species of artificially reducing the size of mangrove islands. Islands 1 and 2 were reduced in size after both the 1969 and 1970 censuses. Island 3 was reduced only after the 1969 census. The control island was not reduced, and the change in its species richness was attributable to random fluctuations. (After Simberloff, 1976.)

richness over the same period. Which of the predictions of island biogeography theory are supported by the results in the figure? What further data would you require to test the other predictions? How would you account for the slight increase in species richness on the control island?

9 ▲ A cascade effect is sometimes proposed to explain the increase in species richness during a community succession. How might this cascade concept apply to the commonly observed gradient of species richness with latitude?

10 Describe how theories of species richness that have been derived on ecological timescales can also be applied to patterns observed in the fossil record.

▲ = Challenge Question

Chapter 11

The Flux of Energy and Matter through Ecosystems

Like all biological entities, ecological communities, at scales from the local to the global, require matter for their construction and energy for their activities. A healthy community, like a healthy body, needs to be adequately fed and watered. We need to understand the routes by which matter and energy enter and leave ecological communities, and the pathways they follow, and the processes that affect them while they are there.

Chapter Contents

11.1 Introduction

11.2 Primary productivity

11.3 The fate of primary productivity

11.4 The process of decomposition

11.5 The flux of matter through ecosystems

11.6 Global biogeochemical cycles

Key Concepts

In this chapter, you will

- recognize that communities are intimately linked with the abiotic environment by fluxes of energy and matter

- understand that net primary productivity is not evenly spread across the Earth

- appreciate that transfer of energy between trophic levels is always inefficient—secondary productivity by herbivores is approximately an order of magnitude less than the primary productivity on which it is based

- recognize that the decomposer system processes much more of a community's energy and matter than the live-consumer system

- appreciate that decomposition results in complex energy-rich molecules being broken down by their consumers (decomposers and detritivores) into carbon dioxide, water, and inorganic nutrients

- understand that in global geochemical cycles, nutrients are moved over vast distances by winds in the atmosphere and in the moving waters of streams and ocean currents

11.1   Introduction

When we focus on the processes driving the fluxes of energy and matter between and within communities, we are concerned more with the ways in which areas of land or water receive and process incident radiation and inorganic nutrients, and with the integrated performance of whole communities and their abiotic environment, and not so much with the behavior of individuals, species, and populations. At this level of organization, we use the term *ecosystem* to denote the biological community together with the abiotic environment in which it is set. Lindemann (1942) laid the foundations for ecological energetics, a science with profound implications both for understanding ecosystem processes and for human food production (Box 11.1).

In order to examine ecosystem processes, it is important to understand some key terms.

the standing crop and primary and secondary productivity

- The bodies of the living organisms within a unit area constitute a *standing crop* of biomass.
- By *biomass* we mean the mass of organisms per unit area of ground (or water) and this is usually expressed in units of energy (e.g., joules per square meter) or dry

---

Box 11.1   *Historical Landmarks*

Ecological Energetics and the Biological Basis of Productivity and Human Welfare

A classic paper by Lindemann (1942) laid the foundations of a science of ecological energetics. He attempted to quantify the concept of food chains and food webs by considering the efficiency of transfer between trophic levels—from incident radiation received by a community through its capture by green plants in photosynthesis to its subsequent use by bacteria, fungi, and animals.

Lindemann's paper was a major catalyst that stimulated the International Biological Programme (IBP). The subject of the IBP was "the biological basis of productivity and human welfare". Given the problem of a rapidly increasing human population, it was recognized that scientific knowledge would be required for rational resource management. Cooperative international research programs focused on the ecological energetics of areas of land, fresh waters, and the seas. The IBP provided the first occasion on which biologists throughout the world were challenged to work together towards a common end.

More recently, another pressing issue has galvanized the ecological community into action. Deforestation, the burning of fossil fuels, and other human influences are expected to cause dramatic changes to global climate and atmospheric composition, and can be expected in turn to influence patterns of productivity and the composition of vegetation on a global scale. The International Geosphere-Biosphere Programme (IGBP) has established a core project on Global Change and Terrestrial Ecosystems (GCTE) with the prime objective of predicting the effects of changes in climate and atmospheric composition on terrestrial ecosystems, including agricultural and production forest systems (Steffen et al., 1992).

organic matter (e.g., metric tons per hectare). In practice we include in biomass all those parts, living or dead, that are attached to the living organism. Thus, it is conventional to regard the whole body of a tree as biomass, despite the fact that most of the wood is dead. Organisms (or their parts) cease to be regarded as biomass when they die (or are shed) and become components of dead organic matter.

- The *primary productivity* of a community is the rate at which biomass is produced *per unit area* by plants, the primary producers. It can be expressed either in units of energy (e.g., joules per square meter per day) or of dry organic matter (e.g., kilograms per hectare per year).
- The total fixation of energy by photosynthesis is referred to as *gross primary productivity* (GPP). A proportion of this, however, is respired away by the plant itself and is lost from the community as respiratory heat (R).
- The difference between GPP and R is known as *net primary productivity* (NPP) and represents the actual rate of production of new biomass that is available for consumption by heterotrophic organisms (bacteria, fungi, and animals).
- The rate of production of biomass by heterotrophs is called *secondary productivity*.

live-consumer systems and decomposer systems

A proportion of primary production is consumed by herbivores which, in turn, are consumed by carnivores. These constitute "the live-consumer system". The fraction of NPP that is not eaten by herbivores is then said to pass through the *decomposer system*. We distinguish two groups of organisms responsible for the decomposition of dead organic matter (detritus); bacteria and fungi are called *decomposers* while animals that consume dead matter are known as *detritivores*.

11.2 Primary Productivity

11.2.1 Geographical patterns in primary productivity

The functioning of the biota of the Earth, and of the communities across its surface, depends crucially on the levels of productivity that the plants are able to achieve. If we wish to understand and protect these communities, and to be able to detect when their functioning is faulty or under threat, then we must understand the underlying patterns of productivity.

the open ocean is, in effect, a marine desert

Global terrestrial net primary productivity is estimated to be about 115×10^9 metric tons dry weight per year, and in the sea 55×10^9 t yr^{-1}. The fact that productivity is not evenly spread across the Earth is emphasized in Table 11.1. A large proportion of the globe produces less than 400 g m^{-2} yr^{-1}. This includes over 30 percent of the land surface and 90 percent of the ocean. At the other extreme, the most productive systems are found amongst swamps and marshlands, estuaries, algal beds and reefs, tropical forests, and cultivated land.

the productivity of forests, grasslands, crops, and lakes follows a latitudinal pattern

In the forest biomes of the world, there is a general latitudinal trend of increasing productivity from northern (boreal), through temperate, to tropical conditions (Table 11.1). The same trend may exist in the productivity of tundra and grassland communities (Figure 11.1a) and in various cultivated crops (Figure 11.1b), but there is much variation. In aquatic communities this trend is clear in lakes (Figure 11.1c) but not in the oceans. The trend with latitude suggests that radiation (a resource)

Table 11.1
Normal ranges of the net annual primary productivity (NNP) and standing crop biomass estimates for contrasting communities of the world. (After Whittaker, 1975.)

Ecosystem type	Area (10^6 km²)	NPP, per unit area (g m^{-2} or t km^{-2})	World NPP (10^9 t)	Biomass per unit area (kg m^{-2})	World biomass (10^9 t)
Tropical forest	24.5	1,000–3,500	49.4	6–80	1,025
Temperate forest	12.0	600–2,500	14.9	6–200	385
Northern coniferous forest	12.0	400–2,000	9.6	6–40	240
Woodland and shrubland	8.5	250–1,200	6.0	2–20	50
Savanna	15.0	200–2,000	13.5	0.2–15	60
Temperate grassland	9.0	200–1,500	5.4	0.2–5	14
Tundra and alpine	8.0	10–400	1.1	0.1–3	5
Desert and semidesert shrub	18.0	10–250	1.6	0.1–4	13
Extreme desert, rock, sand, and ice	24.0	0–10	0.07	0–0.2	0.5
Cultivated land	14.0	100–3,500	9.1	0.4–12	14
Swamp and marsh	2.0	800–3,500	4.0	3–50	30
Lake and stream	2.0	100–1,500	0.5	0–0.1	0.05
Total continental	149		115		1,837
Open ocean	332.0	2–400	41.5	0–0.005	1.0
Upwelling zones	0.4	400 1,000	0.2	0.005–0.1	0.008
Continental shelf	26.6	200–600	9.6	0.001–0.04	0.27
Algal beds and reefs	0.6	500–4,000	1.6	0.04–4	1.2
Estuaries	1.4	200–3,500	2.1	0.01–6	1.4
Total marine	361		55.0		3.9
Full total	510		170		1,841

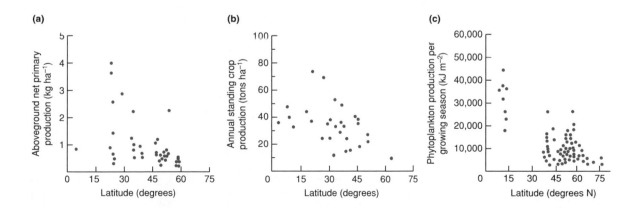

Figure 11.1
Despite much variation, there is a general trend of increasing net primary productivity from the poles (latitude 90°) towards the equator (latitude 0°) for: (a) grassland and tundra ecosystems (after Cooper, 1975); (b) cultivated crops (after Cooper, 1975); and (c) lakes (after Brylinsky & Mann, 1973).

and temperature (a condition) may be the factors usually limiting the productivity of communities. Other factors can, however, constrain productivity within even narrower limits. In the sea, productivity is more often limited by a shortage of nutrients.

11.2.2 Factors limiting primary productivity

What, then, limits primary productivity? In terrestrial communities, solar radiation, carbon dioxide, water, and soil nutrients are the resources required for primary production while temperature, a condition, has a strong influence on the rate of photosynthesis. Carbon dioxide is normally present at around 0.03 percent of atmospheric gases and seems to play no significant role in determining differences between the productivities of different communities (although global increases in carbon dioxide concentration may bring big changes) (Kicklighter et al., 1999). On the other hand, the intensity of radiation, the availability of water and nutrients, and temperature all vary dramatically from place to place. They are all candidates for the role of limiting factor. Which of them actually sets the limit to primary productivity?

terrestrial communities use radiation inefficiently

Depending on location, something between 0 and 5 J of solar energy strikes each square meter of the Earth's surface every minute. If all this were converted by photosynthesis to plant biomass (that is, if photosynthetic efficiency were 100 percent) there would be a prodigious generation of plant material, ten to a 100 times greater than recorded values. However, only about 44 percent of incident shortwave radiation occurs at wavelengths suitable for photosynthesis. Yet, even when this is taken into account, productivity still falls well below the maximum possible. For example, the conifer communities shown in Figure 11.2 had the highest net photosythetic efficiencies, but these were only between 1 and 3 percent. For a similar level of incoming

Figure 11.2
Photosynthetic efficiency (percentage of incoming photosynthetically active radiation converted to aboveground net primary production) for three sets of terrestrial communities in the United States. (After Webb et al., 1983.)

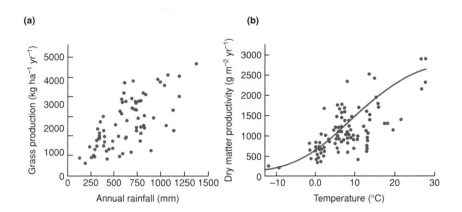

Figure 11.3
Aboveground net primary production of:
(a) grass in savanna regions in relation to
annual rainfall (after Higgins et al., 2000);
and (b) forest in relation to temperature
worldwide (after Reichle, 1970).

radiation, deciduous forests achieved 0.5–1 percent, and, despite their greater energy income, deserts managed only 0.01–0.2 percent. These can be compared with short-term peak efficiencies achieved by crop plants under ideal conditions, when values from 3 to 10 percent can be achieved.

There is no doubt that available radiation would be used more efficiently if other resources were in abundant supply. The much higher values of community productivity from agricultural systems bear witness to this. Shortage of water—an essential resource both as a constituent of cells and for photosynthesis—is often the critical factor. It is not surprising, therefore, that the rainfall of a region is quite closely correlated with its productivity (Figure 11.3a). There is also a clear relationship between the aboveground NPP and mean annual temperature (Figure 11.3b), but note that higher temperatures are associated with rapid transpiration, and thus they increase the rate at which water shortage may become important. Water shortage has direct effects on the rate of plant growth, but it also leads to less dense vegetation. Vegetation that is sparse intercepts less radiation (much of which falls on bare ground). This accounts for much of the difference in productivity noted earlier between desert vegetation and forest. *water and temperature as critical factors*

The productivity of a community can be sustained only for that part of the year when the plants bear photosynthetically active foliage. Deciduous trees have a self-imposed limit on the period of the year during which they bear foliage, while ever-green trees hold a canopy throughout the year (though in some seasons they may barely photosynthesize at all). *NPP increases with the length of the growing season*

No matter how brightly the sun shines and how often the rain falls, no matter how equable is the temperature, productivity will be low if there is no soil in a terrestrial community, or if the soil is deficient in essential mineral nutrients. Of all the mineral nutrients, the one with the strongest influence on community productivity is fixed nitrogen. There is probably no agricultural or forestry system that does not respond to applied nitrogen by increased primary productivity, and this may well be true of natural vegetation as well. The deficiency of other elements, particularly phosphorus, can also hold the productivity of a community far below what is theoretically possible. *NPP may be low because appropriate mineral resources are deficient*

In fact, in the course of a year, the productivity of a terrestrial community may be limited by a succession of factors. The primary productivity of grasslands may be far below the theoretical maximum because the winters are too cold and the intensity of *a succession of factors may limit primary productivity through the year*

Figure 11.4
Relationship between the gross primary productivity of phytoplankton (microscopic plants) in the open water of some Canadian lakes and phosphorus concentration. (After Carignan et al., 2000.)

Figure 11.4
Relationship between the gross primary productivity of phytoplankton (microscopic plants) in the open water of some Canadian lakes and phosphorus concentration. (After Carignan et al., 2000.)

radiation is low, the summers are too dry, the rate of nitrogen mobilization is too slow, or—especially—because overgrazing reduces the standing crop of photosynthetic leaves and much of the incident radiation falls on bare ground.

productive aquatic communities occur where nutrient concentrations are high

In aquatic communities, the factors that most frequently limit primary productivity are the availability of nutrients (particularly nitrate and phosphate) and the intensity of the solar radiation that penetrates the column of water. Productive aquatic communities occur where, for one reason or another, nutrient concentrations are high (for example, Figure 11.4). Lakes receive nutrients by the weathering of rocks and soils in their catchment areas, in rainfall, and as a result of human activity (fertilizers and sewage input—see Chapter 13); they vary considerably in nutrient availability.

In the oceans, locally high levels of primary productivity are associated with high nutrient inputs from two sources. First, nutrients may flow continuously into coastal shelf regions from estuaries. Productivity in the inner shelf region is particularly high because nutrient concentrations are high and the relatively clear water provides a reasonable depth within which net photosynthesis is positive (the euphotic zone). Closer to land, the water is richer in nutrients but is highly turbid and its productivity is less. The least productive zones are in the open ocean where, although the water is clear and the euphotic zone is deep, there are generally extremely low concentrations of nutrients. Local regions of high productivity occur in the open ocean only where there are upwellings from deep nutrient-rich water.

11.3 The Fate of Primary Productivity

Fungi, animals, and most bacteria are heterotrophs: they derive their matter and energy either directly by consuming plant material or indirectly from plants by eating other heterotrophs. Plants, the primary producers, comprise the first trophic level in a community; primary consumers occur at the second trophic level; secondary consumers (carnivores) at the third, and so on.

11.3.1   Relationship between primary and secondary productivity

Since secondary productivity depends on primary productivity, we should expect a positive relationship between the two variables in communities. Figure 11.5 illustrates this general relationship in aquatic and terrestrial examples. Secondary productivity by zooplankton (small animals in the open water), which principally consume phytoplankton cells, is positively related to phytoplankton productivity in a range of lakes in different parts of the world (Figure 11.5a). The productivity of heterotrophic bacteria in lakes and oceans also parallels that of phytoplankton (Figure 11.5b); they metabolize dissolved organic matter released from intact phytoplankton cells or produced as a result of "messy feeding" by grazing animals. Figure 11.5c shows how the number of broods raised by one of Darwin's finches, the seed-eating *Geospiza fortis* (see Figure 2.15), is significantly related to rainfall during the year (primary productivity in many communities is related to rainfall).

> there is a general positive relationship between primary and secondary productivity

In both aquatic and terrestrial communities, secondary productivity by herbivores is approximately one-tenth of the primary productivity upon which it is based. This results in a pyramidal structure in which the productivity of plants provides a broad base upon which a smaller productivity of primary consumers depends, with a still smaller productivity of secondary consumers above that. Trophic levels may also have a pyramidal structure when expressed in terms of density or biomass. But food chains that depend on phytoplankton often have inverted pyramids of biomass: a highly productive but small biomass of short-lived algal cells maintains a larger biomass of longer lived zooplankton.

The productivity of herbivores is invariably less than that of the plants on which they feed. Where has the missing energy gone? First, not all of the plant biomass produced is consumed alive by herbivores. Much dies without being grazed and supports a community of decomposers (bacteria, fungi) and detritivorous animals. Second, not all

> most of the primary productivity does not pass through the grazer system

Figure 11.5
The relationship between primary and secondary productivity: (a) for zooplankton in lakes (after Brylinsky & Mann, 1973); (b) for bacteria in fresh water and seawater (after Cole et al., 1988); and (c) for Darwin's finch (*Geospiza fortis*), measured as clutch size in relation to annual rainfall (open circles are for particularly wet years when El Niño weather events occurred) (after Grant et al., 2000). All regression lines are significant at $P < 0.05$.

the plant biomass eaten by herbivores (nor herbivore biomass eaten by carnivores) is assimilated and available for incorporation into consumer biomass. Some is lost in feces, and this also passes to the decomposers. Third, not all the energy that has been assimilated is actually converted to biomass. A proportion is lost as respiratory heat. This occurs both because no energy conversion process is ever 100 percent efficient (some is lost as unusable random heat, consistent with the second law of thermodynamics) and also because animals do work which requires energy, again released as heat. These three energy pathways occur at all trophic levels and are illustrated in Figure 11.6.

11.3.2   The fundamental importance of energy transfer efficiencies

possible pathways of a joule of energy through a community

A unit of energy (a joule) may be consumed and assimilated by an invertebrate herbivore that uses part of it to do work and loses it as respiratory heat. Or it might be consumed by a vertebrate herbivore and later be assimilated by a carnivore that dies and enters the dead organic matter compartment. Here, what remains of the joule may be assimilated by a fungal hypha and consumed by a soil mite, which uses it to do work, dissipating a further part of the joule as heat. At each consumption step, what remains of the joule may fail to be assimilated and may pass in the feces to dead organic matter, or it may be assimilated and respired, or assimilated and incorporated into the growth of body tissue (or the production of offspring). The body may die and what remains of the joule may enter the dead organic matter compartment, or it may be captured alive by a consumer in the next trophic level where it meets a further set of possible branching pathways. Ultimately, each joule will have found its way out of the community, dissipated as respiratory heat at one or more of the transitions in its path along the food chain. Whereas a molecule or ion may cycle endlessly through the food chains of a community, energy passes through just once.

The possible pathways in the herbivore/carnivore (live-consumer) and decomposer systems are the same, with one critical exception—feces and dead bodies are lost to the former (and enter the decomposer system), but feces and dead bodies from the decomposer system are simply sent back to the dead organic matter compartment at its base. Thus, the energy available as dead organic matter may finally be completely metabolized—and all the energy lost as respiratory heat—even if this requires several circuits through the decomposer system. The exceptions to this are situations (1) where matter is exported out of the local environment to be metabolized elsewhere, for example detritus being washed out of a stream, and (2) where local abiotic conditions have inhibited decomposition and left pockets of incompletely metabolized high-energy matter, otherwise known as oil, coal, and peat.

The proportions of net primary production that flow along each of the possible energy pathways depend on *transfer efficiencies* from one step to the next. We need to know about just three categories of transfer efficiency to be able to predict the pattern of energy flow. These are *consumption efficiency* (CE), *assimilation efficiency* (AE), and *production efficiency* (PE).

consumption, assimilation, and production efficiencies determine the relative importance of energy pathways

Consumption efficiency is the percentage of total productivity available at one trophic level that is consumed ("ingested") by a trophic level above. For primary consumers, CE is the percentage of joules produced per unit time as net primary productivity that finds its way into the guts of herbivores. In the case of secondary consumers, it is the percentage of herbivore productivity eaten by carnivores. The remainder dies without being eaten and enters the decomposer system. Reasonable average figures for consumption efficiency by herbivores are approximately 5 percent in forests, 25 percent in grasslands, and 50 percent in phytoplankton-dominated communities. We know much less about the consumption efficiencies of carnivores; vertebrate predators may consume 50–100 percent of production from vertebrate prey but perhaps only 5 percent from invertebrate prey, while invertebrate predators consume perhaps 25 percent of available invertebrate prey production.

Assimilation efficiency is the percentage of food energy taken into the guts of consumers in a trophic compartment that is assimilated across the gut wall and becomes available for incorporation into growth or is used to do work. The remainder is lost as feces and enters the decomposer system. An "assimilation efficiency" is much less easily ascribed to microorganisms. Food does not pass through a "gut" and feces are not produced. Bacteria and fungi digest dead organic matter externally and,

Figure 11.7
The frequency distribution of trophic-level transfer efficiencies in 48 trophic studies of aquatic communities. There is considerable variation among studies and among trophic levels. The mean is 10.13 percent (standard error = 0.49). (After Pauly & Christensen, 1995.)

between them, typically absorb almost all the product: they are often said to have an "assimilation efficiency" of 100 percent. Assimilation efficiencies are typically low for herbivores, detritivores, and microbivores (20–50 percent) and high for carnivores (around 80 percent). The way that plants allocate production to roots, wood, leaves, seeds, and fruits also influences their usefulness to herbivores. Seeds and fruits may be assimilated with efficiencies as high as 60–70 percent, and leaves with about 50 percent efficiency, while the assimilation efficiency for wood may be as low as 15 percent.

Production efficiency is the percentage of assimilated energy that is incorporated into new biomass—the remainder is entirely lost to the community as respiratory heat. Production efficiency varies mainly according to the taxonomic class of the organisms concerned. Invertebrates in general have high efficiencies (30–40 percent), losing relatively little energy in respiratory heat. Amongst the vertebrates, ectotherms (whose body temperature varies according to environmental temperature; see Section 3.2.6) have intermediate values for PE (around 10 percent), whilst endotherms, with their high energy expenditure associated with maintaining a constant temperature, convert only 1–2 percent of assimilated energy into production. Microorganisms, including protozoa, tend to have very high production efficiencies.

The overall *trophic transfer efficiency* from one trophic level to the next is simply CE × AE × PE. In the period after Lindeman's (1942) pioneering work (see Box 11.1), it was generally assumed that trophic transfer efficiencies were around 10 percent; indeed some ecologists referred to a 10 percent "law". However, there is certainly no law of nature that results in precisely one-tenth of the energy that enters a trophic level transferring to the next. For example, a compilation of trophic studies from a wide range of freshwater and marine environments revealed that trophic-level transfer efficiencies varied between about 2 and 24 percent—although the mean *was* 10.13 percent (Figure 11.7).

11.3.3 Relative roles of the "live-consumer" and "decomposer" systems

Given reasonable values for net primary productivity at a site, and CE, AE, and PE for all the trophic groupings present (herbivores, carnivores, decomposers, detritivores), it is possible to map out the relative importance of different pathways. Figure 11.8 does

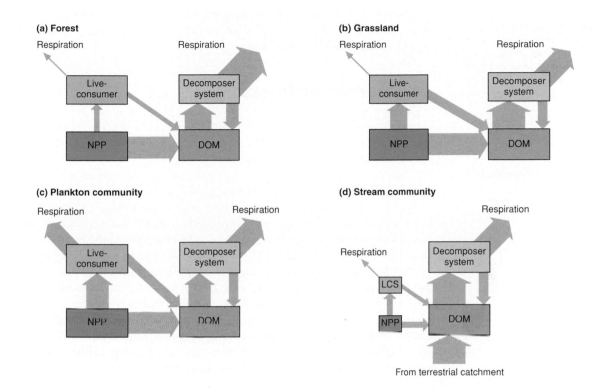

Figure 11.8
General patterns of energy flow for: (a) forest, (b) grassland, (c) a plankton community in the sea, and (d) the community of a stream or small pond. Relative sizes of boxes and arrows are proportional to relative magnitudes of compartments and flows. DOM, dead organic matter; LCS, live-consumer system; NPP, net primary production.

this, in a general way, for a forest, a grassland, a plankton community (of the ocean or a large lake), and the community of a small stream or pond. The decomposer system is probably responsible for the majority of secondary production, and therefore respiratory heat loss, in every community in the world. The "live consumers" have their greatest role in plankton communities, where a large proportion of NPP is consumed alive and assimilated at quite a high efficiency. Even here, though, very high densities of heterotrophic bacteria in the plankton community subsist on dissolved organic molecules excreted by phytoplankton cells, perhaps consuming more than 50 percent of primary productivity as "dead" organic matter in this way.

The "live-consumer" system holds little sway in terrestrial communities because of low herbivore consumption and assimilation efficiencies, and it is almost nonexistent in many small streams and ponds simply because primary productivity is so low. The latter often depend for their energy base on dead organic matter from the terrestrial environment that falls or is washed or blown into the water. The deep-ocean benthic community has a trophic structure very similar to that of streams and ponds. In this case, the community lives in water too deep for photosynthesis to be appreciable or even to take place at all, but it derives its energy base from dead phytoplankton, bacteria, animals, and feces that sink from the autotrophic community in the euphotic zone above. From a different perspective, the ocean bed is equivalent to a forest floor beneath an impenetrable forest canopy.

11.4   The Process of Decomposition

Given the profound importance of the decomposer system, and thus of decomposers (bacteria and fungi) and detritivores, it is important to appreciate the range of organisms and processes involved in decomposition.

decomposition defined

Immobilization is what occurs when an inorganic nutrient element is incorporated into an organic form, primarily during the growth of green plants: for example, when carbon dioxide becomes incorporated into a plant's carbohydrates. Energy (coming, in the case of plants, from the sun) is required for this. Conversely, decomposition involves the release of energy and the *mineralization* of chemical nutrients—the conversion of elements from organic back to inorganic form. Decomposition is defined as the gradual disintegration of dead organic matter (i.e., dead bodies, shed parts of bodies, and feces) and is brought about by both physical and biological agencies. It culminates with complex energy-rich molecules being broken down by their consumers (decomposers and detritivores) into carbon dioxide, water, and inorganic nutrients. Ultimately, the incorporation of solar energy in photosynthesis, and the immobilization of inorganic nutrients into biomass, is balanced by the loss of heat energy and organic nutrients when the organic matter is mineralized.

11.4.1   The decomposers: bacteria and fungi

bacteria and fungi are early colonists of newly dead material

If a scavenging animal, a vulture, or a burying beetle perhaps, does not take a dead resource immediately, the process of decomposition usually starts with colonization by bacteria and fungi. Bacteria and fungal spores are always present in the air and water, and are usually present on (and often in) dead material before it is dead. The early colonists tend to use soluble materials, mainly amino acids and sugars that are freely diffusible. The residual resources, though, are not diffusible and are more resistant to attack. Subsequent decomposition therefore proceeds more slowly, and involves microbial specialists that can break down structural carbohydrates (e.g., celluloses and lignins) and complex proteins such as suberin (cork) and cuticle.

11.4.2   The detritivores and specialist microbivores

most detritivores consume both detritus and its associated bacteria and fungi

The microbivores are a group of animals that operate alongside the detritivores, and which can be difficult to distinguish from them. The name microbivore is reserved for the minute animals that specialize at feeding on bacteria or fungi but are able to exclude detritus from their guts. In fact, though, the majority of detritivorous animals are generalist consumers, of both the detritus itself and the associated bacterial and fungal populations. The invertebrates that take part in the decomposition of dead plant and animal materials are a taxonomically diverse group. In terrestrial environments they

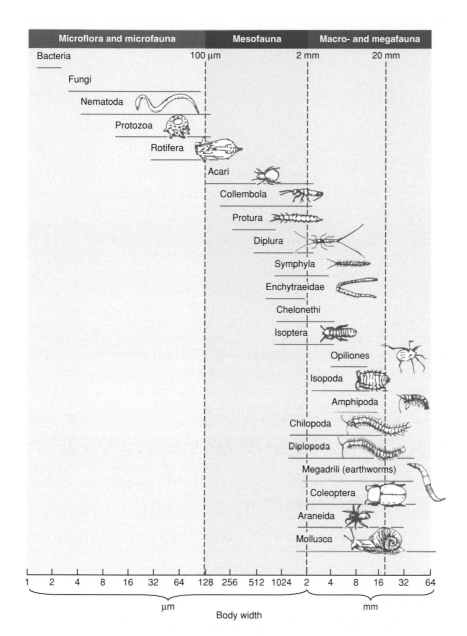

Figure 11.9
Size classification by body width of organisms in terrestrial decomposer food webs. The following groups are wholly carnivorous: Opiliones (harvest spiders), Chilopoda (centipedes), and Araneida (spiders). (After Swift et al., 1979.)

are usually classified according to their size (Figure 11.9). This is not an arbitrary basis for classification, because size is an important feature for organisms that reach their resources by burrowing or crawling among the cracks and crevices of litter or soil.

In freshwater ecology, on the other hand, the study of detritivores has been concerned less with the size of the organisms than with the ways in which they obtain their food (see Figure 4.16). For example, *shredders* are detritivores that feed on coarse

aquatic detritivores are usually classified according to their feeding mode

Figure 11.10
Downstream trends in the Vindel River in Sweden (shown as distance from the confluence with the larger Ume River) in the concentration of fecal pellets (number of fecal pellets per litre $\pm$ standard error) of blackfly larvae (family Simuliidae). The generally lower concentrations in "runs" reflect the higher probability of pellets settling to the riverbed in these sections compared to the "rapids" sections. The numbers above the error bars are percentages of the mass of total organic matter in the flowing water (seston) made up of fecal pellets. (After Malmqvist et al., 2001.)

particulate organic matter, such as tree leaves fallen into a river—these animals fragment the material into finer particles. On the other hand, *collector–filterers*, such as larvae of blackflies in rivers, consume the fine particulate organic matter which otherwise would be carried downstream. Because of very high densities (sometimes as many as 600,000 blackfly larvae per square meter of riverbed) a very large quantity of fine particulate matter is converted by the larvae into fecal pellets (estimated at 429 metric tons dry mass of fecal pellets per day in a Swedish river; Malmqvist et al., 2001). Fecal pellets are much larger than the particulate food of the larvae and so are much more likely to settle out on the riverbed, especially in slower flowing sections of the river (Figure 11.10). Here they provide organic matter as food for many other detritivorous species.

11.4.3 Consumption of plant detritus

UNANSWERED QUESTION: given the abundance of cellulose as a resource why do so few animals possess their own cellulase enzymes?

Two of the major organic components of dead leaves and wood are cellulose and lignin. These pose considerable digestive problems for animal consumers. Digesting cellulose requires *cellulase* enzymes but, surprisingly, cellulases of animal origin have been definitely identified in only one or two species. The majority of detritivores, lacking their own cellulases, rely on the production of cellulases by associated bacteria or fungi or, in some cases, protozoa. The interactions are of a range of types: (1) *obligate mutualisms* between a detritivore and a specific and permanent gut microflora (e.g., bacteria) or microfauna (e.g., protozoa in termite guts); (2) *facultative mutualisms*, where the animals make use of cellulases produced by a microflora which is ingested with detritus as it passes through an unspecialized gut (e.g., woodlice); or (3) "external rumens", where animals simply assimilate the products of the cellulase-producing microflora associated with decomposing plant remains or feces (e.g., springtails—Collembola).

A variety of detritivores may be involved in fragmenting a single leaf. In experiments involving the larvae of shredding stoneflies, three species did not differ much in

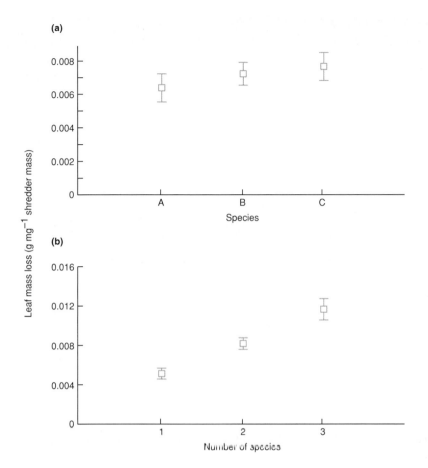

Figure 11.11
Variation in alder leaf mass loss in replicated stream experiments (per gram of leaf per milligram of shredder ± standard error), (a) caused by individual species acting alone (A, *Protonemura meyeri*; B, *Nemoura avicularis*; C, *Taeniopteryx nebulosa*) and (b) averaged for species on their own, for pairs of species in the various possible combinations, and for all three species together. Species' identity did not influence rate of loss of leaf mass ($P = 0.68$) but species richness did ($P = 0.0016$). (After Jonsson & Malmqvist, 2000.)

the efficiency of their contribution to the physical disintegration in a stream of leaves of the alder tree, *Alnus incana* (Figure 11.11a). However, average leaf loss was significantly greater when a pair of species were involved and was faster still when all three species were feeding on the leaf (Figure 11.11b). The same number of stonefly larvae were included in every experiment (12 of a single species, six each in the species pairs, and four each when all three species were present) and the results were expressed in a standard way (leaf mass loss per gram of leaf per milligram of shredder in a 46-day experiment), so the result directly reflects the species diversity present. These results may be indicative of facilitation (feeding by each species benefits from the activity of the others) or complementarity (each species feeds in a slightly different way so their combined effect is enhanced), but the precise mechanism was not determined. Such studies as these have significant implications for the role that biodiversity plays in ecosystem functioning. Given current concerns about the loss of biodiversity worldwide (Chapter 14), we need to know whether or not biodiversity loss will have major consequences for the way ecosystems work. This is an important and controversial area (Box 11.2).

Box 11.2   *Topical ECOncerns*

The Importance of Biodiversity to the Health of Ecosystems

Ecologists agree that some experimental evidence points to a significant role for biodiversity in ecosystem functioning. Figure 11.11, for example, showed how the decomposition rate is slower when fewer species are involved in the process. But some disagree about how much this matters—in other words, whether these kinds of result prove that biodiversity is critical to ecosystem health. This is a significant question at a time when global biodiversity is declining.

The following quotation comes from a commentary by Jocelyn Kaiser that appeared in 2000 in *Science*:

> RIFT OVER BIODIVERSITY DIVIDES ECOLOGISTS—a long-simmering debate among ecologists over the importance of biodiversity to the health of ecosystems has erupted into a full-blown war. Opposing camps are dueling over the quality of key experiments, and some are flinging barbs at meetings and in journals.
> (© 2000 AAAS.)

What lay behind such bellicose language? The disagreement began as part of the normal debate that should occur about any piece of research. To what extent are the conclusions justified from the results and how far can they be generalized from the special circumstances of the experiment to other situations in nature? Various studies around the world seemed to show that the loss of plant or animal species might adversely affect ecosystem function; for example, the productivity of grassland communities appears to be higher when more species are present. This could mean that biodiversity *per se* matters to productivity. But might variables other than species diversity have given rise to increased productivity? For example,

perhaps such a result was a statistical artefact—higher productivity with higher species diversity might be explained simply by the addition of a more productive species to the list (and a more productive species is more likely to be present when more species are included in the experiment).

This kind of debate is healthy, but it took on a new dimension when one of the world's leading learned societies, the Ecological Society of America (ESA), published a pamphlet and sent copies to members of the US Congress. One of a series called "Issues in Ecology", the pamphlet concerned the importance of biodiversity to ecosystem functioning. It summarized the results of several studies but with little discussion of doubts raised by the skeptics. Charged with using inconclusive research to push a policy agenda that plugged the value of biodiversity, the researchers acknowledged their experiments had limitations, but continued to argue that the results were convincing.

The commentator also noted: "Other ecologists safely outside the fray say there is more at stake in this dispute than personalities and egos. Beyond the legitimate scientific question about how much can be learned from experiments is the nagging question—by no means limited to biodiversity—of when scientific data are strong enough to form the basis of policy decisions."

This debate was not really about the quality of the science (since every study has its limitations), but rather the document that ESA sent to Congress, which some said tended to present opinion as fact. Do you think scientists should remain entirely outside the political arena? If not, how would you insure that balanced and generally accepted positions would be presented?

The decomposition of dead material is not simply due to the sum of the activities of decomposers and detritivores; it is largely the result of interactions between the two (Lussenhop, 1992). This can be illustrated by taking an imaginary journey with a leaf fragment through the process of decomposition, focusing attention on a part of the wall of a single cell. Initially, when the leaf falls to the ground, the piece of cell wall is protected from microbial attack because it lies within the plant tissue. The leaf is now chewed and the fragment enters the gut of a woodlouse. Here it meets a new microbial flora in the gut and is acted on by the digestive enzymes of the woodlouse. The fragment emerges, changed by passage through the gut. It is now part of the woodlouse's feces and is much more easily attacked by microorganisms, because it has been fragmented and partially digested. While microorganisms are colonizing the fecal pellet, it may again be eaten, perhaps by a springtail, and pass through the new environment of the springtail's gut. Incompletely digested fragments may again appear, this time in springtail feces, yet more easily accessible to microorganisms. The fragment may pass through several other guts in its progress from being a piece of dead tissue to its inevitable fate of becoming carbon dioxide and minerals.

UNANSWERED QUESTION: where, when, and to what extent will loss of biodiversity cause changes to ecosystem functioning?

11.4.4   Consumption of feces and carrion

The dung of carnivorous vertebrates is relatively poor quality stuff. Carnivores assimilate their food with high efficiency (usually 80 percent or more is digested) and their feces retain only the least digestible components; their decomposition is probably effected almost entirely by bacteria and fungi. In contrast, herbivore dung still contains an abundance of organic matter and is sufficiently thickly spread in the environment to support its own characteristic fauna, consisting of many occasional visitors but with several specific dung feeders. A good example is provided by elephant dung; within a few minutes of dung deposition the area is alive with beetles. The adult dung beetles feed on the dung but they also bury large quantities along with their eggs to provide food for developing larvae (Figure 11.12).

All animals defecate and die, yet feces and dead bodies are not generally very obvious in the environment. This is because of the efficiency of the specialist consumers of these dead organic products. On the other hand, where consumers of feces are absent, a dramatic build up of dung may occur (Box 11.3).

When considering the decomposition of dead bodies, it is helpful to distinguish three categories of organisms that attack carcasses. As before, decomposers (bacteria and fungi) and invertebrate detritivores have roles to play, but, in addition, scavenging vertebrates are often of considerable importance. Many carcasses of a size to make a single meal for one or a few of these scavenging detritivores will be removed completely within a very short time of death, leaving nothing for bacteria, fungi, or invertebrates. This role is played, for example, by Arctic foxes and skuas in polar regions, by crows, gluttons, and badgers in temperate areas, and by a wide variety of birds and mammals, including kites, jackals, and hyenas, in the tropics.

(a)

(b)

11.5 The Flux of Matter through Ecosystems

Chemical elements and compounds are vital for the processes of life. When living organisms expend energy (as they all do, continually), they do so, essentially, in order to extract chemicals from their environment, and to hold on to them and use them for a period before they lose them again. Thus, the activities of organisms profoundly influence the patterns of flux of chemical matter.

The great bulk of living matter in any community is water. The rest is made up mainly of carbon compounds and this is the form in which energy is accumulated and stored. Carbon enters the trophic structure of a community when a simple molecule, carbon dioxide, is taken up in photosynthesis. If it becomes incorporated in net primary productivity, it is available for consumption as part of a sugar, a fat, a protein, or, very often, a cellulose molecule. It follows exactly the same route as energy, being successively consumed and either defecated, assimilated, or used in metabolism, during which the energy of its molecule is dissipated as heat while the carbon is released again to the atmosphere as carbon dioxide. Here, though, the tight link between energy and carbon ends.

Once energy is transformed into heat, it can no longer be used by living organisms to do work or to fuel the synthesis of biomass. The heat is eventually lost to the

Box 11.3   *Topical ECOncerns*

An Australian Dilemma:
Bovine Dung but no Bovine Dung Beetles

Bovine dung has provided an extraordinary and economically very important problem in Australia. During the past 200 years the cow population has risen from just seven individuals (brought over by the first English colonists in 1788) to 30 million or so. These produce some 300 million cow pats per day, covering as much as 6 million acres per year with dung. Deposition of bovine dung poses no particular problem elsewhere in the world, where bovines have existed for millions of years and have an associated fauna that exploits the fecal resources. However, the largest herbivorous animals in Australia until European colonization were marsupials such as kangaroos. The native detritivores, which deal with the dry, fibrous dung pellets of marsupials, cannot cope with cow dung, and the loss of pasture under dung has imposed a huge economic burden on Australian agriculture. The decision was therefore made in 1963 to establish in Australia beetles of African origin, able to dispose of bovine dung, in the most important places and under the most prevalent

conditions where cattle are raised; so far 21 species have been introduced (Doube et al., 1991).

Adding to the problem, Australia is plagued by native bushflies (*Musca vetustissima*) and buffalo flies (*Haematobia irritans exigua*) which deposit eggs on dung pats. The larvae fail to survive in dung that has been buried by beetles, and the presence of beetles has been shown to be effective at reducing fly abundance (Ridsdill-Smith, 1991). Success depends on dung being buried within about 6 days of production, the time it takes for the fly egg (laid on fresh dung) to hatch and develop to the pupal stage. Edwards and Aschenborn (1987) surveyed the nesting behavior in southern Africa of 12 species of dung beetles in the genus *Onitis*. They concluded that *Onitis uncinatus* was a prime candidate for introduction to Australia for fly-control purposes, since substantial amounts of dung were buried on the first night after pat colonization. The least suitable species, *O. viridualus*, spent several days constructing a tunnel and did not commence burying until 6–9 days had elapsed.

atmosphere and can never be recycled: life on Earth is only possible because a fresh supply of solar energy is made available every day. In contrast, the carbon in carbon dioxide can be used again in photosynthesis. Carbon, and all other nutrient elements (nitrogen, phosphorus, etc.), are available to plants as simple organic molecules or ions in the atmosphere (carbon dioxide), or as dissolved ions in water (nitrate, phosphate, potassium, etc.). Each can be incorporated into complex carbon compounds in biomass. Ultimately, however, when the carbon compounds are metabolized to carbon dioxide, the mineral nutrients are released again in simple inorganic form. Another plant may then absorb them, and so an individual atom of a nutrient element may pass repeatedly through one food chain after another.

Unlike the energy of solar radiation, moreover, nutrients are not in unalterable supply. The process of locking some up into living biomass reduces the supply remaining to the rest of the community. If plants, and their consumers, were not eventually decomposed, the supply of nutrients would become exhausted and life on Earth would cease.

energy cannot be cycled and reused—matter can

We can conceive of pools of chemical elements existing in compartments. Some compartments occur in the *atmosphere* (carbon in carbon dioxide, nitrogen as gaseous

biogeochemistry and biogeochemical cycles

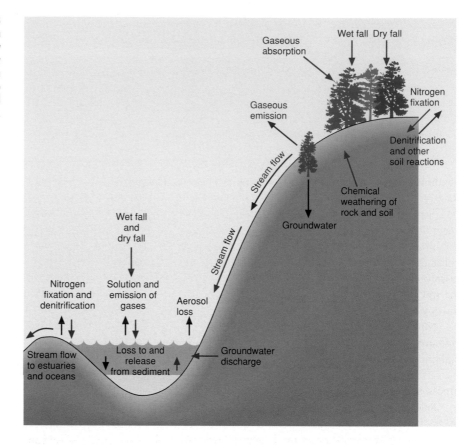

nitrogen, etc.), some in the rocks of the *lithosphere* (calcium as a constituent of calcium carbonate, potassium in feldspar, etc.) and others in the water of soil, streams, lakes, or oceans—the *hydrosphere* (nitrogen in dissolved nitrate, phosphorus in phosphate, carbon in carbonic acid, etc.). In all these cases the elements exist in inorganic form. In contrast, living organisms (the biota) and dead and decaying bodies can be viewed as compartments containing elements in organic form (carbon in cellulose or fat, nitrogen in protein, phosphorus in ATP, etc.). Studies of the chemical processes occurring within these compartments and, more particularly, of the fluxes of elements between them, comprise the science of biogeochemistry.

Nutrients are gained and lost by communities in a variety of ways (Figure 11.13). We can construct a nutrient budget if we can identify and measure all the processes on the credit and debit sides of the equation.

11.5.1   Nutrient budgets in terrestrial ecosystems

nutrient inputs   Weathering of parent bedrock and soil, by both physical and chemical processes, is the dominant natural and ultimate source of nutrients such as calcium, iron, magnesium, phosphorus, and potassium, which may then be taken up via the roots of plants.

Atmospheric carbon dioxide is the source of the carbon content of terrestrial communities. Similarly, gaseous nitrogen from the atmosphere provides most of the nitrogen content of communities. Several types of bacteria and blue–green algae possess the enzyme nitrogenase, which converts gaseous nitrogen to ammonium ions (NH_4^+) that can then be taken up through the roots and used by plants. All terrestrial ecosystems receive some available nitrogen through the activity of free-living nitrogen-fixing bacteria, but communities containing plants such as legumes and alder trees (*Alnus* spp.), with their root nodules containing symbiotic nitrogen-fixing bacteria (Chapter 7), may receive a very substantial proportion of their nitrogen in this way.

Other nutrients from the atmosphere become available to communities in *dryfall* (settling of particles during periods without rain) or *wetfall* (in rain, snow, and fog). Rain is not pure water but contains chemicals derived from a number of sources: (1) trace gases, such as oxides of sulfur and nitrogen; (2) aerosols (produced when tiny water droplets from the oceans evaporate in the atmosphere and leave behind particles rich in sodium, magnesium, chloride, and sulfate); and (3) dust particles from fires, volcanoes, and windstorms, often rich in calcium, potassium, and sulfate. Nutrients dissolved in precipitation mostly become available to plants when the water reaches the soil and can be taken up by plant roots. However, some are absorbed by leaves directly.

Nutrients may circulate within the community for many years. Alternatively, the atom may pass through the system in a matter of minutes, perhaps without interacting with the biota at all. Whatever the case, the atom will eventually be lost through one of the variety of processes that remove nutrients from the system (Figure 11.13). These processes constitute the debit side of the nutrient budget equation.

Release to the atmosphere is one pathway of nutrient loss. In many communities there is an approximate annual balance in the carbon budget; the carbon fixed by photosynthesizing plants is balanced by the carbon released to the atmosphere as carbon dioxide from the respiration of plants, microorganisms, and animals. Plants themselves may be direct sources of gaseous and particulate release. For example, forest canopies produce volatile hydrocarbons (e.g., terpenes) and tropical forest trees appear to emit aerosols containing phosphorus, potassium, and sulfur. Finally, ammonia gas is released during the decomposition of vertebrate excreta. Other pathways of nutrient loss are important in particular instances. For example, fire (either natural, or when, for instance, agricultural practice includes the burning of stubble) can turn a very large proportion of a community's carbon into carbon dioxide in a very short time, and the loss of nitrogen, as volatile gas, can be equally dramatic.

For many elements, the most substantial pathway of loss is in streamflow. The water that drains from the soil of a terrestrial community into a stream carries a load of nutrients that is partly dissolved and partly particulate. With the exception of iron and phosphorus, which are not mobile in soils, the loss of plant nutrients is predominantly in solution. Particulate matter in streamflow occurs both as dead organic matter (mainly tree leaves) and as inorganic particles.

It is the movement of water, under the force of gravity, that links the nutrient budgets of terrestrial and aquatic communities (Figure 11.13). Terrestrial systems lose dissolved and particulate nutrients into streams and ground waters; aquatic systems (including the stream communities themselves, and ultimately the oceans) gain nutrients from streamflow and ground water discharge. Refer to Section 1.3.3 for a discussion

nutrient outputs

of a study (at Hubbard Brook) that explored the chemical linkages at the land–water interface.

11.5.2 Nutrient budgets in aquatic communities

Aquatic systems receive the bulk of their supply of nutrients from stream inflow. In stream and river communities, and also in lakes with a stream outflow, export in outgoing stream water is a major factor. By contrast, in lakes without an outflow (or where this is small relative to lake volume), and also in oceans, nutrient accumulation in permanent sediments is often the major export pathway.

Many lakes in arid regions, lacking a stream outflow, lose water only by evaporation. The waters of these endorheic lakes (internal flow) are thus more concentrated than their freshwater counterparts, being particularly rich in sodium but also in other nutrients such as phosphorus. Saline lakes should not be considered as oddities; globally, they are just as abundant in terms of number and volume as freshwater lakes (Williams, 1988). They are usually very fertile and have dense populations of blue–green algae, and some, such as Lake Nakuru in Kenya, support huge aggregations of plankton-filtering flamingoes (*Phoeniconaias minor*).

The largest of all endorheic "lakes" is the ocean—a huge basin of water supplied by the world's rivers and losing water only by evaporation. Its great size, in comparison to the input from rain and rivers, leads to a remarkably constant chemical composition. Nutrients in the surface waters come from two sources—river inputs and water welling up from the deep (Figure 11.14). A given phosphorus atom, for example, arriving in the surface waters, is taken up by a phytoplankton cell or bacterium, or by microscopic picoplankton, and is passed along the food chain (very much as described above for a lake). Detrital particles are continuously sinking to the deep waters where the majority

Figure 11.14
A simple representation of the major fluxes of phosphorus (P) in the ocean.

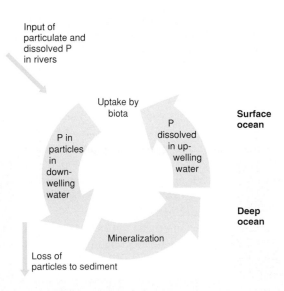

are decomposed, releasing soluble phosphorus that eventually finds its way back to the surface water by upwelling. Only about 1 percent of detrital phosphorus is lost to the sediment in each oceanic mixing cycle.

11.6 Global Biogeochemical Cycles

Nutrients are moved over vast distances by winds in the atmosphere and by the moving waters of streams and ocean currents. There are no boundaries, either natural or political. It is appropriate, therefore, to conclude this chapter by moving to an even larger spatial scale to examine global biogeochemical cycles.

11.6.1 The hydrological cycle

The principal source of water is the oceans; radiant energy makes water evaporate into the atmosphere, winds distribute it over the surface of the globe, and precipitation brings it down to Earth (with a net movement of atmospheric water from oceans to continents), where it may be stored temporarily in soils, lakes, and icefields (Figure 11.15). Loss occurs from the land through evaporation and transpiration or as liquid flow through stream channels and ground water aquifers, eventually to return to the sea. The major pools of water occur in the oceans (97.3 percent of the total for the biosphere; Berner & Berner, 1987), the ice of polar ice caps and glaciers (2.06 percent), deep in the ground water (0.67 percent), and in rivers and lakes (0.01 percent). The proportion that is in transit at any time is very small—water draining through the soil, flowing along rivers, and present as clouds and vapor in the atmosphere constitutes only about 0.08 percent of the total. However, this small percentage plays a crucial role, both by supplying the requirements for the survival of living organisms and for

Figure 11.15
The hydrological cycle showing fluxes and sizes of reservoirs ($\times 10^6$ km^3). (After Berner & Berner, 1987.)

community productivity and because so many chemical nutrients are transported with the water as it moves.

The hydrological cycle would proceed whether or not a biota was present. However, terrestrial vegetation can modify to a significant extent the fluxes that occur. Vegetation can intercept water at two points on this journey, preventing some from reaching the stream and causing it to move back into the atmosphere: (1) by catching some in foliage from which it may evaporate, and (2) by preventing some from draining from the soil water by taking it up in the transpiration stream. We have seen earlier how cutting down the forest in a catchment in Hubbard Brook (see Section 1.3.3) increased the throughput to streams of water together with its load of dissolved and particulate matter. It is small wonder that large-scale deforestation around the globe, usually to create new agricultural land, can lead to loss of top soil, nutrient impoverishment, and increased severity of flooding. Water is a very valuable commodity, and this is reflected in the difficult political exercise of dealing with competing demands of diverting river water for hydroelectric power generation or irrigation for agriculture as opposed to maintaining the intrinsic values of an unmanipulated river.

The world's major abiotic reservoirs for nutrients are illustrated in Figure 11.16. We consider these cycles in turn.

11.6.2   The phosphorus cycle

The principal stocks of phosphorus occur in the water of the soil, rivers, lakes, and oceans and in rocks and ocean sediments. The phosphorus cycle may be described as a sedimentary cycle because of the general tendency for mineral phosphorus to be carried from the land inexorably to the oceans where ultimately it becomes incorporated in the sediments (Figure 11.16a).

the life history of a phosphorus atom

A "typical" phosphorus atom, released from the rock by chemical weathering, may enter and cycle within the terrestrial community for years, decades, or centuries before it is carried via the ground water into a stream. Within a short time of entering the stream (weeks, months, or years), the atom is carried to the ocean. It then makes, on average, about 100 round trips between the surface and deep waters, each lasting perhaps 1,000 years. During each trip, it is taken up by surface-dwelling organisms, before eventually settling into the deep again. On average, on its 100th descent (after 10 million years in the ocean) it fails to be released as soluble phosphorus, but instead enters the bottom sediment in particulate form. Perhaps 100 million years later, the ocean floor is lifted up by geological activity to become dry land. Thus, our phosphorus atom will eventually find its way back via a river to the sea, and to its existence of cycle (biotic uptake and decomposition) within cycle (ocean mixing) within cycle (continental uplift and erosion).

11.6.3   The nitrogen cycle

the nitrogen cycle has an atmospheric phase of overwhelming importance

The atmospheric phase has been widely held to be predominant in the global nitrogen cycle, in which nitrogen fixation and denitrification by microbial organisms are of particular importance (Figure 11.16b). However, recent evidence indicates that nitrogen

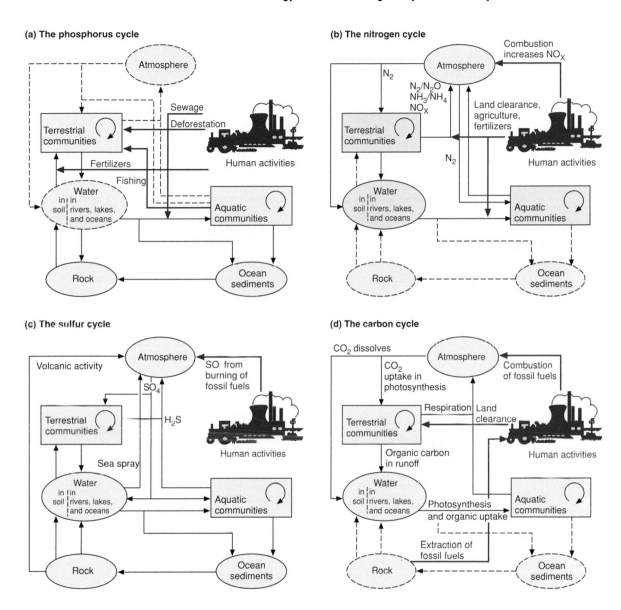

Figure 11.16
The major global pathways of nutrients between the abiotic "reservoirs" of the atmosphere, water (hydrosphere), and rock and sediments (lithosphere), and the biotic "reservoirs" constituted by terrestrial and aquatic communities. Human activities (in red) affect nutrient fluxes through terrestrial and aquatic communities both directly and indirectly, via their effects on global biogeochemical cycling through the release of extra nutrients into the atmosphere and water. Cycles are presented for four important nutrient elements: (a) phosphorus, (b) nitrogen, (c) sulfur, and (d) carbon. Insignificant compartments and fluxes are represented by dashed lines.

from certain geological sources may also be significant in fueling productivity in terrestrial and freshwater communities (Holloway et al., 1998; Thompson et al., 2001). The magnitude of the flux in streamflow from terrestrial to aquatic communities is relatively small, but it is by no means insignificant for the aquatic systems involved. This is because nitrogen is one of the two elements (along with phosphorus) that most often limits plant growth. Finally, there is a small annual loss of nitrogen to ocean sediments.

11.6.4   The sulfur cycle

Three natural biogeochemical processes release sulfur to the atmosphere: the formation of sea-spray aerosols, anaerobic respiration by sulfate-reducing bacteria, and volcanic activity (relatively minor) (Figure 11.16c). Sulfur bacteria release reduced sulfur compounds, particularly H_2S, from waterlogged bog and marsh communities and from marine communities associated with tidal flats. A reverse flow from the atmosphere involves oxidation of sulfur compounds to sulfate, which returns to Earth as both wetfall and dryfall.

the sulfur cycle has an atmospheric phase and a lithospheric phase of similar magnitude

The weathering of rocks provides about half the sulfur draining off land into rivers and lakes, the remainder deriving from atmospheric sources. On its way to the ocean, a proportion of the available sulfur (mainly dissolved sulfate) is taken up by plants, passed along food chains and, via decomposition processes, becomes available again to plants. However, in comparison to phosphorus and nitrogen, a much smaller fraction of the flux of sulfur is involved in internal recycling in terrestrial and aquatic communities. Finally, there is a continuous loss of sulfur to ocean sediments.

11.6.5   The carbon cycle

the opposing forces of photosynthesis and respiration drive the global carbon cycle

Photosynthesis and respiration are the two opposing processes that drive the global carbon cycle. It is predominantly a gaseous cycle, with carbon dioxide as the main vehicle of flux between atmosphere, hydrosphere, and biota. Historically, the lithosphere played only a minor role; fossil fuels lay as dormant reservoirs of carbon until human intervention in recent centuries (Figure 11.16d).

Terrestrial plants use atmospheric carbon dioxide as their carbon source for photosynthesis, whereas aquatic plants use dissolved carbonates (i.e., carbon from the hydrosphere). The two subcycles are linked by exchanges of carbon dioxide between the atmosphere and the oceans. In addition, carbon finds its way into inland waters and oceans as bicarbonate resulting from weathering (carbonation) of calcium-rich rocks such as limestone and chalk. Respiration by plants, animals, and microorganisms releases the carbon locked in photosynthetic products back to the atmospheric and hydrospheric carbon compartments.

11.6.6   Human impacts on biogeochemical cycles

It goes almost without saying that human activities contribute significant inputs of nutrients to ecosystems and disrupt local and global biogeochemical cycles. For example, the amounts of carbon dioxide and oxides of nitrogen and sulfur in the atmosphere have been increased by the burning of fossil fuels and by the exhausts of cars, and the concentrations of nitrate and phosphate in stream water have been raised by agricultural practices and sewage disposal. These changes have far-reaching consequences, which will be discussed in Chapter 13.

Summary

Patterns in primary productivity

Primary production on land is limited by a variety of factors—the quality and quantity of solar radiation, the availability of water, nitrogen, and other key nutrients, and physical conditions, particularly temperature. Productive aquatic communities occur where, for one reason or another, nutrient concentrations are unusually high and the intensity of radiation is not limiting.

The fate of primary productivity

Secondary productivity by herbivores is approximately an order of magnitude less than the primary productivity on which it is based. Energy is lost at each feeding step because consumption efficiencies, assimilation efficiencies, and production efficiencies are all less than 100 percent. The *decomposer system* processes much more of a community's energy and matter than the *live-consumer system*. The energy pathways in the live-consumer and decomposer systems are the same, with one critical exception—feces and dead bodies are lost to the grazer system (and enter the decomposer system), but feces and dead bodies from the decomposer system are simply sent back to the dead organic matter compartment at its base.

The process of decomposition

Decomposition results in complex energy-rich molecules being broken down by their consumers (decomposers and detritivores) into carbon dioxide, water, and inorganic nutrients. Ultimately, the incorporation of solar energy in photosynthesis, and the immobilization of inorganic nutrients into biomass, is balanced by the loss of heat energy and organic nutrients when the organic matter is decomposed. This is brought about partly by physical processes, but mainly by decomposers (bacteria and fungi) and detritivores (animals that feed on dead organic matter).

The flux of matter through ecosystems

Nutrients are gained and lost by communities in a variety of ways. Weathering of parent bedrock and soil, by both physical and chemical processes, is the dominant source of nutrients such as calcium, iron, magnesium, phosphorus, and potassium, which may then be taken up via the roots of plants. Atmospheric carbon dioxide and gaseous nitrogen are the principal sources of the carbon and nitrogen content of terrestrial communities while other nutrients from the atmosphere become available as *dryfall* or in rain, snow, and fog. Nutrients are lost again through release to the atmosphere or in the water that feeds into streams and rivers. Aquatic systems (including the stream communities themselves, and ultimately the oceans) gain nutrients from streamflow and ground water discharge and from the atmosphere by diffusion across their surfaces.

Global biogeochemical cycles

The principal source of water in the hydrological cycle is the oceans; radiant energy makes water evaporate into the atmosphere, winds distribute it over the surface of the globe, and precipitation brings it down to Earth. Phosphorus derives mainly from the weathering of rocks (lithosphere); its cycle may be described as sedimentary because of the general tendency for mineral phosphorus to be carried from the land inexorably to the oceans where ultimately it becomes incorporated in sediments. The sulfur cycle has an atmospheric phase and a lithospheric phase of similar magnitude. The atmospheric phase is predominant in both the global carbon and nitrogen cycles. Photosynthesis and respiration are the two opposing processes that drive the global carbon cycle while nitrogen fixation and denitrification by microbial organisms are of particular importance in the nitrogen cycle. Human activities contribute significant inputs of nutrients to ecosystems and disrupt local and global biogeochemical cycles.

Review Questions

1 A large proportion of the surface of the ocean has a net primary productivity of less than $400 \text{ g m}^{-2} \text{ yr}^{-1}$. The open ocean is, in effect, a marine desert. Why?

2 ▲ Describe the general latitudinal trends in net primary productivity. Suggest reasons why such a latitudinal trend does not occur in the oceans.

3 ▲ Table 11.2 presents the results of a study that contrasted the productivity of a deciduous beech forest (*Fagus sylvatica*) with that of a nearby evergreen spruce forest (*Picea abies*). The beech leaves photosynthesized at a greater rate (per gram dry weight) than those of spruce, and beech "invested" a considerably greater amount of biomass in its leaves each year. Why were these differences not reflected in overall primary productivity? If these species were grown

together, which would you expect to come to dominate the forest? What factors other than productivity might influence the relative competitive status of the two species?

4 What evidence suggests that the productivity of many terrestrial and aquatic communities is limited by nutrients?

5 ▲ In both aquatic and terrestrial communities, secondary productivity by herbivores is approximately one-tenth of the primary productivity upon which it is based. This has led some to suggest the operation of a 10 percent law. Do you subscribe to this view?

6 Account for the observation that in most communities much more energy is processed through the decomposer system than through the live-consumer system.

7 Outline the role played by bacteria and fungi (decomposers) in the flux of energy and matter through a named ecosystem. Imagine what would happen if bacteria and fungi were magically removed—describe the resulting scenario in detail.

8 Energy cannot be cycled and reused—matter can. Discuss this assertion and its significance for ecosystem functioning.

9 Is the ocean simply a large lake in terms of patterns of flux of energy and matter?

10 The hydrological cycle would proceed whether or not a biota was present. Discuss how the presence of vegetation modifies the flow of water through an ecosystem.

▲ = Challenge Question

Table 11.2
Characteristics of representative trees of two contrasting species growing within 1 km of each other on the Solling Plateau, Germany. (After Schulze, 1970; Schulze et al., 1977a; 1977b.)

	Beech	Norway spruce
Age (years)	100	89
Height (m)	27	25.6
Leaf shape	Broad	Needle
Annual production of leaves	Higher	Lower
Photosynthetic capacity per unit dry weight of leaf	Higher	Lower
Length of growing season (days)	176	260
Primary productivity (metric tons of carbon per hectare per year)	8.6	14.9

Web Research Questions

1 Many of the world's forests have been cut down for timber or to open up land for agriculture. Discuss the consequences of deforestation for the communities that live in rivers and the human communities that live in their flood plains.

2 Consider a current topic in the news whose understanding depends on complex population interactions (e.g., prediction of the incidence of Lyme disease based on acorn crop and mouse population increase). Review reports on the topic from a number of commentators and evaluate the variation in their science content. What is needed for a readable *and* well-informed piece on the topic?

3 To an extent that is often underestimated, pathogens (causing infectious disease) may affect populations of their hosts. Consider human populations, and select one or more human pathogens. How many deaths does each pathogen cause per year? Who, exactly, dies? To what extent is the pathogen the primary cause of death—or does it interact with other factors? Which factors?

4 Many pests are pests because they are strong interspecific competitors. This is especially true of weeds competing against crop plants. Consider one or more of the world's important crops— perhaps those growing close to your home. What are the main problem weeds of those crops? What is the range of ways in which those weeds are controlled? To what extent may the different methods be seen as attempts to shift the competitive balance between crop and weed?

5 Many birds undergo migrations, whose study poses considerable problems. Discuss the role of satellite tracking and other technologies in studying bird migration behavior. Contrast the ecological problems of migratory birds whose populations are so large that they may be considered pests in part of their range (e.g., snowgeese) with those whose populations are so small that they are in danger of extinction (e.g., whooping crane).

Part IV

Applied Issues in Ecology

Chapter 12

Sustainability

The sustainability of human activities, and the size and distribution of the human population, have increasingly become preoccupations of the general public and of the politicians who represent them. But attaining or even approaching sustainability requires more than a will to do so—it requires ecological understanding, carefully acquired and even more carefully applied.

Chapter Contents

12.1 Introduction

12.2 The human population "problem"

12.3 Harvesting living resources from the wild

12.4 The farming of monocultures

12.5 Pest control

12.6 Integrated farming systems

12.7 Forecasting agriculturally driven global environmental change

Key Concepts

In this chapter you will

- appreciate the underlying dynamics of human population growth and its relationship to the sustainable (or unsustainable) use of resources
- understand the biological basis of sustainable harvesting of wild populations—particularly in fisheries
- recognize the benefits and costs of farming monocultures
- understand that much agricultural practice has not been sustainable because of loss and degradation of soil
- appreciate that water may be the least sustainable of global resources
- recognize the benefits and costs of different methods of pest control and the importance of devising integrated management practices

12.1   Introduction

what is "sustainability"?

To call an activity "sustainable" means that it can be continued or repeated for the fore-seeable future. Concern has arisen, therefore, precisely because so much human activity is clearly unsustainable. We cannot go on increasing the size of the global human population; we cannot (if we wish to have fish to eat in the future) continue to remove fish from the sea faster than the remaining fish can replace their lost companions; we cannot continue to harvest agricultural crops or forests if the quality and quantity of the soil deteriorates or water resources become inadequate; we cannot continue to use the same pesticides if increasing numbers of pests become resistant to them; we cannot maintain the diversity of nature if we continue to drive species to extinction.

Sustainability has thus become one of the core concepts—perhaps the core concept—in an ever-broadening concern for the fate of Earth and the ecological communities that occupy it. In defining sustainability we used the words "*foreseeable* future". We did so because, when an activity is described as sustainable, it is on the basis of what is known at the time. But many factors remain unknown or unpredictable. Things may take a turn for the worse (as when adverse oceanographic conditions damage a fishery already threatened by overexploitation), or some unforeseen additional problem may be discovered (resistance may appear to some previously irresistible pesticide). On the other hand, technological advances may allow an activity to be sustained that previously seemed unsustainable (new types of pesticide may be discovered that are more finely targeted on the pest itself rather than innocent by-stander species). However, there is a real danger that we observe the many technological and scientific advances that have been made in the past and act on the faith that there will always be a techno-logical "fix" coming along to solve our present problems, too. Unsustainable practices cannot be accepted simply from faith that future advances will make them sustainable after all.

sustainability "comes of age"

The recognition of sustainability's importance as a unifying idea in applied ecology has grown gradually, but there is something to be said for the claim that sustainability really came of age in 1991. This was when the Ecological Society of America published "The sustainable biosphere initiative: an ecological research agenda", "a call-to-arms for all ecologists" with a list of 16 co-authors (Lubchenco, 1991). And in the same year, the World Conservation Union, the United Nations Environment Programme, and the World Wide Fund for Nature jointly published *Caring for the Earth. A Strategy for Sustainable Living* (IUCN/UNEP/WWF, 1991). The detailed contents of these documents are less important than their existence. They indicate a growing preoccupation with sustainability, shared by scientists and pressure groups, and recognition that much of what we do is not sustainable.

12.2   The Human Population "Problem"

12.2.1   Introduction

what is the human population "problem"?

The root of most, if not all, of the environmental problems facing us is the population "problem", the effects of a large and growing population of humans. More people means an increased requirement for energy, a greater drain on nonrenewable resources like

Box 12.1   *Topical ECOncerns*

The Human Population Problem

What is the human population "problem"? This is not an easy question to answer, but what follows are some possible versions of it (Cohen, 1995). The real problem, of course, may be a combination of these— or of these and others. There is little doubt, though, that there *is* a problem—and that the problem is "ours", collectively.

■ **The present size of the global human population is unsustainably high.** Around AD 200, when there were around a quarter of a billion people on Earth, Quintus Septimus Florens Tertullianus wrote that ". . . we are burdensome to the world, the resources are scarcely adequate to us". By 1990 the total had risen to an estimated 5.3 billion.

■ **The present rate of growth in size of the global human population is unsustainably high.** Prior to the widespread agricultural revolution of the 18th century, the human population, very roughly, had taken 1,000 years to double in size. Recently, the total has been doubling about every 43 years.

■ **It is not the size but the distribution over the Earth of the human population that is unsustainable.** The fraction of the population living, highly concentrated, in an urban environment has risen from around 3 percent in 1800 to 29 percent in 1950 and 45 percent in 1995.

■ **It is not the size but the age distribution of the global human population that is unsustainable.** In the "developed" regions of the world, the percentage of the population that was elderly (over 65) rose from 7.6 percent in 1950 to 12.1 percent in 1990. This proportion will jump dramatically after 2010 when the large cohorts born after World War II pass 65.

1950

2000

2050

Box 12.1 (cont'd)

■ **It is not the size but the uneven distribution of resources within the global population that is unsustainable.** In 1992, the 830 million people of the world's richest countries enjoyed an average income equivalent to US$22,000 per annum. The 2.6 billion people in the middle-income countries received $1,600. But the 2 billion in the poorest countries got just $400. These averages themselves hide enormous inequalities.

What role or responsibility does the individual, as opposed to government, have in responding to the human population problem? Which of the variants of the problem, above, pose particular questions of the relationship between the developed and the developing parts of the world or between the "haves" and the "have-nots"?

oil and minerals, more pressure on renewable resources like fish and forests (Section 12.3), more need for food production through agriculture (Section 12.4), and so on. The issue is undoubtedly one of sustainability: things cannot go on the way they are. Yet it is not clear exactly what the "problem" is (Box 12.1). Here, therefore, we examine first the size and growth rate of the global human population and how we reached our current state, then how successful we can expect to be in projecting forward into the future, before finally addressing the "problem" more directly by asking the question—"How many people can the Earth support?"

12.2.2 Population growth up to the present

past population growth: "more than exponential"

When the finger is pointed at human population *growth* as the key issue, it is often said that what is wrong is that the global population has been growing "exponentially". But in an exponentially growing population (Chapter 5) the rate of increase per individual is constant. The population as a whole grows at an accelerating rate (a plot of numbers against time sweeps upwards), because the population growth rate is a product of the individual rate (constant) and the accelerating number of individuals. In Chapter 5, such exponential growth was contrasted with a population limited by intraspecific competition (such as one described by the "logistic" equation), where the rate of increase per individual *decreases* as population size increases. In the case of the global human population, however, the rate of increase per individual (and also therefore the annual percentage increase in size: the rate of increase per 100 individuals) has certainly not been decreasing—but neither has it remained constant (Cohen, 1995) (Box 12.2). Rather, the individual rate has itself been accelerating. Even exponential growth would be unsustainable; but the more-than-exponential growth that we have been witnessing would, if continued, become unsustainable even sooner.

Box 12.2   *Quantitative Aspects*

The Growth of Human Populations

Figure (a) shows estimates, from a variety of sources (indicated by different symbols) of the size of the global human population from 2,000 years ago to the present. Apart from the occasional hesitation and even rarer downturn (such as that caused by the ravages of the Black Death towards the end of the 14th century) the overall picture is clearly one of ever more rapid population growth: the slope of the curve gets steeper and steeper.

But is this exponential growth? The answer is a conclusive "No". Figure (b) shows this same graph (with the individual estimates omitted), but also shows: (1) what an exponentially growing population would have looked like that started at the same point 2,000 years ago and finished at the present population size; and (2) for the sake of contrast, a population anchored at the same start and finish points but growing according to the logistic equation.

Disregarding the logistic as utterly unrealistic (we have not approached a global "carrying capacity"), it is also clear that exponential growth is much more "gradual" than what has actually been observed. The crux of the difference between these three graphs is shown in Figure (c), which uses the same information, but this time plots the changing individual rate of increase against time (the percentage increases would look the same except that all figures would be 100 times greater). For the logistic, under the influence of increasingly intense intraspecific competition, this declines in a straight line down to zero—as it always does for the logistic. For exponential growth, the rate is constant—again "by definition". But the actual growth curve gives rise to an individual rate which not only increases with time as the global population has increased, but it increases more than linearly—it accelerates.

(a)

Box 12.2 (cont'd)

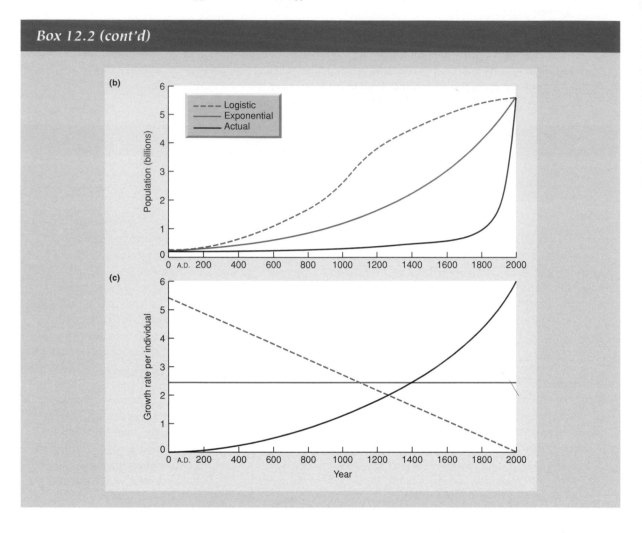

12.2.3   Predicting the future

prediction is more than projection

It is interesting to see what has happened to the total human population in the past—and to do so alerts us to the scale of the problem we face—but the major practical importance of such a survey lies in the opportunity it might provide to predict future population sizes and rates of growth. There is an enormous difference, however, between projection and prediction. Simply to project forwards would be to make the almost certainly false assumption that things will go on in the future just as they have in the past.

the global population is heterogeneous

Prediction, by contrast, requires an *understanding* of what has happened in the past, as well as how the present differs from the past, and finally how these differences might translate into future patterns of population growth. In particular, it is essential to recognize that the global population of humans is a collection of smaller populations, each with its own often very different characteristics. Like all ecological populations, the human population is heterogeneous.

One common way in which subpopulations have been distinguished has been in terms of the "demographic transition". The term "demographic transition" refers both to an idealized pattern of changing birth and death rates as they have occurred during the history of populations, but also to a hypothesis proposed to explain this pattern (Cohen, 1995). Three groups of nations can be recognized: those that passed through the demographic transition "early" (pre-1945), "late" (since then), or "not yet" ("pretransition" countries). The pattern is as follows. Initially, both the birth rate and the death rate are high, but the former is only slightly greater than the latter, so the overall rate of population increase is only moderate or small. (This is presumed to have been the case in all human populations at some time in the past.) Next, the death rate declines while the birth rate remains high, so the population growth rate increases. Subsequently, however, the birth rate also declines until it is similar to or perhaps even lower than the death rate. The population growth rate therefore declines again and may even become negative—though at a population size far greater than before the transition began (Figure 12.1).

The hypothesis commonly proposed to explain this transition, put simply, is that it is an inevitable consequence of industrialization, education, and general modernization leading, first, through medical advances, to the drop in death rates, and then, through the choices people make (delaying having children and so on) to the drop in birth rates. The question remains, though: is the demographic transition a real repeatable phenomenon or just an academic ideal? This is an important question. If we can

early, late, and future demographic transitions

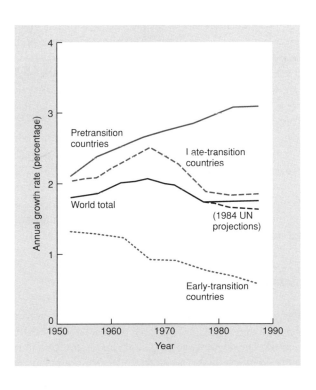

Figure 12.1
The annual rates of population growth, 1950–90, according to United Nations estimates, for early-transition, late-transition, and pretransition countries and for the global total. (After Cohen, 1995.)

Figure 12.2
The decline in the annual rate of population
growth in Europe since 1850 has been
associated with a decline in the death rate,
followed by a decline in the birth rate, and
an overall narrowing of the gap between the
two. (After Cohen, 1995.)

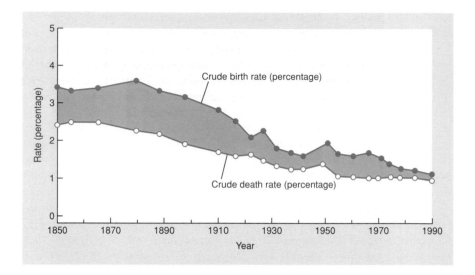

UNANSWERED QUESTION: can
demographic transitions be
understood?—and
manipulated?

understand the demographic transition where it has occurred, we might predict how it
will occur in those countries that have yet to pass through it, and hence predict future
rates of growth much more accurately than by extrapolation. Future patterns may even
be amenable to manipulation by appropriately enlightened policies.

There is some support for the idealized pattern over the past 150 years if the
"early-transition" countries of Europe are taken as a whole (Figure 12.2) and this
apparent demographic transition was, as we have seen, associated with a steady decline
in population growth rate in the early-transition countries (Figure 12.1). But there has
been enormous variation in the speed with which rates have declined, in the length
of the delay between the drop in death rates and birth rates, in the present rates themselves, and so on. It is clear that neither the pattern, nor the underlying mechanisms,
are either simple or always the same. To take just one example, fertility only began to
fall in England and Wales once literacy levels were high and a mere 15 percent of men
were employed in agriculture; whereas fertility began to fall in Bulgaria, and long
before that in France, when literacy levels were low and the populations were still
largely agricultural (van de Walle & Knodel, 1980; Cohen, 1995). Comparable variations exist amongst countries that are just entering or have just started demographic
transitions—the result, perhaps, of differences in government policy, religion, and so
on. There is still much to be learned about demographic transitions before the future
can be predicted with any degree of certainty, much less manipulated.

12.2.4 Two future inevitabilities

unsustainable age structures?

If it were possible to effect some kind of demographic transition in all countries of the
world, so that the birth rates were no more than the death rates, would the population
"problem" be solved? The answer, sadly, is "No", for at least two important reasons.
First, there is a big difference in age structure between a population with equal birth

and death rates when both those rates are high and one when both are low. When "life tables" were described in Chapter 5, we made the point that the net reproductive rate of a population was a reflection of the age-related patterns of survival and birth. A given net reproductive rate, though, can be arrived at through a literally infinite number of different birth and death patterns, and these different combinations themselves give rise to different age structures within the population. If birth rates are high but survival rates low ("pretransition") then there will be many young, and relatively few old, individuals in the population. But if birth rates are low but survival rates high—the "ideal" to which we might aspire "post-transition"—then relatively few young, productive individuals will be called upon to support many who are old, unproductive, and dependent (see Box 12.1). The size and growth rates of the human population are not the only problems: the age structure of a population adds yet another.

Moreover, suppose that our understanding was so sophisticated, and our power so complete, that we could establish equal birth and death rates tomorrow. Would the human population stop growing? The answer, once again, is "No". Population growth has its own momentum, which would still have to be contended with. Even with a birth rate matched to the death rate, there would be many years before a stable age structure was established, and in the mean time there would be considerable further population growth before numbers leveled off. A recent estimate suggests that the global population of 5.3 billion in 1990 is not likely to level off until around 2070, at a population size of 9.0 billion (Lutz et al., 2001). The reason, simply, is that there are, for example, many more babies now than there were 25 years ago, and so even if birth rate per capita drops considerably now, there will still be many more births in 25 years time, when these babies grow up, than at present; and these children, in turn, will continue the momentum effect before an approximately stable age structure is eventually established.

the momentum of population growth

12.2.5 A global carrying capacity?

The current rate of increase in the size of the global population is unsustainable even though it is lower now than it has been; in a finite space and with finite resources, no population can continue to grow for ever. What is an appropriate response to this? To suggest an answer, it is necessary to have some sense of a target, and thus it is interesting, and may be important, to know how large a population of humans could be sustained on Earth. What is the global carrying capacity?

Many estimates have been proposed over the last 300 or so years, a number of which are surveyed in Figure 12.3. There is astonishing variation in these estimates— we are no nearer an agreed carrying capacity now than we have ever been—and there is as much variation amongst estimates now as at any time in the past: even the estimates in the figure since 1970 span three orders of magnitude—from 1 to 1,000 billion.

To help make sense of the apparent confusion of Figure 12.3, a few examples of how carrying capacity estimates have been arrived at are described below (Cohen, 1995).

In 1967, C.T. de Wit, a plant population ecologist, asked the question "How many people can live on Earth if photosynthesis is the limiting process?" The answer he arrived at was roughly 1,000 billion. He built into his calculation the fact that the

some estimates of "the global carrying capacity" . . .

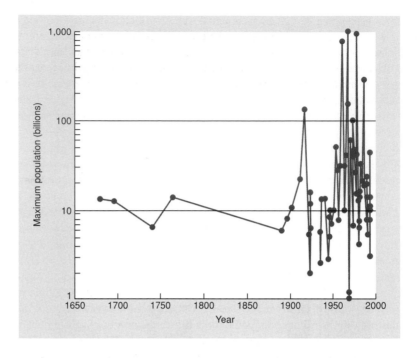

length of the potential growing season varied with latitude, but assumed, amongst other things, that neither water nor minerals were limiting. He acknowledged "This is how many people could live *from* the Earth; not *on* the Earth." In other words, if people wanted to eat meat, or wanted what most of us consider a reasonable amount of living space, and so on, then the estimate would be much less.

. . . ranging from 1 billion to 1,000 billion

By contrast, H.R. Hulett in 1970 assumed that United States levels of affluence and consumption were "optimal" for the whole world, and that not only food but requirements for renewable resources like wood and nonrenewable resources like steel and aluminum needed to be brought into the calculations. The figure he came up with was no more than 1 billion (less than one-fifth of the 1990 total). He further assumed, however, that the global supply of the various limiting resources would remain the same as the number of producers changed (in his scenario, declined) and the distribution of consumption amongst them became equal. This seems unlikely.

This last assumption was made by R.W. Kates and others in a series of reports from 1988, although they worked from global rather than United States averages for consumption rates. With these they estimated support for 5.9 billion people on a basic diet (principally vegetarian), 3.9 billion on an "improved" diet (about 15 percent of calories from animal products), and 2.9 billion on a diet with 25 percent of calories from animal products.

It is clear from the examples above that there is a difference between the number the Earth can support and the number that can be supported with an acceptable standard of living. The higher estimates come closer to the concept of a carrying capacity that we normally apply to other organisms (Chapter 5)—a number "imposed" by the limiting resources of the environment. But it is unlikely that many of us would choose to live crushed up against an environmental ceiling or wish it on our descendants. In any case, it is a big step to assume that the human population is limited "from below" by its resources rather than "from above" by its natural enemies. Infectious disease in particular, which not long ago was considered to be an enemy largely vanquished, is now once again perceived, for example by the World Health Organization, as a major threat to human welfare. In a growing global tuberculosis epidemic, for example, 2.7 million people died in 1993 but an estimated 1.7 billion were infected but had not yet developed the disease (Dobson & Carper, 1996). We saw in Chapter 8 that many infectious diseases thrive best in the densest populations.

Any suggestions we make about a global carrying capacity clearly depend on choices we make both for ourselves and for others. Most of us would choose to live at least as well as we do at present, but can the global population afford to choose for the whole world to live at least as well as those in developed countries do now? The answer to any question depends on what is meant by the question—defining what we mean by "the global carrying capacity" is far from straightforward.

UNANSWERED QUESTION: what do we mean by "the global carrying capacity"?

12.3   Harvesting Living Resources from the Wild

A major limit to the number of people the Earth can support is the food that can be obtained. Populations of many species living freely in the wild are exploited for food by humans, who "cull" or "harvest" a proportion of the population, leaving some individuals behind to grow and reproduce material for future harvests. Primitive human societies obtained all their resources like this, by hunting and gathering from nature, and humans continue to garner some resources in this way. The resources may be fish from the sea, deer from a moorland, or timber from a forest. There is an important difference between resources obtained in this way and those that are farmed (see Sections 12.4 and 12.5). Farmed resources are obtained by taking chosen species of plant or animal, domesticating them (often changing them genetically), and growing or rearing them in more or less controlled monocultures. These resources tend to be owned and managed by a farmer or organization. In contrast, most of the oceans and forests that are fished and hunted have at one time been common property, open to free-for-all unsustainable looting by all-comers. Recently, though, fishing and hunting have also come under increasing national and international regulation and national claims to "ownership". Many of our examples in this section are of fish or fisheries, but the principles apply to the harvesting of any natural resource.

12.3.1   Fisheries—maximum sustainable yields

Whenever a natural population is exploited there is a risk of overexploitation: too many individuals are removed and the population is driven into biological jeopardy or economic insignificance—perhaps even to extinction. Global catches of marine fish rose five-fold between 1950 and 1989. However, many of the world's fish stocks are now at or near the point of overexploitation (Figure 12.4). But harvesters also want to avoid underexploitation: if fewer individuals are removed than the population can bear, the harvested crop is smaller than necessary, potential consumers are deprived,

aiming for the narrow path between over- and underexploitation

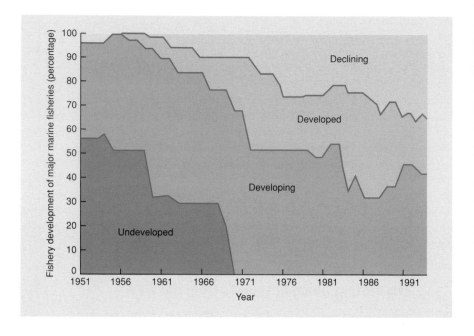

and those who do the harvesting are underemployed. It is not easy to tread the narrow path between under- and overexploitation. It is asking a great deal of a management policy that it combine the well-being of the exploited species, the profitability of the harvesting enterprise, continuing employment for the workforce, and the maintenance of traditional lifestyles, social customs, and natural biodiversity.

population dynamics in the absence of exploitation—humped net recruitment curves

The most fundamental aspects of ecology that we need to understand here were introduced in Chapter 5 when the effects of intraspecific competition on populations were discussed. To determine the best way to exploit a population, it is necessary to know what the consequences will be of different exploitation "strategies". But in order to know these consequences, we first need some understanding of the dynamics of the population in the absence of, or prior to, exploitation. It is usual to assume that, before it is exploited, a harvestable population is crowded and intraspecific competition is intense. Summarizing from Chapter 5, and remembering that these are broad generalizations:

1 Populations in the absence of exploitation can be expected to settle around their carrying capacity, but exploitation will reduce numbers to less than this.
2 Exploitation, by reducing the intensity of competition, moves the population "leftwards" along the humped net recruitment curve, increasing the net number of recruits to the population per unit time (Figure 12.5).

MSY—the narrow path?

In fact, we can go further with Figure 12.5, since it is clear from the shape of the curve that there must be an "intermediate" population size at which the rate of net recruitment is highest. Consider a timescale of years. The peak of the curve might be "10 million new fish each year". This is then also the highest number of new fish that

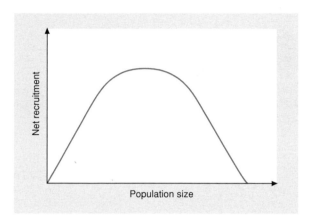

Figure 12.5
The humped relationship between the net recruitment into a population (births minus deaths) and the size of that population, resulting from the effects of intraspecific competition (Chapter 5). Population size increases from left to right, but increasing rates of exploitation take the population from right to left.

could be removed from the population each year which the population itself could replenish. It is known as the maximum sustainable yield (MSY): the largest harvest that can be removed from the population regularly and indefinitely. It looks as though a fishery could tread the narrow path between under- and overexploitation if the fishers could find a way to achieve this maximum sustainable yield.

the MSY concept has shortcomings

The MSY concept has been the guiding principle in resource management for many years in fisheries, forestry, and wildlife exploitation, but it is very far from being the perfect answer for a variety of reasons.

1 By treating the population as a number of similar individuals, it ignores all aspects of population structure such as size or age classes and their differential rates of growth, survival, and reproduction.
2 By being based on a single recruitment curve, it treats the environment as unvarying.
3 In practice, it may be impossible to obtain a reliable estimate of the MSY.
4 Achieving an MSY is by no means the only, nor necessarily the best, criterion by which success in the management of a harvesting operation should be judged. (It may, for example, be more important to provide stable, long-term employment for the workforce.)

12.3.2 Obtaining MSYs through fixed quotas

the fragility of fixed quota harvesting . . .

There are two simple ways of obtaining an MSY on a regular basis: through a "fixed quota" and through a "fixed effort". With fixed quota MSY harvesting (Figure 12.6), the same amount, the MSY, is removed from the population every year. If (and it is a big if) the population stayed exactly at the peak of its net recruitment curve, then this could work: each year the members of the population, through their own growth and reproduction, would add exactly what the harvesting removed. But if by chance numbers fell even slightly below those at which the curve peaked, then the numbers harvested would exceed those recruited. Population size would then decline to below the peak of the curve, and if a fixed quota at the MSY level was maintained the population would carry on declining until it was extinct (Figure 12.6). Furthermore, if the MSY was even

Figure 12.6
Fixed quota harvesting. The figure shows a single recruitment curve (solid line) and two fixed quota harvesting curves (dashed lines) with a high quota (h_h) and MSY quota (h_m). Arrows in the figure refer to changes to be expected in abundance under the influence of the harvesting rate to which the arrows are closest. The black dots indicate equilibria. At h_h the only "equilibrium" is when the population is driven to extinction. The MSY is obtained at h_m because it just touches the peak of the recruitment curve (at a density N_m): populations greater than N_m are reduced to N_m, but populations smaller than N_m are driven to extinction.

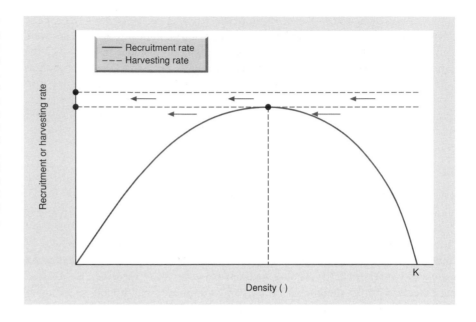

Figure 12.7
The catch history of the Peruvian anchoveta fishery. (After Hilborn & Walters, 1992.)

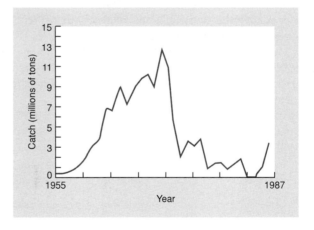

slightly overestimated (and reliable estimates are hard to come by) then the harvesting rate would always exceed the recruitment rate and extinction would again follow. In short, a fixed quota at the MSY level might be desirable and reasonable in a wholly predictable world about which we had perfect knowledge. But in the real world of fluctuating environments and imperfect data sets, these fixed quotas are open invitations to disaster (Clark, 1981).

... borne out in practice

Nevertheless, a fixed quota strategy has frequently been used, where a management agency formulates an estimate of the MSY, which is then adopted as the annual quota. On a specified day in the year, the fishery is opened and the accumulated catch is logged. A fairly typical example is provided by the Peruvian anchoveta (*Engraulis ringens*) fishery (Figure 12.7). From 1960 to 1972 this was the world's largest single fishery, and it constituted a major sector of the Peruvian economy. Fisheries experts advised that the MSY was around 10 million metric tons annually, and catches were limited accordingly. But the fishing capacity of the fleet expanded, and in 1972 the catch crashed. Overfishing seems, at the least, to have been a major cause of the collapse, although its effects were compounded by the influences of profound environmental fluctuations, discussed below. A moratorium on fishing might have allowed the stocks to recover, but this was not politically feasible: 20,000 people were dependent on the anchoveta industry for employment. The Peruvian government has therefore allowed fishing to continue. The catches have never recovered.

12.3.3   Obtaining MSYs through fixed effort

An alternative to trying to maintain a constant harvest is to maintain a constant "harvesting effort" (e.g., the number of "trawler days" in a fishery or the number of "gun days" with a hunted population). With such a regime the amount harvested should increase with the size of the population being harvested (Figure 12.8). Now, in contrast to Figure 12.6, if density drops below the peak, new recruitment exceeds the amount harvested and the population recovers. The risk of extinction is much reduced. The disadvantages, however, are first, because there is a fixed effort, the yield varies with population size (there are good, but, more to the point, bad years), and second, steps need to be taken to insure that nobody makes a greater effort than they are supposed to. Nonetheless, there are many examples of harvests being managed by legislative regulation of effort. Harvesting of the important Pacific halibut (*Hippoglossus stenolepis*), for example, is limited by seasonal closures and sanctuary zones, though heavy investment in fisheries protection vessels is needed to control law-breakers.

the relative robustness of fixed effort harvesting

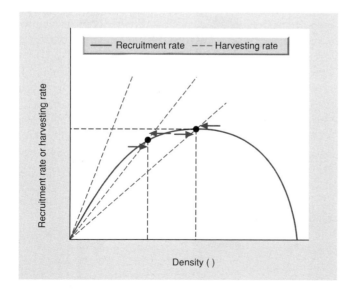

Figure 12.8
Fixed effort harvesting. Curves, arrows, and dots as in Figure 12.6. The MSY is obtained with an effort of E_m, leading to a stable equilibrium at a density of N_m with a yield of h_m. At a somewhat higher effort (E_h), the equilibrium density and the yield are both lower than with E_m but the equilibrium is still stable. Only at a much higher effort (E_0) is the population driven to extinction.

Density ( )

Recruitment rate or harvesting rate

—— Recruitment rate   - - - Harvesting rate

12.3.4   Beyond MSY

There is no doubt that fishing pressure often exerts a great strain on populations. But the collapse of fish stocks in one year rather than any other is often the result of an occurrence of unusually unfavorable environmental conditions, rather than simply overfishing.

environmental fluctuations—the anchoveta and El Niño

Harvests of the Peruvian anchoveta (Figure 12.7) collapsed from 1972 to 1973, but a previous steady rise in catches had already dipped in the mid-1960s as a result of an "El Niño event"—this happens when warm tropical water from the north reduces the upwelling, and hence the productivity, of the nutrient-rich cold Peruvian current coming from the south. By 1973, however, commercial fishing had so greatly increased that the subsequent El Niño event had even more severe consequences. There were some signs of recovery from 1973 to 1982, but a further collapse occurred in 1983 associated with yet another El Niño event. It is unlikely that the El Niño events would have had such severe effects if the anchoveta had only been lightly fished. It is equally clear, though, that the history of the Peruvian anchoveta fishery cannot be explained simply in terms of overfishing.

population structure and the Arctic cod (Gadus morhua)

So far, this account has ignored population structure of the exploited species. This is a bad fault for two reasons. First, most harvesting practices are primarily interested in only a portion of the harvested population (mature trees, fish that are large enough to be saleable, etc.). Second, "recruitment" is, in practice, a complex process incorporating adult survival, adult fecundity, juvenile survival, juvenile growth, and so on, each of which may respond in its own way to changes in density and harvesting strategy. An example of a model that takes some of these variables into account was that developed for the Arcto-Norwegian cod fishery, the most northerly fish stock in the Atlantic ocean. The numbers of fish in different age classes were known for the late 1960s and this information was used to predict the tonnage of fish likely to be caught with different intensities of harvesting and with different net mesh sizes. The model predicted that the long-term prospects for the fishery were best insured with a low intensity of fishing (less than 30 percent) and a large mesh size. These gave the fish more opportunity to grow and reproduce before they were caught (Figure 12.9). The recommendations from the model were ignored and, as predicted, the stocks of cod fell disastrously.

precautionary management, closed areas, and "data-less" management

Managing most marine fisheries to achieve optimum yields is an unattainable dream. There are generally too few researchers to do the work and, in many parts of the world, no researchers at all. In these situations, a precautionary approach to fisheries management might involve locking away a proportion of a coastal or coral community in marine protected areas (Hall, 1998). The term *data-less management* has been applied to situations where local villagers follow simple prescriptions to make sustainability more likely—for example, locals on the Pacific island of Vanuatu were provided with some simple principles of management for their trochus (*Tectus niloticus*) shellfishery (stocks should be harvested every 3 years and left unfished in between) with an apparently successful outcome (Johannes, 1998).

12.4   The Farming of Monocultures

Globally there is abundant food. Between 1961 and 1996 the per capita food supply in developing countries increased by 32 percent (Figure 12.10a) and the proportion of the world's population that was undernourished fell from 35 to 21 percent though this was very unevenly distributed (Figure 12.10b). Yet 840 million people remain hungry worldwide, and the rate of increase in per capita food production is falling.

Fishing and hunting (Section 12.3) have been human activities since our early history as hunter–gatherers. But the harvest that can be taken from nature was totally inadequate to support the main phases of human population growth. Increasingly, both animals and plants have been domesticated and managed in ways that allowed much greater rates of production. The great bulk of the human food resource is now farmed—produced usually as dense populations of single species (monocultures). This allows them to be managed in specialized ways that can maximize their productivity—

Figure 12.10
(a) Changes in world food production and food produced per capita over the period 1961–96. The values in 1989–91 are adjusted to 100 to make comparisons easier. (b) Changes in agricultural food production in the major regions of the world over the period 1961–96. The values in 1989–91 are adjusted to 100 to make comparisons easier. The developed countries category includes the former Soviet Union. (After United Nations, 1998.)

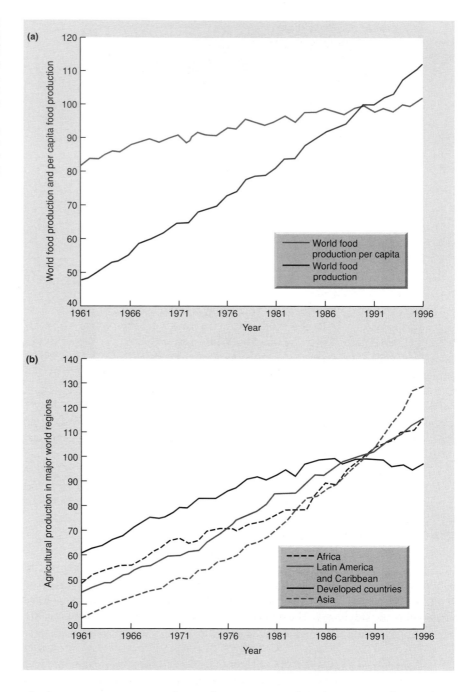

whether as immense monocultures of rice, corn, or wheat (Figure 12.11), or as live-stock factories producing beef, pork, or poultry. Fish, indeed, are increasingly managed in the same way (aquaculture)—reared in enclosures, fed with controlled diets, and harvested in mass production. Nearly one-quarter of the fish supply in Asia is already produced in this way.

Figure 12.11
Agricultural monoculture: wheat as far as the eye can see.

Only monoculture can maximize the rate of food production. This is because it allows the farmer to control and optimize with high precision the density of the populations (livestock or crop plants), the quantity and quality of their resources (food supplied to livestock, and fertilizer and water to the crops), and often even the physical conditions of temperature and humidity. With many animals, the monocultures extend to segregating livestock or poultry into narrow age bands or age classes. There need be none of the uneconomic mixing of calves with cows or chickens with hens; fish eggs and fry can be segregated from potentially cannibalistic adults; the grossly uneconomic equality of the sex ratio that is common in nature can be distorted by culling to give efficient all-female dairy herds of cattle, or all-hen populations in batteries for egg production. This is a far cry from the ecology of the primitive human hunter–gatherers, who subsisted on their gleanings from the tangled web of wild nature!

monoculture—and beyond . . .

To what extent, though, are modern farming methods sustainable? There is abundant evidence that a high price has to be paid to sustain the high rates of food production achieved by farmed monocultures. For example, they offer ideal conditions for the epidemic spread of diseases such as mastitis, brucellosis, and swine fever among livestock and coccidiosis among poultry. Farmed animals are normally kept at densities far higher than their species would meet in nature with the result that disease transmission rates are magnified (Chapter 8). In addition, high rates of transmission between herds occur as animals are sold from one farming enterprise to another, and it is easy for the farmers themselves, with mud on their boots and their vehicles, to act as vectors of pests and disease. The dramatic spread of foot and mouth disease in 2001 among British livestock provides a graphic example.

. . . but disease spreads in monocultures

Crop plants, too, provide illustrations of the fragility of human dependence on monocultures. The potato, for example, was not introduced across the Atlantic to Europe until the second half of the 16th century, but three centuries later other foods had given way to it, and it had become the almost exclusive food crop of the poorer half of the population of Ireland. Dense monoculture, though, provided ideal conditions for the devastating spread of "late blight" (the fungal pathogen *Phytophthora infestans*) when it also crossed the Atlantic in the 1840s. The disease spread rapidly, dramatically reducing potato yields and also decomposing the tubers in storage. Out of the Irish population of about 8 million, 1.1 million died in the resulting famine and another 1.5 million emigrated to the UK and the USA.

In more modern history, an outbreak of "southern corn leaf blight" (caused again by a fungus, *Helminthosporium maydis*) developed in southeastern America in the late 1960s and spread rapidly after 1970. Most of the corn grown in the area had been derived from the same stock and was genetically almost uniform. This extreme mono-culture allowed one specialized race of the pathogen to have devastating consequences. The damage was estimated as at least 1 billion dollars in the USA and had repercussions on grain prices worldwide.

12.4.1   Degradation and erosion of soil

A United Nations report (1998) stated "Agricultural intensification in recent decades has taken a heavy toll on the environment. Poor cultivation and irrigation techniques and excessive use of pesticides and herbicides have led to widespread soil degradation and water contamination." Around 300 million hectares are now severely degraded around the world, and a further 1.2 billion hectares—10 percent of the Earth's vege-tated surface—can be described as moderately degraded. Clearly much of agricultural practice has not been sustainable.

agriculture and forestry require soil

Land without soil can support only very small primitive plants such as lichens and mosses that can cling onto a rock surface. The rest of the world's terrestrial vegetation has to be rooted in soil that gives it physical support. The soil also serves as a store of essential mineral nutrients and water that are extracted by the roots during plant growth. Soil develops by the accumulation of finely divided mineral products of rock weathering and decomposing organic residues from previous vegetation. The char-acteristics of the soil under natural vegetation in any particular climatic region and on any particular rock type depends on the balance between these processes of accumu-lation and forces that degrade and remove the soil.

soil forms ... and is lost

The formation and persistence of soil in a region depend on local natural checks and balances. Soil may be lost by being washed or blown away, perhaps to be rede-posited as an accumulation of fine textured "loess" somewhere else. Soil is best pro-tected when it contains organic matter, is always wholly covered with vegetation, is finely interwoven with roots and rootlets, and is on horizontal ground. Natural soil systems are probably *always* too fragile to be fully sustained when land is brought into cultivation. Dramatic evidence of unsustainable land use came from the "Dust Bowl" disaster in the Great Plains of the United States and a similar disaster happening cur-rently in China (Box 12.3).

soil maintenance

In an ideal sustainable world, new soil would be formed as fast as the old was lost. In Britain about 0.2 t of new soil is produced naturally per hectare per year and it has been suggested that a tolerable (although not sustainable) rate of soil erosion might be about 2.0 t ha^{-1} yr^{-1}. However, rates of erosion have been recorded up to 48 t ha^{-1} yr^{-1}!

Almost all (perhaps all) agricultural land will support higher yields if artificial fer-tilizers are applied to supplement the nitrogen, phosphorus, and potassium supplied naturally by the soil. Fertilizers are cheap, easy to handle, of a guaranteed composition, allow even and accurate application, and higher and more predictable yields. When

Box 12.3 *Historical Landmarks*

Soil Erosion, America's Historic "Dust Bowl", and China's Current Problem

Large areas of southeastern Colorado, southwestern Kansas, and parts of Texas, Oklahoma, and northeastern New Mexico had been used to support rangeland management of livestock. The vegetation consisted largely of native perennial grasses and had been neither ploughed nor sown with seed.

At the time of World War I, much of the land was ploughed and annual crops of wheat were grown. There were poor crops in the early 1930s due to severe drought and the top soil was exposed and carried away by the wind. Black blizzards of wind-blown soil blocked out the sun and piled the dirt in drifts. Occasionally the dust storms swept completely across the country to the East Coast. Thousands of families were forced to leave the region at the height of the Great Depression in the early and mid-1930s. The wind erosion was gradually halted with federal aid: windbreaks were planted and much of the grassland was restored. By the early 1940s the area had largely recovered.

The story is being repeated today in northwest China, where the need to feed 1.3 billion people has

Dust bowl field and abandoned farm (© Visuals Unlimited).

led to the raising of too many cattle and sheep, and the use of too many ploughs. This is more than the land can stand and 2,300 km² are turning to desert each year. A huge dust storm blanketed areas from Canada to Arizona in April 2001—the dust originated in China.

---

there is an overwhelming reliance on them, however, maintaining the organic matter capital of the soil tends to be neglected and has declined everywhere.

The degradation of soil by agriculture can be prevented, or at least slowed down, by: (1) incorporating farmyard manure, crop residues, and animal wastes, (2) alternating years under cultivation with years of fallow, or (3) returning the land to grazed pasture or rangeland. Such practices conserve soil quality in technologically sophisticated agricultures in temperate regions.

But soil degradation is most serious and least easily prevented in Third World countries. The problems are greatest in high rainfall areas and on steeply sloping ground in the tropics where organic matter in the soil also decomposes more rapidly. The United Nations soil conservation strategy of "Agenda 21" (in Rio de Janeiro, 1992) recommended measures to prevent soil erosion and promote erosion control.

contour plowing and terracing— Agenda 21

The most cost-effective technology used in reducing soil erosion is considered to be contour-based cultivation (Figure 12.12). In India, contour ditches have helped to quadruple the survival chances of tree seedlings and quintuple their early growth in height. Deeply rooted, hedge-forming vetiver grass, planted in contour strips across

Figure 12.12
Terracing of hill and mountainous land
(© D. Cavagnaro, Visuals Unlimited).

hill slopes, slows water runoff dramatically, reduces erosion, and increases the moisture available for crop growth. Currently 90 percent of soil conservation efforts in India are based on such biological systems. Simple technologies involving rock embankments constructed along contour lines for soil and water conservation have also been successful. Embanked fields in Burkina Faso (West Africa) yielded an average of 10 percent more than traditional fields in a normal year and, in drier years, almost 50 percent more (United Nations, 1998). Such terracing provides a very high level of soil conservation but is possible only where labor is cheap. On lesser slopes, by ploughing and cultivating in strips along the contours, runoff of soil can be significantly reduced.

desertization and salinization

Agricultural land is also highly susceptible to degradation in arid and semiarid regions. Both overgrazing and excessive cultivation expose the soil directly to erosion by the wind and to rare but fierce rainstorms. In the process of "desertization", land that is arid or semiarid but has supported subsistence or nomadic agriculture gives way to desert. The process has often been slowed down for a time by irrigating the land. This gives a temporary remission but lowers the water table and salts accumulate in the top soil (salinization). Once salts have started to accumulate, the process of salinization tends to spread and leads to an expansion of sterile white salt deserts. This has been a particular hazard in irrigated areas of Pakistan.

forests protect . . . except when harvested by clear-felling

Forests protect soil from erosion because the canopy absorbs the direct impact of the rain on the soil surface, the perennial root systems bind the soil, and leaf fall continually adds organic matter. But when forests are clear-felled and then replanted, there is an open "window of opportunity" for soil erosion until the forest canopy closes again. Cultivation and replanting along contours gives some control over soil erosion during this danger period, but the best precaution is to avoid clear-felling and extract only a proportion of a forest stand at each harvest. This can often be technically difficult and more expensive.

12.4.2 Sustainability of water as a resource

water is a finite global resource

In the 1960s and 1970s, the main worry about the sustainability of global resources concerned energy supplies that were recognized to be finite and exhaustible. While

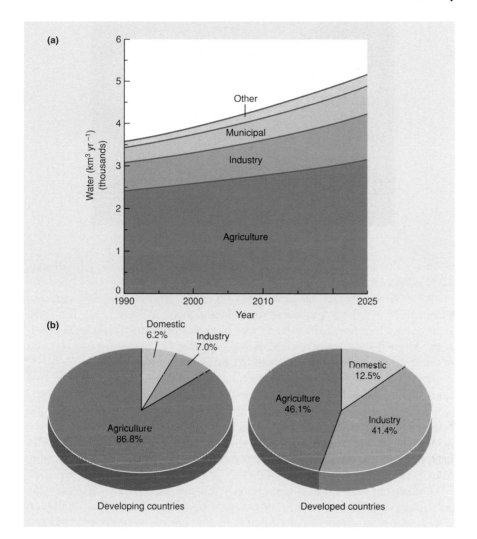

Figure 12.13
(a) Changes in global withdrawal of water for agriculture, industry, municipal (largely domestic), and other users during the period from 1990 and projected to the year 2025. (b) The contrast between the extraction of water for agriculture, industry, and domestic use in developing and developed countries. (After United Nations, 1998.)

energy resources remain finite, concern has shifted because exploration has revealed much larger reserves of oil, gas, and even coal than had been entered into earlier environmental balance sheets. Water has now come into sharper focus. Fresh water, which is used in crop irrigation and for domestic consumption, is of crucial importance (Figure 12.13). On a global scale, agriculture is the largest consumer of fresh water, taking about 70 percent of available supplies and more than 90 percent in parts of South America, central Asia, and Africa (Figure 12.14).

There is a fixed stock of water on the globe and it is continually recycled as it evaporates from vegetation, land, and sea and is then condensed and redistributed as precipitation. The human species now uses, directly or indirectly, more than half of the world's accessible water supply. The fresh water available per capita worldwide has fallen from 17,000 m^3 in 1950 to 7,300 m^3 in 1995. Many assessments of the problems of water supply suggest that countries with less than 1,000 m^3 per person per year

water—the resource that future wars will be fought over?

Figure 12.14
The soils around the Nile in Egypt can support plant growth where they are irrigated, but where irrigation stops (note the background in the photo) there is desert.

experience chronic scarcity. Water is widely thought to be the resource that future wars will be fought over. Even at a national level, the allocation of water resources can cause political problems, as occur for example in conflicts in California between urban and agricultural demands for water from the Colorado River. At the international level, conflict arises between countries that are upstream of their neighbors and are in a position to dam and divert water supplies. There are bitter cross-border disputes in South America, Africa, and the Middle East between nations that share river basins.

One response to chronic scarcity of water is to pump it from underground aquifers—but this often happens faster than the aquifers can be recharged. Such activity is clearly profligate and unsustainable. It is also the main cause of the loss of land from agriculture due to salinization. The demand for accessible supplies of water for both agriculture and domestic use has led to the plumbing of the Earth's river systems on a vast scale. The number of river dams more than 15 m high has increased from about 5,000 in 1950 to 38,000 in the 1990s.

contamination and conservation

In Chapter 13, we discuss the pollution of water by excreta and by the pesticides and fertilizers applied in agriculture. Water that is uncontaminated by disease, nitrates, or pesticides is an especially valuable commodity, but contamination is easy and removing contaminants (e.g., nitrates) is very expensive. Major dams built to control and conserve water in North and West Africa create large bodies of open water in which contamination spreads easily; one consequence has been the rapid spread of schistosomiasis (a flatworm disease of humans) along rivers with infection rates rising from 1–10 percent to 98–100 percent.

Maintaining water supplies for human use also creates problems for the conservation of wildlife (Chapter 14). The water flow in many of the world's larger rivers is now very heavily controlled—in some cases little water now reaches the sea and wetlands have been lost or are at risk. Moreover, silt accumulates up river instead of spreading into deltas and flood plains. The results may be catastrophic for wildlife areas and for human communities as well. For example, there is reason to believe that failure of silt deposition in the Nile delta (together with rising sea levels) may cause Egypt to lose up to 19 percent of its habitable land and displace 16 percent of its population within 60 years.

12.5   Pest Control

what is a pest?

Pest control is another area in which the sustainability of agricultural practice may be threatened. A pest species is one that humans consider undesirable. Estimates suggest that there are around 67,000 species of pests that attack agricultural crops worldwide: 8,000 weeds that compete with crops; 9,000 insects and mites, and 50,000 plant pathogens that prey on them (Pimentel, 1993). Here we consider the sustainability of insect pest control in agriculture to illustrate the types of problems that arise in managed monocultures. We could equally well have chosen the control of weeds or molluscs, or of the pests and diseases of farmed livestock, poultry, or fish.

12.5.1   Aims of pest control: economic injury levels and action thresholds

EILs for pests, non-pests, and potential pests

Economics and sustainability are intimately tied together. Market forces insure that uneconomic practices are not sustainable. One might imagine that the aim of pest control is total eradication of the pest, but this is not the general rule. Rather, the aim is to reduce the pest population to a level at which it does not pay to achieve yet more control (the economic injury level or EIL). The EIL for a hypothetical pest is illustrated in Figure 12.15a. It is greater than zero (eradication is not profitable) but it is also below the typical, average abundance of the species. If the species was naturally self-limited to a density below the EIL, then it would never make economic sense to apply "control" measures, and the species could not, by definition, be considered a "pest" (Figure 12.15b). There are other species, though, which have a carrying capacity (Chapter 5) in excess of their EIL, but have a typical abundance which is kept below the EIL by natural enemies (Figure 12.15c). These are potential pests. They can become actual pests if their enemies are removed.

When a pest population has reached a density at which it is causing economic injury, however, it is generally too late to start controlling it. More important, then, is the control action threshold (CAT): the density of the pest at which action should be taken to prevent it reaching the EIL. CATs are predictions based on detailed studies of past outbreaks or sometimes on correlations with climatic records. They may take into account the numbers not only of the pest itself but also of its natural enemies. As an example, in order to control the spotted alfalfa aphid (*Therioaphis trifolii*) on hay alfalfa in California, control measures have to be taken at these times and under these circumstances:

1 In spring when the aphid population reaches 40 aphids per stem.
2 In summer and fall when the population reaches 20 aphids per stem, but the first three cuttings of hay are not treated if the ratio of ladybirds (beetle predators of the aphids) to aphids is one adult per 5–10 aphids, or three larvae per 40 aphids on standing hay, or one larva per 50 aphids on stubble.
3 During winter when there are 50–70 aphids per stem (Flint & van den Bosch, 1981).

Figure 12.15
(a) The population fluctuations of a hypothetical pest. Abundance fluctuates around an "equilibrium abundance" set by the pest's interactions with its food, predators, etc. It makes economic sense to control the pest when its abundance exceeds the economic injury level (EIL). Being a pest, its abundance exceeds the EIL most of the time (assuming it is not being controlled). (b) By contrast, a species that cannot be a pest fluctuates always below its EIL. (c) "Potential" pests fluctuate normally below their EIL but rise above it in the absence of one or more of their natural enemies.

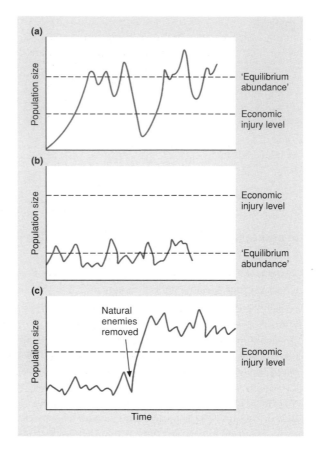

12.5.2　Problems with chemical pesticides—and their virtues

A pesticide gets a bad name if, as is usually the case, it kills more species than just the one at which it is aimed. It may then become a pollutant (Chapter 13). However, in the context of the sustainability of agriculture, its bad name is especially justified if it kills the pests' natural enemies and so contributes to undoing what it was employed to do. Thus, the numbers of a pest sometimes increase rapidly some time after the application of a pesticide. This is known as "target pest resurgence" and occurs when treatment kills both large numbers of the pest *and* large numbers of their natural enemies. Pest individuals that survive the pesticide or that migrate into the area later find themselves with a plentiful food resource but few, if any, natural enemies. The pest population may then explode.

The after effects of applying a pesticide may involve even more subtle reactions. When a pesticide is applied, it may not be only the target pest that resurges. Alongside the target are likely to be a number of potential pest species that had been kept in check by their natural enemies (Figure 12.15c). If the pesticide destroys these, the potential pests become real ones—and are called secondary pests. A dramatic example concerns the insect pests of cotton in Central America. In 1950, when mass dissemination of organic insecticides began, there were two primary pests: the Alabama leafworm and the boll weevil (Smith, 1998). Organochlorine and organophosphate insecticides were applied fewer than five times a year and initially had apparently miraculous results— yields soared. By 1955, however, three secondary pests had emerged: the cotton bollworm, the cotton aphid, and the false pink bollworm. The pesticide applications rose to 8–10 per year. This reduced the problem of the aphid and the false pink bollworm, but led to the emergence of five further secondary pests. By the 1960s, the original two pest species had become eight and there were, on average, 28 applications of insecticide per year. Clearly, such a rate of pesticide application is not sustainable.

Chemical pesticides lose their role in sustainable agriculture if the pests evolve resistance. The evolution of pesticide resistance is simply natural selection in action (Chapter 2). It is almost certain to occur when vast numbers of a genetically variable population are killed. One or a few individuals may be unusually resistant (perhaps because they possess an enzyme that can detoxify the pesticide). If the pesticide is applied repeatedly, each successive generation of the pest will contain a larger propor- tion of resistant individuals. Pests typically have a high intrinsic rate of reproduction, and so a few individuals in one generation may give rise to hundreds or thousands in the next, and resistance spreads very rapidly in a population.

evolved resistance

This problem was often ignored in the past, even though the first case of DDT resistance was reported as early as 1946 (house flies in Sweden). Now, the scale of the problem is illustrated in Figure 12.16, which shows the exponential increases in the number of insect species resistant to insecticides. The house fly has developed resistance worldwide to virtually every chemical that has been employed against it. The evolution of pesticide resistance can be slowed, though, by changing from one pesticide to another, in a repeated sequence that is rapid enough that resistance does not have time to emerge (Roush & McKenzie, 1987).

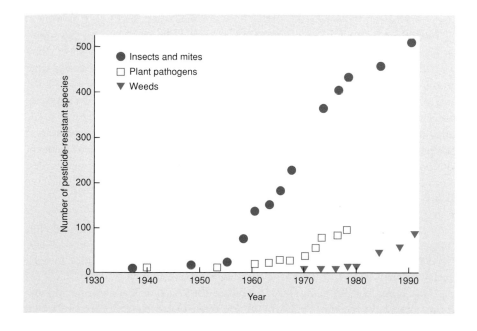

Figure 12.16
The increase in the number of arthropod (insects and mites), plant pathogens, and weed species reported to be resistant to at least one pesticide. (After Gould, 1991.)

Figure 12.17
Trend in global pesticide production rates, measured as millions of metric tons per year. (From World Health Organization, 1990.)

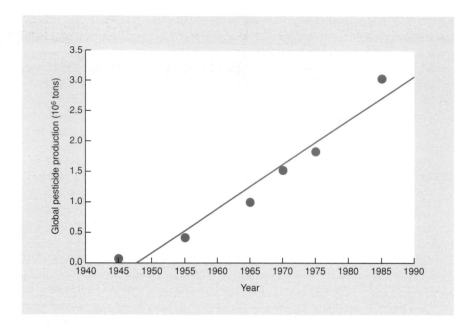

pesticides work

If chemical pesticides brought nothing but problems, however—if their use was intrinsically and acutely unsustainable—then they would already have fallen out of widespread use. This has not happened. Instead, their rate of production has increased rapidly (Figure 12.17). The ratio of cost to benefit for the individual producer has remained in favor of pesticide use: they do what is asked of them. In the United States, insecticides are estimated to benefit the agricultural producer to the tune of around $5 for every $1 spent (Pimentel et al., 1978).

Moreover, in many poorer countries, the prospect of imminent mass starvation, or of an epidemic disease, are so frightening that the social and health costs of using pesticides have to be ignored. In general, the use of pesticides is justified by objective measures such as "lives saved", "economic efficiency of food production", and "total food produced". In these very fundamental senses, their use may be described as sustainable. In practice, sustainability depends on continually developing new pesticides that keep at least one step ahead of the pests; pesticides that are less persistent, biodegradable, and more accurately targeted at the pests.

12.5.3 Biological control

Outbreaks of pests occur repeatedly and so does the need to apply pesticides. But biologists can sometimes replace chemicals with other tools that do the same job and cost a great deal less. These involve the manipulation of the natural enemies of pests (biological control) and the controlled evolution of host resistance.

Most crops have been introduced to the areas in which they are grown and some of their pests have accompanied them or have arrived later. However, the pest's own

Box 12.4   *Historical Landmarks*

The Cottony Cushion Scale: a Classic Case of Importation

The cottony cushion scale insect, *Icerya purchasi*, was first discovered as a pest of Californian citrus orchards in 1868. By 1886 it had brought the citrus industry close to the point of destruction. Would-be pest controllers initiated a worldwide correspondence to try and discover the natural home and natural enemies of the scale. In 1887 they received a report from Adelaide, Australia, of apparently natural populations of the scale along with a dipteran parasitoid, *Cryptochaetum* spp. Thus, in 1888, an expert on insect taxonomy, Koebele, was despatched to collect parasitoids for importation into California. He made a widespread tour of Australia but both scale and parasitoid were very hard to find. Eventually he was successful—but he also found a ladybird beetle, the vedalia *Rodolia cardinalis*, feeding on the scale insect. Then, on his way home, he visited New

Zealand and found the vedalia doing very well on the scale insects. He despatched around 12,000 *Cryptochaetum* and 500 vedalia back to California.

Initially, the parasitoids seemed simply to have disappeared, but the predatory beetles underwent such a population explosion that all infestations of the scale insects in California were controlled by the end of 1890. In fact, although the beetles have usually taken most or all of the credit, the long-term outcome has been that the vedalia has been largely instrumental in keeping the scale in check inland, but *Cryptochaetum* is the main agent of control on the coast (Flint & van den Bosch, 1981).

Biological control of the cottony cushion scale has subsequently been transferred to 50 other countries and savings have been incalculable: the outlay, including salaries, was no more than $5,000.

natural enemies have often been left behind. The pest populations have then exploded with disastrous consequences. A search of the area where the crop originated might reveal predators or diseases of the pest that can be introduced to control it. There is then a hope that the introduced enemy will persist and keep the pest population below an economically damaging threshold indefinitely.

There are three main types of biological control. The natural enemy of a pest may be *imported* from another geographical area—very often the area in which the pest originated. Probably the best example of this has become a classic: the control of the cottony cushion scale (Box 12.4). Sometimes the natural enemy is unable to persist throughout the year but is repeatedly reintroduced (*inoculated*). Finally, a crop may be deliberately *inundated* with large numbers of a natural enemy of the pest, for example a bacterium or virus, which does not persist, but acts, like a chemical pesticide, to kill those pests that are present at the time.

three types of biological control

Inoculation is widely practiced in the biological control of pests in glasshouses, where crops are removed, along with the pests and their natural enemies, at the end of the growing season. The two species of natural enemy that are used most widely in this way are *Phytoseiulus persimilis*, a mite which preys on a spider mite pest of cucumbers and other vegetables, and *Encarsia formosa*, a parasitoid wasp that attacks a whitefly pest of tomatoes and cucumbers. By 1985, around 500 million individuals of each species were being produced for release each year in Western Europe.

control by inoculation

Table 12.1
The record of insects as biological control agents against insect pests and weeds

	Insect pests	Weeds
Control agent species	563	126
Pest species	292	70
Countries	168	55
Cases where agent has become established	1,063	367
Substantial successes	421	113
Successes as a percentage of establishments	40	31

biological control: excellent when it works ...

Insects have been the main agents of biological control against both insect pests and weeds. Table 12.1 summarizes the extent to which they have been used and the proportion of cases where the establishment of an agent has greatly reduced or eliminated the need for other control measures.

... but sometimes nontarget organisms are affected

Biological control may appear to be a particularly environmentally friendly approach to pest control, but examples are coming to light where even carefully chosen and apparently successful introductions of biological control agents have impacted on nontarget species. *Cactoblastis* moths, which were introduced to Australia and were dramatically successful at controlling exotic cactuses, were accidentally introduced to Florida where they have been attacking several native cacti (Cory & Myers, 2000). Similarly, a seed-feeding weevil (*Rhinocyllus conicus*), introduced to North America to control exotic *Carduus* thistles, attacks several native thistles and has adverse impacts on populations of a native picture-winged fly (*Paracantha culta*) that feeds on the thistle seeds (Louda et al., 1997). Such ecological effects need to be better evaluated in future assessments of potential biocontrol agents.

12.6   Integrated Farming Systems

integrated pest management

The desire for sustainable agriculture has increasingly led to more ecological approaches to food production, which are often given the label "integrated farming systems". Part of this, and something that preceded it historically, is a similar approach to pest control: integrated pest management (IPM).

IPM is a practical philosophy of pest management. It combines physical control (for example, simply keeping pests away from crops), cultural control (for example, rotating the crop planted in a field so pests cannot build up their numbers over several years), biological and chemical control, and the use of resistant varieties. It has come of age as part of the reaction against the unthinking use of chemical pesticides in the 1940s and 1950s.

IPM is ecologically based but uses all methods of control—including chemicals—where appropriate. It relies heavily on natural mortality factors, such as enemies and weather, and seeks to disrupt them as little as possible. It aims to control pests below an economically damaging level (the EIL), andds on monitoring the abundance of pests and their natural enemies and using various control methods as complementary parts of an overall program. IPM therefore calls for specialist pest managers or advisers. Broad-spectrum pesticides in particular, although not excluded, are used only very sparingly, and if chemicals are used at all it is in ways that minimize the costs and the quantities used. The essence of the IPM approach is to make the control measures fit the pest problem, and no two pest problems are the same—even in adjacent fields.

One example of IPM concerns the crops of cotton in the San Joaquin Valley, California. The crops had been afflicted by target pest resurgence (Figure 12.18a), secondary pest outbreaks (Figure 12.18b,c), and the evolution of resistance to chemical insecticides (Figure 12.18d). The key pest was the lygus bug *Lygus hesperus*, which feeds primarily on fruiting cotton buds. Another major pest was the cotton bollworm (*Heliothis zea*). However, in the San Joaquin Valley this seems largely to have been a secondary pest induced by insecticides, as were two other caterpillar pests, the beet army worm (*Spodoptera exigua*) and the cabbage looper (*Trichoplusia ni*). An IPM program was aimed at reducing insecticide usage to prevent secondary outbreaks, and also to avoid certain commonly used insecticides that decrease the yield of cotton plants.

Prior to the development of the program, a CAT had been established for the lygus bug, but based on little information. Hence, the threshold remained the same throughout the year—a serious error. Further research showed that the lygus bug could inflict serious damage only during the budding season (broadly, early June to mid-July). Thus, in the IPM program the threshold was set at ten bugs per 50 sweep-net sweeps on two consecutive sampling dates between June 1 and July 15. After that time, insecticide application for lygus bug control was avoided. In fact, the threshold was subsequently refined to one based on a bug : bud ratio, which was more dynamic and more accurate. Another aspect of control within the system centered on manipulation of alfalfa, a favored host plant of the lygus bug. In particular, the interplanting of narrow alfalfa strips (5–10 m) in the cotton fields proved effective in attracting the bugs out of the cotton. In addition, attempts were made to augment the populations of natural enemies by providing them with supplementary nutrition.

The key to a successful IPM program is a good field monitoring system. Here, each field was sampled twice a week from the beginning of budding (roughly mid-May) to the end of August. Plant development, and pest and natural enemy population data were all collected. Additional sampling for the cotton bollworm began around August 1 and continued until mid-September. Overall, the program represents a harmonious integration of naturally occurring (but augmented) biological controls and carefully timed insecticide treatments. The success of the program for cotton and also for citrus in the San Joaquin Valley is illustrated in Table 12.2.

It has increasingly become apparent, in an agricultural context at least, that implicit in the philosophy of IPM is the idea that pest control cannot be isolated from other aspects of food production and is especially bound up with the means by which soil fertility is maintained and improved. Thus, a number of programs have been initiated to develop and put into practice sustainable food production methods that incorporate IPM, including not only IFS (integrated farming systems) but also LISA (low input sustainable agriculture) in the United States, and LIFE (lower input farming and environment) in Europe (International Organisation for Biological Control,

IPM of cotton pests in California

integrated farming systems—LISA, IFS, and LIFE

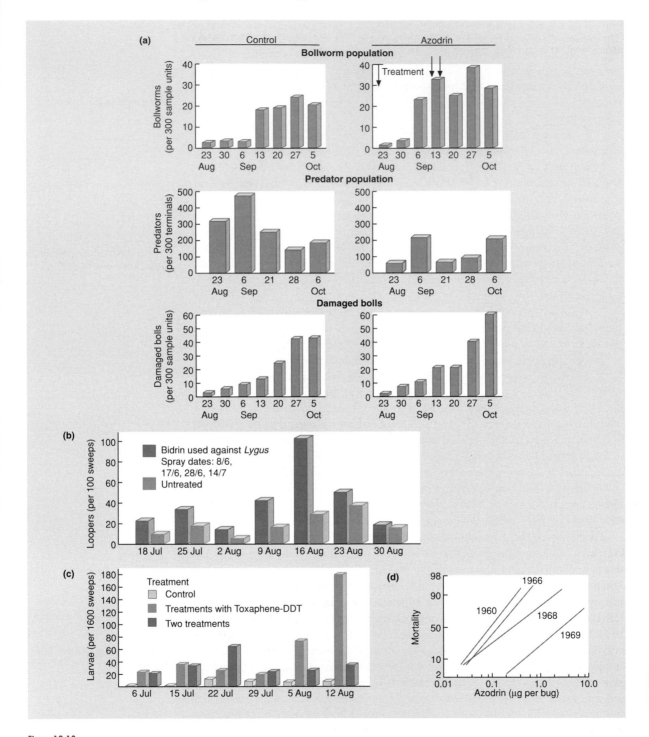

Figure 12.18
Pesticide problems amongst cotton pests in San Joaquin Valley, California. (a) Target pest resurgence: bollworms resurged because the abundance of their natural predators was reduced and the number of damaged bolls was therefore higher. (b) An increase in cabbage loopers and (c) in beet army worms were seen when plots were sprayed against lygus bugs—both are examples of secondary pest outbreaks. (d) Increasing resistance of lygus bugs to Azodrin®. (After van den Bosch et al., 1971.)

Table 12.2
The success of integrated pest management (IPM) in San Joaquin Valley, California 1970–1. With IPM the outlay on insecticides was cut drastically yet the financial yield was not reduced. (Figures in brackets are standard deviations.)

	IPM users	Nonusers
Average insecticide cost per acre of cotton (US$)	4.9 (± 3.9)	12.0 (± 7.4)
Average yield per acre of cotton (US$)	270.2 (± 27.5)	247.8 (± 55.1)
Average insecticide cost per acre of citrus (US$)	20.5 (± 15.0)	42.4 (± 18.3)
Average yield per acre of citrus (US$)	515.8 (± 260.6)	502.9 (± 157.0)

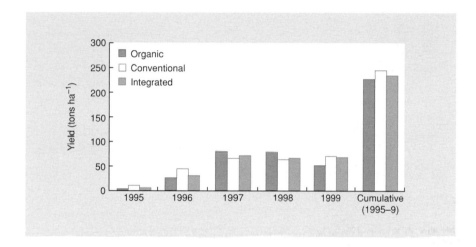

Figure 12.19
The fruit yields of three apple production systems. (From Reganold et al., 2001.)

1989; National Research Council, 1990). All share a commitment to the development of sustainable agricultural systems.

    These approaches have advantages in terms of reduced environmental hazards. Even so, it is unreasonable to suppose that they will be adopted widely unless they are also sound in economic terms. As we have already noted, in an industry such as agriculture, practices that are economically unsustainable are, ultimately, unsustainable overall. In this context, Figure 12.19 shows the yields of apples from organic, conventional, and integrated production systems in Washington State from 1995 to 1999 (Reganold et al., 2001). Organic management excludes such conventional inputs as synthetic pesticides and fertilizers whilst integrated farming uses reduced amounts of chemicals by integrating organic and conventional approaches. All three systems gave similar apple yields but the organic and integrated systems had higher soil quality and potentially lower environmental impacts. When compared with conventional and integrated systems, the organic system produced sweeter apples, higher profitability, and greater energy efficiency. Note, however, that despite some widely held beliefs, organic farming is not totally free of adverse environmental consequences (Box 12.5).

environmental and economic
sustainability

Box 12.5   *Topical ECOncerns*

Myths About Organic Farming

"Organic farming" is distinguished from other farming methods mainly by the prohibition of soluble mineral inputs and by the rejection of synthetic herbicides and pesticides in favor of natural pesticides. Anthony Trewavas (2001) reviewed the pros and cons of organic farming in a commentary in the influential journal *Nature*, extracts of which are reproduced below (refer to the full article for more detail and relevant references).

Urban myths of organic farming
There is a widespread belief that low-yielding organic agricultural systems are more friendly to the environment and more sustainable than high-yielding farming systems. The current aims of organic systems—maintenance of soil fertility, avoidance of pollution, use of crop rotation, animal welfare concerns and wider environmental aspects—would be hard to quarrel with. But the rules and regulations that have to be followed to achieve these ends caused one leading organic researcher to admit that in organic farming "there is very little science" and "this gives rise to a great deal of illogicality and confusion particularly in some areas of production."

Confusion of principles: Organic proponents assert that better plants are produced from minerals derived from manure breakdown, so that organic food is superior and improves human health. Hundreds of rigorous tests have failed to reveal better-tasting properties or improved nutritional value.

Problems with pesticides and chemicals: Approved pesticides for organic farmers include *copper sulfate*, which has caused liver damage in vineyard workers, kills worms and is persistent in soil and produce (to be banned by the European Commission after 2002), *rotenone*, recently shown to induce Parkinson's disease, and *Bacillus thurigiensis* spores, causing fatal lung infections in mice.

Uses and misuses of manure: Soluble minerals are not used on organic farms. Although crude rock phosphate may be allowed, potassium chloride is banned; sylvanite, another form of potassium chloride, may be permitted. The main alternative mineral source for crop nutrients is animal or green manure. Manure treatment used on any mixed farm improves soil quality, but conventional crop rotation seems equally effective. Manure breakdown cannot be synchronized with crop canopy growth, as is desirable, but continues throughout the growing season. Ploughing in of legume crops (a necessary part of the organic method to build soil fertility) and continued manure breakdown leads to nitrate leaching into aquifers and waterways at identical rates to conventional farms.

(All content © *Nature* 2001 and may not be republished without permission.)

Trewavas concludes his commentary by noting that organic agriculture was originally formulated as an ideology, but today's global problems—such as climate change and population growth—need agricultural pragmatism and flexibility, not ideology. Do you agree? How would you determine whether the frequent mechanical weeding used by many organic farmers to keep their fields clear of weeds would be less or more damaging to farmland birds, earthworms, and insects than cautious use of a herbicide, coupled with no-till conventional farming?

12.7 Forecasting Agriculturally Driven Global Environmental Change

Much attention has been focused on the predicted far-reaching consequences of global climate change caused by human activities such as the burning of fossil fuels. We deal with this in Section 13.4. However, very significant threats are also posed to ecosystems around the world by increasing agricultural development. In this chapter we have considered the problems of the more-than-exponential increase in the human population, and of the associated impacts of increased erosion, unsustainability of water supplies, salinization and desertification, excess plant nutrients finding their way into waterways, and the unwanted consequences of chemical pesticides. Model projections suggest that all these will increase over the next 50 years as more land is converted to grow crops and pasture (Tilman et al., 2001) (Figure 12.20). This can be expected to place biodiversity at high risk, particularly because population increases are predicted to be greatest in species-rich tropical areas. To control the environmental impacts of agricultural expansion, we will need scientific and technological advances as well as the implementation of effective government policies.

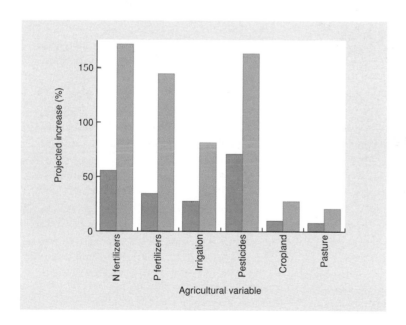

Figure 12.20
Projected increases in nitrogen (N) and phosphorus (P) fertilizers, irrigated land, pesticide use, and total areas under crops and pasture for the years 2020 (pink bars) and 2050 (blue bars). (From Laurance, 2001, after data in Tilman et al., 2001.)

Summary

The human population problem

Resource use by humans is defined as sustainable if it can be continued for the forseeable future. The root of most environmental problems is the population "problem": a large human population that has been growing at a more-than-exponential rate.

Three groups of nations can be recognized: those that passed through the demographic transition "early", "late", or "not yet". Even if it were possible instantaneously to bring about the transition in all remaining countries of the world, the population problem would not be solved, partly because population growth has its own momentum.

The global carrying capacity is variously estimated at between 1 and 1,000 billion, depending mainly on what is deemed to constitute an acceptable standard of living.

Harvesting living resources from the wild

Whenever a natural population is exploited by harvesting there is a risk of overexploitation. But harvesters also want to avoid underexploitation, where potential consumers are deprived and those who do the harvesting are underemployed.

The concept of the maximum sustainable yield (MSY) has been a guiding principle in the exploitation of natural populations. There are two simple ways of obtaining an MSY on a regular basis: through a "fixed quota" and through a "fixed effort". Two limitations of the MSY approach are (1) that it treats populations as a number of similar individuals, and (2) that it treats the environment as unvarying. Improved harvesting strategies take both these into account. Lack of knowledge of most fisheries around the world means that management is often based on the precautionary principle, often in the absence of data.

The farming of monocultures

Increasingly, animals and plants have been domesticated and managed in ways that have allowed much greater harvests—usually as monocultures. But a high price may be paid to sustain these high rates of food production. Monocultures offer ideal conditions for the epidemic spread of diseases and lead to widespread degradation of land.

Sustainability of soil and of water supplies

In an ideal sustainable world, new soil would be formed as fast as the old was lost, but in most agricultural systems this is not achieved. When there is an overwhelming reliance on artificial fertilizers, maintaining the organic matter capital of the soil tends to be neglected and this has declined worldwide.

Soil degradation can be slowed down by incorporating manures and residues, alternating cultivation and fallow, or returning the land to grazed pasture. In tropical regions, terracing is widely practiced over hilly and mountainous terraine. In arid regions, overgrazing and excessive cultivation can lead to desertization and salinization.

Water is widely thought to be the resource that future wars will be fought over. On a global scale, agriculture is the largest consumer of fresh water. Pumping water from underground aquifers is the main cause of loss of agricultural land through salinization.

Pest control

The aim of pest control is to reduce the pest population to its economic injury level, but control action thresholds may be of more immediate importance.

Pesticides may kill species other than their target and may give rise to target pest resurgence or secondary pest outbreaks. Pests may also evolve pesticide resistance.

Biologists may manipulate the natural enemies of pests (biological control) through importation, inoculation, or inundation, but even biocontrol agents can have unwanted effects on nontarget species.

Integrated farming systems

Integrated pest management (IPM) is a practical philosophy of pest management that is ecologically based but uses all methods of control where appropriate. It relies heavily on natural mortality factors and calls for specialist pest managers or advisers.

Implicit in the philosophy of IPM is the idea that pest control cannot be isolated from other aspects of food production. A number of programs have been initiated to develop and put into practice sustainable food production methods that incorporate IPM. Evidence has been accumulating that this sustainable farming approach can yield improved economic returns too.

Agriculturally induced global change

It is clear that very significant threats are posed to ecosystems around the world by the increasing human population and concomitant increases to agricultural development; these are expected to have a particularly damaging effect on biodiversity because most agricultural growth is predicted to occur in the species-rich tropics.

Review Questions

1 ▲ What is sustainability? Is it possible to have sustainable population growth? Sustainable use of fossil fuels? Sustainable use of forest trees? Justify your answers.

2 Describe what is meant by "the demographic transition" in a human population. Explain why it might be important, for future management of human population growth, to discover whether the demographic transition is an academic ideal or a process through which all human populations necessarily pass.

3 ▲ The number of people that Earth can support depends on their standard of living. Argue the case either for or against developing nations having the right to expect standards of living those in the developed world take for granted.

4 Contrast the ways in which "fixed quota" and "fixed effort" harvesting strategies seek to extract maximum sustainable yields from natural populations.

5 Discuss the pros and cons of agricultural monocultures.

6 One of the main bodies regulating the production of organic food (food produced without synthetic fertilizers or pesticides) in the United Kingdom is the Soil Association. Explain why you think it has adopted this name.

7 Explain the meaning and importance of the terms "economic injury level" and "control action threshold".

8 Weigh up the advantages and disadvantages of the chemical and biological control of pests.

9 Explain why methods of pest control and methods of soil fertility maintenance need to be considered together in integrated farming systems.

10 ▲ Hilborn and Walters (1992) have suggested that there are three attitudes that ecologists can take when they enter the public arena. The first is to claim that ecological interactions are too complex, and our understanding and our data too poor, for definite pronouncements to be made (for fear of being wrong). The second possibility is for ecologists to concentrate exclusively on ecology and arrive at a recommendation designed to satisfy purely ecological criteria. The third is for ecologists to make ecological recommendations that are as accurate and realistic as possible, but to accept that these will be incorporated with a broader range of factors when management decisions are made—and may be rejected. Which of these do you favor, and why?

▲ = Challenge Question

Chapter 13

Pollution

As populations grow, organisms become more crowded and both deplete their environment of the essentials for life and contaminate it with products that hinder their activities. The human species is unique in using fire (and recently nuclear fission) to provide the energy for doing work. This has given the species remarkable power to change the environment. When we do not like what has changed in our environment we call it pollution.

Chapter Contents

13.1   Introduction

13.2   Urban pollution

13.3   Agricultural pollution

13.4   Atmospheric pollution

13.5   Nuclear radiation

13.6   Mining and quarrying

13.7   Restoration ecology

Key Concepts

In this chapter you will

- recognize that *Homo sapiens* is just one species among many whose activities can reduce the quality of their environment

- understand that social and urban living exaggerate the problems of waste disposal and that corpses, feces, and urine can be serious pollutants

- appreciate that human use of fire and fuel makes our species a unique polluter and that there are a multitude of consequences in atmospheric pollution

- understand that nuclear radiation caused by human activities adds to that already present

- realize that quarrying and mining, smelting, and disposal of metals are seriously polluting

- appreciate that most terrestrial pollutants ultimately pollute rivers, lakes, oceans, or the atmosphere

- understand that applied ecology has a role in the restoration of polluted environments

13.1   Introduction

The verb *pollute* derives from the Latin *polluere,* meaning "to soil". To most people, environmental pollution means the contamination of the environment by human waste, especially sewage, and by the unwanted products of human activities, including gases from power plants, leaks of radioactive material, car exhaust fumes, and runoff from agricultural land of pesticides and fertilizers.

The human species is not unique in depleting and contaminating its environment. However, we are certainly unique in using fire, fossil fuels, and nuclear fission to provide the energy for doing work. This power generation has extremely far-reaching consequences for the state of the land, aquatic ecosystems, and the atmosphere. More-over, the generated energy has given *Homo sapiens* remarkable power to transform landscapes through activities such as urbanization, industrial agriculture, and forestry; to mine, smelt, and use metals (causing further pollution); to create new chemical diversity with attendant dangers; to add to the hazard of natural radiation; and to alter the atmosphere and even the nature of the climate.

13.2   Urban Pollution

As human populations grew and were increasingly concentrated in cities, two sources of pollution became particularly important: feces and corpses.

13.2.1   Burial and cremation

Both Jews and Romans had recognized that cemeteries were sources of pollution and disease and they placed their graveyards outside the city walls of Jerusalem and Rome. But in other areas bodies were buried within cities. In the Middle Ages in London coffins were placed tier upon tier until sometimes they were within a few centimeters of the surface; to make room for more burials, the gravediggers moved bones and partly decomposed remains to nearby pits. In Western societies the safe and hygienic disposal of corpses in religious rites has increasingly become a civic responsibility.

cremation   Cremation has become increasingly acceptable in many societies, partly in reaction to the massive loss of countryside that cemeteries represent—itself a dramatic example of environmental pollution. In Hindu communities corpses are disposed of by burning on an open fire. At first sight this appears to be much less damaging to the environment than burial, but enormous volumes of wood are used in the funeral pyres. More than 50 percent of human corpses (and those of many pets) are now cremated in England and nearly 100 percent in Japan, but in the United States the figure is only about 10 percent. Modern cremation contributes its small share to environmental pollution in the form of carbon dioxide (and mercury from tooth fillings).

the use of scavengers to deal with corpses   Perhaps the least polluting methods for disposing of corpses are burial at sea and direct exposure on land, as practiced by Parsis. The naked corpses are exposed in special isolated sites for vultures to strip off the flesh. This takes only 1 or 2 hours and the bared bones then dry in the sun and are swept away. The "ecosystem service" provided by vultures provides an extra dimension to the need to conserve them (Box 13.1).

Box 13.1 *Topical ECOncerns*

An Ecosystem Service Provided by Vultures

The following article by Luke Harding appeared in the *Guardian Weekly*.

Threat to Parsi rites as India's vultures face extinction

From the outside, the Tower of Silence at the top of Bombay's prestigious Malabar Hill appears unremarkable. The only hint of what goes on inside are the vultures that perch on its whitewashed outer walls.

Inside the secret complex, surrounding a Roman-style amphitheatre, banyan and casarina trees flourish in the tropical sun. And there are, of course, the dead—two or three corpses most days, exposed to the elements and left for the vultures to pick clean.

But this ancient, ecologically impeccable method of disposing of the dead, practised by India's Parsi community for hundreds of years, is under threat. The problem is not with the method, but with the vultures. Over the last three or four years India's vulture population has suffered a dramatic and mysterious decline. They are, literally, falling off their perches. In some areas the vulture has vanished, with extinction possibly five years away.

For the Parsis, the vulture shortage poses an immediate difficulty: what to do with the dead? The community, which fled Persia 1,200 years ago to escape Arab persecution, believes earth and fire to be sacred. Cremation or burial is regarded as an offence.

. . . Conservationists and scientists from all over the world gathered in Delhi last month to debate how the vulture can be saved. Classified as "critically endangered", the most likely explanation for its decline is a virus, the conference heard. Across India the vultures are showing the same symptoms: a drooping neck and lethargy. Weak birds eventually fall out of the trees and die. Numbers have fallen by 90%. Two species have been devastated—the white-backed and long-billed vultures.

(All content © *Guardian* and may not be published without permission.)

Native organisms and natural communities provide a wide range of "ecosystem services" for people (see Box 14.1), of which this is one of the most unusual. What other ecosystem services can you think of that are provided by vertebrates? Is the case for conservation action made stronger by the fact that an endangered species is of material value to people?

Humans are not unique in regarding corpses as pollutants that must be removed from the community—the process has even been given a name, *necrophoresis*. Worker ants recognize corpses and remove them from the nest. If material from a dead ant is sprayed onto a live one, workers will quickly carry it out of the nest, as if it were a corpse.

disposal of corpses in nonhuman species . . .

13.2.2 Feces, urine, and other urban waste

Feces and urine are unlikely to cause problems in natural communities of nonsocial species—indeed both feces and urine are sometimes put to use to mark territories. However, they may act as pollutants in the environment of social animals. For example,

. . . and the problems of feces in nonhuman species

cattle avoid grass near their dung for more than a month, responding to certain volatile chemicals (Dohi et al., 1991). Most mammals that live as social families in burrows avoid soiling them and deposit their urine and feces outside, even in special latrines. Birds avoid fouling their nests and may carry away nestlings' feces in special fecal sacs.

All human body products, but most notably feces and urine, can be regarded as pollutants. The Greeks were probably the first to control the accumulation of pollution within towns, and a law of 320 BC forbade dumping of waste in the streets. The Romans were also very pollution conscious, dumping city waste in pits outside the city walls. When the Roman and Greek civilizations perished, their quite sophisticated control of urban pollution collapsed. Medieval castles, for example, were often designed with latrines projecting from castle walls that simply dumped waste at the base of the walls (the accumulated wastes give archaeologists a direct record of historic diets and infestation with intestinal worms!). Until the 14th and 15th centuries the open streets again became the main, and often the only, destination for human and animal feces and urine, from which they were washed during rainstorms into rivers, lakes, and oceans. Horses, used for transport, and farm animals, brought into cities and kept alive for milk or until they were slaughtered for meat, contributed their share of feces to the burden of pollution. A special trade developed, that of the scavenger, who was paid to carry waste to dumps outside the cities; in 1714 every city in England had an official scavenger (the forerunner of the Environmental Protection Agency!). Intense city pollution developed later in America, but municipal collection of garbage had started in Boston before the end of the 18th century.

Even when water closets (invented by Thomas Crapper) began to be installed early in the 19th century, the underground reservoirs (cesspools) into which they emptied often overflowed and contaminated drinking water. Outbreaks of cholera in the middle of the 19th century were traced directly to this source of contamination, and this discovery led to the connection of household waste directly to sewers in both Britain and the United States.

At first glance, the easiest way to cope with accumulated feces and urine might appear to be to dilute them in large bodies of water. However, it is not easy to dispose of human waste and at the same time provide healthy drinking water. In addition to health issues there can also be profound ecological effects of disposing of sewage in water bodies.

when natural ecosystems cannot cope with the extra supply of organic matter and nutrients in human waste

All natural ecosystems have an inherent capacity to decompose feces (see Section 11.4.4) and up to a point natural decomposition processes in rivers, lakes, and oceans may cope with increased organic matter from human sewage without obvious changes to the nature of the biological communities they contain. However, problems arise when the rate of input of sewage exceeds this capacity. First, excessively high rates of decomposition of dead organic matter in rivers and lakes can lead to anaerobic conditions (causing the death of fish and invertebrates). This is because oxygen is consumed by the decomposer microorganisms faster than it is replenished from photosynthesis by aquatic plants and diffusion from the air. Second, the supply of nutrients such as phosphate and nitrate that normally limit plant growth in water bodies may be increased to a level where algal growth is so great that it shades and kills other aquatic plants. This leads to high rates of algal decomposition and consequent oxygen-poor conditions (a process known as *cultural eutrophication*).

Modern sewage systems were developed as ecological devices for pollution management. They aim to capture pollutants from wastewater and to clean it, usually in a drainage system separate from the one that carries heavy flows of storm water. Ideally a sewage system cleans polluted water to a state suitable for drinking before discharging it back into rivers, lakes, and the sea. The full treatment of sewage has three stages, though in many places only the first or second stages are actually used before discharge into the environment (Figure 13.1). After paper, rags, and plastic have been removed by passing the sewage through screens, *primary treatment* is a physical process in which much of the solid organic sewage waste is allowed to settle to the bottom of settlement tanks, from which it is removed as sludge. *Secondary treatment* is a biological process in which the process is designed to mimic (and indeed enhance) natural decomposition. In its simplest version, the partly cleaned water is sprayed onto a layer of crushed rock

sewage treatment systems are needed

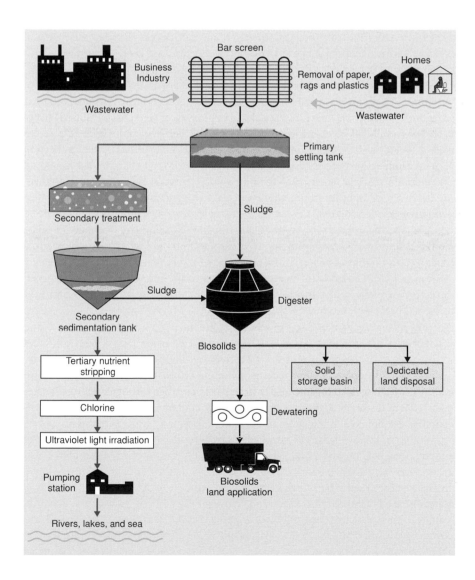

Figure 13.1
The sequence of treatments commonly applied to the sewage waste from a modern urban community.

within which microorganisms have been encouraged to grow; as the water trickles down through these *percolating* or *trickling filters*, natural decomposition mineralizes much of the remaining organic matter, releasing carbon dioxide to the atmosphere. A more sophisticated and efficient method of secondary treatment is the *activated sludge method*, in which the sewage is passed into aerated tanks containing sludge that is activated, or seeded with microorganisms. After secondary treatment the remaining solids are settled to yield more sludge. The wastewater now appears clean, but it still contains two types of "impurities", namely, disease organisms and high concentrations of mineral nutrients that may have health consequences (see Section 13.3.2) but also cause eutrophication if released into rivers and lakes. A final "polishing" stage usually includes chlorination, and sometimes ultraviolet light irradiation to kill bacteria; full *tertiary treatment* involves the stripping of nutrients, largely by artificial and expensive chemical processes.

products of sewage treatment are themselves pollutants

Untreated sewage is obviously a pollutant, with adverse health and ecological consequences for the water bodies into which it is discharged. However, discharge of sewage that has only been subject to primary treatment is still likely to cause eutrophication because it remains rich in organic matter and nutrients. Moreover, even secondary treatment removes only the organic matter, leaving wastewater rich in plant nutrients. The sludge that accumulates in the settling tanks is itself a pollutant that has to be disposed of, usually by dumping at sea or burying in landfill sites. Buried sludge decomposes anaerobically, sometimes taking more than 20 years to mineralize completely, and it produces methane, which contributes to atmospheric pollution (Section 13.4.2). Sludge can be more appropriately used as a fertilizer, either dried or as a liquid sprayed onto the land; in this way the nutrient cycle can be reconstituted by returning nutrients, assimilated from crops by people, to agricultural land to be taken up by future crops.

The range of products of sewage treatment (wastewater, nutrients, sludge, methane) illustrates very elegantly how each solution to a problem of environmental pollution leads almost inevitably to a new type of pollution. Moreover, almost every stage in pollution control is demanding of energy and therefore causes more pollution. An environmentally attractive fate for sludge involves digesting it in a controlled microbial process to produce methane, which is then burned to produce the energy that powers the mechanical stages of sewage treatment.

other urban waste

Urban waste includes a very wide range of materials from widely different sources—e.g., the chemical industry, the military, and households—and includes plastics, wood, and metal. Such waste is commonly deposited in landfill sites, which are then capped with soil and passed to agriculture or urban development. There have been many cases in which these have subsequently caused problems as toxic chemicals start to leach into neighboring watercourses and nauseating gases are released.

Love Canal: a classic example of industrial urban pollution

Pollution is not necessarily stopped by burying the pollutants out of sight. An attempt to build a canal joining the upper and lower Niagara River in 1892 was abandoned and the Love Canal excavation was sold in 1920 as a site for disposing of chemical waste. The Hooker Chemical Corporation, the city of Niagara, and the United States Army all deposited material at the site (including perhaps materials for chemical warfare). The canal was then filled with about 20,000 tons of toxic waste and was covered with soil and sold to the Niagara Board of Education (with a warning that chemical wastes had been deposited). A school for about 400 students was completed in 1955 on the land near the center of the landfill and a community of 800 single-family homes and 240 low-income apartments had been built around the site by 1978. Residents complained of foul odors and smelly materials emerging from the soil and consultants were called in. They found storage drums just beneath the surface and high concentrations of polychlorinated biphenyls (PCBs) in the storm sewage system. Little was done except to place window fans in a few homes where the residues were especially high. A health order was issued on August 2, 1978, recommending that the school be closed, that pregnant women and children below the age of 2 years be evacuated, and that residents eat nothing from their gardens and avoid spending time in their basements. The state agreed to buy 239 homes closest to the canal. In the period from January 1979 to February 1980 there were 22 pregnancies among women from Love Canal, 18 of which ended in miscarriage, stillbirth, or a child with birth defects (for more details see Gibbs, 1998).

13.3   Agricultural Pollution

When intensive livestock production forces animals to live the equivalent of urban life, their waste is produced faster than natural decomposers and detritivores can handle it. All the problems of human urban overpopulation then affect domestic livestock. Intensive agriculture is also associated with an increase in the level of nitrates that runs into rivers and lakes (and into drinking water) and problems associated with the use of insecticides and herbicides.

13.3.1   Intensive livestock management

Pigs, cattle, and poultry are the three major contributors to pollution in industrialized agriculture feedlots. The waste from factory-farmed poultry is easily dried and forms a transportable, inoffensive, and valuable crop and garden fertilizer. In contrast, the excreta from cattle and pigs are 90 percent water and have an unpleasant smell (Table 13.1). A commercial unit for fattening 10,000 pigs produces as much pollution as a town of 18,000 inhabitants.

refuse from poultry farming is easy to transport and use as fertilizer . . .

. . . but excreta from cattle and pigs are much more bulky (and smelly)

Table 13.1
The average percentage nitrogen and dry matter content of manures and slurries. (From Gostick, 1982.)

Source	Nitrogen (%)	Dry matter (%)
Cattle farmyard manure	0.6	25
Pig farmyard manure	0.6	25
Poultry broiler litter	2.4	70
Cattle slurry	0.5	10
Pig slurry	0.6	10

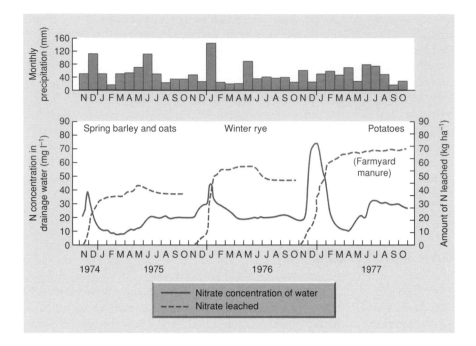

The law in the United States and Europe increasingly restricts the discharge of agricultural slurry into watercourses. The simplest and often the most economically sound practice returns the material to the land as semisolid manure or as sprayed slurry. This dilutes its concentration in the environment to what might have occurred in a more primitive and sustainable type of agriculture and converts pollutant into fertilizer. In turn, of course, new pollution is created—in this case the smell of applied slurry can be offensive. Soil microorganisms decompose the organic components of sewage and slurry and release carbon dioxide into the atmosphere, while most of the mineral nutrients become immobilized in the soil, available to be absorbed as nutrients by the vegetation. Nitrogen is a special case: nitrate ions are not adsorbed in the soil and rainfall leaches them into drainage (and therefore potential drinking) water (Figure 13.2). The nitrate becomes a new pollutant.

13.3.2 Nitrate

nitrates in drinking water are a hazard to health . . .

Excess nitrate in drinking water is a health hazard—the Environmental Protection Agency in the United States recommends a maximum concentration of 10 mg l^{-1} in drinking water. Nitrates may contribute to the formation of carcinogenic nitrosamines and in young children may reduce the oxygen-carrying capacity of the blood. Human and animal wastes are just two of the contributors to the nitrate pollution of water; agriculture and forestry are also major culprits.

. . . and leach readily from agricultural and forest soils

Most of the fixed nitrogen in natural communities is present in the vegetation and in the organic fraction of the soil. As organisms die they contribute organic matter to the soil, and this decomposes to release carbon dioxide so that the ratio of carbon to nitrogen falls; when the ratio approaches 10 : 1, nitrogen begins to be released from the soil organic matter as ammonium ions. In aerated regions of the soil, the ammonium ions become oxidized to nitrite and then to nitrate ions, which are leached by rainfall down the soil profile dissolved in the drainage waters and then enter aquifers or rivers *en route* to the sea.

Both the processes of organic matter decomposition and the formation of nitrates are usually fastest in the summer, when natural vegetation is growing most quickly. Nitrates are then absorbed by the growing vegetation as fast as they are formed. They are not present in the soil long enough for significant quantities to be leached out of the plants' rooting zone and lost to the community. Natural vegetation is a "nitrogen-tight" ecosystem.

In contrast, there are several reasons why nitrates leach more easily from agricultural land and managed forests than from natural vegetation.

1 For part of the year agricultural land carries little or no living vegetation to absorb nitrates (and for many years forest biomass is below its maximum).

2 Crops and managed forests are usually monocultures that can capture nitrates only from their own rooting zones, whereas natural vegetation often has a diversity of rooting systems and depths.

3 When straw and forestry waste are burned, the organic nitrogen within them is returned to the soil as nitrates.

4 When agricultural land is used for grazing animals their metabolism speeds up the rate at which carbon is respired, reduces the C : N ratio, and increases nitrate formation and leaching.

5 Nitrogen in agriculture fertilizer is usually applied only once or twice a year rather than being steadily released as it is during the growth of natural vegetation; it is therefore more readily leached and finds its way into drainage waters.

Because nitrogen is not efficiently recycled on agricultural land or in managed forests, repeated cropping leads to losses of nitrogen from the ecosystem and thus to decreasing crop productivity. The growth of human populations places heavy demands on agriculture to produce ever greater yields; to maintain crop yields the available nitrogen has to be supplemented with fertilizer nitrogen.

Some of the nitrogen used in agricultural fertilizers is obtained by mining potassium nitrate in Chile and Peru, but the majority comes from the energy-expensive industrial process of nitrogen fixation, in which nitrogen is catalytically combined with hydrogen under high pressure to form ammonia and, in turn, nitrate. Nitrogen fertilizers are applied in agriculture either as nitrates or as urea or ammonium compounds (which are oxidized to nitrates). However, it is wrong to regard artificial fertilization as the only practice that leads to nitrate pollution; nitrogen fixed by crops of legumes such as alfalfa, clover, peas, and beans also finds its way into nitrates that leach into drainage water.

The problem of nitrate pollution in drinking water is tackled mainly by encouraging wetland plants to use it in their growth or by modifying agricultural and forestry practices to ease the problem. This may involve maintaining ground cover of vegetation year-round, by practicing mixed cropping rather than monoculture, by returning

most agricultural crops depend on fertilizer nitrogen . . .

. . . or nitrogen fixation by legumes

UNANSWERED QUESTION: can intensive agriculture or forestry be practiced without creating nitrate pollution of the drainage water?

organic matter to the soil, by maintaining low stocking levels, and by applying nitrogen fertilizers only in the season when they are most rapidly absorbed by the crop. However, all these restrictions reduce crop yields and generate other costs. Nitrate pollution of drainage waters appears to be an inevitable consequence of intensive agriculture or forestry.

13.3.3   Pesticides

Many of the manufactured chemicals that are used to kill pests have become important environmental pollutants. The most widely polluting pesticides are those used to control pests and weeds that damage crops in agriculture, horticulture, and forestry or to kill pests that transmit diseases of livestock and humans. All are sprayed or dusted onto the areas in which the pests live, and only a very small proportion hit the target—most land on the (resistant) crop or on bare ground. Such pesticides are therefore used in much larger quantities than strictly necessary. Characteristics of the most widely used pesticides are described in Chapter 12.

In the early industrial development of pesticides, manufacturers were not much concerned with the specificity of their product: they could damage anything, provided it did not harm the crop, humans, or domestic animals. The potential for disaster is illustrated by the occasion when massive doses of the insecticide dieldrin were applied to large areas of Illinois farmland from 1954 to 1958 to "eradicate" a grassland pest, the Japanese beetle. Cattle and sheep on the farms were poisoned, 90 percent of the cats on the farms and a number of dogs were killed, and among the wildlife 12 species of mammals and 19 species of birds suffered losses (Luckman & Decker, 1960).

pesticides are most polluting when they are unselective, are persistent, and accumulate in food chains . . .

Chemical insecticides are generally intended to control particular target pests at particular places and times. Problems arise when they are toxic to many more species than just the target and particularly when they drift beyond the target areas and persist in the environment beyond the target time. The organochlorine insecticides have caused particularly severe problems because they are *biomagnified*. Biomagnification happens when a pesticide is present in an organism that becomes the prey of another and the predator fails to excrete the pesticide. It then accumulates in the body of the predator. The predator may itself be eaten by a further predator, and the insecticide becomes more and more concentrated as it passes up the food chain. Top predators in aquatic and terrestrial food chains, which were never intended as targets, can then accumulate extraordinarily high doses (Figure 13.3 and Box 13.2).

. . . but some species reabsorb and concentrate the pollutants to dangerous levels

most terrestrial pollution finds its way via rivers and estuaries to the sea— the ultimate dilution?

The oceans receive whatever drains and leaches from the land, and the seemingly ideal fate for a human pollutant is to be diluted and carried away in the vastness of the oceans. However, what has been diluted in the sea may subsequently be accumulated by biological processes, and, like their terrestrial counterparts, predators at the top of marine food chains can accumulate damaging concentrations from what appear to be innocuously dilute sources (Figure 13.3). Indeed, persistent chlorinated biphenyls can accumulate to concentrations 1,000 times that found in invertebrates lower in the food chain in the livers of whales, dolphins, and porpoises, which are at particular risk because they cannot metabolize these pollutants efficiently. Such top predators may not only suffer from the direct effects of pesticide residues; the

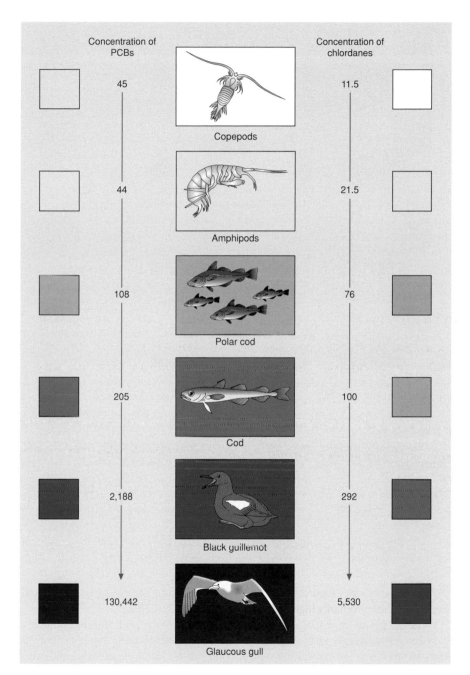

Concentration of PCBs

45

44

108

205

2,188

130,442

Copepods

Amphipods

Polar cod

Cod

Black guillemot

Glaucous gull

Concentration of chlordanes

11.5

21.5

76

100

292

5,530

Figure 13.3
Organochlorines, applied as pesticides on land, are transported to the Arctic through river runoff and oceanic and atmospheric circulation. A study in the Barents Sea shows how two classes of pesticide are biomagnified during passage through the marine food chain. Concentrations in seawater are very low. Herbivorous copepods (that feed on phytoplankton) have higher concentrations (measured in nanograms per gram of lipid in the organisms), and predatory amphipods have higher concentrations still. The polar cod (*Boreogadus saida*), which feeds on the invertebrates, and cod (*Gadus morhua*), which also includes polar cod in its diet, show further evidence of biomagnification. However, it is the higher steps in the food chain where biomagnification is most marked, because the seabirds that feed on the fish (black guillemots—*Cephus grylle*) or on fish and other seabirds (glaucous gull—*Larus hyperboreus*) have much less ability to eliminate the chemicals than fish or invertebrates. Note how chlordanes are biomagnified to a lesser extent than polychlorinated biphenyls (PCBs). This results from the birds' greater ability to metabolize and excrete the former class of pesticide. (After data in Borga et al., 2001.)

compounds may also damage their immune systems and lay the organisms open to infection from pathogens.

Herbicides are used in very large amounts and on a worldwide scale. They are active against pest plants and when used at commercial rates appear to have few significant effects on animals. Herbicide pollution of the environment has not aroused

herbicides and pollution

Box 13.2   *Topical ECOncerns*

Pollution and the Thickness of Birds' Eggshells

The peregrine falcon (*Falco peregrinus*) is a particularly distinctive and beautiful bird of prey with an almost worldwide range. Until the 1940s about 500 pairs bred regularly in the eastern states of the United States and about 1,000 pairs in the west and Mexico. In the late 1940s their numbers started a rapid decline, and by the mid-1970s the bird had disappeared from almost all the eastern states and its numbers had fallen by 80–90 percent in the west. Similar dramatic declines were occurring in Europe. The peregrine falcon was listed as an *endangered species* (at risk of extinction). The decline occurred in many other birds of prey and was traced to failure to hatch normal broods. There was a very high breakage of eggs in the nest. The cause was eventually identified as the accumulation of DDT in the parent birds. The pesticide had apparently contaminated seeds and insects which had then been eaten by small birds and had accumulated in their tissues. In turn these had been caught and eaten by birds of prey and the pesticide interfered with their reproduction—in particular causing the eggs to have thin shells and be more likely to break. The use of DDT was banned in the United States in 1972. Programs were developed to breed peregrines in captivity and at least 4,000 peregrines were bred and released to the wild. Peregrines are now breeding successfully over much of the United States and are no longer considered an endangered species. In Britain recovery has been so successful that the peregrine is becoming regarded as a pest by pigeon fanciers and lovers of the smaller songbirds.

It was possible to identify DDT pollution as a cause of eggshell thinning because egg shells had been collected as dated specimens in museums and private collections. A measure of eggshell thickness in collections of eggs of the sparrow hawk (*Accipiter nisus*) showed a sudden stepwise fall of 17 percent in 1947, when DDT began to be used widely in agriculture, and a steady increase in thickness since DDT was banned (see graph).

It was a surprise to ornithologists in Britain to find evidence of a decline in eggshell thickness of 2–10 percent in four species of thrush (*Turdus*) since the mid-19th century (Green, 1998). This seemed to have started long before the development of organic pesticides and there was no sudden change when DDT was introduced. Snails are an important part of the diet of thrushes and they derive much of the calcium for their egg shells from the snails. There is convincing evidence that acid rain (see Section 13.4.3) since the Industrial Revolution has acidified leaf litter and reduced its calcium content, leading to a reduction in snail populations and to the calcium content of their shells. The shells of wild birds' eggs have therefore recorded two of the major, but quite different, forces of environmental pollution—pesticides and acid rain.

the passions associated with insecticides. However, conservationists worry about the loss of many species of "weed" that are the food hosts for larvae of butterflies and other insects and whose seeds form the main diet of many birds. A recent development has been the genetic modification of crops such as soybean to produce resistance to the nonselective herbicide glyphosate (see Section 1.3.4). When these specialized crops are sown, the farmer can use glyphosate to kill all species of weed within the crop, a strategy that is particularly worrying to conservationists because it removes even minor species that would not harm crop yield.

In Europe, many of the most beautiful cornfield weeds can now be found only in

Box 13.2 (cont'd)

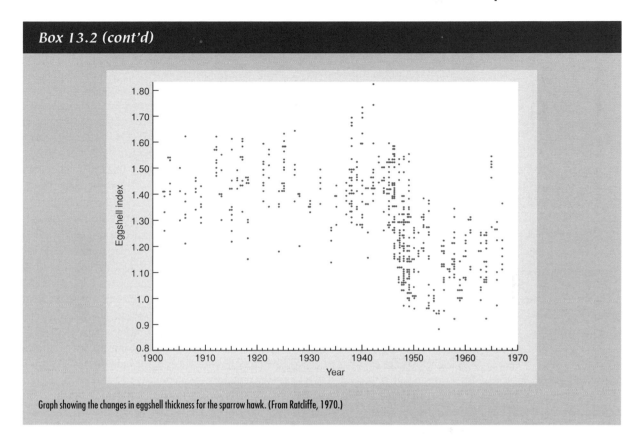

Graph showing the changes in eggshell thickness for the sparrow hawk. (From Ratcliffe, 1970.)

the more primitive agricultures of Central and Eastern Europe, while the glorious floristic diversity of Texan roadsides contrasts vividly with the crop monocultures at their side. It may be important to maintain roadsides, hedges, and field margins in an unpolluted condition if a residue of the natural diversity of fauna and flora is to persist.

13.4   Atmospheric Pollution

In many of the ways in which humans pollute their environment they are just another social animal, facing essentially the same problems of accumulating waste as ants, bees, and termites. But humans have added a type of pollution of a quite different order—we are unique in the way we obtain and manage external sources of energy. In its simplest manifestation this takes the form of using fire to cook food (rendering a wide variety of plants, and particularly seeds, digestible). The development of cooking was also the first step by which humans contributed more than just their own metabolic carbon dioxide to the atmosphere: this was now augmented by the carbon dioxide produced from burned fuel. To cooking was added in turn the burning of fuel to give warmth, bake clay, smelt ores, work metals, fix atmospheric nitrogen, drive machinery, and power transport. All these operations are peculiar to the human species and have given it unique powers to pollute the environment.

The rich biodiversity of agricultural weeds has been largely destroyed by the use of selective herbicides. This photograph shows a range of arable weeds that still persist in primitive peasant farms in Europe but are nearing extinction elsewhere.

humans are unique among animals
in using fuel to release energy for
work

The history of humans as a unique polluter can be written in terms of their primary sources of energy as they have changed from using wood to coal to oil to nuclear fuel. In theory, wood is a sustainable fuel, but in practice this is only true if the population that uses it remains small. As peasant populations grow, trees must be felled farther and farther from home. In parts of Ethiopia the women of a village may spend a day walking to the nearest source of wood and a second day carrying it back to the village. A major element of the Industrial Revolution was the switch from using wood to coal (and later oil) as a source of power. No fossil fuel is sustainable in the long term—like wood, both coal and oil are exhaustible resources as well as pollutants. Each type of energy source, including nuclear power, makes its own special contribution to pollution. Even hydroelectric, wind, and wave power, and photoelectric processes (which use photovoltaic cells to convert solar radiation directly into usable electric power) use machinery that has been made by polluting industries.

The use of fossil fuels has many adverse effects. For example, more than 4 million tons of oil enters the world's waterways every year; some as seepage from the ocean floor, some from industry (e.g., gasoline solvents), and more than 1 million tons from oil tankers or from wells drilled into the seabed. Oil in and on the sea affects wildlife in many ways. It reduces the level of aeration of the water, and it prevents light from penetrating the surface. Damage to invertebrates can be widespread, affecting chitons, mussels, and crustaceans, and encrusting species of hydroids and bryozoans as well as seaweeds and kelps. Feathers become choked with oil so that seabirds cannot fly and fish gills become coated and cease to function. The largest accident in the United States occurred on March 24, 1989, when the oil tanker *Exxon Valdez* ran aground in Prince William Sound, Alaska. It spilled nearly 50,000 tons of crude oil, which spread along the coast for nearly 1,000 km, contaminating the shores of a national forest, five state parks, four state critical habitat areas, and a state game sanctuary. The episode is believed to have killed 300 harbor seals, 2,800 sea otters, 250,000 birds, and possibly 13 killer whales. Many commercial fisheries were closed for a year or more because of the concern that fish caught in the area might find their way into the human food chain. By 1996, 28 species and resources were still listed as having failed to recover.

However, the most profound and far-reaching consequences of the use of fossil fuels are those of atmospheric pollution. From the mid-19th to the mid-20th century the burning of fossil fuels (together with deforestation) contributed about 9×10^{10} metric tons of carbon dioxide to pollute the atmosphere and a further 9×10^{10} metric tons has been added since 1950.

13.4.1 Carbon dioxide—a major atmospheric pollutant

The concentration of carbon dioxide (CO_2) in the atmosphere has increased from about 280 parts per million (ppm) in 1750 to about 370 ppm today, and it is projected to continue to rise, to 700 ppm by the year 2100 unless there are rather profound changes in human behavior. A remarkable census of atmospheric CO_2 was started in 1958 at Mauna Loa Observatory in Hawaii and this detected the extraordinary pattern shown in Figure 13.4. The principal cause of this increase has been the burning of fossil fuels, which in 1980 released about 5.2×10^9 metric tons of carbon into the atmosphere (Table 13.2).

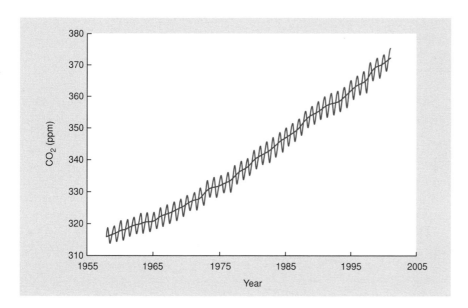

Figure 13.4
The concentration of atmospheric CO_2 measured at the Mauna Loa Observatory, Hawaii showing the seasonal cycle (resulting from changes in photosynthetic rate) and the long-term increase that is due largely to the burning of fossil fuels. (Courtesy of the Climate Monitoring and Diagnostics Laboratory [CMDL] of the National Oceanic and Atmospheric Administration [NOAA].)

Table 13.2

Balancing the global carbon budget (in metric tons per year) in 1980 to account for increases in atmospheric carbon caused by human activities. In the row labeled "Missing" the minus sign indicated the need to identify an unknown uptake of carbon of the size shown. This has now been identified as fertilization of terrestrial vegetation by atmospheric CO_2 so that an increase of the order of what was estimated as "missing" can be accounted for by an increase in the amount of carbon locked in extra vegetation biomass (Kicklighter et al., 1999). (After Detwiler & Hall, 1988.)

	Extreme low estimate	Median estimate	Extreme high estimate
Release to atmosphere			
Fossil fuel combustion	4.7	5.2	5.7
Cement production	0.1	0.1	0.1
Tropical forest clearance	0.4	1.0	1.6
Nontropical forest clearance	−0.1	0.0	0.1
Total release	5.1	6.3	7.5
Accounted for			
Atmospheric increase	−2.9	−2.9	−2.9
Ocean uptake	−2.5	−2.2	−1.8
Missing?	−0.3	+1.2	+2.8

The clearing and burning of tropical forest to make way for agriculture or timber production and the decay of the residues make a further contribution to atmospheric CO_2 (Table 13.2). A considerable amount of this is recaptured in photosynthesis by the succeeding vegetation (Kicklighter et al., 1999), but this is least when forest is converted to grassland, which has a much lower biomass. In total about 1.0×10^9 metric tons per year has been estimated to be released through changes in tropical land use

... due to the burning of fossil fuels and deforestation

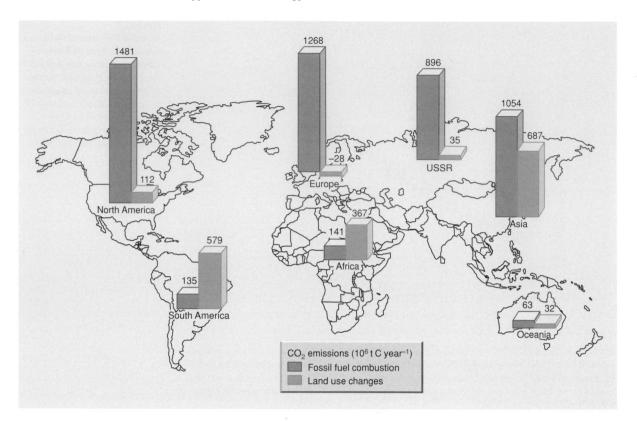

Figure 13.5
Emissions of CO_2 in 1980 resulting from fossil fuel consumption and land-use changes (mainly forest clearance). (After UNEP, 1991.)

(Detwiler & Hall, 1988). This calculation was made for 1980, and the figure for tropical forest clearance must now be significantly greater as a result of the uncontrollable spread of forest fires in Indonesia and in South America following the droughts associated with the El Niño phenomenon of 1997–8.

Figure 13.5 shows the overwhelming contribution made to atmospheric CO_2 by the combustion of fossil fuel in northern industrial countries and the relatively smaller contribution from changes in land use in the tropics and subtropics. Note that estimates of the relative contribution of industrial countries varies considerably, both in total and in terms of the use of power by cars and other transportation, in manufacturing, in the service sector (buildings not involved in manufacturing), and in homes (Schipper et al., 2001). On a per capita basis, the United States, Australia, and Canada emit the most carbon (more than 3 tons of carbon per head of population per year) and Norway the least (about 1 ton per capita) (Figure 13.6).

It should be possible to draw up a balance sheet that shows how the CO_2 produced by human activities translates into the changes in concentration in the atmosphere. Table 13.2 is an attempt to do this, but it exposes a problem. Human activities release $5.1–7.5 \times 10^9$ metric tons of carbon to the atmosphere every year. But the increase in atmospheric CO_2 (2.9×10^9 metric tons) accounts for only 60 percent of

Figure 13.6
International comparisons of CO$_2$ emissions estimated for 1994 in four activity categories, expressed as tons of carbon emitted per head of population. (Courtesy of the International Energy Agency, after Schipper et al., 2001.)

this, a percentage that has remained remarkably constant for 40 years (Hansen et al., 1999). The oceans absorb CO$_2$ from the atmosphere, and it is estimated that they may absorb 1.8–2.5×10^9 metric tons of the carbon released by human activities. The balance sheets in Table 13.2 for the year 1980 did not add up. More recent analyses indicate that terrestrial vegetation has been "fertilized" by the increased atmospheric CO$_2$ so that a considerable amount of extra carbon has become locked up in vegetation biomass (Kicklighter et al., 1999).

13.4.2 The greenhouse effect

The atmosphere behaves as a greenhouse: during the day solar radiation warms up the Earth's surface, which then reradiates energy outward, principally as infrared radiation. Carbon dioxide, nitrous oxide, methane, ozone, and chlorofluorocarbons (CFCs) absorb infrared radiation and, like the glass on a greenhouse, prevent the radiation from escaping and keep the temperature high. The concentrations of all these gases have increased rapidly since the Industrial Revolution. It appears that the present air temperature at the land surface is $0.6 \pm 0.2°C$ warmer than in preindustrial times, and temperatures are predicted to continue to rise by a further 1.4–$5.8°C$ by 2100 (IPCC, 2001). Such changes will probably lead to a melting of the ice caps, a consequent rising of sea level, and large changes in the pattern of global climates and the distribution of species. In addition, changes in temperature and global climate can be expected to influence world patterns of famine and disease.

Global warming is not evenly distributed over the surface of the Earth. Figure 13.7 shows the measured global change as the trends in surface temperature over the 46 years from 1951 to 1997. Areas in North America (Alaska) and Asia experienced rises of 1.5–$2°C$ in that period, and these areas are predicted to continue experiencing the fastest warming in the first half of the next century. In some regions the temperature has apparently not changed during this period (New York, for example) and should not change greatly in the next 50 years. There are also some areas, notably Greenland and the northern Pacific Ocean, where surface temperatures have fallen.

industrial gases accumulate in the atmosphere—and this accumulation leads to global warming

Figure 13.7
Change in the surface temperature of the globe expressed as the linear trend over 46 years from 1951 to 1997. The colour bar gives temperatures in °C. (From Hansen et al., 1999.)

vegetation has migrated since postglacial warming but has failed to keep pace with changing climates

the greenhouse effect warms 50–100 times faster than postglacial warming

Global temperatures have changed naturally in the past, most notably during the recurring ice ages. We are currently approaching the end of one of the warming periods that started 20,000 years ago and during which global temperatures have risen by about 8°C. The greenhouse effect adds to global warming at a time when temperatures are already higher than they have been for 400,000 years. Buried pollen gives us evidence of past changes in vegetation and shows that North American forest boundaries have migrated north at rates of 100–500 m yr^{-1} since the last ice age. However this rate of advance has not been fast enough to keep pace with postglacial warming, and the present pattern of forest boundaries is still catching up with postglacial warming. The rate of global warming forecast to result from the greenhouse effect is 50–100 times faster than that of postglacial warming!

Of all the types of environmental pollution caused by human activities none may have such profound effects. We must expect latitudinal and altitudinal changes to species' distributions and widespread extinctions as floras and faunas fail to track and keep up with the rate of change in global temperatures (Hughes, 2000). Extinctions will almost certainly be much faster than after glacial cycles because humans have now fragmented the landscapes and destroyed the continuity of migration routes.

The growth rate at which greenhouse gases are forcing climatic change may have reached its peak in the late 1970s. Since then the damaging effects of CFCs have been recognized and substitutes have been found. The growth rate of methane pollution has also fallen quite sharply—perhaps because the area of land under rice cultivation has not been increasing so fast. Moreover, the growth rate of carbon dioxide in the atmosphere has fallen slightly below prediction, perhaps because the oceans absorb more than had been estimated (Hansen et al., 1999). However, there is no reason for complacency—greenhouse levels are still increasing, though perhaps not quite so fast.

13.4.3 Acid rain

Of the pollutants that humans release into the atmosphere most are returned to Earth, about half as gases or particles and half dissolved or suspended in rain, snow, and fog. They may be carried in the wind for hundreds of kilometres across state and national

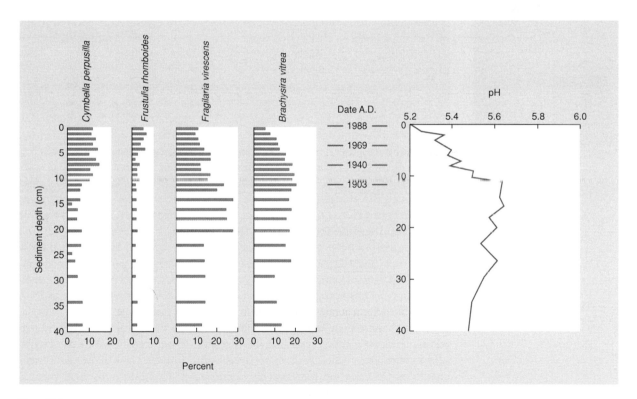

Figure 13.8
The history of the diatom flora of an Irish lake (Lough Maam, County Donegal) can be traced by taking cores from the sediment at the bottom of the lake. The percentage of various diatom species at different depths reflects the flora present at various times in the past (four species are illustrated). The age of layers of sediment can be determined by the radioactive decay of lead-210 (and other elements). We know the pH tolerance of the diatom species from their present distribution and this can be used to reconstruct what the pH of the lake has been in the past. Note how the waters have been acidified since about 1900. The diatoms *Fragilaria virescens* and *Brachysira vitrea* have declined markedly during this period while the acid-tolerant *Cymbella perpusilla* and *Frustulia rhomboides* have increased after 1900. (After Flower et al., 1994.)

oxides of sulfur and nitrogen form
acids in the atmosphere ... and fall
as acid rain

borders, and when they cause harm they can be the source of bitter international dispute. Atmospheric pollution by sulfur dioxide (SO_2) and oxides of nitrogen (NO_X) interact with water and oxygen in the atmosphere to form dilute sulfuric and nitric acids, which fall as "acid rain".

Rain water has a pH of about 5.6, but pollutants lower it to below 5.0 and values as low as 2.4 have been recorded in Britain, 2.8 in Scandinavia, and even 2.1 in the United States. Acid rain acidifies the water in lakes and streams, especially when the composition of the underlying soil and rock does not help to neutralize the acidity. A high concentration of hydrogen ions may itself be toxic, but changes in the availability of nutrients and other toxins are usually more important. At a pH below 4.0–4.5 the concentrations of aluminum (Al^{3+}), iron (Fe^{3+}), and manganese (Mn^{2+}) become toxic to most plants and to aquatic animals that expose delicate tissues directly to the water (such as the gills of fish). Acid rain is most damaging in water that is already naturally acidic: it may then lower the pH so far that it sterilizes the environment for many of the native species (e.g., Figure 13.8).

Many of the most visibly dramatic effects of acid rain have been observed in the forests of Central Europe where industry has depended on low-quality coal with a high sulfur content and forest die-back on a massive scale has resulted. For example, in the Czech Republic nearly 60 percent of the forests have been damaged or destroyed. Even in the United States, high elevation spruce forests have been affected, including the Shenandoah and Great Smoky Mountain National Parks (Figure 13.9).

There are two main responses to the acid rain problem: reduce its origin or soften its consequences. The burning of fossil fuel to generate electricity in the United States accounts for about 70 percent of the SO_2 and 30 percent of the NO_X emitted to the atmosphere. SO_2 emissions from electric generators can be reduced by switching from coal (1–5 percent sulfur) and oil (2–3 percent sulfur) to gas, which has a lower sulfur content. Alternatively, when coal is burned the sulfur can be extracted before or after burning but in either case the extracted sulfur (e.g., as calcium sulfate) is itself a potential pollutant. The process uses limestone, and when this started to be used on a large scale to reduce sulfur acid emissions there were angry protests at the quarrying that damaged landscapes and conservation sites!

The Clean Air Act in the United States, which is designed to reduce annual emissions of SO_2 (and NO_X) by power plants, sets an interesting, peculiarly American, model for pollution control called *allowance trading*. Each power plant was allocated an allowance of units based on its past record of fuel consumption and emission. A unit allows 1 metric ton of SO_2 to be emitted in a year. The units can be bought, sold, or banked, and this system gives a financial incentive to the power company to control its emissions. If it curtails them, there will be units of allowance that can be sold to another company with more lax controls. In effect, this allowance system confers market costs and values on environmental pollution. In most other parts of the world, when governments intervene to reduce emissions, they do so by taxing or fining the polluter.

The acidification of lakes can be reduced by applying lime, but like almost all measures that control pollution this creates environmental problems of its own. The necessary lime, which has to be quarried from limestone, requires a great deal of heat and contributes CO_2 to the atmosphere.

Figure 13.9
Acid rain damage to spruce forest.
(© Rob and Ann Simpson, Visuals
Unlimited.)

liming can reduce lake acidity—but
causes new problems

13.4.4   Thinning of the ozone layer

Ozone is produced by the influence of sunlight on oxygen and during the oxidation of carbon monoxide and hydrocarbons such as methane. It has three very different roles in environmental pollution. The first two are negative, in the sense that undesirable "polluting" consequences occur as the concentration of ozone increases. First, in atmospheres polluted with methane, industrial hydrocarbons, NO_X gases, and carbon monoxide, ozone can reach concentrations that are toxic to plants and that contribute to smog. Second, ozone is also a greenhouse gas, though it is not particularly significant in this respect.

ozone can have adverse consequences locally . . .

However, ozone also accumulates as a layer in the upper atmosphere; this "ozone layer" is beneficial. It absorbs most of the ultraviolet (UV) radiation (wavelength 200–300 nm) incident on the Earth's upper atmosphere and so makes the Earth habitable for plants and animals. The frequency of skin cancer among humans has focused attention on the damage caused by exposure to the sun and on the importance of the stability of the ozone layer.

. . . but in the upper atmosphere it shields the Earth from damaging UV radiation

Evidence that nitric oxide produced by supersonic aircraft might contribute massively to reduce the level of atmospheric ozone led to the halting of their large-scale development. However, that has by no means been the end of the story. CFCs had been developed as aerosols and refrigerants and have been used on a very large international scale. It became clear that these posed the threat that their chlorine content could interact with and destroy atmospheric ozone. International agreements to phase them out have already had effects. Similarly, methyl bromide, a soil fumigant used on a large scale, especially in southern Europe, was recognized to be an important pollutant because it destroys ozone in the upper atmosphere. The plan to ban its use after 2001 could make the largest single contribution to protecting the ozone layer.

Ozone chemical processes are very complicated, and methane, nitrous oxide, and carbon monoxide may all play a part in its decomposition. The upper atmosphere is not the easiest place in which to study the chemical characteristics of gases! But pollution of the upper atmosphere poses questions of the greatest significance for environmental scientists, especially since the discovery that the concentration of ozone in the atmosphere over Antarctica (Figure 13.10) had started to decline by 1978 and was doing so very rapidly after 1982. It is clearly in the interests of humans and probably most other organisms that ozone concentrations should remain low close to home (e.g., minimizing smog) but high in the upper atmosphere, and that we should find out how to insure this.

chlorine compounds and other pollutants decompose ozone in the atmosphere and need to be phased out

UNANSWERED QUESTION: is it possible to minimize ozone concentrations close to home but to stabilize them high in the atmosphere?

13.5   Nuclear Radiation

Fossil fuels are exhaustible, are increasingly costly to extract, pollute the atmosphere, and contribute to global warming. Humans have been slow to develop the full potential for hydroelectric power or power harvested from nonpolluting wind, wave, and tidal energy (but note again that the engineering work needed to harvest such sources of energy is certainly not pollution free). We have been quicker to exploit energy from nuclear fission, a source that in the beginning was viewed as an almost ideal, long-term

Figure 13.10
Thinning of the ozone layer over the
Antarctic.

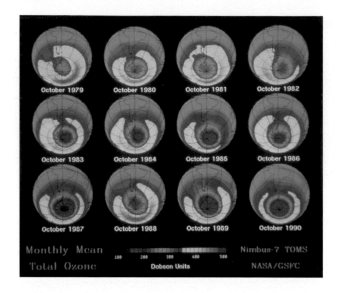

source of industrial, domestic, and military power. However, the view that the release of radiation could be readily controlled faded rapidly. Some leakage occurs from nuclear power reactors, and it is doubtful whether the reprocessing of waste nuclear fuel can ever be made completely clean. Moreover, the polluting power of radioactive wastes has a timescale that may be orders of magnitude greater than that of other human pollutants. For example, plutonium-239 has a half-life of about 25,000 years. Plutonium is separated and recovered from the spent fuel in nuclear reactors and stocks are expected to have risen to more than 100 metric tons by 2010. Ways have to be found to protect against the risks of leakage over this sort of timescale, perhaps by burial in deep mines after incorporation in glass. In the longer term it may become necessary to devise reactions that can burn it more efficiently or technology that can fire such pollutants into outer space or into the sun (the ultimate dilution!).

natural background radiation and that produced by human activities are of similar magnitudes

The radiation received by an organism arises from human activities (nuclear warfare, leakage from nuclear plants, accidents such as the Chernobyl meltdown, and medical use) together with a very similar-sized contribution from "background radiation"—mainly cosmic rays, the radioactive decay of materials such as radium and thorium in the Earth's crust, and radioactive isotopes that are present in food and that decay within the body (Table 13.3). It is a sobering thought that the total radiation given to a cancer tumor in a patient is many thousand times greater than the total normal exposure from the combined natural and artificial background radiation. Background radiation is not evenly distributed. Cosmic rays are more intense at high altitudes, and the radioactive gas radon, which is formed by the decay of radium in the soil, can be a significant source of radiation over certain rock formations, from which it may leak into households.

different isotopes follow different pathways and become concentrated in different tissues

Some radioisotopes become concentrated in particular tissues. Iodine-131, a product of fallout of nuclear explosions that emits dangerous β- and γ-radiation, is preferentially absorbed and accumulated by the thyroid gland, where it reaches concentrations 100 times higher than in the rest of the body. It has a half-life of only 8 days,

Table 13.3
Estimates of the average annual dose from various sources of irradiation received per human per year by people in the United States.
(From Encyclopaedia Britannica, © 1994–8.)

Source of radiation	Average dose rates (microsievert/year)	% of annual total
Natural environmental		
Cosmic radiation	0.27–1.30	13.7
Terrestrial radiation	0.28–1.15	14.2
Internal radioactive isotopes (K^{40}, C^{14}, etc.)	0.36	18.3
Subtotal	0.91	46.2
Human-made environmental		
Technologically enhanced	0.04	2.0
Global fallout	0.04	2.0
Nuclear power	0.002	0.1
Medical diagnostic	0.79	39.6
Radiopharmaceutical	0.14	7.0
Occupational	0.01	0.5
Miscellaneous	0.05	2.5
Subtotal	1.06	53.7
Total	**1.97**	**100**

but the high local concentration can lead to thyroid cancer, particularly in children. In contrast, cesium-137 and strontium-90 have much longer half-lives (37 and 28 years, respectively). Released into the environment in nuclear warfare and by leakage from nuclear industry, they are similar to major plant nutrients (cesium is similar to potassium and strontium to calcium) and as a result are readily absorbed by plants from the soil. They then pass into meat and milk and so into the diet of humans, to whom they can be harmful. Strontium, for example, may be laid down in bone as if it was calcium. Radioactive isotopes that enter rivers, lakes, and oceans are also absorbed and concentrated up food chains and may ultimately find their way into human diets.

A major accident in 1986 at a nuclear power station at Chernobyl in the Ukraine released 50–185 million curies of radionucleides into the atmosphere. Close to the explosion, 32 deaths occurred within a very short time. Farther away, individuals contracted radiation sickness and some died. Effects in the locality have continued to appear—livestock have been born deformed, and thousands of radiation-induced illnesses and deaths of cancer are expected in the longer term. Farther afield, wind-dispersed atmospheric pollution from Chernobyl was detected in Sweden 3 days after the accident. Fallout also reached the British Isles. Figure 13.11 shows the persistence of cesium-137 in the acid soils of the northwest of Britain, where it has been absorbed by plants and eaten by sheep. The sale of sheep for food was still banned more than 10 years after the accident because of the persistence of the isotope at dangerous levels.

Chernobyl is the only major nuclear disaster that has occurred to date, excepting the atomic bombs dropped on Hiroshima and Nagasaki, which released less than half as much radiation (though with far greater loss of human life). However, there have

Chernobyl—the worst nuclear pollution disaster so far

Figure 13.11
An example of long-distance environmental
pollution. The distribution in Great Britain of
fallout of cesium-137 from the Chernobyl
nuclear accident in the Soviet Union in 1986.
The contours show the persistence of the
cesium on acid upland soils where it is
recycled through soil, plants, and animals.
On typical lowland soils cesium does not
persist in food chains. (After NERC, 1990.)

been "near misses", most dramatically in 1979 at Three Mile Island, near Harrisburg, Pennsylvania, when a reactor core was partly melted and large amounts of radioactive material were released—but into a special containment structure that had been specifically designed to deal with such a happening.

species differ in their sensitivity to nuclear radiation

The effects of environmental pollution are usually described in terms of the harm they cause humans. Taking a broader, ecological view, a benchmark experiment was performed in an oak/pine forest at the Brookhaven National Laboratory, on Long Island, New York (Woodwell, 1970) (Figure 13.12). A 9,500 Ci radiation source of cesium-137 was exposed in the forest in the fall of 1961. Trees close to the source soon died, and over time a concentric pattern of vegetation developed around the source,

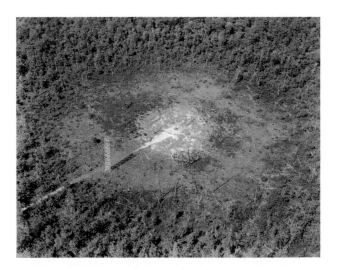

Figure 13.12
Photograph of the Brookhaven experiment,
showing the nuclear pile in the woodland
and the changes in the vegetation around it
(Courtesy of George M. Woodwell).

reflecting different sensitivities to radiation damage. Some mosses colonized ground
close to the source. Farther away annuals and then perennials persisted and shrubs and
trees formed the most distant zone. It was very noticeable that the species most tolerant
to radiation were also those that were quickest to establish after other natural hazards
in the forest, such as fire and hurricane. Moreover, the order of tolerance to radiation
was generally similar to the order of tolerance to drought, frost, and fire. The most
tolerant organisms were usually small, rapidly reproducing species—the ones generally regarded as r-strategists (see the discussion in Section 5.6).

13.6 Mining and Quarrying

Many animals are ecological engineers: they dig, burrow, and build and can change the
physical character of their environment (Jones et al., 1997). Human engineering is on a
grander scale: quarrying, building, and even manipulating whole landscapes as in
contour agriculture, the Great Wall of China, the Egyptian pyramids, and Stonehenge.
These examples of human engineering on a massive scale were powered, like the
ecological engineering of other species, solely by metabolic energy. The work was done
by the muscles of humans and their draft livestock using only energy derived from their
food. It was, in a sense, no more polluting of the environment than the natural burrowing activities of gophers and moles or the architectural creations of termites. The ability
of humans to use nonmetabolic energy sources is responsible for the unique scale of
modern engineering. In particular the ability to control and use fire made it possible for
humans to extract metals from their ores and so ultimately made the Industrial
Revolution possible.

Metals were first used by humans in the late Stone Age, about 6,500 years ago.
Gold, silver, and copper were the first metals used; they are easy to extract because they
exist in nature as the metals themselves rather than as chemical compounds. Nuggets
of pure metallic gold were found in riverbeds and were beaten and molded for decora-
tion. Once such metals were valued it was an obvious step to dig and mine for them,

*quarrying and mining are "ecological
engineering" on a vast scale*

and from that point almost every phase in the extraction and industrial use of metals involves a sequence of phases of environmental pollution.

Each type of metal has its own peculiarities. In this chapter the mining and purification of copper are used to illustrate pollution by the extraction of metal. Copper is present in deposits either as the metal or as copper sulfide or oxide. Like most metal deposits it usually exists in a mixture with other metals, some of which may be worth saving (e.g., gold), whereas others are discarded in more or less hazardous waste.

The mining industry may pollute at every stage of extraction, purification, and disposal:

■ Mining or quarrying exposes the metal and its ores. Many of the world's copper reserves are close to the surface and are easily extracted by open-cut mining: the copper mines of Bougainville (Solomon Islands, Papua New Guinea) and of Utah are among the largest human scars on the Earth's surface (Figure 13.13).
■ The ores are crushed and finely ground. This processing immediately exposes ores to the elements, and even after the best has been extracted the residues are copper-rich and the metal leaches as toxic waste into rivers and lakes. Waters close to copper mines are commonly brilliantly blue–green with copper salts and are quite sterile.
■ The finely ground ore is agitated in water, and the metal becomes concentrated in the froth and dried to a cake. The remainder, which is still rich in copper, may be further concentrated to recover more of the metal. Ultimately water and solid "tailings" have insufficient copper to warrant further extraction but contain sufficient copper to form a hazardous and polluting waste.

Figure 13.13
Binyon Canyon Mine, Utah. A toxic and sterile environment created by the world's largest excavation (© David R. Frazier).

- The concentrate is then roasted to 1,230–1,300°C, polluting the atmosphere by burning of the necessary fuel. The roasting drives off a host of pollutants such as arsenic, mercury, and sulfur into the atmosphere. Further refining at high temperatures removes more of the sulfur (as sulfur dioxide) and yields metal copper of 99.5 percent purity.
- The copper can now be purified by electrolysis, which leaves most of the other metals in a sludge that may be further purified (to remove gold, for example) but ultimately contributes yet more toxic waste.

There are many variations in the commercial process of copper extraction, smelting, and refining, and some have been designed to reduce pollution. For example, some of the SO_2 can be removed and converted to sulfuric acid that can then be used in the electrolytic phase.

Each type of metal presents its own peculiar problems to the mining and smelting engineer, and increasingly the levels of environmental pollution are controlled by law to insure low rates of pollution and efficient clean-up. There is, however, a history of accumulating hazardous waste from before the time when communities became sensitive to environmental pollution. The Environmental Protection Agency (EPA) administers these historic problem sites in the United States (by means of the Superfund program) and funds pollution control through a tax on the chemical and petroleum industries.

The major role of some metals as environmental pollutants occurs after they have been purified and used industrially and are then released into the environment as industrial waste. Lead and mercury are particularly striking examples. Lead became an environmental pollutant from the moment that the Romans started to use it to make water pipes and so started to pollute their drinking water. Though lead is ranked by the EPA as number 1 in their list of 275 hazardous substances, 3.8 million children in the United States still drink lead-contaminated water. Lead poses a particular risk for the development of the nervous system in young children and in the fetus and is being phased out of many commercial uses. Where lead tetraethyl is still used as an additive to gasoline, lead pollutes the atmosphere, especially in cities.

It is not clear whether lead pollution has significant consequences for wildlife on the land or in aquatic environments, but it does not appear to become concentrated along food chains. This is a major contrast with mercury.

Mercury is used in a variety of specialized applications in industry and medicine, in electric switches, batteries, fluorescent and mercury vapor lights, thermometers, barometers, and dental amalgams. The main culprits in releasing mercury to the atmosphere are, in order of importance, coal-fired power plants, medical waste incinerators, municipal waste incinerators, and industrial boilers. In the natural environment mercury can be converted by microbial activity to methylmercury, a form that is readily absorbed and accumulated up food chains, especially in lakes and estuaries. Fish, the top predators, may accumulate concentrations of mercury 10,000–100,000 times that in the surrounding water (Bowles et al., 2001). Native peoples who hunt and eat wildlife can accumulate even higher concentrations. Mercury is a serious poison that can cause permanent damage to the human brain and kidneys, and particularly to the developing fetus. It may also damage the immune system.

the disposal of hazardous waste may be regulated by legislation

lead and mercury can be especially dangerous pollutants

13.7 Restoration Ecology

Ecologists are concerned to understand the relationship between organisms and their environment, and as ecological science matures they are becoming better able to manage this relationship. *Restoration ecology* is a discipline directed at managing communities that have been damaged by pollution. Management may aim to restore communities to their prepollution state or to create new types of community that can tolerate the pollution.

After an oil spill, for example, the aim of the restoration ecologist is usually to remove the polluting oil and help the marine communities to recover to their original condition. Cleaning up after a major spill is peculiarly difficult. Physical removal is the preferred option, but it can be enormously costly. The oil can sometimes be skimmed off, but only in calm water. Straw and ash can be spread on the water to soak up the oil, but then the problem of dealing with the oil-soaked residue arises! In Europe, detergents have been sprayed onto oil slicks to disperse the oil, but the detergent probably does more harm than good to the biota of the polluted environment. Seaweeds and kelps and their fauna cannot be sown like a grass sward, and so recovery of the flora and fauna has to await dispersal from undamaged parts of the coastline.

Restoration ecologists have been particularly successful at reintroducing species that have been damaged by pollution (e.g., Box 13.2) and revegetating open scars on the landscape caused by mining and quarrying. The latter may involve deliberately sowing species or varieties of plants that tolerate toxic materials (see the discussion in Section 2.3.2 of the evolution of genetic strains of grasses and other plants that tolerate high concentrations of copper and other heavy metals). Often, however, it is the sheer absence of soil and its nutrients (particularly nitrogen) that prevents mines and quarries from being naturally colonized by vegetation.

Ecologists and managers now appreciate the role played by wetlands and riparian (bankside) vegetation in reducing the transport from land to river of soil and organic particles and, more particularly, in absorbing nutrients from enriched agricultural, forestry, and urban runoff. Economists are beginning to make some tentative valuations of such natural water-purifying "ecosystem services", in multimillion dollar annual estimates. Innovative water managers are restoring marshy areas (drained long ago for agricultural or urban development) by building artificial wetlands consisting of a linked series of treatment and enhancement marshes, through which secondarily treated sewage is transformed by natural vegetation into a progressively cleaner form. In essence, these artificial wetlands are being used for tertiary treatment of sewage (Section 13.2.2), and in the process provide an environment for wildlife and recreation in an urban setting.

Prevention of pollution is, of course, better than cure. Most of the major disasters of environmental pollution have been caused by human decisions to take high risks and shortcuts. But it may be as unrealistic to imagine that such human behavior can change as to suppose that we will learn to prevent earthquakes, volcanoes, and hurricanes.

Summary

Waste from humans, their livestock, and other species

Almost all functions of the human body have been regarded historically as polluting the environment. Feces, urine, and corpses are primary pollutants in the crowded environments of most social animals where waste is produced faster than it can be dealt with by natural decomposition. Patterns of behavior have evolved that reduce pollution within the community, for example, fecal sacs of nesting birds, latrines of social mammals, removal of corpses and other waste from the nests of colonial insects. All of these activities have their parallels in human societies, in rituals of burial and in the technology for disposing of feces and urine, that contribute to community hygiene and help to prevent the spread of disease.

Sewage systems use water to carry human excreta away from crowded communities and accelerate its decomposition by managed microbial populations. The resultant sludge is dumped at sea, buried, or used as fertilizer. The intensive production of livestock in factory farming is seriously polluting, and agricultural slurry may need to be thinly dispersed over extensive farmland to dilute it to a level that the natural microbial flora can decompose and then release as fertilizer.

Nitrates and pesticides

Nitrates released from human or agricultural waste may leach into potential drinking waters and create a health hazard; they are expensive to remove. Natural vegetation can absorb nitrates and maintain low concentrations in soil and water. But both agriculture and forestry make ecosystems "leaky" and allow nitrates to leach more readily into drainage (potential drinking) water.

Pesticides can be serious environmental pollutants, especially to wildlife. The greatest damage is caused by accidents and misuse and by those pesticides (like chlorinated biphenyls) that accumulate along food chains and reach very high concentrations in top predators.

Fossil fuels and atmospheric pollution

Humans are unique polluters of the environment because they burn fossil fuels to give warmth, cook, smelt metals, fix atmospheric nitrogen, drive machinery, and power transport. Each of these activities pollutes the environment. Deforestation for wood, mining for coal, and drilling (and spillage) of oil all harm the environment and contribute in various ways to pollute the atmosphere with CO_2, SO_2, oxides of nitrogen (NO_X), and hydrocarbons.

Atmospheric pollutants contribute to the "greenhouse effect" by which the Earth's atmosphere hinders radiation away from the Earth and forces global warming. Predicted rates of warming are 50–100 times faster than that experienced during the past 20,000 years of global warming and are expected to lead to widespread migration and extinctions of species.

Atmospheric pollution by SO_2 and NO_X leads to the deposition of acid rain, which can be toxic to vegetation either directly or by changing soil pH and the toxicity of mineral ions. The Clean Air Act aims to control such pollution in the United States.

Atmospheric pollution by industrial products, especially the fumigant methyl bromide and chlorine compounds, such as chlorofluorocarbons, reduces the ozone concentration in the upper atmosphere. This allows increased carcinogenic UV radiation to reach the Earth's surface. Thinning of the ozone layer was first observed in 1978 and has increased rapidly. Use of the more guilty chemicals is being phased out internationally.

Nuclear radiation

Fossil fuels are exhaustible: nuclear fission provides an alternative source of industrial energy as well as a powerful military weapon. Theoretically it produces little pollution, but in practice power stations leak and human failure has lead to major disaster (e.g., Chernobyl). Bioaccumulation of radioactive isotopes such as iodine, cesium, and strontium intensifies the radiation hazard above that present in the environment.

Mining, quarrying, and restoration

Mining and quarrying for metals and their extraction and purification result in severe environmental pollution at every stage of processing. Copper, lead, and mercury are especially severe pollutants of land and water. Mercury (which bioaccumulates) and lead pose severe health hazards.

A science of restoration ecology in which fundamental ecological science is applied to minimize environmental pollution and restore a fauna and flora to polluted land, mines, shores, and estuaries is developing.

Review Questions

1 What are the features that distinguish human pollution of the environment from that by other social organisms?

2 Explain why it may be impossible to achieve increasing agricultural production without creating unacceptable levels of nitrate in drinking water.

3 ▲ Consider the toilet that you most frequently use. Find out where your sewage goes and how it is treated. What pollution problems are you contributing to as a result of your sewage disposal?

4 Describe the causes of acid rain and the way in which it damages terrestrial and aquatic communities.

5 ▲ Hydroelectric schemes provide one of the least polluting ways of generating power. Nevertheless, they have a number of negative effects on natural systems. What are they?

6 Define the characteristics that make some pesticides particularly dangerous pollutants.

7 Describe the ways in which the use of metals by humans has created problems of environmental pollution.

8 Define the greenhouse effect and list the pollutants that contribute to it.

9 ▲ Is it sensible to treat environmental pollution as more than just another density-dependent effect of population growth? Discuss why you consider it different—or do not.

10 ▲ It is often argued that environmental pollution can be prevented only by making it costly to the polluter. Discuss the ways in which this is, or might be, done.

▲ = Challenge Question

Chapter 14

Conservation

Natural ecosystems have been placed at threat by a plethora of human influences, particularly in the face of a burgeoning human population. Conservation is the science concerned with increasing the probability that the Earth's species and communities (or, more generally, its biodiversity) will persist into the future. We need to appreciate the scale of the problem, understand the threats posed by human activities, and consider how our knowledge of ecology can be brought to bear to provide remedies.

Chapter Contents

14.1 Introduction

14.2 Threats to species

14.3 Threats to communities

14.4 Conservation in practice

Key Concepts

In this chapter you will

- appreciate that in seeking to conserve the Earth's species and communities, we are often woefully ignorant of what there is to conserve
- understand that endangered species are usually rare but not all rare species are endangered
- recognize that a variety of human influences, including overexploitation, habitat disruption, and the introduction of exotic species, have made species rarer and increased their probability of extinction
- appreciate that populations that become very small may experience genetic problems
- understand that some species are at risk for a single reason but often a combination of factors is at work
- appreciate that conservation involves the development of species management plans but also often requires a broader, community perspective

14.1 Introduction

The term "biodiversity" makes frequent appearances in both the popular media and the scientific literature—but it often does so without an unambiguous definition. At its simplest, it is species richness, the number of species present in a defined geographical unit (Chapter 10). Biodiversity, though, can also be viewed at scales smaller and larger than the species. For example, we may include genetic diversity within species, perhaps seeking to conserve genetically distinct subpopulations and subspecies. Above the species level, we may wish to insure that species without close relatives are afforded special protection, so that the overall evolutionary variety of the world's biota is maintained as large as possible. At a larger scale still, we may include in biodiversity the variety of community types present in a region—swamps, deserts, early and late stages in a woodland succession, and so on. Thus, "biodiversity" may itself, quite reasonably, have a diversity of meanings. Yet it is necessary to be specific if the term is to be of any practical use. Ecologists and conservationists must define precisely what it is they mean to conserve in their particular circumstances, and how to measure whether this has been achieved.

Most often the focus of concern of conservation biologists is the rate of extinction of species in the face of human influence. To judge the scale of this problem, we need to know the total number of species that occur in the world, the rate at which these are going extinct, and how this rate compares with that of prehuman times. Unfortunately, there are considerable uncertainties in our estimates of all these things. About 1.8 million species have so far been named (Figure 14.1), but the real number must be much larger. Estimates have been derived in a variety of ways. One approach, for example, uses information on the rate of discovery of new species to project forward, group by taxonomic group, to a total estimate of up to 6–7 million species in the world. However, the uncertainties in estimating global species richness are profound and our best guesses range from 3 to 30 million or more (Gaston, 1998).

An important lesson from the fossil record is that the vast majority of (probably all) species eventually become extinct—more than 99 percent of species that ever existed are now extinct. However, given that individual species are believed, on average, to have lasted about 1–10 million years, and if we estimate conservatively that the total number of species on Earth is 10 million, we would predict that only an average of between 100 and 1,000 species (0.001–0.01 percent) would go extinct each century. The current observed rate of extinction of birds and mammals of about 1 percent per century is 100–1,000 times this "natural" background rate. Furthermore, the scale of the most powerful human influence, that of habitat destruction, continues to increase, and the list of endangered species in many taxa is alarmingly long (Table 14.1).

The evidence, then, while inconclusive to a degree because of the unavoidable difficulty of making accurate estimates, suggests that our children and grandchildren may live through a period of species extinction comparable to the "natural" mass extinctions evident in the geological record (see Section 10.6). But should we care? To many, the answer is a resounding and unhesitating "yes". Whether the answer seems obvious or debatable, however, it is important to consider *why* we should care—*why* biodiversity is valuable (Box 14.1).

Figure 14.1
Globally, about 1.8 million species have so far been named—these are unevenly distributed across taxonomic groups, with insects providing by far the greatest number. (After Alonso et al., 2001.)

Known spieces

4,000 species of bacteria

4,500 species of mammals

5,000 species of viruses

10,000 species of birds

12,000 species of amphibians and reptiles

22,000 species of fish

70,000 species of fungi

270,000 species of plants

400,000 species of invertebrates (excluding insects)

960,000 species of insects, approximately 600,000 of which are beetles

Conservation biology relies on an understanding of the threats faced by individual species (Section 14.2) and by whole communities (Section 14.3). After presenting this background, we consider in Section 14.4 the options open to conservation biologists to maintain or restore biodiversity.

14.2 Threats to Species

A basic aim of conservation is to prevent species from becoming extinct, either regionally or globally. But how do we define the risk of extinction that a species faces? A species can be described as:

classification of threat to species

- *vulnerable* if there is considered to be a 10 percent probability of extinction within 100 years,

Table 14.1

The numbers and percentages of named animal and plant species in major taxa judged to be threatened with extinction. The higher values associated with plants, birds, and mammals may reflect our greater knowledge of these taxa. (After Smith et al., 1993.)

Taxons	Number of threatened species	Approximate total species	Percentage endangered
Animals			
Molluscs	354	1.0×10^3	0.4
Crustaceans	126	4.0×10^3	3
Insects	873	1.2×10^6	0.07
Fishes	452	2.4×10^4	2
Amphibians	59	3.0×10^3	2
Reptiles	167	6.0×10^3	3
Birds	1,029	9.5×10^3	11
Mammals	505	4.5×10^3	11
Total	3,565	1.35×10^6	0.3
Plants			
Gymnosperms	242	758	32
Monocotyledons	4,421	5.2×10^4	9
Monocotyledons: palms	925	2,820	33
Dicotyledons	17,474	1.9×10^5	9
Total	22,137	2.4×10^5	9

Figure 14.2
Levels of threat as a function of time and probability of extinction. (After Akçakaya, 1992.)

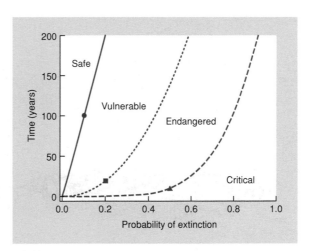

- *endangered* if the probability is 20 percent within 20 years or ten generations, whichever is longer, and
- *critical* if within 5 years or two generations the risk of extinction is at least 50 percent (Figure 14.2).

Box 14.1   *Topical ECOncerns*

What is the Value of Biodiversity?

To most people, biological diversity is undeniably of value. But it does not always lend itself to the economic valuation on which political decisions are usually based. Standard economics has generally failed to assign value to ecological resources, so that the costs of environmental damage or depletion of living resources have usually been disregarded. A major challenge for the future is the development of a new *ecological economics* (Costanza et al., 1997) in which the worth of species, communities, and ecosystems can be assigned financial value to be set against the gains to be made in industrial and other human projects that may damage them.

Many species are recognized, now, as having actual *direct value* as living resources; many more species are likely to have a potential value that as yet remains untapped. For example, wild meat, fish, and plants remain a vital resource in many parts of the world, while most of the world's food is derived from plants that were originally domesticated from wild plants in tropical and semiarid regions. In future, wild strains of these species may be exploited for their genetic diversity, and quite different species of plants and animals may be found that are appropriate for domestication. Secondly, as we saw in Chapter 12, the potential benefits that could come from natural enemies if they could be used as biological control agents for pest species are enormous; most natural enemies of most pests remain unstudied and often unrecognized. Finally, about 40 percent of the prescription and nonprescription drugs used throughout the world have active ingredients extracted from plants and animals. Aspirin, probably the world's most widely used drug, was derived

Box 14.1 (cont'd)

originally from the leaves of the tropical willow, *Salix alba*. The nine-banded armadillo (*Dasypus novemcinctus*) has been used to study leprosy and prepare a vaccine for the disease; the Florida manatee (*Trichechus manatus*), an endangered mammal, is being used to help understand hemophilia; while the rose periwinkle (*Catharanthus roseus*), a marine snail from Madagasgar, has yielded two potent drugs effective in treating blood cancer.

Other species have *indirect economic value*. For example, a multitude of wild insect species are responsible for pollinating crop plants. In a different context, the monetary value of ecotourism, which depends on biodiversity, is becoming ever more considerable. Each year, nearly 200 million adults and children in the United States take part in nature recreation and spend about $4 billion on fees, travel, lodging, food, and equipment. Moreover, ecotourists, who visit a country wholly or partly to experience its biological diversity, spend approximately $12 billion a year worldwide on their enjoyment of the natural world (Primack, 1993). On a smaller scale, a multitude of natural history films, books, and educational programs are "consumed" annually without harming the wildlife upon which they are based. More ingenuity is required to find ways to measure the indirect economic benefits that accrue as a result of natural biodiversity; for example, biological communities can be of vital importance by maintaining the chemical quality of natural waters, in buffering ecosystems against floods and droughts, in protecting and maintaining soils, in regulating local and even global climate, and in breaking down or immobilizing organic and inorganic wastes.

The final category is *ethical value*. Many people believe that there are ethical grounds for conservation, arguing that every species is of value in its own right,

and would be of equal value even if people were not here to appreciate or exploit it. From this perspective even species with no conceivable economic value require protection.

The value of all the "*ecosystem services*" described above has been estimated globally at US$33 trillion per year, nearly double the gross national product of all the world's economies (Costanza et al., 1997).

It would be wrong, though, to see things only from the point of view of conservation—not that there are really arguments *against* conservation as such, but there are arguments in favor of the human activities that make conservation a necessity: agriculture, the felling of trees, the harvesting of wild animal populations, the exploitation of minerals, the burning of fossil fuels, irrigation, the discharge of wastes, and so on. To be effective, it is likely that the arguments of conservationists must ultimately be framed in cost–benefit terms because governments will always determine their policies against a background of the money they have to spend and the priorities accepted by their electorates.

A government conservation authority is considering a proposal to designate a marine reserve at a rocky promontory of great scenic beauty. The site is very diverse in species, including a few that are rare. Commercial and recreational fishers wish to continue fishing at this unusually productive site, local people have mixed feelings about an expected influx of tourists, while conservationists (who mostly live a long way from the site) believe that the conservation value is such that no fishing should be permitted and visitor numbers should be strictly controlled. Imagine that you are an arbitrator chairing a meeting of all interested parties. What arguments do you think they will put forward? What decision would you reach and why?

The giant panda (*Ailuropoda melanoleuca*) is rare on three counts—it has a small geographical range, requires a specialized habitat, and where it does occur local populations are small.

Based on the above criteria, for example, 43 percent of vertebrate species have been classified as threatened (i.e., they fell into one of the above categories) (Mace, 1994).

Species that are at high risk of extinction are almost always rare, but not all rare species are at risk. We need to ask what precisely we mean by rare. A species may be rare:

there are several ways of being rare

- in the sense that its geographical range is small,
- in the sense that its habitat range is narrow, or
- in the sense that local populations, even where they do occur, are small.

Species that are rare on all three counts, such as the giant panda (*Ailuropoda melanoleuca*) and the giant cave weta (*Gymnoplectron giganteum*, a primitive insect from New Zealand) are intrinsically vulnerable to extinction. However, species need only be rare in one sense in order to become endangered. For example, the peregrine falcon (*Falco peregrinus*) is broadly distributed across habitats and geographical regions, yet, because it exists always at low densities, local populations in the United States have become extinct and have had to be re-established with individuals bred in captivity (see Box 13.2). The grand Otago skink (*Leiolopisma grande*), on the other hand, is found in only one part of New Zealand and is restricted to schist gullies containing native tussock grasses—but where it occurs, it does so at relatively high densities. It is at risk because changes to pastoral land use threaten its special habitats.

Nevertheless, rare species, just by virtue of their rarity, are not necessarily at risk of extinction. In fact it seems that many, probably most, species are naturally rare. We have already said (Chapter 2) that everything is almost always absent from almost everywhere. Succinctly: many species are born rare—but others have rarity thrust upon them. Other things being equal, it is easier to make a rare species extinct, simply because a localized effect may be sufficient to push it to the brink. Next, therefore, we deal with the various categories of human influence that increase the chances of species extinction.

some species have rarity thrust upon them

14.2.1 Overexploitation

large animals are prone to overexploitation

The essence of overexploitation is that populations are harvested at a rate that is unsustainable, given their natural rates of mortality and capacities for reproduction (see Section 12.3). We have already discussed the idea that in prehistoric times humans were responsible for the extinction of many large animals, the so-called megaherbivores, by overexploiting them (see Section 10.6). In more recent times, the history of the great whales has followed a similar pattern, while today we are still taking our toll of other vulnerable giants. Sharks provide an interesting example. Among the most feared of species (although attacks are much rarer than held in the popular imagination), large numbers are taken for sport, many others to make shark fin soup, while a large proportion of the estimated annual 200 million shark kills are accidental by-catches of commercial fishing. Evidence is mounting that many species of shark have been declining in abundance, a trend that should come as no surprise given their late ages of maturity, slow reproductive cycles, and low fecundities (Manire & Gruber, 1990). Sharks are among the most important predators in the marine environment, and their enforced rarity may have widespread repercussions in ocean communities.

UNANSWERED QUESTION: what are the consequences for ocean communities of the increased rarity of sharks as top predators?

the threat posed by collectors

A feature of animals that are collected for ornamentation, whether for their body parts or as exotic pets, is that their value to collectors goes up as they become rarer. Thus, instead of the normal safeguard of a density-dependent reduction in consumption rate at low density (see Section 8.5), the very opposite occurs. The phenomenon is not restricted to animals. New Zealand's endemic mistletoe (*Trilepidia adamsii*), for example, parasitic on a few forest understorey shrubs and small trees, was undoubtedly overcollected to provide herbarium specimens. Always a rare species, its extinction (recorded from 1867 to 1954 but not seen since) was due to overcollecting combined with forest clearance and perhaps an adverse effect on fruit dispersal because of reductions in bird populations.

14.2.2 Habitat disruption

Habitats may be adversely affected by human influence in three main ways. First, a proportion of the habitat available to a particular species may simply be destroyed, for urban and industrial development or for the production of food and other natural resources such as timber. Second, habitat may be degraded by pollution (Chapter 13) to the extent that conditions become untenable for certain species. Third, habitat may be disturbed by human activities to the detriment of some of its occupants.

habitat may be destroyed . . .

Forest clearance has been, and is still, the most pervasive of the forces of habitat destruction. Much of the native temperate forest in the developed world was destroyed long ago, while current rates of deforestation in the tropics are 1 percent or more per annum. As a consequence, more than half of the wildlife habitat has been destroyed in most of the world's tropical countries. The process of habitat destruction often results in the habitat available to a particular species being more fragmented than was historically the case. This can have several repercussions for the populations concerned, a point we take up again in Section 14.2.6.

Degradation by pollution can take many forms, from the application of pesticides that harm nontarget organisms, to acid rain with its adverse effects on organisms as diverse as forest trees, amphibians in ponds, and fish in lakes, to global climate change that may turn out to have the most pervasive influence of all. Aquatic environments are particularly vulnerable to pollution. Water, inorganic chemicals, and organic matter enter from drainage basins, with which streams, rivers, lakes, and continental shelves are intimately connected. Land-use changes, waste disposal, and water impoundment and abstraction can profoundly affect the patterns of water flow and the quality of their water (Allan & Flecker, 1993).

... or degraded

Habitat disturbance is not such a pervasive influence as destruction or degradation but certain species are particularly sensitive. For example, diving and snorkeling on coral reefs, even in marine protected areas, can cause damage through direct physical contact with hands, body, equipment, and fins. Often the disturbance is minor, but this can amount to cumulative damage and a reduction in the populations of vulnerable branching corals. In one analysis of 214 divers in a marine park on Australia's Great Barrier Reef, 15 percent of divers damaged or broke corals, mostly by fin flicks (Rouphael & Inglis, 2001). Impacts were much more likely to be caused by male than female divers, whilst specialist underwater photographers caused more damage on average (1.6 breaks per 10 minutes) than divers without cameras (0.3 breaks per 10 minutes). Nature recreation, ecotourism, and even ecological research are not without risk of disturbance and a decline of the populations concerned.

... or disturbed

14.2.3 Introduced species

Invasions of exotic species into new geographical areas sometimes occur naturally and without human agency. However, human actions have increased this trickle to a flood. Human-caused introductions may occur either accidentally as a consequence of human transport, or intentionally but illegally to serve some private purpose, or legitimately to procure some hoped-for public benefit by bringing a pest under control, producing new agricultural products, or providing novel recreational opportunities. Many introduced species are assimilated into communities without much obvious effect. However, some have been responsible for dramatic changes to native species and natural communities.

For example, the accidental introduction of the brown tree snake *Boiga irregularis* onto Guam, an island in the Pacific, has, through nest predation, reduced ten endemic forest bird species to the point of extinction. The gradual spread of the snake from its bridgehead population in the center of the island has been paralleled by the timing of the loss of bird species to the north and south (Figure 14.3). Similarly, the introduction as a source of human food of the predaceous Nile perch (*Lates nilotica*) to the enormously species-rich Lake Victoria in East Africa has driven most of its 350 endemic species of fish to extinction or near extinction (Kaufman, 1992).

introduced predators

Conservation biologists are particularly concerned about the effects of introduced species wherever there are communities of native organisms that are largely endemic (that is, live nowhere else in the world). Indeed, one of the major reasons for the world's huge diversity is the occurrence of centers of endemism so that similar habitats in different parts of the world are occupied by different groups of species that happen

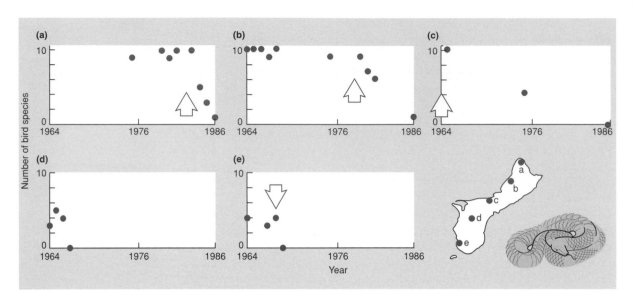

Figure 14.3
Decline in the number of forest bird species at five locations on the island of Guam. The large arrows indicate the first sightings of the brown tree snake at each location (in location d, the snake was first sighted in the early 1950s). (After Savidge, 1987.)

introductions leading to homogenization

to have evolved there. If every species naturally had access to everywhere on the globe, we might expect a relatively small number of successful species to become dominant in each biome. The extent to which this homogenization can happen naturally is restricted by the limited powers of dispersal of most species in the face of the physical barriers to dispersal that exist. By virtue of the transport opportunities offered by humans, these barriers have been breached by an ever-increasing number of exotic species (Mack et al., 2000). The effects of introductions have been to convert a hugely diverse range of local community compositions into something much more homogeneous.

It would be wrong, however, to conclude that introducing species to a region will inevitably cause a decline in species richness there. For example, there are numerous species of plants, invertebrates, and vertebrates found in continental Europe but absent from the British Isles (many because they have so far failed to recolonize after the last glaciation). Their introduction would be likely to augment British biodiversity. The significant detrimental effect noted above arises where aggressive species provide a novel challenge to endemic biotas ill equipped to deal with them.

14.2.4   Possible genetic problems in small populations

loss of evolutionary potential

Theory tells conservation biologists to beware genetic problems in small populations that may arise through loss of genetic variation (Box 14.2). The preservation of genetic diversity is important in the first place because of the long-term evolutionary potential it provides. Rare forms of a gene (alleles), or combinations of alleles, may confer no

Box 14.2   *Quantitative Aspects*

What Determines Genetic Variation?

Genetic variation is determined primarily by the joint action of natural selection and genetic drift (where the frequency of genes in a population is determined by chance rather than evolutionary advantage). The relative importance of genetic drift is higher in small isolated populations that, as a consequence, are expected to lose genetic variation. The rate at which this happens depends on the *effective population size* (N_e). This is the size of the "genetically idealized" population to which the actual population (N) is equivalent in genetic terms.

N_e is usually less, often much less, than N, for a number of reasons (detailed formulae can be found in Lande & Barrowclough, 1987):

1 If the sex ratio is not 1 : 1; for instance, with 100 breeding males and 400 breeding females $N = 500$, but $N_e = 320$.

2 If the distribution of progeny from individual to individual is not random; for instance, if 500 individuals each produce one individual for the next generation on average ($N = 500$), but the variance in progeny production is 5 (with random variation this would be 1), then $N_e = 100$.

3 If population size varies from generation to generation, then N_e is disproportionately influenced by the smaller sizes; for instance, for the sequence 500, 100, 200, 900, 800, the mean is $N = 500$ but $N_e = 258$.

How many individuals are needed to maintain genetic variability? Franklin (1980) suggested that an effective population size of about 50 would be unlikely to suffer from inbreeding depression, while 500 might be needed to maintain longer term evolutionary potential. Such rules of thumb should be applied cautiously and, bearing in mind the relationship between N_e and N, the minimum population size N should probably be set at five to ten times N_e, that is, 2,500–5,000 individuals (Nunney & Campbell, 1993).

immediate advantage but could turn out to be well suited to changed environmental conditions in the future. Small populations tend to have less variation and hence lower evolutionary potential.

A more immediate potential problem is inbreeding depression. When populations are small there is a tendency for individuals breeding with one another to be related. The offspring of inbreeding parents tend to have levels of heterozygosity far below that of the population as a whole. Individuals with low levels of heterozygosity often have reduced fitness. There are many examples of inbreeding depression—breeders of domesticated animals and plants, for example, have long been aware of reductions in fertility, survivorship, growth rates, and resistance to disease. Furthermore, as well as the problems of a generally low level of heterozygosity, all populations carry specific recessive alleles which are deleterious, or even lethal, when homozygous (human examples include sickle-cell anemia and cystic fibrosis). Close relatives are more likely to carry the same deleterious recessives. Hence, individuals that breed with close relatives are more likely to produce offspring where the same harmful allele is derived from both parents and where the deleterious effect is therefore expressed.

the risk of inbreeding depression

Figure 14.4
Relationships for 23 populations of
Gentianella germanica between population
size and (a) mean number of fruits per
plant, (b) mean number of seeds per fruit,
and (c) mean number of seeds per plant.
(d) The relationship between the population
growth rate from 1993 to 1995 (ratio of
population sizes) and the population size (in
1994). (From Fischer & Matthies, 1998.)

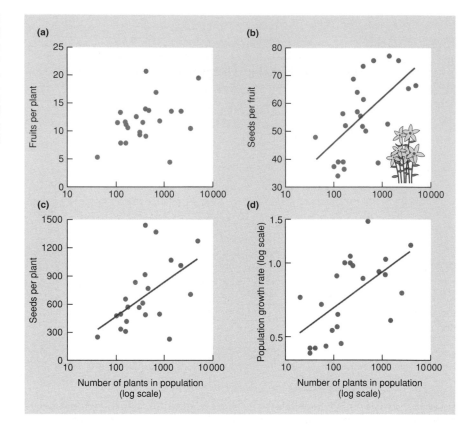

In their study of 23 local populations of the rare plant *Gentianella germanica* in grasslands in the Jura mountains (Swiss–German border), Fischer and Matthies (1998) found a negative correlation between reproductive performance and population size (Figure 14.4a–c). Furthermore, population size decreased between 1993 and 1995 in most of the studied populations, but population size decreased more rapidly in the smaller populations (Figure 14.4d). These results are consistent with the hypothesis that genetic effects resulted in a reduction in fitness in the small populations. However, they may equally have been caused by differences in local habitat conditions (small populations may be small because they have low fecundity resulting from low-quality habitat) or because of the disruption of plant–pollinator interactions (small populations may have low fecundity because of low frequencies of visitation by pollinators). To determine whether genetic differences were, indeed, responsible, seeds from each population were grown under standard conditions in a common garden experiment. We have already witnessed the power of such transplant experiments to reveal intrinsic differences among populations of a species within its geographical range (see Section 2.3). After 17 months in the common garden, there were significantly more flowering plants and more flowers (per planted seed) for seeds from large populations than those from small populations. We can conclude that genetic effects are of importance for population persistence in this rare species.

14.2.5   A review of risks

Some species are at risk for a single reason, but often, as in the case of the New Zealand mistletoe discussed earlier, a combination of factors is at work. A review of the factors responsible for recorded vertebrate extinctions showed that habitat loss, overexploitation, and species introductions were all of considerable significance, although habitat loss was less prominent in the case of reptiles and overexploitation hardly figured in the case of fishes (Table 14.2). As far as threatened extinctions are concerned, habitat loss is most commonly the major hazard, while risk of overexploitation remains very high, especially for mammals and reptiles.

It is interesting that no example of extinction due to genetic problems has so far come to light. Perhaps inbreeding depression has occurred, although undetected, as part of the "death rattle" of some dying populations (Caughley, 1994). Thus, a population may have been reduced to a very small size by one or more of the processes described above, and this may have led to an increased frequency of matings among relatives and the expression of deleterious recessive alleles in offspring, leading to reduced survivorship and fecundity and causing the population to become smaller still—the so-called extinction vortex (Figure 14.5). Perhaps the small populations of *Gentianella germanica* (Figure 14.4) have entered an extinction vortex.

extinction vortices

Table 14.2
A review of the factors responsible for the recorded extinctions of vertebrates and an assessment of the risks currently facing species categorized globally as endangered, vulnerable, or rare by the International Union for the Conservation of Nature (IUCN). (After Reid & Miller, 1989.)

Group	Habitat loss	Overexploitation†	Species introduction	Predators	Other	Unknown
Extinctions						
Mammals	19	23	20	1	1	36
Birds	20	11	22	0	2	37
Reptiles	5	32	42	0	0	21
Fishes	35	4	30	0	4	48
Threatened extinctions						
Mammals	68	54	6	8	12	–
Birds	58	30	28	1	1	–
Reptiles	53	63	17	3	6	–
Amphibians	77	29	14	–	3	–
Fishes	78	12	28	–	2	–

Percentage due to each cause*

* The values indicated represent the percentage of species that are influenced by the given factor. Some species may be influenced by more than one factor, thus, some rows may exceed 100 percent.
† Overexploitation includes commercial, sport, and subsistence hunting and live animal capture for any purpose.

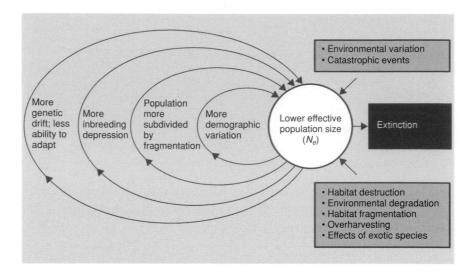

14.2.6 The dynamics of small populations

Much of conservation biology is a crisis discipline. Inevitably, conservationists are confronted with too many problems and too few resources. Should we focus attention on the various forces that bring species to extinction and attempt to persuade governments to act to reduce their prevalence? Or should we restrict our activities to identifying areas of high species richness where we can set up and protect reserves? Or should we identify species at most risk of extinction and work out ways of keeping them in existence? Ideally, we should do all these things. The greatest pressure, however, is often in the area of species preservation; the remaining population of giant pandas in China (or yellow-eyed penguins in New Zealand or spotted owls in North America) has become so small that if nothing is done the species may become extinct within a few years or decades. There is a pressing need to understand the dynamics of small populations.

These are governed by a high level of uncertainty—whereas large populations can be described as being governed by the law of averages (Caughley, 1994). Three kinds of uncertainty or variation can be identified that are or particular importance to the fate of small populations.

1 *Demographic uncertainty.* Random variations in the number of individuals that are born male or female, or in the number that happen to die or reproduce in a given year, or in the genetic "quality" of the individuals in terms of survival/reproductive capacities can matter very much to the fate of small populations. Suppose a breeding pair produces a clutch consisting entirely of females—such an event would go unnoticed in a large population but would be the last straw for a species down to its last pair.

2 *Environmental uncertainty.* Unpredictable changes in environmental factors, whether "disasters" (such as floods, storms, or droughts of a magnitude that occur very

rarely) or more minor (year to year variation in average temperature or rainfall) can also seal the fate of a small population. A small population is more likely than a large one to be reduced by adverse conditions to zero or to numbers so low that recovery is impossible.

3 *Spatial uncertainty*. Many species consist of an assemblage of subpopulations that occur in more or less discrete patches of habitat (habitat fragments). Since the subpopulations are likely to differ in terms of demographic uncertainty, and the patches they occupy in terms of environmental uncertainty, the dynamics of extinction and local recolonization can be expected to have a large influence on the chance of extinction of the overall metapopulation (see Section 9.3).

To illustrate some of these ideas, take the demise in North America of the heath hen (*Tympanuchus cupido cupido*). This bird was once extremely common from Maine to Virginia. Being highly edible and easy to shoot (and also susceptible to introduced cats and affected by conversion of its grassland habitat to farmland), it had by 1830 disappeared from the mainland and was only found on the island of Martha's Vineyard. In 1908 a reserve was established for the remaining 50 birds and by 1915 the population had increased to several thousand. However, 1916 was a bad year. Fire (a disaster) eliminated much of the breeding ground; there was a particularly hard winter coupled with an influx of goshawks (environmental uncertainty); and finally poultry disease arrived on the scene (another disaster). At this point the remnant population was likely to have become subject to demographic uncertainty; for example, of the 13 birds remaining in 1928 only two were females. A single bird was left in 1930 and the species went extinct in 1932.

the case of the heath hen

Of the high risk factors associated with local extinctions of plant and animal species, habitat or island area is probably the most pervasive. Figure 14.6 shows the negative relationships for a variety of taxa between annual extinction rate and area. No doubt the main reason for the vulnerability of populations in small areas is the fact that the populations themselves are small. This is illustrated in Figure 14.7 for bird species on islands and for bighorn sheep in various desert areas in southwest United States.

the importance of habitat or island area

In fact, loss of habitat frequently results not only in a reduction in the absolute size of a population but also the division of the original population into a metapopulation of

habitat fragmentation

Figure 14.6
Percentage extinction rates as a function of habitat area for (a) zooplankton in lakes in northeastern USA, (b) birds on northern European islands, and (c) vascular plants in southern Sweden. (After data in Pimm, 1991.)

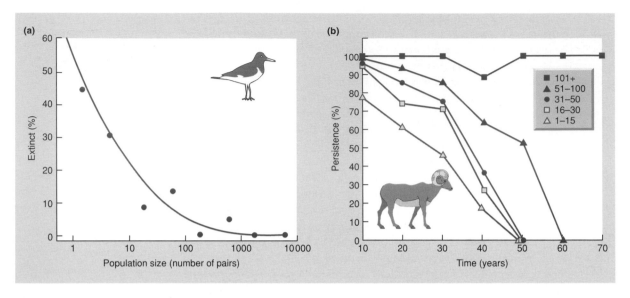

Figure 14.7
(a) The extinction rate of island birds is higher for small populations. (b) The percentage of populations of bighorn sheep in North America that persist over a 70-year period is lowest where the initial population size was small (open triangles: 1–15 individuals) and is highest where initial population size was large (solid squares: 101+ individuals). (After Berger, 1990.)

semi-isolated subpopulations. Further fragmentation can result in a decrease in the average size of fragments, an increase in the distance between them, and an increase in the proportion of edge habitat (Burgman et al., 1993). A question of fundamental importance, then, is whether a species is more at risk simply because its population is subdivided. In other words, would a single population of a given size be less or more at risk than one divided into a number of subpopulations in habitat fragments?

The general conclusion from theoretical explorations is that the answer lies in the balance between the connectedness of different subpopulations on the one hand, and the correlation between the dynamics of different subpopulations on the other. Thus, where the probability of dispersal between fragments (that is, connectedness) is high, metapopulations will tend to persist for longer than unfragmented populations. The reason is that when individual subpopulations go extinct, there is a good chance that they will be restarted by a colonist from another subpopulation. However, where extinction events in different subpopulations are strongly correlated (because environmental variation acts identically in all fragments) metapopulations will be more at risk than unfragmented populations. The reason here is that individual subpopulations, being small, are vulnerable to extinction, and when one goes extinct, they all tend to.

A long-term study of the checkerspot butterfly (*Euphydryas editha*) at Jasper Ridge (western United States) illustrates some of these phenomena. Figure 14.8a shows that if we view the whole population (metapopulation) as a unit, there is a fair degree of consistency in overall abundance, despite a certain amount of year to year variation. However, when the population is examined as spatially distinct subpopulations, we see evidence, first, that environmental variation is not well correlated between subpopulations (patterns of increase and decrease are not very similar), and also that in area G the

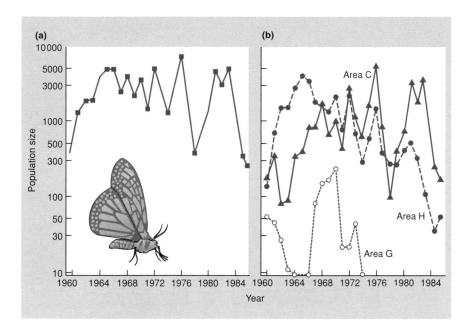

Figure 14.8
The abundance of checkerspot butterflies (a) when viewed as a single population, and (b) when subdivided into three spatially distinct units. (After Ehrlich & Murphy, 1987).

subpopulation went extinct twice, recolonizing the first time but not the second (Figure 14.8b).

14.3 Threats to Communities

So far in this chapter, attention has been focused on individual species, treating them as though they were largely independent entities and applying what we know about population dynamics. However, it hardly needs to be pointed out that conservation of biodiversity also requires a broader perspective in which we apply our knowledge of communities and ecosystems. There are many reasons for this.

If we ignore community interactions, a chain of extinctions may follow inexorably from the extinction of particular native species, which therefore deserve special attention. Flying foxes in the genus *Pteropus*, which occur on South Pacific islands, are the major, and sometimes the only, pollinators and seed dispersers for hundreds of native plants (many of which are of considerable economic importance, providing medicines, fiber, dyes, prized timber, and foods). Flying foxes are highly vulnerable to human hunters and there is widespread concern about declining numbers. On Guam, for example, the two indigenous species are either extinct, or virtually so, and there are already indications of reductions in fruiting and dispersal (Cox et al., 1991).

A similarly unwanted outcome may occur if a particularly influential exotic species is introduced into a community without regard to food web consequences. Our example is provided by the introduction into Flathead Lake and its tributaries in Montana, USA, of opossum shrimp (*Mysis relicta*) to provide food for kokanee salmon (*Oncorhynchus nerka*) (itself a previous introduction). Instead of benefitting the fishery

chains of extinctions—the case of the flying fox

new species for old

Figure 14.9
As the opposum shrimp population increased and exploited zooplankton in Flathead Lake, the kokanee salmon declined, as did the eagle population that relied on the salmon. (After Spencer et al., 1991.)

Grizzly bears (*Ursus arctos*) and bald eagles (*Haliaeetus leucocephalus*) feed on spawning salmon when these are available. The collapse of the kokanee salmon population in Flathead Lake and its tributaries means that these are much rarer sights.

as intended, the shrimps heavily exploited the zooplankton food of the salmon while avoiding predation themselves because of their behavioral adaptation of migrating to deep water during the day. The collapse of the kokanee population has apparently dramatically reduced the numbers of bald eagles and grizzly bears that gather each fall to feed on spawning salmon and their carcasses (Figure 14.9). Ecotourism has also suffered a serious decline because the charismatic megafauna are now less in evidence.

We have already noted that species richness declines with habitat area when previously continuous habitat becomes fragmented. But how does trophic organization change in the process of fragmentation? If we are not careful, we may devise a plan to conserve some endangered species but forget to allow for the ecological functional links required to keep the system going. Conservation managers need to be wary about the scale of their plans. Where the focus is on a particular species, for example the elephants discussed below in Section 14.4.1, the recommended minimum viable population may, by chance, define a minimum viable area that is big enough to contain all the necessary functional links. But ecologically defined boundaries may extend further than required from just the perspective of any particular species. For example, there is little point in setting up a wetland reserve unless it extends sufficiently far into the surrounding landscape to provide the necessary habitat for terrestrial phases of aquatic insects and to insure that terrestrial energy and nutrient fluxes provide appropriate inputs to the aquatic ecosystem.

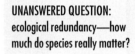

In the community and ecosystem contexts, we can pose the question: Can we lose a few, several, or many species and still maintain, for example, productivity, nutrient cycling, or community resilience in the face of disturbance? Three possible scenarios are illustrated in Figure 14.10. In each case, the community process operates at a "natural" level at the high (natural) level of species richness and the process reduces to zero when no species are left. But there are three curves between these two extremes. Curve 1 shows little change to the process until many species have been lost, corresponding to a situation where there is high "ecological redundancy". Loss of biodiversity is of least significance in this scenario. Curve 2 has every species contributing to community functioning; if this corresponded to reality we should beware any loss of biodiversity but the system may be able to cope with a certain level of loss. Curve 3 has the system collapsing with even the loss of one or a few (perhaps keystone) species. Which of these models is closest to reality? The simple answer is that we have almost no idea.

UNANSWERED QUESTION: ecological redundancy—how much do species really matter?

In one study, though, Mikkelson (1993) gathered together information from five contrasting situations where fragments of different sizes were available—one set of mountain tops and two sets of islands (each of which comprised continuous habitat

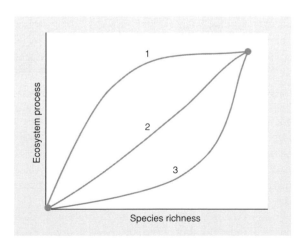

Figure 14.10
Hypothetical relationships between a basic ecosystem process (such as productivity, nutrient cycling, stability in composition) and species richness. For further details see text.

6,000–14,000 years ago) and two groups of nature reserves in areas of Australia where habitat fragmentation began at the turn of the 20th century. The overall picture was that species richness declined as a result of fragmentation; but trophic organization (the proportional representation of plants, herbivores, and carnivores) varied very little between different-sized fragments. This, and other studies, therefore, indicate that large fragments are indispensible if we wish to conserve high biodiversity. On the other hand, Mikkelson's conclusion provides some cheer since it implies that, up to a point at least, fragmentation need not lead to the dislocation of all fundamental ecological processes.

14.4   Conservation in Practice

14.4.1   Species management plans

Given the environmental circumstances and species characteristics of a particular rare species, what is the chance it will go extinct in a specified period? Alternatively, how big must its population be to reduce the chance of extinction to an acceptable level? These are frequently the crunch questions in conservation biology.

trying to determine the minimum viable population . . .

The ideal classic approach of experimentation, which might involve setting up and monitoring for several years a number of populations of various sizes, is unavailable to those concerned with species at risk because the situation is usually too urgent and there are inevitably too few individuals to work with. How then are we to decide what constitutes the minimum viable population? Three approaches will be discussed in turn: (1) a search for patterns in evidence already gathered in long-term studies, (2) subjective assessment based on expert knowledge, and (3) development of population models. Each has its limitations, which we will explore by looking at particular examples.

. . . from available biogeographical data—a risky approach

Data sets such as those in Figure 14.7b are unusual because they depend on a long-term commitment to monitoring a number of populations—in this case, bighorn sheep in desert areas. If we set an arbitrary definition of the necessary minimum viable population (MVP) as one that will give at least a 95 percent probability of persistence for 100 years, we can explore data like these to provide an approximate estimate of MVP. Populations of fewer than 50 individuals all went extinct within 50 years while only 50 percent of populations of 51–100 sheep lasted for 50 years. Evidently, for an MVP here we require more than 100 individuals. Indeed, for these sheep such populations demonstrated close to 100 percent persistence over the maximum period studied of 70 years. The value for conservation of studies like this, however, is limited because they deal with species that are generally not at risk. It is only safe to use them to produce recommendations for the management of endangered species if the species of concern and the ones in the study are sufficiently similar in their vital statistics, and if the environmental regimes are similar.

. . . by subjective expert assessment

Information that may be relevant to a conservation crisis exists not only in the scientific literature but also in the minds of experts. By bringing experts together in a conservation workshop, well-informed decisions can be reached. The recommended actions may not be entirely correct, but when a decision about an endangered species has to be made without the possibility of further research, this approach has much to

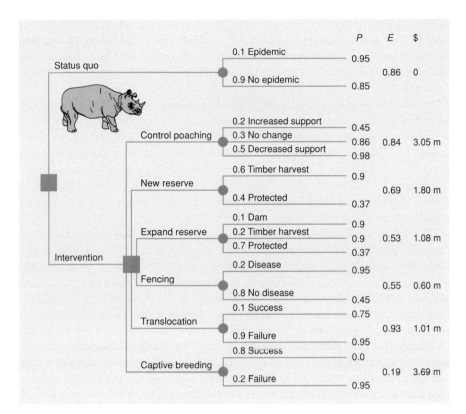

Figure 14.11
A decision tree for management of the Sumatran rhino. For the *status quo* option, there was estimated to be a probability of 0.1 that a disease epidemic will occur in the next 30 years, and hence a probability of 0.9 that no epidemic will occur. If there is an epidemic, the probability of extinction, *P*, was estimated to be 0.95 (i.e., 95 percent probability of extinction in 30 years), whereas with no epidemic *P* was 0.85. The overall estimate of species extinction for an option, *E*, is then given by: *E* = (probability of first scenario × *P* for first scenario) + (probability of second scenario × *P* for second scenario). In the case of the *status quo* option, *E* = (0.1 × 0.95) + (0.9 × 0.85) = 0.86. The values of *P* and *E* for the other options are calculated in a similar way. The final column in the figure then lists the estimated costs of the various options. (From Maguire et al., 1987.)

commend it. It makes use of available data, knowledge, and experience in a situation when a decision is needed and time for further research is unavailable. Moreover, it explores the various options in a systematic manner and does not duck the regrettable but inevitable truth that unlimited resources will not be available. However, it also runs a risk. In the absence of all necessary data, the recommended best option may simply be wrong.

The Sumatran rhino (*Dicerorhinus sumatrensis*) persists only in small, isolated sub-populations in an increasingly fragmented habitat in Sabah (East Malaysia), Indonesia, and West Malaysia, and perhaps also in Thailand and Burma. Unprotected habitat is threatened by timber harvest, human resettlement, and hydroelectric development. There are only a few designated reserves, which are themselves subject to exploitation, and only two individuals were held in captivity at the time of a workshop to discuss their fate (Maguire et al., 1987).

The vulnerability of the Sumatran rhino, the way this vulnerability varies with different management options, and the most appropriate management option given various criteria were assessed by a technique known as *decision analysis*. A decision tree is shown in Figure 14.11, based on the estimated probabilities of the species becoming extinct (*P*) within a 30-year period (equivalent to approximately two rhino generations). The tree was constructed in the following way. The two squares are decision points: the first distinguishes between intervention on the rhino's behalf and nonintervention (*status quo*); the second distinguishes the various management options. For each option, the line branches at a small circle. The branches represent alternative scenarios that might occur, and the numbers on each branch indicate the probabilities estimated for the alternative scenarios (for further details see the legend to Figure 14.11).

Consider two of the management options in a little more detail. The first is to fence an area in an existing or new reserve, managing the resulting high density of rhinos with supplemental feeding and veterinary care. Disease here is a major risk: the probability of an epidemic was estimated to be higher than in the *status quo* option (0.2 as opposed to 0.1) because the density would be higher, and P, if there was an epidemic, was again 0.95. On the other hand, if fencing was successful, P was expected to fall to 0.45, giving an overall estimate of extinction (E) of 0.55. The fenced area would cost around US$0.6 million over 30 years.

For the establishment of a captive breeding program, animals would have to be captured from the wild, increasing P if the program failed to an expected 0.95. However, P would clearly drop to zero if the program succeeded. The cost, though, would be high, since it would involve the development of facilities and techniques in Malaysia and Indonesia (around $2.06 million) and the extension of those that already exist in the United States and Great Britain ($1.63 million). The probability of success was estimated to be 0.8, and the overall E was therefore 0.19.

Where do these various probability values come from? The answer is from a combination of data, the educated use of data, educated guesswork, and experience with related species. Which would be the best management option? The answer depends on what criteria are used in the judgement of "best". Suppose we wanted simply to minimize the chances of extinction, irrespective of cost. The proposal for the best option would then be captive breeding. In practice, though, costs are most unlikely to be ignored. We would then need to identify an option with an acceptably low E but also an acceptable cost.

With the benefit of hindsight, Caughley (1994; data of N. Leader-Williams) reported that about $2.5 million was spent catching Sumatran rhinos; three died during capture, six died postcapture, and of the 21 rhinos now in captivity only one has given birth and she was pregnant when captured. Leader-Williams suggests that those $2.5 million might have been used more effectively to protect 700 km^2 of prime rhino habitat for nearly two decades. This could in theory hold a population of 70 Sumatran rhinos which, with a rate of increase of 0.06 per individual per year (shown by other rhino species given adequate protection), might have given birth to 90 calves during that period.

... by simulation modeling: the case of the African elephant

Simulation models provide another way of gauging viability. In such models, every individual is treated separately in terms of the probability (with an attached degree of uncertainty) that it will survive or produce a certain number of offspring within each of a succession of time periods. The program is then run many times, each giving a different population trajectory because of the random elements involved. The outputs, for each set of model parameters used, are estimates of the probability of extinction (the proportion of simulated populations that go extinct) over a specified time interval.

Such a population viability analysis (PVA) was performed for African elephants (*Loxodonta africana*). Overall numbers are in decline, and few are expected to survive

Table 14.3
The survivorship for 12 elephant age classes in normal years (occur in 0.47 of 5-year periods), and in years with 10-year drought cycles (0.41 of 5-year periods), 50-year, and 250-year drought (0.1 and 0.02 of 5-year periods, respectively). (After Armbruster & Lander, 1992.)

Age class (years)	Female survivorship			
	Normal years	10-year droughts	50-year droughts	250-year droughts
0–5	0.500	0.477	0.250	0.01
5–10	0.887	0.877	0.639	0.15
10–15	0.884	0.884	0.789	0.20
15–20	0.898	0.898	0.819	0.20
20–25	0.905	0.905	0.728	0.20
25–30	0.883	0.883	0.464	0.10
30–35	0.881	0.881	0.475	0.10
35–40	0.875	0.875	0.138	0.05
40–45	0.857	0.857	0.405	0.10
45–50	0.625	0.625	0.086	0.01
50–55	0.400	0.400	0.016	0.01
55–60	0.000	0.000	0.000	0.00

over the next few decades outside high-security areas, mainly because of habitat loss and poaching for ivory. The elephant population was modeled in twelve 5-year age classes through discrete 5-year time steps. Values for age-specific survivorship and density-dependent reproductive rates were derived from a thorough data set from Tsavo National Park in Kenya, because its semiarid nature has the general characterisics of the land planned for game reserves now and in the future. Environmental stochasticity was incorporated as possible drought events affecting sex- and age-specific survivorship—again, realistic data from Tsavo were used, based on a mild drought cycle of approximately 10 years superimposed on a more severe 50-year drought cycle and an even more severe 250-year cycle. Table 14.3 gives the survivorship of females under "normal" conditions and the three drought conditions. The relationship between habitat area and the probability of extinction was examined in 1,000-year simulations; at least 1,000 replicates were performed for each model. Extinctions were taken to have occurred when no individuals remained or when only a single sex was represented.

The results imply that an area of 500 square miles is required to yield a 99 percent probability of persistence for 1,000 years (Figure 14.12). In fact, the researchers recommended to managers an even more conservative minimum area of 1,000 square miles for reserves, recognizing in particular the unreliability of their estimates for survivorship in the youngest age class. Of the parks and game reserves in central and southern Africa, only 35 percent are larger than 1,000 square miles.

In an ideal world, a PVA would enable us to produce a specific and reliable recommendation for an endangered species of the population size, or reserve area, that would permit persistence for a given period with a given level of probability. But this is rarely if ever achievable because the biological data are hardly ever good enough. The

Figure 14.12
The cumulative probability of elephant
population extinction over 1,000 years for
six differently sized habitat areas. (After
Armbruster & Lande, 1992.)

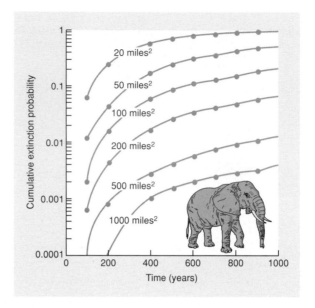

modelers know this. It is important that conservation managers appreciate it too. Nevertheless, we have seen how models allow us to make the very best use of available data and may well give us the confidence to make a choice between various possible management options and to identify the relative importance of different risk factors. It is also heartening that a retrospective analysis of 21 PVAs (concerning bird, mammal, reptile, and fish populations) indicated that predictions relating to population declines were remarkably accurate (Brook et al., 2000). The sorts of management interventions that may then be recommended in the light of results of a PVA include translocation of individuals to augment the numbers and/or genetic diversity of target populations, raising the carrying capacity by artificial feeding, restricting dispersal by fencing, fostering of young (or cross-fostering of young by related species), reducing mortality by controlling predators or poachers, or through vaccination, and of course habitat preservation.

14.4.2 *Ex situ* conservation

Captive propagation is attractive in its directness, its immediacy, and the opportunity it provides to preserve options. It can be used to provide demographic or genetic reservoirs for enhancing existing natural populations or establishing new ones, or to provide a final refuge for species with no immediate hope of survival in the wild. Thus the Pere David deer (*Elaphuurs davidianus*) and the Mongolian or Przewalski's wild horse (*Equus przewalski*), well established in captivity, are believed extinct in nature; while the endangered but still surviving wild populations of blackbuck (*Antilope cervicapra*), the Arabian oryx (*Oryx leucoryx*), the peregrine falcon (*Falco peregrinus*), the giant tortoise (*Geochelone elephantopus*), and the European otter (*Lutra lutra*), among others, have all been augmented from captive-bred populations.

Examples of species that are well established in captivity but are extinct or endangered in the wild—the Pere David deer (*Elaphuurs davidianus*), the Mongolian or Przewalski's wild horse (*Equus przewalski*), the Arabian oryx (*Oryx leucoryx*), the giant tortoise (*Geochelone elephantopus*), and the blackbuck (*Antilope cervicapra*).

zoo or ark?

Zoos have recently become the center of a spirited debate. At one extreme, some see zoological and botanical gardens (and germbanks and seedbanks) as small arks that provide refuge for endangered species from a flood of species extinctions; others see them as living museums—once a species enters a zoo it is essentially dead (Ginsberg, 1993). The cases above demonstrate the merit of the ark perspective. However, captive breeding for conservation should involve three stages—(1) planning a conservation strategy, (2) captive breeding, and (3) reintroduction—and these need to be integrated with habitat protection and *in situ* conservation. Regrettably, many species are simply inappropriate for captive breeding and eventual reintroduction (e.g., breeding technology unavailable, appropriate habitat unavailable, causes for endangerment irreversible) and available resources are insufficient to build an "ark" big enough for a significant proportion of the world's biota.

14.4.3   Protected areas

getting the most from the conservation dollar

Producing survival plans for individual species may be the best way to deal with species recognized to be in deep trouble and identified to be of special importance (e.g., keystone species, evolutionarily unique species, charismatic large animals that are easy to "sell" to the public). However, there is no possibility that all endangered species can be dealt with one at a time. For instance, the US Fish and Wildlife Service calculated it would need to spend about $4.6 billion over 10 years to fully recover all gazetted species in the United States (US Department of the Interior, 1990), whereas the annual budget for 1993 was $60 million (Losos, 1993). We can, though, expect to conserve the greatest biodiversity if we protect whole communities by setting aside protected areas.

Protected areas of various kinds (national parks, nature reserves, multiple-use management areas, etc.) grew in number and area through the 20th century, with the greatest expansion occurring since 1970. However, the 4,500 protected areas in existence in 1989 still only represented 3.2 percent of the world's land area. At best, and given the political will, perhaps 6 percent of land area may eventually be provided with protection—the rest would be considered necessary to provide the natural resources needed by the human population (Primack, 1993).

marine protected areas—conservation and sustainable use can go hand in hand

Priorities for marine conservation, which have lagged behind terrestrial efforts, are now being urgently addressed. Most of the world's biota is found in the sea (32 of the 33 known phyla are marine, 15 exclusively so) and marine communities are subject to a number of potentially adverse influences, including overfishing, habitat disruption, and, in particular, pollution from land-based activities. Many of the new generation of marine protected areas are designed as multiple-use reserves, accommodating many different users (environmentalists, cultural harvesters, recreational and commercial fishers, etc.) (Agardy, 1994). It is clear, too, that conservation and sustainable use on land (forestry, agriculture) can often proceed hand in hand as long as planning has a scientific basis and the negotiated objectives are clear (Margules & Pressey, 2000).

centers of diversity, endemism, or extinction

It is important to devise priorities so that the restricted number of new protected areas, in terrestrial and marine settings, can be evaluated systematically and chosen with care. We know that the biotas of different locations vary in species richness (centers of diversity), the extent to which the biota is unique (centers of endemism), and the extent to which the biota is endangered (hotspots of extinction, for example because of imminent habitat destruction). One or more of these criteria could be used to prioritize potential areas for protection (Figure 14.13).

design of nature reserves

A perhaps rather surprising application of island biogeography theory (see Section 10.5.1) is in nature conservation. This is because many conserved areas and nature reserves are surrounded by an "ocean" of habitat made unsuitable, and therefore hostile, by people. Can the study of islands in general provide us with "design principles"

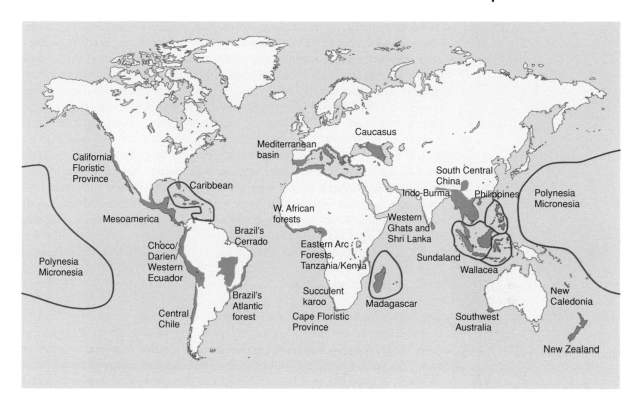

Figure 14.13
The global distribution of biodiversity hotspots where exceptional concentrations of endemic species are undergoing exceptional loss of habitat. As many as 44 percent of all species of the Earth's vascular plants and 35 percent of its vertebrates are confined to 25 hotspots that make up only 1.4 percent of its land surface. (After Myers et al., 2000b.)

that can be used in the planning of nature reserves? The answer is a cautious "yes" and some general points can be made.

1 One problem that conservation managers sometimes face is whether to construct one large reserve or several small ones adding up to the same total area. If each of the small reserves supported the same species, then it would be preferable to construct the larger reserve in the expectation of conserving more species (this recommendation derives from the species–area relationships discussed in Section 10.5.1).

2 On the other hand, if the region as a whole is heterogeneous, then each of the small reserves may support a different group of species and the total conserved might exceed that in a large reserve. In fact, collections of small islands tend to contain more species than a comparable area composed of one or a few large islands. The pattern is similar for habitat islands and, most significantly, for national parks. Thus, several small parks contained more species than larger ones of the same area in studies of mammals and birds in East African parks, of mammals and lizards in Australian reserves, and of large mammals in national parks in the United States. It seems likely that habitat heterogeneity is a general feature of considerable importance in determining species richness.

3 A point of particular significance is that local extinctions are common events, therefore recolonization of habitat fragments is critical for the survival of fragmented populations. Thus, we need to pay particular attention to the spatial relationships amongst fragments, including the provision of dispersal corridors. There are potential disadvantages—for example, corridors could increase the correlation among fragments of catastrophic effects such as the spread of fire or disease—but the arguments in favor are persuasive. Indeed, high recolonization rates (even if this means conservation managers themselves moving organisms around) may be indispensable to the success of conservation of endangered metapopulations. Note especially that human fragmentation of the landscape, producing subpopulations that are more and more isolated, is likely to have had the strongest effect on populations with naturally low rates of dispersal. Thus, the widespread declines of the world's amphibians may be due in significant measure to their poor potential for dispersal (Blaustein et al., 1994).

14.4.4 Restoration ecology

Conservation is an appropriate aim when there is something to conserve. In many cases, however, matters have gone too far. Species, or, quite likely, whole communities have disappeared, been destroyed, or have been altered beyond recognition. Restoration, rather than conservation, may then seem appropriate (Box 14.3).

restoring a "natural" community

But what, precisely, is to be restored? Firstly, there may be an attempt to restore the plants and animals that are considered to be the natural inhabitants of that area of land—"natural" in the sense that they were the species living there prior to the disturbance that ultimately led to the need for restoration. There are three main problems. First, we may have no clear idea what the authentic, original species are. Second, their natural pattern of establishment may take years. And third, short-cutting these natural processes, even if possible in principle, would require detailed knowledge of the ecologies of many species, which we rarely if ever have.

removing unwanted species

A community may have been changed dramatically because of the effects of one or more introduced species. In such cases, restoration involves removing and not just adding species (Myers et al., 2000a). New Zealand's offshore islands provide some significant success stories. Over the centuries, domestic stock were purposely introduced to many islands. Successful removals of cattle, goats, pigs, and cats occurred between 1916 and 1936. The much more daunting task of tackling the smaller, faster reproducing, and less conspicuous species, such as Norway rats, ship rats, and mice, has occupied teams of New Zealand ecologists for the past two decades, and with considerable success. Over 120 pest eradication campaigns have been conducted to restore island habitats in a way that improves them for endangered species already present and that opens up opportunities to translocate individuals from more vulnerable settings on the mainland (Towns & Ballantine, 1993). Some successful campaigns have also been staged to reduce exotic predators in forest patches on New Zealand's mainland areas (Saunders & Norton, 2001).

community rehabilitation or replacement

An alternative approach is necessary where abiotic conditions have been profoundly, perhaps irreversibly, altered. This occurs, for instance, when mine wastes of various sorts are deposited above ground. Here, restoration as such may not be

Box 14.3   *Topical ECOncerns*

People's Conflicting Demands on Ecosystems

This extract from an article in the *Boston Globe*, written by Beth Daley, shows graphically the conflicting demands people place on natural ecosystems.

Saving the "River of Grass"—the world's biggest environmental restoration faces one looming obstacle: thirsty cities

Weston, Florida. Few talk about it, but the restoration of the Everglades is as much about watering lawns in this planned community of peach-colored homes as it is about saving the fabled River of Grass and its wildlife.

The most ambitious environmental restoration project ever attempted is underway, rebuilding a swamp once considered so vile the federal government built a maze of levees, canals, and pumps to drain it. The damage from 50 years is everywhere: sixty-eight threatened plants and animals, 1.5 million acres infested with invasive weeds, and almost 2 billion gallons of water lost to the sea every day. Now, with much fanfare, engineers are rewiring nature to recapture that water in a daring attempt to stop the Everglades from dying of thirst. It's expected to take 40 years.

But as laudable as saving mangrove swamps and Cape Sable seaside sparrows may sound, the gleaming skyscrapers of Miami and sprawling suburbs like Weston stand to benefit enormously, too: about one-fifth of the water recovered by the Everglades restoration will go to human uses, from irrigation on farms to drinking water for fast-growing cities and towns.

Conservationists fear that dwindling bird populations and fragile vegetation will have a harder time making a comeback if water is siphoned off to communities first. They want a law spelling out exactly what the Everglades will get and when so politics in the future can't alter the formula.

How would you weigh up the costs and benefits of a restoration scheme such as this? The environment is set to benefit, enhancing natural values and recreational opportunities, some communities will be winners too by gaining access to more precious water (but possibly at the expense of the rate of environmental improvement); other communities will lose out as papaya orchards and horse ranches disappear under water as the swamps are reclaimed. Is it realistic to expect an economic cost : benefit analysis (where ecological economic values—see Box 14.1—are introduced to the normal mix of market forces) to provide the right balance?

realistic. Bradshaw (1984) has suggested the terms "rehabilitation" for the establishment of a community that is similar to but not the same as the original, and "replacement" for the establishment of a quite different community. The pragmatic approach of replacement will often be necessary, establishing whichever species can be established, in order to provide aesthetic or recreational facilities, or perhaps even a productive environment in terms, say, of forestry or grassland.

Whether the aim is pragmatic or romantic (pristine restoration), the exercise is one in community rather than population ecology. There is a need, in other words, for an understanding of the principles and processes described throughout this book.

restoration ecology is an acid test of ecological understanding

14.4.5 Finale—a healthy approach to conservation

the "triage" approach to setting
priorities

In desperate times, painful decisions have to be made about priorities. Thus, wounded soldiers arriving at field hospitals in World War I were subjected to a *triage* evaluation: priority 1 (those who were likely to survive but only with rapid intervention), priority 2 (those who were likely to survive without rapid intervention), and priority 3 (those who were likely to die with or without intervention). Conservation managers are often faced with the same kind of choices and need to demonstrate some courage in giving up on hopeless cases, and prioritizing those species and habitats where something can be done.

the biodiversity healthcare
compendium

Some argue that the species approach is very much an emergency room strategy when what is needed is better preventative and primary care—something that can most appropriately be accomplished by focusing on communities and ecosystems. We have also seen that intensive care (captive breeding in zoos or botanical gardens) and reconstructive surgery (restoration ecology of communities) need to figure in the biodiversity healthcare compendium. All these approaches have their place in protecting and enhancing the health of the living world.

taking a balanced view

The spectrum of opinions on conservation is complete. It ranges from the environmental extremist, who is prepared to destroy property and put human life at risk for what is seen as unacceptable exploitation of animals, to the other extreme of the exploitational extremist, who is prepared to destroy a rare habitat just as it is about to achieve protected status. There are zealots on both sides of the spectrum too. On the one hand, there are the industrialists, fishers, farmers, and foresters who accept none of the conservationist case and are not prepared to look objectively at the scientific evidence, while, on the other, are the environmental zealots—preservationists who seem unwilling to accept any exploitation of the natural world, some even pronouncing that fishing, hunting, or logging are intrinsically wrong. The middle ground is occupied by both exploiters and conservationists whose basic philosophy holds that natural resources can be used, but this should be done in a sustainable and balanced manner. A thorough understanding of the principles and applications of ecological science should enable all to pay healthy regard to the scientific aspects of what, in its broader context, is very much an ethical and sociopolitical problem.

Summary

In seeking to conserve the Earth's species and communities, we are often woefully ignorant of what there is to conserve

Conservation is the science concerned with increasing the probability that the Earth's species and communities (or, more generally, its biodiversity) will persist into the future. Biodiversity is, at its most basic, the number of species present, but it can also be viewed at smaller scales (e.g., genetic variation within populations) and larger scales (e.g., the variety of community types present in a region). About 1.8 million species have so far been named but the real number is probably between 3 and 30 million. The current observed rate of extinction may be as much as 100–1,000 times the background rate indicated by the fossil record.

Endangered species are usually rare but not all rare species are endangered

A species may be rare in the sense that its geographical and or habitat ranges are small or in the sense that local populations, even where they do occur, are small. Many species are naturally rare, but just by virtue of their rarity species are not necessarily at risk of extinction. However, other things being equal, it is easier to make a rare species extinct. Some species are born rare, others have rarity thrust upon them as a result of the actions of humans.

A variety of human influences have made species rarer and increased their probability of extinction

The principal causes of decline are overexploitaion, habitat degradation, and the introduction of exotic species. Overexploitation occurs when people harvest a population (for food or trophies) at a rate that is unsustainable. Humans adversely affect habitat in three ways—a proportion of available habitat may simply be destroyed, or it may be degraded by pollution, or it may be disturbed by human activities to the detriment of some of its occupants. Human-caused introductions of exotic species, which may occur accidentally or intentionally, have sometimes been responsible for dramatic changes to native species and natural communities.

Populations whose size is reduced by human agency may experience genetic problems

Rare alleles of a gene may confer no immediate advantage but could turn out to be well suited to changed environmental conditions in the future—small populations that have lost rare alleles through genetic drift have less potential to adapt. A more immediate potential problem is inbreeding depression; when populations are small there is a tendency for individuals breeding with one another

to be related, and this may lead to reductions in fertility, survivorship, growth rates, and resistance to disease.

Some species are at risk for a single reason but often a combination of factors is at work

A given population may have been reduced to a very small size by one or more of the processes described above, and this may have led to an increased frequency of matings among relatives and the expression of deleterious recessive alleles in their offspring, leading to reduced survivorship and fecundity and causing the population to become smaller still—the so-called extinction vortex.

Conservation biologists need to understand the dynamics of both small and fragmented populations

Much of conservation biology is a crisis discipline concerned with small populations in immediate danger of extinction. The dynamics of small populations are governed by a high level of uncertainty, whereas large populations can be described as being governed by "the law of averages". Three kinds of uncertainty or variation can be identified that are of particular importance to the fate of small populations: demographic uncertainty, environmental uncertainty, and spatial uncertainty. Moreover, loss of habitat frequently results not only in a reduction in the absolute size of a population but also the division of the original population into a number of fragments.

Conservation also often requires a broader, community perspective

A chain of extinctions may follow inexorably from the extinction of particular native species. A similarly unwanted outcome may occur if an exotic species is introduced into a community without regard to food web consequences. Moreover, when habitats become fragmented we need to pay attention not only to the fate of particular species but also to whether food webs maintain their functional integrity.

Practical conservation often requires the development of species' management plans, by a variety of means

Given the environmental circumstances and species characteristics of a particular rare species, what is the chance it will go extinct in a specified period? There are three approaches to determine the minimum viable population: a search for patterns in evidence already gathered in long-term studies, subjective assessment based on expert knowledge, and the development of simulation population models.

Captive breeding has a limited role to play in conservation

Captive propagation is attractive in its directness and the opportunity it provides to preserve options. However, many species are simply inappropriate for

captive breeding and/or eventual reintroduction, and available resources are insufficient to build an "ark" big enough for a significant proportion of the world's biota.

Protected areas need to be chosen and designed with care

Given limited funds to purchase protected areas, it is important to devise priorities so that they can be evaluated systematically and chosen with care. We know that the biotas of different locations vary in species richness, the extent to which the biota is unique, and the extent to which the biota is endangered; one or more of these criteria could be used to prioritize potential areas for protection. The principles of island biogeography theory provide some clues about the most appropriate shape and disposition of protected areas. A balanced approach to conservation requires a compendium of ecological "healthcare" measures: preventative and primary care (setting aside protected areas), emergency care for critically endangered species, intensive care (captive breeding in zoological and botanical gardens), and reconstructive surgery (restoration ecology).

Review Questions

1 ▲ Of the estimated 3–30 million species on Earth, only about 1.8 million have so far been named. How important is it for the conservation of biodiversity that we can name the species involved?

2 Species may be "rare" on three counts: what are these? From your own experience, provide examples of three "rare" species and explain the nature of their rarity.

3 ▲ Researchers collected data on the relative abundance of 16 Peruvian mammal species in forest areas that contrasted in whether they were subject to light or heavy hunting by local people. As an index of vulnerability to hunting they used the reduction in relative abundance in the heavily versus lightly hunted areas. This is plotted against intrinsic rate of population increase (r_{max}), age of first reproduction, and longevity in Figure 14.14.

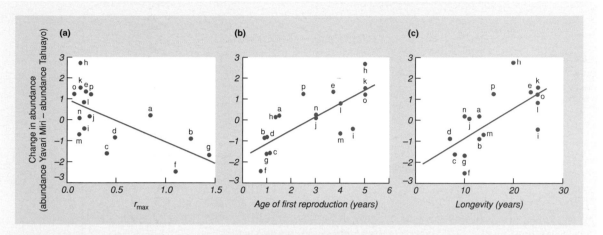

Figure 14.14
Relationships between (a) r_{max}, (b) age of first reproduction, (c) longevity and the vulnerability of mammals to population declines measured as the change in abundance between lightly and heavily hunted areas of forest. The mammals are represented by the following letters: a, white-lipped peccary; b, collared peccary; c, red brocket deer; d, grey brocket deer; e, lowland tapir; f, black agouti; g, green acouchy; h, woolly monkey; i, howler monkey; j, red wakari monkey; k, brown capuchin; l, white-fronted capuchin; m, monk saki monkey; n, titti monkey; o, spider monkey, p, squirrel monkey. (After Bodmer et al., 1997.)

Provide explanations for the relationships shown in the figure. Would you expect the variables r_{max}, age of first reproduction, and longevity to be intercorrelated? If so, how? Many species of large animals have gone extinct in the last 50,000 years. What light do the results of this study shed on the possible role of overexploitation by humans in historic extinctions? On the basis of these results, what advice would you give wildlife managers about conserving mammals in Peruvian forests?

4 Are there any circumstances where the intentional introduction of an exotic species can be considered a good thing because it enhances biodiversity?

5 Unpredictable temporal variability is a feature of most ecosystems. How can conservation biologists allow for such uncertainty when they devise species management plans?

6 Explain, with examples, how the the loss or introduction of a single species can have conservation consequences throughout a whole ecological community.

7 Discuss the triage approach to conservation assessment. List some highly endangered species of which you are aware and propose priorities for conservation action. Are any so hopeless that they should be allowed to go extinct?

8 Discuss the value of zoos and botanical gardens in nature conservation.

9 Why is restoration ecology said to be an "acid test" of ecological understanding?

10 ▲ The famous ecologist of the early 20th century, A.G. Tansley, when asked what he meant by nature conservation, said it was maintaining world in the state he knew as a child. From your perspective, as we enter the new millenium, how would you define the aims of conservation biology?

▲ = Challenge Question

Web Research Questions

1 Before global warming became number one on the environmental agenda, acid rain received frequent mention in the media. Have the problems associated with acid rain got better or worse, or are they now simply the least of our worries?

2 Biodiversity has become a buzzword in the media. How do ecologists define biodiversity and how well do those in the media understand its meaning?

3 Explore any lists of endangered species that you can find and evaluate the relative importance of different categories of risks of extinction to the group(s) you come up with.

4 Sharks. Most people fear them and might prefer to see them dead. Others see them as a valuable resource to exploit (for shark fin soup or big game hunting). Environmentalists see them as peculiarly susceptible to overexploitation and in need of conservation. Consider the diverse range of views and propose an overall strategy for their exploitation and/or conservation.

5 Discuss the pros and cons of traditional tertiary treatment (stripping of nutrients by chemical means) at sewage works as opposed to passing secondarily treated sewage through artificial wetlands.

References

Abramsky, Z. & Rosenzweig, M.L. (1983) Tilman's predicted productivity—diversity relationship shown by desert rodents. *Nature*, 309, 150–151.

Agardy, M.T. (1994) Advances in marine conservation: the role of marine protected areas. *Trends in Ecology and Evolution*, 9, 267–270.

Akçakaya, H.R. (1992) Population viability analysis and risk assessment. In: *Proceedings of Wildlife 2001: Populations* (D.R. McCullough, ed.). Elsevier, Amsterdam.

Albrecht, M. & Gotelli, N.J. (2001) Spatial and temporal niche partitioning in grassland ants. *Oecologia*, 126, 134–141.

Al-Hiyaly, S.A., McNeilly, T. & Bradshaw, A.D. (1988) The effects of zinc contamination from electricity pylons—evolution in a replicated situation. *New Phytologist*, 110, 571–580.

Allan, J.D. & Flecker, A.S. (1993) Biodiversity conservation in running waters. *Bioscience*, 43, 32–43.

Alliende, M.C. & Harper, J.L. (1989) Demographic studies of a dioecious tree. I. Colonization, sex and age-structure of a population of *Sarex cinerea*. *Journal of Ecology*, 77, 1029–1047.

Alonso, A., Dallmeier, F., Granek, E. & Raven, P. (2001) *Connecting with the Tapestry of Life*. Smithsonian Institution/ Monitoring and Assessment of Biodiversity Program and President's Committee of Advisors on Science and Technology, Washington, DC.

Ameczua, A.B. & Holyoak, M. (2000) Empirical evidence for predator–prey source–sink dynamics. *Ecology*, 81, 3087–3098.

Anderson, R.M. (1982) Epidemiology. In: *Modern Parasitology* (F.E.G. Cox, ed.), pp. 205–251. Blackwell Scientific Publications, Oxford.

Anderson, R.M. & May, R.M. (1991) *Infectious Diseases of Humans: Dynamics and Control*. Oxford University Press, Oxford.

Andrewartha, H.G. (1961) *Introduction to the Study of Animal Populations*. Methuen, London.

Andrewartha, H.G. & Birch, L.C. (1954) *The Distribution and Abundance of Animals*. University of Chicago Press, Chicago.

Andrews, P., Lorde, J.M. & Nesbit Evans, E.M. (1979) Patterns of ecological diversity in fossil and mammalian faunas. *Biological Journal of the Linnean Society*, 11, 177–205.

Angel, M.V. (1994) Spatial distribution of marine organisms: patterns and processes. In: *Large Scale Ecology and Conservation Biology* (P.J. Edwards, R.M. May & N.R. Webb, eds), pp. 59–109. Blackwell, Oxford.

Armbruster, P. & Lande, R. (1992) A population viability analysis for African elephant (*Loxodonta africana*): how big should reserves be? *Conservation Biology*, 7, 602–610.

Ashworth, S.T. & Kennedy, C.R. (1999) Density-dependent effects on *Anguillicola crassus* (Nematoda) within its European eel definitive host. *Parasitology*, 118, 289–296.

Aston, J.L. & Bradshaw, A.D. (1966) Evolution in closely adjacent plant populations. II. *Agrostis stolonifera* in maritime habitats. *Heredity*, 21, 649–664.

Audesirk, T. & Audesirk, G. (1996) *Biology: Life on Earth*. Prentice Hall, Upper Saddle River, NJ.

Ayre, D.J. (1985) Localized adaptation of clones of the sea anemone *Actinia tenebrosa*. *Evolution*, 39, 1250–1260.

Ayre, D.J. (1995) Localized adaptation of sea anemone clones: evidence from transplantation over two spatial scales. *Journal of Animal Ecology*, 64, 186–196.

Bach, C.E. (1994) Effects of herbivory and genotype on growth and survivorship of sand-dune willow (*Salix cordata*). *Ecological Entomology*, 19, 303–309.

Baltensweiler, W., Benz, G., Bovey, P. & Delucchi, V. (1977) Dynamics of larch budmoth populations. *Annual Review of Ecology and Systematics*, 22, 79–100.

Batzli, G.O. (1983) Responses of arctic rodent populations to nutritional factors. *Oikos*, 40, 396–406.

Bazzaz, F.A. (1979) The physiological ecology of plant succession. *Annual Review of Ecology and Systematics*, 10, 351–371.

Bazzaz, F.A. (1996) *Plants in Changing Environments*. Cambridge University Press, Cambridge.

Beaver, R.A. (1979) Host specificity of temperate and tropical animals. *Nature*, 281, 139–141.

Becker, P. (1992) Colonization of islands by carnivorous and herbivorous Heteroptera and Coleoptera: effects of island area, plant species richness, and 'extinction' rates. *Journal of Biogeography*, 19, 163–171.

Begon, M., Sait, S.M. & Thompson, D.J. (1995) Persistence of a predator–prey system: refuges and generation cycles? *Proceedings of the Royal Society of London, Series B*, 260, 131–137.

Bentley, S. & Whittaker, J.B. (1979) Effects of grazing by a chrysomelid beetle, *Gastrophysa viridula*, on competition between *Rumex obtusifolius* and *Rumex crispus*. *Journal of Ecology*, 69, 79–90.

Berger, J. (1990) Persistence of different-sized populations: an empirical assessment of rapid extinctions in bighorn sheep. *Conservation Biology*, 4, 91–98.

Berner, E.K. & Berner, R.A. (1987) *The Global Water Cycle: Geochemistry and Environment*. Prentice-Hall, Englewood Cliffs, NJ.

Berven, K.A. (1995) Population regulation in the wood frog, *Rana sylvatica*, from three diverse geographic localities. *Australian Journal of Ecology*, 20, 385–392.

Blaustein, A.R., Wake, D.B. & Sousa, W.P. (1994) Amphibian declines: judging stability, persistence, and susceptibility of populations to local and global extinctions. *Conservation Biology*, 8, 60–71.

Bodmer, R.E., Eisenberg, J.F. & Redford, K.H. (1997) Hunting and the likelihood of extinction of Amazonian mammals. *Conservation Biology*, 11, 460–466.

Borga, K., Gabrielsen, G.W. & Skaare, J.U. (2001) Biomagnification of organochlorines along a Barents Sea food chain. *Environmental Pollution*, 113, 187–198.

Bowles, K.C., Apte, S.C., Maher, W.A., Kawei, M. & Smith, R. (2001) Bioaccumulation and biomagnification of mercury in Lake Murray, Papua New Guinea. *Canadian Journal of Fisheries and Aquatic Science*, 58, 888–897.

Bradshaw, A.D. (1984) Ecological principles and land reclamation practice. *Landscape Planning*, 11, 35–48.

Breznak, J.A. (1975) Symbiotic relationships between termites and their intestinal biota. In: *Symbiosis* (D.H. Jennings & D.L. Lee, eds), pp. 559–580. Symposium 29, Society for Experimental Biology, Cambridge University Press, Cambridge. 13.5.2.

Briand, F. (1983) Environmental control of food web structure. *Ecology*, 64, 253–263.

Brook, B.W., O'Grady, J.J., Chapman, A.P., Burgman, M.A., Akçakaya, H.R. & Frankham, R. (2000) Predictive accuracy of population viability analysis in conservation biology. *Nature*, 404, 385–387.

Brookes, M. (1998) The species enigma. *New Scientist*, June 13, 1998.

Brown, J.H. & Davidson, D.W. (1977) Competition between seed-eating rodents and ants in desert ecosystems. *Science*, 196, 880–882.

Brown, K.M. (1982) Resource overlap and competition in pond snails: an experimental analysis. *Ecology*, 63, 412–422.

Brown, V.K. & Southwood, T.R.E. (1983) Trophic diversity, niche breadth, and generation times of exopterygote insects in a secondary succession. *Oecologia*, 56, 220–225.

Brunet, A.K. & Medellín, R.A. (2001) The species–area relationship in bat assemblages of tropical caves. *Journal of Mammalogy*, 82, 1114–1122.

Brylinski, M. & Mann, K.H. (1973) An analysis of factors governing productivity in lakes and reservoirs. *Limnology and Oceanography*, 18, 1–14.

Burgman, M.A., Ferson, S. & Akçakaya, H.R. (1993) *Risk Assessment in Conservation Biology*. Chapman & Hall, London.

Carignan, R., Planas, D. & Vis, C. (2000) Planktonic production and respiration in oligotrophic shield lakes. *Limnology and Oceanography*, 45, 189–199.

Caughley, G. (1994) Directions in conservation biology. *Journal of Animal Ecology*, 63, 215–244.

Charnov, E.L. (1976) Optimal foraging: attack strategy of a mantid. *American Naturalist*, 110, 141–151.

Clark, C.W. (1981) Bioeconomics. In: *Theoretical Ecology: Principles and Applications*, 2nd edn. (R.M. May, ed.), pp. 387–418. Blackwell Scientific Publications, Oxford.

Clements, F.L. (1905) *Research Methods in Ecology*. University of Nevada Press, Lincoln.

Cohen, J.E. (1995) *How Many People Can the Earth Support?* W.W. Norton & Co., New York.

Cole, J.J., Findlay, S. & Pace, M.L. (1988) Bacterial production in fresh and salt water ecosystems: a cross-system overview. *Marine Ecology Progress Series*, 4, 1–10.

Comins, H.N., Hassell, M.P. & May, R.M. (1992) The spatial dynamics of host–parasitoid systems. *Journal of Animal Ecology*, 61, 735–748.

Connell, J.H. (1978) Diversity in tropical rainforests and coral reefs. *Science*, 199, 1302–1310.

Connell, J.H. (1983) On the prevalence and relative importance of interspecific competition: evidence from field experiments. *American Naturalist*, 122, 661–696.

References **503**

Cook, J.A., Chubb, J.C. & Veltkamp, C.J. (1998) Epibionts of *Asellus aquaticus* (L.) (Crustacea, Isopoda): an SEM study. *Freshwater Biology*, 39, 423–438.

Cooper, J.P. (ed.) (1975) *Photosynthesis and Productivity in Different Environments*. Cambridge University Press, Cambridge.

Cory, J.S. & Myers, J.H. (2000) Direct and indirect ecological effects of biological control. *Trends in Ecology and Evolution*, 15, 137–139.

Costanza, R., D'Arge, R., de Groot, R. et al. (1997) The value of the world's ecosystem services and natural capital. *Nature* 387, 253–260.

Cox P.A., Elmquist, T., Pierson, E.D. & Rainey, W.E. (1991) Flying foxes as strong interactors in South Pacific island ecosystems: a conservation hypothesis. *Conservation Biology*, 5, 448–454.

Crawley, M.J. (1986) The structure of plant communities. In: *Plant Ecology* (M.J. Crawley, ed.), pp. 1–50. Blackwell, Oxford.

Crisp, M.D. & Lange, R.T. (1976) Age structure distribution and survival under grazing of the arid zone shrub *Acacia burkitii*. *Oikos*, 27, 86–92.

Currie, D.J. (1991) Energy and large-scale patterns of animal and plant species richness. *American Naturalist*, 137, 27–49.

Currie, D.J. & Paquin, V. (1987) Large-scale biogeographical patterns of species richness in trees. *Nature*, 39, 326–327.

Davidson, D.W. (1977) Species diversity and community organization in desert seed-eating ants. *Ecology*, 58, 711–724.

Davidson, J. & Andrewartha, H.G. (1948) The influence of rainfall, evaporation and atmospheric temperature on fluctuations in the size of a natural population of *Thrips imaginis* (Thysanoptera). *Journal of Animal Ecology*, 17, 200–222.

Davis, M.B. (1976) Pleistocene biogeography of temperate deciduous forest. *Geoscience and Management*, 13, 13–26.

Davis, M.B. & Shaw, R.G. (2001) Range shifts and adaptive responses to quarternary climate change. *Science*, 292, 673–679.

Davis, M.B., Sugita, S., Calcote, R.R., Ferrari, J.B. & Frelich, L.E. (1994) Historical development of alternate communities in a hemlock-hardwood forest in northern Michigan, USA. In: *Large-scale Ecology and Conservation Biogoy* (R.M. May, N. Webb & P. Edwards, eds), pp. 19–39. Blackwell Scientific Publications, Oxford.

de Wit, C.T., Tow, P.G. & Ennik, G.C. (1966) Competition between legumes and grasses. *Verslagen van landbouwkundige onderzoekingen*, 112, 1017–1045.

Deevey, E.S. (1947) Life tables for natural populations of animals. *Quarterly Review of Biology*, 22, 283–314.

Denno, R.F., McClure, M.S. & Ott, J.R. (1995) Interspecific interactions in phytophagous insects: competition reexamined and resurrected. *Annual Review of Entomology*, 40, 297–331.

Deshmukh, I. (1986) *Ecology and Tropical Biology*. Blackwell Scientific Publications, Oxford.

Detwiler, R.P. & Hall, C.A.S. (1988) Tropical forests and the global carbon cycle. *Science*, 239, 42–47.

Diamond, J.M. (1972) Biogeographic kinetics: estimation of relaxation times for avifaunas of South-West Pacific islands. *Proceedings of the National Academy of Science of the USA*, 69, 3199–3203.

Diamond, J.M. (1983) Taxonomy by nucleotides. *Nature*, 305, 17–18.

Dirzo, R. & Harper, J.L. (1982) Experimental studies of slug–plant interactions. IV. The performance of cyanogenic and acyanogenic morphs of *Trifolium repens* in the field. *Journal of Ecology*, 70, 119–138.

Dobson, A.P. & Carper, E.R. (1996) Infectious diseases and human population history. *Bioscience*, 46, 115–126.

Dohi, H., Yamada, A. & Entsu, S. (1991) Cattle feeding deterrents emitted from cattle feces. *Journal of Chemical Ecology*, 17, 1197–1203.

Doube, B.M., Macqueen, A., Ridsdill-Smith, T.J. & Weir, T.A. (1991) Native and introduced dung beetles in Australia. In: *Dung Beetle Ecology* (I. Hanski & Y. Cambefort, eds), pp. 255–278. Princeton University Press, Princeton, NJ.

Duynisveld, W.H.M., Strebel, O. & Bottcher, J. (1988) Are nitrate leaching from arable and nitrate pollution of groundwater avoidable? *Ecological Bulletins*, 39, 116–125.

Ebert, D., Zschokke-Rohringer, C.D. & Carius, H.J. (2000) Dose effects and density-dependent regulation in two microparasites of *Daphnia magna*. *Oecologia*, 122, 200–209.

Edwards, P.B. & Aschenborn, H.H. (1987) Patterns of nesting and dung burial in *Onitis* dung beetles: implications for pasture productivity and fly control. *Journal of Applied Ecology*, 24, 837–851.

Ehrlich, P.R. & Murphy, D.D. (1987) Conservation lessons from long-term studies of checkerspot butterflies. *Conservation Biology*, 1, 122–131.

Eis, S., Garman, E.H. & Ebel, L.F. (1965) Relation between cone production and diameter increment of douglas fir (*Pseudotsuga menziesii* (Mirb). Franco), grand fir (*Abies grandis* Dougl.) and western white pine (*Pinus monticola* Dougl.), *Canadian Journal of Botany*, 43, 1553–1559.

Elliott, J.M. (1994) *Quantitative Ecology and the Brown Trout*. Oxford University Press, Oxford.

Elton, C. (1927) *Animal Ecology*. Sidgwick & Jackson, London.

Elton, C. (1933) *The Ecology of Animals*. Methuen, London.

Elton, C. (1940) *The Ecology of Animals*, 2nd edn. Methuen, London.

Elton, C.S. (1958) *The Ecology of Invasion by Animals and Plants*. Methuen, London.

Emiliani, C. (1966) Isotopic palaeotemperatures. *Science*, 154, 851–857.

Endler, J.A. (1980) Natural selection on color patterns in *Poecilia reticulata*. *Evolution*, 34, 76–91.

Erwin, T.L. (1982) Tropical forests: their richness in Coleoptera and other arthropod species. *Coleopterists Bulletin*, 36, 74–75.

Fenner, F. (1983) Biological control, as exemplified by smallpox eradication and myxomatosis. *Proceedings of the Royal Society, Series B*, 218, 259–285.

Fenner, F. & Ratcliffe, R.N. (1965) *Myxomatosis*. Cambridge University Press, London.

Ferguson, R.G. (1933) The Indian tuberculosis problem and some preventative measures. *National Tuberculosis Association Transactions*, 29, 93–106.

Fischer, M. & Matthies, D. (1998) Effects of population size on performance in the rare plant *Gentianella germanica*. *Journal of Ecology*, 86, 195–204.

Fitter, A.H. (1991) The ecological significance of root system architecture. In: *Plant Root Growth: an Ecological Perspective* (D. Atkinson, ed.), pp. 229–243. Blackwell Scientific Publications, Oxford.

FitzGibbon, C.D. (1990) Anti-predator strategies of immature Thomson's gazelles: hiding and the prone response. *Animal Behaviour*, 40, 846–855.

FitzGibbon, C.D. & Fanshawe, J. (1989) The condition and age of Thomson's gazelles killed by cheetahs and wild dogs. *Journal of Zoology*, 218, 99–107.

Flecker, A.S. & Townsend, C.R. (1994) Community-wide consequences of trout introduction in New Zealand streams. *Ecological Applications*, 4, 798–807.

Flessa, K.W. & Jablonski, D. (1995) Biogeography of recent marine bivalve mollusks and its implications of paleobiogeography and the geography of extinction: a progress report. *Historical Biology*, 10, 25–47.

Flint, M.L. & van den Bosch, R. (1981) *Introduction to Integrated Pest Management*. Plenum Press, New York.

Flödder, S. & Sommer, U. (1999) Diversity in planktonic communities: an experimental test of the intermediate disturbance hypothesis. *Limnology and Oceanography*, 44, 1114–1119.

Flower, R.J., Rippey, B., Rose, N.L., Appleby, P.G. & Battarbee, R.W. (1994) Palaeolimnological evidence for the acidification and contamination of lakes by atmospheric pollution in western Ireland. *Journal of Ecology*, 82, 581–596.

Ford, E.B. (1975) *Ecological Genetics*, 4th edn. Chapman & Hall, London.

Ford, M.J. (1982) *The Changing Climate: Responses of the Natural Fauna and Flora*. George Allen & Unwin, London.

Fox, C.J. (2001) Recent trends in stock-recruitment of blackwater herring (*Clupea harengus* L.) in relation to larval production. *ICES Journal of Marine Science*, 58, 750–762.

Frank, D.A. & McNaughton, S.J. (1991) Stability increases with diversity in plant communities: empirical evidence from the 1988 Yellowstone drought. *Oikos*, 62, 360–362.

Franklin, I.A. (1980) Evolutionary change in small populations. In: *Conservation Biology, an Evolutionary-Ecological Perspect-ive* (M.E. Soulé & B.A. Wilcox, eds), pp. 135–149. Sinauer Associates, Sunderland, MA.

Fridriksson, S. (1975) *Surtsey: Evolution of Life on a Volcanic Island*. Butterworths, London.

Gadgil, M. (1971) Dispersal: population consequences and evolution. *Ecology*, 52, 253–261.

Galloway, L.F. & Fenster, C.B. (2000) Population differentiation in an annual legume: local adaptation. *Evolution*, 54, 1173–1181.

Gaston, K.J. (1998) *Biodiversity*. Blackwell Science, Oxford.

Gathreaux, S.A. (1978) The structure and organization of avian communities in forests. In: *Proceedings of the Workshop on Management of Southern Forests for Nongame Birds* (R.M. DeGraaf, ed.), pp. 17–37. Southern Forest Station, Asheville, NC.

Gende, S.M., Quinn, T.P. & Willson, M.F. (2001) Consumption choice by bears feeding on salmon. *Oecologia*, 127, 372–382.

Gibbs, L. (1998) *Love Canal: the Story Continues*. New Society Publishers.

Ginsberg, J.R. (1993) Can we build an Ark? *Trends in Ecology and Evolution*, 8, 4–6.

Godfray, H.C.J. & Crawley, M.J. (1998) Introductions. In: *Conservation Science and Action* (W.J. Sutherland, ed.), pp. 39–65.

Gostick, K.G. (1982) Agricultural Development and Advisory Service (ADAS) recommendations to farmers on manure disposal and recycling. *Philosophical Transactions of the Royal Society of London, Series B*, 296, 329–332.

Gotthard, K., Nylin, S. & Wiklund, C. (1999) Seasonal plasticity in two satyrine butterflies: state-dependent decision making in relation to daylength. *Oikos*, 84, 453–462.

Gough, L., Shaver, G.R., Carroll, J., Royer, D.L. & Laundre, J.A. (2000) Vascular plant species richness in Alaskan arctic tundra: the importance of soil pH. *Journal of Ecology*, 88, 54–66.

Gould, F. (1991) The evolutionary potential of crop pests. *American Scientist*, 79, 496–507.

Gould, W.A. & Walker, M.D. (1997) Landscape-scale patterns in plant species richness along an arctic river. *Canadian Journal of Botany*, 75, 1748–1765.

Grant, P.R., Grant, B.R., Keller, L.F. & Petren, K. (2000) Effects of El Nino events on Darwin's finch productivity. *Ecology*, 81, 2442–2457.

Green, R.E. (1998) Long-term decline in the thickness of eggshells of thrushes, *Turdus* spp., in Britain. *Proceedings of the Royal Society of London, Series B*, 265, 679–684.

Groves, R.H. & Williams, J. D. (1975) Growth of skeleton weed (*Chondrilla juncea* L.) as affected by growth of subterranean clover (*Trifolium subterraneum* L.) and infection by *Puccinia chondrilla* Bubak and Syd. *Australian Journal of Agricultural Research*, 26, 975–983.

Haefner, P.A. (1970) The effect of low dissolved oxygen concentrations on temperature–salinity tolerance of the sand shrimp, *Crangon septemspinosa*. *Physiological Zoology*, 43, 30–37.

Hairston, N.G., Smith, F.E. & Slobodkin, L.B. (1960) Community structure, population control, and competition. *American Naturalist*, 44, 421–425.

Halaj, J., Ross, D.W. & Moldenke, A.R. (2000) Importance of habitat structure to the arthropod food-web in Douglas-fir canopies. *Oikos*, 90, 139–152.

Hall, D.C., Norgard, R.B. & True, P.K. (1975) The performance of independent pest management consultants. *California Agriculture*, 29, 12–14.

Hall, S.J. (1998) Closed areas for fisheries management—the case consolidates. *Trends in Ecology and Evolution*, 13, 297–298.

Hall, S.J. & Raffaelli, D.G. (1993) Food webs: theory and reality. *Advances in Ecological Research*, 24, 187–239.

Hansen, J., Ruedy, R., Glascoe, J. & Sato, M. (1999) GISS analysis of surface temperature change. *Journal of Geophysical Research*, 104, 30997–31022.

Hanski, I. (1994) Metapopulation ecology. In: *Spatial and Temporal Aspects of Population Processes* (O.E. Rhodes Jr., ed.), pp. 13–43. University of Georgia, Georgia.

Hanski, I., Pakkala, T., Kuussaari, M. & Lei, G. (1995) Metapopulation persistence of an endangered butterfly in a fragmented landscape. *Oikos*, 72, 21–28.

Harcourt, D.G. (1971) Population dynamics of *Leptinotarsa decemlineata* (Say) in eastern Ontario. III. Major population processes. *Canadian Entomologist*, 103, 1049–1061.

Harper, J.L. (1955) The influence of the environment on seed and seedling mortality. VI. The effects of the interaction of soil moisture content and temperature on the mortality of maize grains. *Annals of Applied Biology*, 43, 696–708.

Harper, J.L. (1977) *The Population Biology of Plants*. Academic Press, London

Harper, J.L. & White, J. (1974) The demography of plants. *Annual Review of Ecology and Systematics*, 5, 419–463.

Hassell, M.P., Latto, J. & May, R.M. (1989) Seeing the wood for the trees: detecting density dependence from existing life-table studies. *Journal of Animal Ecology*, 58, 883–892.

Heal, O.W., Menault, J.C. & Steffen, W.L. (1993) *Towards a Global Terrestrial Observing System (GTOS): Detecting and Monitoring Change in Terrestrial Ecosystems*. MAB Digest 14 and IGBP Global Change Report 26, UNESCO, Paris and IGBP, Stockholm.

Higgins, S.I., Bond, W.J. & Trollope, W.S.W. (2000) Fire, resprouting and variability: a recipe for grass–tree coexistence in savanna. *Journal of Ecology*, 88, 213–229.

Hilborn, R. & Walters, C.J. (1992) *Quantitative Fisheries Stock Assessment*. Chapman & Hall, New York.

Hildrew, A.G. & Townsend, C.R. (1980) Aggregation, interference and the foraging by larvae of *Plectrocnemia conspersa* (Trichoptera: Polycentropodidae). *Animal Behaviour*, 28, 553–560.

Holloway, J.M., Dahlgren, R.A., Hansen, B. & Casey, W.H. (1998) Contribution of bedrock nitrogen to high nitrate concentrations in stream water. *Nature*, 395, 785–788.

Holt, R.D. & Hassell, M.P. (1993) Environmental heterogeneity and the stability of host–parasitoid interactions. *Journal of Animal Ecology*, 62, 89–100.

Hoyer, M.V. & Canfield, D.E. (1994) Bird abundance and species richness on Florida lakes: influence of trophic status, lake morphology and aquatic macrophytes. *Hydrobiologia*, 297, 107–119.

Hudson, P.J., Dobson, A.P. & Newborn, D. (1992) Do parasites make prey vulnerable to predation? Red grouse and parasites. *Journal of Animal Ecology*, 61, 681–692.

Hudson, P.J., Dobson, A.P. & Newborn, D. (1998) Prevention of population cycles by parasite removal. *Science*, 282, 2256–2258.

Huffaker, C.B. (1958) Experimental studies on predation: dispersion factors and predator–prey oscillations. *Hilgardia*, 27, 343–383.

Hughes, L. (2000) Biological consequences of global warming: is the signal already apparent. *Trends in Ecology and Evolution*, 15, 56–61.

Hunter, M.L. & Yonzon, P. (1992) Altitudinal distributions of birds, mammals, people, forests, and parks in Nepal. *Conservation Biology*, 7, 420–423.

Huryn, A.D. (1998) Ecosystem-level evidence for top-down and bottom-up control of production in a grassland stream system. *Oecologia*, 115, 173–183.

Hutchinson, G.E. (1957) Concluding remarks. *Cold Spring Harbour Symposium on Quantitative Biology*, 22, 415–427.

Inouye, D.W. (1978) Resource partitioning in bumblebees: experimental studies of foraging behaviour. *Ecology*, 59, 672–678.

Inouye, R.S. & Tilman, D. (1995) Convergence and divergence of old-field vegetation after 11 yr of nitrogen addition. *Ecology*, 76, 1872–1877.

Inouye, R.S., Huntly, N.J., Tilman, D., Tester, J.R., Stillwell, M. & Zinnel, K.C. (1987) Old-field succession on a Minnesota sand plain. *Ecology*, 68, 12–26.

Interlandi, S.J. & Kilham, S.S. (2001) Limiting resources and the regulation of diversity in phytoplankton communities. *Ecology*, 82, 1270–1282.

International Organisation for Biological Control (1989) *Current Status of Integrated Farming Systems Research in Western Europe* (P. Vereijken & D.J. Royle, eds). IOBC WPRS Bulletin 12(5).

IPCC (2001) *Third Assessment Report*. Working Group 1, Intergovernmental Panel on Climate Change.

Irvine, R.J., Stien, A., Dallas, J.F., Halvorsen, O., Langvatn, R. & Albon, S.D. (2001) Contrasting regulation of fecundity in two abomasal nematodes of Svarlbard reindeer (*Rangifer tarandus platyrynchus*). *Parasitology*, 122, 673–681.

IUCN/UNEP/WWF (1991) *Caring for the Earth. A Strategy for Sustainable Living*. Gland, Switzerland.

Ives, A.R. (1992) Continuous-time models of host–parasitoid interactions. *American Naturalist*, 140, 1–29.

Jackson, S.T. & Weng, C. (1999) Late quaternary extinction of a tree species in eastern North America. *Proceedings of the National Academy of Sciences*, 96, 13847–13852.

Jain, S.K. & Bradshaw, A.D. (1966) Evolutionary divergence among adjacent plant populations. I. The evidence and its theoretical analysis. *Heredity*, 21, 407–411.

Janis, C.M. (1993) Tertiary mammal evolution in the context of changing climates, vegetation and tectonic events. *Annual Review of Ecology and Systematics*, 24, 467–500.

Johannes, R.E. (1998) Government-supported village-based management of marine resources in Vanuatu. *Ocean Coastal Management*, 40, 165–186.

Jones, C.G., Lawton, J.H. & Shaachak, M. (1997) Positive and negative effects of organisms as physical ecosystem engineers. *Ecology*, 78, 1946–1957.

Jones, M. & Harper, J.L. (1987) The influence of neighbours on the growth of trees. I. The demography of buds in *Betula pendula*. *Proceedings of the Royal Society of London, Series B*, 232, 1–18.

Jonsson, M. & Malmqvist, B. (2000) Ecosystem process rate increases with animal species richness: evidence from leaf-eating, aquatic insects. *Oikos*, 89, 519–523.

Kaiser, J. (2000) Rift over biodiversity divides ecologists. *Science*, 89, 1282–1283.

Karlsson, P.S. & Jacobson, A. (2001) Onset of reproduction in *Rhododendron lapponicum* shoots: the effect of shoot size, age, and nutrient status at two subarctic sites. *Oikos*, 94, 279–286.

Kaufman, L. (1992) Catastrophic change in a species-rich freshwater ecosystem: lessons from Lake Victoria. *Bioscience*, 42, 846–858.

Keddy, P.A. (1982) Experimental demography of the sand-dune annual, *Cakile edentula*, growing along an environmental gradient in Nova Scotia. *Journal of Ecology*, 69, 615–630.

Keeley, E.R. (2001) Demographic responses to food and space competition by juvenile steelhead trout. *Ecology*, 82, 1247–1259.

Keith, L.B. (1983) Role of food in hare population cycles. *Oikos*, 40, 385–395.

Keith, L.B., Cary, J.R., Ronstad, O.J. & Brittingham, M.C. (1984) Demography and ecology of a declining snowshoe hare population. *Wildlife Monographs*, 90, 1–43.

Kerbes, R.H., Kotanen, P.M. & Jefferies, R.L. (1990) Destruction of wetland habitats by lesser snow geese: a keystone species on the west coast of Hudson Bay. *Journal of Applied Ecology*, 27, 242–258.

Kettlewell, H.B.D. (1955) Selection experiments on industrial melanism in the Lepidoptera. *Heredity*, 9, 323–342.

Kicklighter, D.W., Bruno, M., Donges, S. et al. (1999) A first-order analysis of the potential role of CO_2 fertilization to affect the global carbon budget: a comparison of four terrestrial biosphere models. *Tellus*, 51B, 343–366.

Kingston, T.J. & Coe, M.J. (1977) The biology of a giant dung-beetle (*Heliocorpis dilloni*) (Coleoptera: Scarabaeidae). *Journal of Zoology*, 181, 243–263.

Kodric-Brown, A. & Brown, J.M. (1993) Highly structured fish communities in Australian desert springs. *Ecology*, 74, 1847–1855.

Koella, J.C., Sörensen, F.L. & Anderson, R.A. (1998) The malaria parasite, *Plasmodium falciparum*, increases the frequency of multiple feeding of its mosquito vector, *Anopheles gambiae*. *Proceedings of the Royal Society of London, Series B*, 265, 763–768.

Kratz, T.K., Webster, K.E., Bowser, C.J. et al. (1997) The influence of landscape position on lakes in Northern Wisconsin. *Freshwater Biology*, 37, 209–217.

Krebs, C.J. (1972) *Ecology*. Harper & Row, New York.

Krebs, C.J., Boonstra, R., Boutin, S. et al. (1992) What drives the snowshoe hare cycle in Canada's Yukon. In: *Wildlife 2001: Populations* (D.R. McCullough & R.H. Barrett, eds), pp. 886–896. Elsevier, New York.

Krebs, J.R. (1978) Optimal foraging: decision rules for predators. In: *Behavioural Ecology: an Evolutionary Approach* (J.R. Krebs & N.B. Davies, eds), pp. 23–63. Blackwell Scientific Publications, Oxford.

Krebs, J.R., Erichsen, J.T., Webber, M.I. & Charnov, E.L. (1977) Optimal prey selection in the great tit (*Parus major*). *Animal Behaviour*, 25, 30–38.

Kullberg, C. & Ekman, J. (2000) Does predation maintain tit community diversity? *Oikos*, 89, 41–45.

Lande, R. & Barrowclough, G.F. (1987) Effective population size, genetic variation, and their use in population management. In: *Viable Populations for Conservation* (M.E. Soulé, ed.), pp. 87–123. Cambridge University Press, Cambridge.

Larcher, W. (1980) *Physiological Plant Ecology*, 2nd edn. Springer-Verlag, Berlin.

Laurance, W.F. (2001) Future shock: forecasting a grim fate for the Earth. *Trends in Ecology and Evolution*, 16, 531–533.

Lawlor, L.R. (1980) Structure and stability in natural and randomly constructed competitive communities. *American Naturalist*, 116, 394–408.

Lawrence, W.H. & Rediske, J.H. (1962) Fate of sown douglas-fir seed. *Forest Science*, 8, 211–218.

Lawton, J.H. & May, R.M. (1984) The birds of Selborne. *Nature*, 306, 732–733.

Lawton, J.H. & Woodroffe, G.L. (1991) Habitat and the distribution of water voles: why are there gaps in a species' range? *Journal of Animal Ecology*, 60, 79–91.

Le Cren, E.D. (1973) Some examples of the mechanisms that control the population dynamics of salmonid fish. In: *The Mathematical Theory of the Dynamics of Biological Populations* (M.S. Bartlett & R.W. Hiorns, eds), pp. 125–135. Academic Press, London.

Lennartsson, T., Nilsson, P. & Tuomi, J. (1998) Induction of overcompensation in the field gentian, *Gentianella campestris*. *Ecology*, 79, 1061–1072.

Leroy, F. & de Vuyst, L. (2001) Growth of the bacteriocin-producing *Lactobacillus sakei* strain CTC 494 in MRS broth is strongly reduced due to nutrient exhaustion: a nutrient depletion model for the growth of lactic acid bacteria. *Applied and Environmental Microbiology*, 67, 4407–4413.

Leverich, W.J. & Levin, D.A. (1979) Age-specific survivorship and reproduction in *Phlox drummondii*. *American Naturalist*, 113, 881–903.

Levins, R. (1969) Some demographic and genetic consequences of environmental heterogeneity for biological control. *Bulletin of the Entomological Society of America*, 15, 237–240.

Likens, G.E. (1989) Some aspects of air pollutant effects on terrestrial ecosystems and prospects for the future. *Ambio*, 18, 172–178.

Likens, G.E. (1992) *The Ecosystem Approach: its Use and Abuse*. Excellence in Ecology, Book 3. Ecology Institute, Oldendorf-Luhe, Germany.

Likens, G.E. & Bormann, F.G. (1975) An experimental approach to New England landscapes. In: *Coupling of Land and Water Systems* (A.D. Hasler, ed.), pp. 7–30. Springer-Verlag, New York.

Likens, G.E. & Bormann, F.H. (1995) *Biogeochemistry of a Forested Ecosystem*, 2nd edn. Springer-Verlag, New York.

Likens, G.E., Bormann, F.H., Pierce, R.S. & Fisher, D.W. (1971) Nutrient–hydrologic cycle interaction in small forested watershed ecosystems. In: *Productivity of Forest Ecosystems* (P. Duvogneaud, ed.). UNESCO, Paris.

Likens, G.E., Driscoll, C.T. & Buso, D.C. (1996) Long-term effects of acid rain: response and recovery of a forest ecosystem. *Science*, 272, 244–245.

Lindeman, R.L. (1942) The trophic–dynamic aspect of ecology. *Ecology*, 23, 399–418.

Lobo, J.M., Castro, I. & Moreno, J.C. (2001) Spatial and environmental determinants of vascular plant species richness distribution in the Iberian Peninsular and Balearic Islands. *Biological Journal of the Linnean Society*, 73, 233–253.

Losos, E. (1993) The future of the US Endangered Species Act. *Trends in Ecology and Evolution*, 8, 332–336.

Lotka, A.J. (1932) The growth of mixed population: two species competing for a common food supply. *Journal of the Washington Academy of Sciences*, 22, 461–469.

Louda, S.M. (1982) Distributional ecology: variation in plant recruitment over a gradient in relation to insect seed predation. *Ecological Monographs*, 52, 25–41.

Louda, S.M. (1983) Seed predation and seedling mortality in the recruitment of a shrub, *Haplopappus venetus* (Asteraceae), along a climatic gradient. *Ecology*, 64, 511–521.

Louda, S.M., Kendall, D., Connor, J. & Simberloff, D. (1997) Ecological effects of an insect introduced for the biological control of weeds. *Science*, 277, 1088–1090.

Lövei, G.L. (1997) Global change through invasion. *Nature*, 388, 627–628.

Lubchenco, J. (1978) Plant species diversity in a marine intertidal community: importance of herbivore food preference and algal competitive abilities. *American Naturalist*, 112, 23–39.

Lubchenco, J. (1991) The sustainable biosphere initiative: an ecological research agenda. *Ecology*, 72, 371–412.

Luckman, W.H. & Decker, G.C. (1960) A 5-year report on observations in the Japanese beetle control area of Sheldon, Illinois. *Journal of Economic Entomology*, 53, 821–827.

Lukens, R.J. & Mullany, R. (1972) The influence of shade and wet on southern corn blight. *Plant Disease Reporter*, 56, 203–206.

Lussenhop, J. (1992) Mechanisms of microarthropod–microbial interactions in soil. *Advances in Ecological Research*, 23, 1–33.

Lutz, W., Sanderson, W. & Scherbov, S. (2001) The end of world population growth. *Nature*, 412, 543–545.

MacArthur, J.W. (1975) Environmental fluctuations an species diversity. In: *Ecology and Evolution of Communities* (M.L. Cody & J.M. Diamond, eds), pp. 74–80. Belknap, Cambridge, MA.

MacArthur, R.H. (1955) Fluctuations of animal populations and a measure of community stability. *Ecology*, 36, 533–536.

MacArthur, R.H. (1972) *Geographical Ecology*. Harper & Row, New York.

MacArthur, R.H. & Pianka, E.R. (1966) On optimal use of a patchy environment. *American Naturalist*, 100, 603–609.

MacArthur, R.H. & Wilson, E.O. (1967) *The Theory of Island Biogeography*. Princeton University Press, Princeton, NJ.

Mace, G.M. (1994) An investigation into methods for categorizing the conservation status of species. In: *Large-Scale Ecology and Conservation Biology* (P.J. Edwards, R.M. May & N.R. Webb, eds), pp. 293–312. Blackwell Scientific Publications, Oxford.

Mack, R.N., Simberloff, D., Lonsdale, W.M., Evans, H., Clout, M. & Bazzaz, F.A. (2000) Biotic invasions: causes, epidemiology, global consequences, and control. *Ecological Applications*, 10, 689–710.

MacLulick, D.A. (1937) Fluctuations in numbers of the varying hare (*Lepus americanus*). *University of Toronto Studies, Biology Series*, 43, 1–136.

Maguire, L.A., Seal, U.S. & Brussard, P.F. (1987) Managing critically endangered species: the Sumatran rhino as a case study. In: *Viable Populations for Conservation* (M.E. Soulé, ed.), pp. 141–158. Cambridge University Press, Cambridge.

Magurran, A.E. (1998) Population differentiation without speciation. *Philosophical Transactions of the Royal Society of London, Series B*, 353, 275–286.

Malmqvist, B., Wotton, R.S. & Zhang, Y. (2001) Suspension feeders transform massive amounts of seston in large northern rivers. *Oikos*, 92, 35–43.

Manire, C.A. & Gruber, S.H. (1990) Many sharks may be headed toward extinction. *Conservation Biology*, 4, 10–11.

Margules, C.R. & Pressey, R.L. (2000) Systematic conservation planning. *Nature*, 405, 243–253.

Marshall, I.D. & Douglas, G.W. (1961) Studies in the epidemiology of infectious myxomatosis of rabbits. VIII. Further observations on changes in the innate resistance of Australian wild rabbits exposed to myxomatosis. *Journal of Hygiene*, 59, 117–112.

Martin, P.S. (1984) Prehistoric overkill: the global model. In: *Quaternary Extinctions: a Prehistoric Revolution* (P.S. Martin & R.G. Klein, eds). University of Arizona Press, Tuscon, AZ.

Marzusch, K. (1952) Untersuchungen über di Temperaturabhängigkeit von Lebensprozessen bei Insekten unter besonderer Berücksichtigung winter-schlantender Kartoffelkäfer. *Zeitschrift für vergleicherde Physiologie*, 34, 75–92.

May, R.M. (1981) Patterns in multi-species communities. In: *Theoretical Ecology: Principles and Applications*, 2nd edn (R.M. May, ed.), pp. 197–227. Blackwell Scientific Publications, Oxford.

May, R.M. & Anderson, R.M. (1983) Epidemiology and genetics in the coevolution of parasites and hosts. *Proceedings of the Royal Society of London, Series B*, 219, 281–313.

McIntosh, A.R. & Townsend, C.R. (1994) Interpopulation variation in mayfly antipredator tactics: differential effects of contrasting predatory fish. *Ecology*, 75, 2078–2090.

McIntosh, A.R. & Townsend, C.R. (1996) Interactions between fish, grazing invertebrates and algae in a New Zealand stream: a trophic cascade mediated by fish-induced changes to grazer behavior. *Oecologia*, 108, 174–181.

McKay, J.K., Bishop, J.G., Lin, J-Z., Richards, J.H., Sala, A. & Mitchell-Olds, T. (2001) Local adaptation across a climatic gradient despite small effective population size in the rare sapphire rockcress. *Proceedings of the Royal Society of London, Series B*, 268, 1715–1721.

McNaughton, S.J. (1977) Diversity and stability of ecological communities: a comment on the role of empiricism in ecology. *American Naturalist*, 111, 515–525.

Mikkelson, G.M. (1993) How do food webs fall apart? A study of changes in trophic structure during relaxation on habitat fragments. *Oikos*, 67, 539–547.

Mittelbach, G.G., Steiner, C.F., Scheiner, S.M. et al. (2001) What is the observed relationship between species richness and productivity? *Ecology*, 82, 2381–2396.

Morris, D.W. & Davidson, D.L. (2000) Optimally foraging mice match patch use with habitat differences in fitness. *Ecology*, 81, 2061–2066.

Morrison, L.W. (1997) The insular biogeography of small Bahamian cays. *Journal of Ecology*, 85, 441–454.

Murdoch, W.W. & Stewart-Oaten, A. (1975) Predation and population stability. *Advances in Ecological Research*, 9, 1–131.

Murdoch, W.W., Briggs, C.J., Nisbet, R.M., Gurney, W.S.C. & Stewart-Oaten, A. (1992) Aggregation and stability in metapopulation models. *American Naturalist*, 140, 41–58.

Murton, R.K., Westwood, N.J. & Isaacson, A.J. (1974) A study of wood-pigeon shooting: the exploitation of a natural animal population. *Journal of Applied Ecology*, 11, 61–81.

Mwendera, E.J., Saleem, M.A.M. & Woldu, Z. (1997) Vegetation response to cattle grazing in the Ethiopian Highlands. *Agriculture, Ecosystems and Environment*, 64, 43–51.

Myers, J.H., Simberloff, D., Kuris, A.M. & Carey, J.R. (2000a) Eradication revisited: dealing with exotic species. *Trends in Ecology and Evolution*, 15, 316–320.

Myers, N., Mittermeier, R.A., Mittermeier, C.G., da Fonseca, G.A.B. & Kent, J. (2000b) Biodiversity hotspots for conservation priorities. *Nature*, 403, 853–858.

Myers, R.A. (2001) Stock and recruitment: generalizations about maximum reproductive rate, density dependence, and variability using meta-analytic approaches. *ICES Journal of Marine Science*, 58, 937–951.

National Research Council (1990) *Alternative Agriculture*. National Academy of Sciences, Academy Press, Washington, DC.

Nedergaard, J. & Cannon, B. (1990) Mammalian hibernation. *Philosophical Transactions of the Royal Society, Series B*, 326, 669–686; also in *Life at Low Temperatures* (R.M. Laws & F. Franks, eds), pp. 153–170. The Royal Society, London.

Neilson, R.P., Prentice, I.C., Smith, B., Kittel, T. & Viner, D. (1998) Simulated changes in vegetation distribution under global warming. Available as Annex C at http://www.epa.gov/globalwarming/reports/pubs/ipcc/annex/index.html.

NERC (1990) *Our Changing Environment*. Natural Environment Research Council, London. (NERC acknowledges the significant contribution of Fred Pearce to the document.)

Nicholson, A.J. (1954) An outline of the dynamics of animal populations. *Australian Journal of Zoology*, 2, 9–65.

Nielsen, B.O. & Ejlerson, A. (1977) The distribution of herbivory in a beech canopy. *Ecological Entomology*, 2, 293–299.

Niklas, K.J., Tiffney, B.H. & Knoll, A.H. (1983) Patterns in vascular land plant diversification. *Nature*, 303, 614–616.

Norton, I.O. & Sclater, J.G. (1979) A model for the evolution of the Indian Ocean and the breakup of Gondwanaland. *Journal of Geophysical Research*, 84, 6803–6830.

Nunney, L. & Campbell, K.A. (1993) Assessing minimum viable population sizes: demography meets population genetics. *Trends in Ecology and Evolution*, 8, 234–239.

Ogden, J. (1968) *Studies on reproductive strategy with particular reference to selected composites*. PhD. thesis, University of Wales.

Owen-Smith, N. (1987) Pleistocene extinctions: the pivotal role of megaherbivores. *Paleobiology*, 13, 351–362.

Paine, R.T. (1966) Food web complexity and species diversity. *American Naturalist*, 100, 65–75.

Paine, R.T. (1979) Disaster, catastrophe and local persistence of the sea palm *Postelsia palmaeformis*. *Science*, 205, 685–687.

Pauly, D. & Christensen, V. (1995) Primary production required to sustain global fisheries. *Nature*, 374, 255–257.

Pavia, H. & Toth, G.B. (2000) Inducible chemical resistance to herbivory in the brown seaweed *Ascophyllum nodosum*. *Ecology*, 81, 3212–3225.

Pearl, R. (1927) The growth of populations. *Quarterly Review of Biology*, 2, 532–548.

Pearl, R. (1928) *The Rate of Living*. Knopf, New York.

Perrins, C.M. (1965) Population fluctuations and clutch size in the great tit, *Parus major* L. *Journal of Animal Ecology*, 34, 601–647.

Petren, K. & Case, T.J. (1996) An experimental demonstration of exploitation competition in an ongoing invasion. *Ecology*, 77, 118–132.

Petren, K., Grant, B.R. & Grant, P.R. (1999) A phylogeny of Darwin's finches based on microsatellite DNA variation. *Proceeding of the Royal Society of London, Series B*, 266, 321–329.

Pianka, E.R. (1983) *Evolutionary Ecology*, 3rd edn. Harper & Row, New York.

Pimentel, D. (1993) Cultural controls for insect pest management. In: *Pest Control and Sustainable Agriculture* (S. Corey, D. Dall & W. Milne, eds), pp. 35–38. CSIRO, East Melbourne.

Pimentel, D., Krummel, J., Gallahan, D. et al. (1978) Benefits and costs of pesticide use in U.S. food production. *Bioscience*, 28, 777–784.

Pimentel, D., Lach, L., Zuniga, R. & Morrison, D. (2000) Environmental and economic costs of nonindigenous species in the United States. *BioScience*, 50, 53–65.

Pimm, S.L. (1991) *The Balance of Nature: Ecological Issues in the Conservation of Species and Communities*. University of Chicago Press, Chicago.

Pisek, A., Larcher, W., Vegis, A. & Napp-Zin, K. (1973) The normal temperature range. In: *Temperature and Life* (H. Precht, J. Christopherson, H. Hense & W. Larcher, eds), pp. 102–194. Springer-Verlag, Berlin.

Pitcher, T.J. & Hart, P.J.B. (1982) *Fisheries Ecology*. Croom Helm, London.

Playfair, J.H.L. (1996) *Immunology at a Glance*, 6th edn. Blackwell Science, Oxford.

Power, M.E. (1990) Effects of fish in river food webs. *Science*, 250, 411–415.

Price, P.W. (1980) *Evolutionary Biology of Parasites*. Princeton University Press, Princeton, NJ.

Primack, R.B. (1993) *Essentials of Conservation Biology*. Sinauer Associates, Sunderland, MA.

Pyke, G.H. (1982) Local geographic distributions of bumblebees near Crested Butte, Colorado: competition and community structure. *Ecology*, 63, 555–573.

Raffaelli, D. & Hawkins, S. (1996) *Intertidal Ecology*. Kluwer, Dordrecht.

Randall, M.G.M. (1982) The dynamics of an insect population throughout its altitudinal distribution: *Coleophora alticolella* (Lepidoptera) in northern England. *Journal of Animal Ecology*, 51, 993–1016.

Ratcliffe, D.A. (1970) Changes attributable to pesticides in egg breakage frequence and eggshell thickness in some British birds. *Journal of Applied Ecology*, 7, 67–107.

Rätti, O., Dufva, R. & Alatalo, R.V. (1993) Blood parasites and male fitness in the pied flycatcher. *Oecologia*, 96, 410–414.

Raunkiaer, C. (1934) *The Life and Form of Plants*. Oxford University Press, Oxford. (Translated from the original published in Danish, 1907.)

Reganold, J.P., Glover, J.D., Andrews, P.K. & Hinman, H.R. (2001) Sustainability of three apple production systems. *Nature*, 410, 926–929.

Reichle, D.E. (1970) *Analysis of Temperate Forest Ecosystems*. Springer-Verlag, New York

Reid, W.V. & Miller, K.R. (1989) *Keeping Options Alive: the Scientific Basis for Conserving Biodiversity*. World Resources Institute, Washington, DC.

Richards, O.W. & Waloff, N. (1954) Studies on the biology and population dynamics of British grasshoppers. *Anti-Locust Bulletin*, 17, 1–182.

Rickards, J., Kelleher, M.J. & Storey, K.B. (1987) Strategies of freeze avoidance in larvae of the goldenrod gall moth *Epiblema scudderiana*: winter profiles of a natural population. *Journal of Insect Physiology*, 33, 581–586.

Ricklefs, R.E. (1973) *Ecology*. Nelson, London.

Ricklefs, R.E. & Lovette, I.J. (1999) The role of island area per se and habitat diversity in the species–area relationships of four Lesser Antillean faunal groups. *Journal of Animal Ecology*, 68, 1142–1160.

Ridley, M. (1993) *Evolution*. Blackwell Science, Boston.

Ridsdill-Smith, T.J. (1991) Competition in dung-breeding insects. In: *Reproductive Behaviour of Insects* (W.J. Bailey & T.J. Ridsdill-Smith, eds), pp. 264–294. Chapman & Hall, London.

Robinson, R.A. & Sutherland, W.R. (1999) *Ecography*, 22, 447–454.

Rohr, D.H. (2001) Reproductive trade-offs in the elapid snakes *Austrelap superbus* and *Austrelap ramsayi*. *Canadian Journal of Zoology*, 79, 1030–1037.

Rosenzweig, M.L. (1971) Paradox of enrichment: destabilization of exploitation ecosystems in ecological time. *Science*, 171, 385–387.

Rouphael, A.B. & Inglis, G.J. (2001) "Take only photographs and leave only footprints"? An experimental study of the impacts of underwater photographers on coral reef dive sites. *Biological Conservation*, 100, 281–287.

Roush, R.T. & McKenzie, J.A. (1987) Ecological genetics of insecticide and acaricide resistance. *Annual Review of Entomology*, 32, 361–380.

Ruiters, C. & McKenzie, B. (1994) Seasonal allocation and efficiency patterns of biomass and resources in the perennial geophyte *Sparaxis grandiflora* subspecies *fimbriata* (Iridaceae) in lowland coastal Fynbos, South Africa. *Annals of Botany*, 74, 633–646.

Sale, P.F. & Douglas, W.A. (1984) Temporal variability in the community structure of fish on coral patch reefs and the relation of community structure to reef structure. *Ecology*, 65, 409–422.

Salisbury, E.J. (1942) *The Reproductive Capacity of Plants*. Bell, London.

Sarukhán, J. & Harper, J.L. (1973) Studies on plant demography: *Ranunculus repens* L., *R. bulbosus* L. and *R. acris* L.I. Population flux and survivorship. *Journal of Ecology*, 61, 675–716.

Saunders, A. & Norton, D.A. (2001) Ecological restoration at Mainland Islands in New Zealand. *Biological Conservation*, 99, 109–119.

Savidge, J.A. (1987) Extinction of an island forest avifauna by an introduced snake. *Ecology*, 68, 660–668.

Schall, J.J. (1992) Parasite-mediated competition in *Anolis* lizards. *Oecologia*, 92, 58–64.

Schipper, L., Murtishaw, S. & Unander, F. (2001) International comparisons of sectoral carbon dioxide emissions using a cross-country decomposition technique. *Energy Journal*, 22, 35–75.

Schluter, D. & McPhail, J.D. (1993) Character displacement and replicate adaptive radiation. *Trends in Ecology and Evolution*, 8, 197–200.

Schoener, T.W. (1983) Field experiments on interspecific competition. *American Naturalist*, 122, 240–285.

Schoenly, K., Beaver, R.A. & Heumier, T.A. (1991) On the trophic relations of insects: a food-web approach. *American Naturalist*, 137, 597–638.

Schulze, E.D. (1970) Dre CO_2-Gaswechsel de Buche (*Fagus sylvatica* L.) in Abhäbgigkeit von den Klimafaktoren in Freiland. *Flora, Jena*, 159, 177–232.

Schulze, E.D., Fuchs, M.I. & Fuchs, M. (1977a) Spatial distribution of photosynthetic capacity and performance in a mountain spruce forest in northern Germany. I. Biomass distribution and daily CO_2 uptake in different crown layers. *Oecologia*, 29, 43–61.

Schulze, E.D., Fuchs, M.I. & Fuchs, M. (1977b) Spatial distribution of photosynthetic capacity and performance in a mountain spruce forest in northern Germany. III. The significance of the evergreen habit. *Oecologia*, 30, 239–249.

Shankar Raman, T., Rawat, G.S. & Johnsingh, A.J.T. (1998) Recovery of tropical rainforest avifauna in relation to vegetation succession following shifting cultivation in Mizoram, north-east India. *Journal of Applied Ecology*, 35, 214–231.

Shaw, D.J. & Dobson, A.P. (1995) Patterns of macroparasite abundance and aggregation in wildlife populations: a quantitative review. *Parasitology*, 111, S111–S133.

Silvertown, J.W. (1982) *Introduction to Plant Population Ecology*. Longman, London.

Simberloff, D.S. (1976) Experimental zoogeography of islands: effects of island size. *Ecology*, 57, 629–648.

Simberloff, D., Dayan T., Jones, C. & Ogura, G. (2000) Character displacement and release in the small Indian mongoose, *Herpestes javanicus*. *Ecology*, 91, 2086–2099.

Simon, K.R. & Townsend, C.R. (in press) The impacts of freshwater invaders at different levels of ecological organisation, with emphasis on ecosystem consequences. *Freshwater Biology*.

Sinclair, A.R.E. & Norton-Griffiths, M. (1982) Does competition or facilitation regulate migrant ungulate populations in the Serengeti? A test of hypothesis. *Oecologia*, 53, 354–369.

Slobodkin, L.B., Smith, F.E. & Hairston, N.G. (1967) Regulation in terrestrial ecosystems, and the implied balance of nature. *American Naturalist*, 101, 109–124.

Smith, F.D.M., May, R.M., Pellew, R., Johnson, T.H. & Walter, K.R. (1993) How much do we know about the current extinction rate? *Trends in Ecology and Evolution*, 8, 375–378.

Smith, J.N.M., Krebs, C.J., Sinclair, A.R.E. & Boonstra, R. (1988) Population biology of snowshoe hares. II. Interactions with winter food plants. *Journal of Animal Ecology*, 57, 269–286.

Smith, J.W. (1998) Boll weevil eradication: area-wide pest management. *Annals of the Entomological Society of America*, 91, 239–247.

Sousa, M.E. (1979a) Experimental investigation of disturbance and ecological succession in a rocky intertidal algal community. *Ecological Monographs*, 49, 227–254.

Sousa, M.E. (1979b) Disturbance in marine intertidal boulder fields: the nonequilibrium maintenance of species diversity. *Ecology*, 60, 1225–1239.

Spencer, C.N., McClelland, B.R. & Stanford, J.A. (1991) Shrimp stocking, salmon collapse, and eagle displacement. *Bioscience*, 41, 14–21.

Spiller, D.A. & Schoener, T.W. (1994) Effects of a top and intermediate predators in a terrestrial food web. *Ecology*, 75, 182–196.

Sprent, J.I. & Sprent, P. (1990) *Nitrogen Fixing Organisms: Pure and Applied Aspects*. Chapman & Hall, London.

Stauffer, R.C. (1975) *Charles Darwin's Natural Selection: Being the second part of his big species book written from 1856 to 1858*. Cambridge University Press, Lodnon.

Steffen, W.L., Walker, B.H., Ingram, J.S.I. & Koch, G.W. (eds) (1992) *Global Change and Terrestrial Ecosystems: the Operational Plan*. IGBP/ICSU, Stockholm, Sweden.

Stenseth, N.C., Falck, W., Bjornstad, O.N. & Krebs, C.J. (1997) Population regulation in snowshoe hare and lynx populations: asymmetric food web configurations between the snowshoe hare and the lynx. *Proceedings of the National Academy of Science of the USA*, 94, 5147–5152.

Stevens, C.E. (1988) *Comparative Physiology of the Vertebrate Digestive System*. Cambridge University Press, London.

Stoll, P. & Prati, D. (2001) Intraspecific aggregation alters competitive interactions in experimental plant communities. *Ecology*, 82, 319–327.

Stone, G.N. & Cook, J.M. (1998) The structure of cynipid oak galls: patterns in the evolution of an extended phenotype. *Proceedings of the Royal Society of London, Series B*, 265, 979–988.

Strauss, S.Y. & Agrawal, A.A. (1999) The ecology and evolution of plant tolerance to herbivory. *Trends in Ecology and Evolution*, 14, 179–185.

Stromberg, P.C., Toussant, M.J. & Dubey, J.P. (1978) Population biology of *Paragonimus kellicotti metacercariae* in central Ohio. *Parasitology*, 77, 13–18.

Strong, D.R. (1992) Are trophic cascades all wet? Differentiation and donor-control in speciose ecosystems. *Ecology*, 73, 747–754.

Strong, D.R. Jr., Lawton, J.H. & Southwood, T.R.E. (1984) *Insects on Plants: Community Patterns and Mechanisms*. Blackwell Scientific Publications, Oxford.

Sutton, S.L. & Collins, N.M. (1991) Insects and tropical forest conservation. In: *The Conservation of Insects and Their Habitats* (N.M. Collins & J.A. Thomas, eds), pp. 405–424. Academic Press, London.

Swift, M.J., Heal, O.W. & Anderson, J.M. (1979) *Decomposition in Terrestrial Systems*. Blackwell Scientific Publications, Oxford.

Symonides, E. (1979) The structure and population dynamics of psammophytes on inland dunes. II. Loose-sod populations. *Ekologia Polska*, 27, 191–234.

Symonides, E. (1983) Population size regulation as a result of intra population interactions. I. The effect of density on the survival and development of individuals of *Erophila verna* (L.). *Ekologia Polska*, 31, 839–881.

Taniguchi, Y. & Nakano, S. (2000) Condition-specific competition: implications for the altitudinal distribution of stream fishes. *Ecology*, 81, 2027–2039.

Tansley, A.G (1904) The problems of ecology. *New Phytologist*, 3, 191–200.

Thomas, C.D. & Harrison, S. (1992) Spatial dynamics of a patchily distributed butterfly species. *Journal of Applied Ecology*, 61, 437–446.

Thomas, C.D. & Jones, T.M. (1993) Partial recovery of a skipper butterfly (*Hesperia comma*) from population refuges: lessons for conservation in a fragmented landscape. *Journal of Animal Ecology*, 62, 472–481.

Thomas, C.D., Thomas, J.A. & Warren, M.S. (1992) Distributions of occupied and vacant butterfly habitats in fragmented landscapes. *Oecologia*, 92, 563–567.

Thompson, R.M., Townsend, C.R., Craw, D., Frew, R. & Riley,

R. (2001) (Further) links from rocks to plants. *Trends in Ecology and Evolution*, 16, 543.

Tilman, D. (1982) *Resource Competition and Community Structure*. Princeton University Press, Princeton, NJ.

Tilman, D. (1986) Resources, competition and the dynamics of plant communities. In: *Plant Ecology* (M.J. Crawley, ed.), pp. 51–74. Blackwell Scientific Publications, Oxford.

Tilman, D., Mattson, M. & Langer, S. (1981) Competition and nutrient kinetics along a temperature gradient: an experimental test of a mechanistic approach to niche theory. *Limnology and Oceanography*, 26, 1020–1033.

Tilman, D., Fargione, J., Wolff, B. et al. (2001) Forecasting agriculturally driven global environmental change. *Science*, 292, 281–284.

Tjallingii, W.F. & Hogen Esch, Th. (1993) Fine structure of aphid stylet routes in plant tissues in correlation with EPG signals. *Physiological Entomology*, 18, 317–328.

Tokeshi, M. (1993) Species abundance patterns and community structure. *Advances in Ecological Research*, 24, 112–186.

Tonn, W.M. & Magnuson, J.J. (1982) Patterns in the species composition and richness of fish assemblages in northern Wisconsin lakes. *Ecology*, 63, 137–154.

Towns, D.R. & Ballantine, W.J. (1993) Conservation and restoration of New Zealand island ecosystems. *Trends in Ecology and Evolution*, 8, 452–457.

Townsend, C.R. (1996) Concepts in river ecology: pattern and process in the catchment hierarchy. *Archiv fur Hydrobiologie* 113 (Suppl. 10), 3–21.

Townsend, C.R. & Crowl, T.A. (1991) Fragmented population structure in a native New Zealand fish: an effect of introduced brown trout? *Oikos*, 61, 348–354.

Townsend, C.R. & Hildrew, A.G. (1980) Foraging in a patchy environment by a predatory net-spinning caddis larva: a test of optimal foraging theory. *Oecologia*, 47, 219–221.

Townsend, C.R., Hildrew, A.G. & Francis, J.E. (1983) Community structure in some southern English streams: the influence of physiochemical factors. *Freshwater Biology*, 13, 521–544.

Townsend, C.R., Scarsbrook, M.R. & Dolédec, S. (1997) The intermediate disturbance hypothesis, refugia and bio-diversity in streams. *Limnology and Oceanography*, 42, 938–949.

Townsend, C.R., Thompson, R.M., McIntosh, A.R. et al. (1998) Disturbance, resource supply, and food-web architecture in streams. *Ecology Letters*, 1, 200–209.

Trewavas, A. (2001) Urban myths of organic farming: organic agriculture began as an idealogy, but can it meet today's needs? *Nature*, 410, 409–410.

Tripet, F. & Richner, H. (1999) Density-dependent processes in the population dynamics of a bird ectoparasite *Ceratophyllus gallinae*. *Ecology*, 80, 1267–1277.

Turkington, R. & Harper, J.L. (1979) The growth, distribution and neighbour relationships of *Trifolium repens* in a permanent pasture. IV. Fine scale biotic differentiation. *Journal of Ecology*, 67, 245–254.

Turner, J.R.G., Lennon, J.J. & Greenwood, J.J.D. (1996) Does climate cause the global biodiversity gradient? In: *Aspects of the Genesis and Maintenance of Biological Diversity* (M. Hochberg, J. Claubert & R. Barbault, eds). Oxford University Press, London, New York.

UNEP (1991) *Environmental Data Report*, 3rd edn. Basil Blackwell, Oxford.

United Nations (1998) *Global Change and Sustainable Development: Critical Trends*. Report of the Secretary General, United Nations, New York. Also available on the world wide web at www.un.org/esa/sustdev/trends.html

US Department of the Interior (1990) *Audit Report: the Endangered Species Program*. US Fish and Wildlife Service Report 90–98.

Valentine, J.W. (1970) How many marine invertebrate fossil species? A new approximation. *Journal of Paleontology*, 44, 410–415.

van de Walle, E. & Knodel, J. (1980) Europe's fertility transition: new evidence and lessons for today's developing world. *Population Bulletin*, 34, 1–43.

van den Bosch, R., Leigh, T.F., Falcon, L.A., Stern, V.M., Gonzales, D. & Hagen, K.S. (1971) The developing program of integrated control of cotton pests in California. In: *Biological Control* (C.B. Huffaker, ed.), pp. 377–394. Plenum Press, New York.

Vannotte, R.L., Minshall, G.W., Cummins, K.W., Sedell, J.R. & Cushing, C.E. (1980) The river continuum concept. *Canadian Journal of Fisheries and Aquatic Sciences*, 37, 130–137.

Vázquez, G.J.A. & Givnish, T.J. (1998) Altitudinal gradients in tropical forest composition, structure, and diversity in the Sierra de Manantlán. *Journal of Ecology*, 86, 999–1020.

Vivas-Martinez, S., Basanez, M-G., Botto, C. et al. (2000) Amazonian onchocerciasis: parasitological profiles by host-age, sex, and endemicity in southern Venezuela. *Parasitology*, 121, 513–525.

Volterra, V. (1926) Variations and fluctuations of the numbers of individuals in animal species living together. (Reprinted in 1931. In: *Animal Ecology* (R.N. Chapman, ed.), pp. 409–448. McGraw Hill, New York.)

Waage, J.K. & Greathead, D.J. (1988) Biological control: challenges and opportunities. *Philosophical Transactions of the Royal Society of London, Series B*, 318, 111–128.

Walsh, J.A. (1983) Selective primary health care: strategies for control of disease in the developing world. IV. Measles. *Reviews of Infectious Diseases*, 5, 330–340.

Warren, P.H. (1989) Spatial and temporal variation in the structure of a freshwater food web. *Oikos*, 55, 299–311.

Watkinson, A.R. (1984) Yield-density relationships: the influence of resource availability on growth and self-thinning in populations of *Vulpia fasciculata*. *Annals of Botany*, 53, 469–482.

Watkinson, A.R. & Harper, J.L. (1978) The demography of a sand dune annual: *Vulpia fasciculata*. I. The natural regulation of populations. *Journal of Ecology*, 66, 15–33.

Watkinson, A.R., Freckleton, R.P., Robinson, R.A. & Sutherland, W.J. (2000) Predictions of biodiversity response to genetically modified herbicide-tolerant crops. *Science*, 289, 1554–1557.

Webb, S.D. (1987) Community patterns in extinct terrestrial invertebrates. In: *Organization of Communities: Past and Present* (J.H.R. Gee & P.S. Giller, eds), pp. 439–468. Blackwell Scientific Publications, Oxford.

Webb, W.L., Lauenroth, W.K., Szarek, S.R. & Kinerson, R.S. (1983) Primary production and abiotic controls in forests, grasslands and desert ecosystems in the United States. *Ecology*, 64, 134–151.

Wegener, A. (1915) *Entstehung der Kontinenter und Ozeaner*. Samml. Viewig, Braunschweig. (English translation 1924. *The Origins of Continents and Oceans*, translated by J.G.A. Skerl. Methuen, London.)

Werner, H.H. & Hall, D.J. (1974) Optimal foraging and the size selection of prey by the bluegill sunfish *Lepomis macrochirus*. *Ecology*, 55, 1042–1052.

Westover, K.M., Kennedy, A.C. & Kelley, S.E. (1997) Patterns of rhizosphere community structure associated with co-occurring plant species. *Journal of Ecology*, 85, 863–874.

Whitehead, A.N. (1953) *Science and the Modern World*. Cambridge University Press, Cambridge.

Whittaker, R.H. (1975) *Communities and Ecosystems*, 2nd edn. Macmillan, London.

Williams, W.D. (1988) Limnological imbalances: an antipodean viewpoint. *Freshwater Biology*, 20, 407–420.

Winemiller, K.O. (1990) Spatial and temporal variation in tropical fish trophic networks. *Ecological Monographs*, 60, 331–367.

Woiwod, I.P. & Hanski, I. (1992) Patterns of density dependence in moths and aphids. *Journal of Animal Ecology*, 61, 619–629.

Woodwell, G.M. (1970) Effects of pollution on the structure and physiology of ecosystems. *Science*, 168, 429–433.

Worland, M.R. & Convey, P. (2001) Rapid cold hardening in Antarctic microarthropods. *Functional Ecology*, 15, 515–524.

World Health Organization (1990) *Public Health Impacts of Pesticides Used in Agriculture*. WHO/UN Environment Program, Geneva.

Wootton, J.T. (1992) Indirect effects, prey susceptibility, and habitat selection: impacts of birds on limpets and algae. *Ecology*, 73, 981–991.

Wurtsbaugh, W.A. (1992) Good-web modification by an invertebrate predator in the Great Salt Lake (USA). *Oecologia*, 89, 168–175.

Yodzis, P. (1986) Competition, mortality and community structure. In: *Community Ecology* (J. Diamond & T.J. Case, eds), pp. 480–491. Harper & Row, New York.

Index

Page numbers in **bold** refer to tables, in *italic* to figures.

Abies balsamea (balsam fir) 116
abundance
  biotic/abiotic factor correlations 295
  crowding effects 298
  density-dependent regulatory processes
    297, 297, 298, 302, *303*, 304, 306
  estimates 295
  key-factor analysis 299–304, *303*, 306
    key phase mortality 302, 304, 306
  predation effects 264–5
  theoretical determinants 296–9, *299*
  see also cycles of abundance
abyssal regions 141, 356
Acacia–ant interactions 235
Acacia burkittii 168, *168*
Acacia cornigera (bulls's horn acacia) 235,
  236
acclimatization/acclimation 81, 81, 85
Acer saccharum (sugar maple) *119*, 131,
  313–14
acid rain 30–2, *31*, 448, *455*, 455–6, *456*
acidic soils 116
  plant tolerance mechanisms 341–2
acidophiles 88
Aconitum columbianum (monkshood)
  194, *195*
Actinia tenebrosa 47–8, **48**
Actinomyces scabies 241
activated sludge 442
activity levels 18, *19*, 20, 21
actual evapotranspiration (AET)
  measurements 335, *336*
adaptation 53, *54*
  local 44, *46*, 46–8, 47
aerial mammals *126*, 126

aerosols, mineral nutrient sources 387
African elephant (*Loxodonta africana*)
  488–9, **489**, *490*
African ground squirrel 102–3
age-specific fecundity schedules 163
age structure of population 168
aggregated (clumped) distribution *171*,
  171, *172*
  interspecific competition 210, *211*
  parasites/mutualists 245, *246*
aggressive behavior, interspecific
  competition 190–2, *191*
agricultural slurries 444
agriculture 415–22
  food production levels *416*
  forecasting global environmental impact
    433, 433, 435
  integrated farming systems 428–31
  intensive livestock management 443–4
  pest control *see* pest control
  pollution 443–9
    nitrates 444 6, 465
    pesticides 446–9
  soil degradation/erosion 418–20
  temperate grasslands 130
  water resource consumption *421*, 421,
    422
  see also monoculture farming
Agriolimax reticulatus 102
Agrobacterium tumefaciens 241
agromyzid flies 241
Agrostis capillaris 53
Agrostis stolonifera (creeping bent grass)
  44, *46*
AIDS (acquired immunodeficiency
  syndrome) 264
Alabama leafworm 425
Alauda arvensis (skylark) 32–3, *33*, *34*

alder (*Alnus* spp.) 387
alderfly 243
algal symbionts 228, 240, 252
  lichens 252
  regulation of numbers 246
alkaline soils 116
allowance trading 456
Alnus incana 381, 381
Alnus spp. (alder) 387
alternate hosts 247, 248
alternative hosts 247–8
altitude gradients 115, *117*, 120, 354–6,
  355, 364
Ambrosia artemisifolia 313
American beech (*Fagus grandifolia*) 314
ammonia 387
Amoeba dysenterica 232
amoebic dysentery 232
amphibians 350, 494
  fossil record *358*
analagous structures 67, *68*
Andricus 242
Androsace septentrionalis *304*, 304, 306
Anguillicola crassus 247
animals
  aquatic
    adaptations 133–4
    oxygen supply 133
  biomes characterization 125–7, *126*
  defenses against consumption 101–3,
    103
  nutrition 98–100
  resource-use patterns 96–103, 109
annual life cycles 158, 159, *160*
  life tables **165**, 165–8, **167**
annual plants 25, 125
  cohort life table construction 166–7,
    167

annual plants (*cont.*):
  dry environments 90
  pioneer species 313
  seed dormancy/seed banks 159
  semelparous/iteroparous species 159
Anolis 60
Anolis gingivinus 261
Anolis wattsi 261
Anopheles annulipes 251
Anopheles gambiae 243, 243
Anoxanthum odoratum 54
Antilope cervicapra (blackbuck) 490
Anthrax 241
antifreeze compounds 81, 82
Aphanes avensis 93
aphids 171, 172, 229
  honeydew excretion 100
  intracellular/algal symbionts 240, 252
  plant parasite transmission 248
  stylet 97, 97
Apodemus sylvaticus (wood mouse) 224, 224
aposematism 102, *103*
apple thrip 298
apples, organic farming *431*, 431
applied ecology 5, 35
  see also conservation; sustainability
aquaculture 416
aquatic environments 118, 133–43, 145–6
  animal adaptations 133–4
  depth gradients in species richness *356*, 356
  nutrient budgets *388*, 388–9
  pollution 475
    chemical insecticides 446, *447*
  primary productivity *372*, 372
aquifers exploitation 422
Arabian oryx (*Oryx leucoryx*) 490, *491*
Arabis fecunda (sapphire rockcress) 44, *46*
arboreal mammals *126*, 126
Arctic tern 87
armadillo 103
Artemisia (wormwood) 131
artificial feeding programs 490
artificial fertilizer 418–19
  nitrate leaching from soil 445
  production 445
Arvicola terrestris (water vole) 307
Ascaris (roundworm) 42, 232
Ascophyllum nodosum 230, 262, *264*
Asellus aquaticus (water louse) 231

aspen (*Populus*) 132
aspirin 471
assimilation efficiency 375–6
Asterionella formosa 104, 189–90, *190*, 192, 194–5, *196*
atmospheric elements 385, 387
  dryfall 387
  wetfall 387
atmospheric pollution 449–57, 465
  acid rain *455*, 455–6, *456*
  allowance trading in USA 456
  carbon dioxide 450–3, *451*, *451*, *452*, *453*, 455, 456
  greenhouse effect 453–5, *454*
  lead 463
  mining activities 463
  ozone layer destruction 457
autotrophs 96
Azalea 116

Bacillus thurigiensis 432
background radiation 458, **459**
bacteria
  decomposers 378
  intracellular symbionts 240
  pathogens 225
badger 282
Balanus glandula 339, 340
bald eagle (*Haliaetus leucocephalus*) *484*, 484
baleen whale 257
balsam fir (*Abies balsamea*) 116
bamboo 162
barnacle 141, 339, 340
basic reproductive rate (*R*) 166
bat 350
Bates, H.W. 39
beet army worm (*Spodoptera exigua*) 429
beetles 350
  latitudinal gradients *354*, 354
behavior
  defenses against consumption 102–3
  host response to parasites 242–3, *243*
benthic communities 356
Berberis 247
Betula (birch) 132
Betula pendula (silver birch) 174, *175*
Betula pubescens 81
biennial plants 159, 161
biodiversity 39
  conservation *see* conservation
  definition 468
  ecosystem function 381, *382*
  genetically modified crops impact 32–5
  hotspots *492*
  northern coniferous forest (taiga) 132
  parasites 224, 224, 227, 228
  tropical 6, 63, *64*, 127–9, 346–7, 353–4
  value 468, 471
  see also species richness

biogeochemical cycles 389–92, 393
  human activities impact 392
biogeochemistry 386
biological pest control 260, 426–8, 471
  inoculation procedure 427
  insect control agents **428**, 428
  nontarget species effects 428
  species introductions 427
biological scale 8
biomagnification 446, *447*
biomass 367–8
biomes 114, 120–33, 145
  classification 120–1, *122*, *123*, 124–7
  distribution 114, *115*
    altitude/latitude effects 115, *117*
    climate change impact 143, *144*
  faunas 125–7, *126*
  local variations in conditions 114–15
  rainfall/temperature ranges *116*
  vegetation types 121, *123*, *124*, 124–5
biomonotonous communities 132
  disease outbreaks 132
biospecies 55–6
birch (*Betula*) 81, 132
bird flea (*Ceratophyllus gallinae*) *283*, 283
bird louse 230–2
birds 257, 350
  altitude gradients *355*, 355
  cellulose digestion 235
  desert 131
  eggshell thickness following DDT accumulation 448, *449*
  GM crops impact 32–3, *33*, *34*
  latitudinal gradients 353–4
  limpets predation 318, *319*
  migration 174
  northern coniferous forest/tundra 133
  temperate forest 132
  vegetation-related distribution *125*, 125
birth rate 152, 185
  density dependence 174–6, *176*, 179
  human demographic transition 405
  measurement 154, 157
  population dynamics 297
Biston betularia (peppered moth) 51–2, *52*, *53*, 55
black bear (*Ursos americanus*) 272
black stem rust (*Puccinia graminis tritici*) *248*
blackbuck (*Antilope cervicapra*) 490
Blackwater herring (*Clupea harengus*) *181*
blowfly (*Lucilia*) 241
blue tit (*Parus caeruleus*) 215, 215, *283*, 283
bluegill sunfish (*Lepomis macrochirus*) 272, *273*
body size 87–8
  reproductive activity relationship 162–3, *163*

body temperature regulation 85–6
Boiga irregularis (brown tree snake) 475, *476*
boll weevil 425
Bombus appositus 194, *195*, 209, 210
Bombus bifarius 209, 210
Bombus flavifrons 194, *195*, 210
Bombus frigidus 209, 210
Bombus kirbyllus 209, 210
Bombus occidentalis 210
Bombus sylvicolla 209, 210
bonito 141
boron 93
Bothriochloa insculpta 288, *289*
Botrytis anthophila 243
Botrytis cinerea (gray mould) 241
Bowerbankia 230
bromeliads 228
brown bear (*Ursos arctos*) 272
brown tree snake (*Boiga irregularis*) 475, *476*
brown trout (*Salmo trutta*) 18–23, *19*, 36, 134, *181*
   community structure changes following introduction 21, *22*
      primary production 21, *23*
   invertebrate feeding behavior response 18, *19*, 20
   population distribution determinants **20**, *20*
brucellosis 417
Buchnera 240, 252
buds 121
   Raunkiaer's "life forms" 121, *124*, 124–5
buffalo fly (*Haematobia irritans exigua*) 385
bull's horn acacia (*Acacia cornigera*) 235, *236*
bumble bees
   coexistence/exclusion 194, *195*
   interspecific competition
      community structure effects 208–10, *209*
      niche complementarity 210
   proboscis length 194, 209, 210
burial of human corpses 438
burrowing animals 103
butterflies 350
   latitudinal gradients 352, *353*

C3 pathway 91
C4 pathway 91
cabbage looper (*Trichoplusia ni*) 429
cabbage root fly 102
cacti 131
Cactoblastis 428
caddis fly (*Plectrocnemia conspersa*) 243, 267, *268*
Cakile edentula 172, *173*, 173
calcareous soil 116
calcicole/calcifuge flora 116

calcium 92, 386
   soil levels 32
Californian mussel (*Mytilus californianus*) 201, *202*, 318, 339, 340
Calluna vulgaris (heather) 119, *120*
Calvus multicornis 230
CAM pathway 91
Camarhynchus (*Cactospiza*) *pallida* 58
Camarhynchus psittacula 58
camel 235
Canada lynx (*Lynx canadensis*) 278, *279*
canopy communities 228
   tropical rain forest 127, *128*
Capsella bursa-pastoris 210, *211*
captive breeding programs 488, 490–2, 498
carbon 385, 386
   terrestrial ecosystem budgets 387
carbon compounds 384, 385
carbon cycle *391*, 392
carbon dioxide 384, 385, 387
   atmospheric concentrations 94, 444, 445
      daily flux 95–6
      global increases 94, 95, 143, 450–3, *451*, *451*, *452*, *453*, 455, 456
      green plant requirements 88, 90, 94–6, 370
      greenhouse gases 453
      primary productivity 370
Carcinus maenus 290
Cardamine hirsuta 210, *211*
Cardulus 428
Carex spp. 319
caribou (*Rangifer tarandus*) 133
carnivores 257, 372
   assimilation efficiency 376
   biomass characterization 125, *126*, 126
   consumption efficiency 375
   desert 131
   digestive processes 100
   excretory products 100
   nutrient composition of animal prey 98, 100
   prey age/fitness characteristics 266, 266–7
   secondary productivity 374
   top carnivore removal experiments 339, 339–40
carp introductions 23
carrion 202
   decomposition 383
carrying capacity (*K*) 175, *176*, 176, 180
   fisheries exploitation strategies 410
   global human populations 407–9, *408*
Catharanthus roseus (rose periwinkle) 471
cecidomyid wasps 241
cellulases 380
cellulose-digesting microflora
   detritivore mutualisms 380–1, *381*
   herbivore mutualisms 100, 234–5
      invertebrates 235

Centaurea solstitalis (yellow star thistle) 23, *24*
Ceratophyllus gallinae (bird flea) 283, *283*
Certhidea olivacea (warbler finch) 58
cesium-137 459
Chagas' disease 247
chain of extinctions 483
Chamaecrista fasciculate 44
chamaephytes 124
chaparral 121
character displacement
   Indian mongoose (*Herpestes* sp.) canine teeth 204, *205*, 206
   three-spined stickleback (*Gasterosteus aculeatus*) 206, *206*–7
cheat grass 213
checkerspot butterfly (*Euphydryas editha*) 482, *483*
cheetah 266
chemical defenses
   animals 101–2
   plants 101
      desert species 131
      responses to herbivory 262, *264*
chemical pesticides/insecticides 424–6
   evolved resistance 425, *425*, 429, *430*
   food chain biomagnification 446, *447*
      marine food chains 446–7, *447*
      thickness of birds' eggshells 448, *449*
   pollution 446–9
   scale of use 426, *426*
   secondary pests emergence 424, 425
   target pest resurgence 424
chemical waste pollution 442–3
chemosynthesis 76
Chen caerulescens (lesser snow goose) 319, *320*
Chenopodium album (fat hen) 32–3, 159
Chernobyl nuclear disaster 459, *460*
Cherokee darter 138
chilling injury 79
chiton 339
Chlorella 240
chlorofluorocarbons (CFCs) 143, 453, 455
cholera 232, 440
Chondrilla juncea (skeleton weed) 212, 212
Chondrus crispus 289, 290
Choristoneura fumiferana (spruce budworm) 132
Chorthippus brunneus (common field grasshopper) 159, **165**, 165–6
cinnebar moth 96
clay soil 91
Clean Air Act 31, 456
Clements, F.E. 5
climate 112
   ocean current influences 112–14, *113*
   rainfall *see* rainfall
   solar radiation 112, *113*

climate (*cont.*):
  species richness effects 343, *344*, 363
  topographic influences *114*, 114
climate change 61–4, 70, 85
  biome boundary changes 143, *144*
  fossil record 359
  glaciation cycles *62*, 62, 454
  greenhouse effect 95, 453–5, *454*
  pollen record analysis 62–3, *63*
  predicted consequences 64
  timescale 64
climax community 311, 317
climbing mammals *126*, 126
cline 56–7
clumped distribution *see* aggregated
    distribution
Clupea harengus (Blackwater herring)
    181
coal tit (*Parus ater*) 215, 288
coastal environments 141–2
  disturbance-related gaps 309–10
  zonation 141, *142*
coccidiosis 417
cod fisheries 414, *415*
coevolution 55
  parasites 249–52, 254
    gene-for-gene relationship 249
coexistence 6, 190–1, *192*, 194–5,
    213
  Competitive Exclusion Principle
      (Gause's Principle) 197, 201
  environmental heterogeneity effects
      201–2
  ephemeral environments 202
  evolutionary aspects 203–4
  intraspecific aggregates 210, *211*
  niche differentiation 211
  predator-mediated 288, 339
cohort life tables 163, *164*, 164–5
  annual life cycles **165**, 165–8, **167**
  survivorship curves 166–7, *167*
Coleophora alticolella (rush moth) 82
collector–filterers
  detritivores 380
  stream invertebrates 136
collector–gatherer stream invertebrates
    136
colonization
  disturbance-created gaps 310
  island of Surtey 350, *352*
  metapopulation dynamics 307, 308
  rapid population growth 182–3
Colorado potato beetle (*Leptinotarsa*
    decemlineata) **300**, 300, *301*, **302**,
    303
Columba livia 230–1
Columba palumbus (woodpigeon) 264
commensals 227, 228
common field grasshopper (*Chorthippus*
    brunneus) 159, **165**, 165–6
common garden experiments 44, *45*, 46,
    478

communities 8
  classification 134
  experimental approaches 10–11
  patchiness/patches 117
community matrix *294*
community structure
  chains of extinctions 483
  interspecific competition 207–13
  patch dynamics 308
    dominance-controlled communities
        310, 311
    founder-controlled communities
        310–11, *311*
  predator–prey relationships 288–91,
      292
  species introductions 21, 483–4
  stability 323–7
    complexity relationship 323–7, *327*
    definition 323
  temporal patterns 308–12, 328
  top carnivore removal experiments *339*,
      339–40
  see also food webs
compensation 105, *106*
competition 6
  dominance-controlled communities
      311
  environmental conditions 83
  optimal foraging behavior *270*, 270
  predation disturbance interactions
      288
    predator interference 282
  selection pressures 55
  species richness effects 337
  see also interspecific competition;
      intraspecific competition; resource
      competition
Competitive Exclusion Principle (Gause's
    Principle) 197, 201, 204
competitive lottery 310
confidence intervals 14–15
confidence levels 11, 12, 14
Connochaetes taurinus (wildebeest) *180*
conservation 6, 138, 467–98
  captive breeding programs 488, 490–2
  practical aspects 486–96, 498
  protected areas 492–4
  restoration ecology 494–5
  species management plans 486–90
consumers 96–8, 372
  energy transfer efficiencies 374–6
  generalists 96, 97
  life spans 97
  patterns of energy flow 373–4, *374*
  prey relationships 256–92
  specialists 96, 97
consumption efficiency 375
contest-like competition 105
continental drift 64, *65*, 66–7, 70
continuous breeders *161*, 161
  life table construction 168
contour-based cultivation 419–20, *420*

control action threshold (CAT) 423
convergent evolution 67, *68*, 70
copper 93
copper mining 461, *462*, 462–3
copper sulfate 432
coral reef 118, 310–11, *311*
  habitat disturbance 475
corals, algal symbionts 240
cormorant 141
correlations 26, 295
cosmic rays 458
cotton aphid 425
cotton bollworm (*Heliothis zea*) 425,
    429
cotton crop pest control 429, *430*
cottony cushion scale (*Icerya ourchasi*)
    427
courtship rituals 56
crabs 141, 174
Crangon septemspinosa (sand shrimp)
    107, 108
creeping bent grass (*Agrostis stolonifera*)
    44, *46*
creeping buttercup (*Ranunculus repens*)
    296
cremation 438
creosote plant (*Larrea mexicana*) 131
crested tit (*Parus cristatus*) 288
crop production 416, *417*, 417
crowding 171–2
  abundance determination 298
  fisheries exploitation strategies 410
  K-selecting habitats 183
  predator–prey relationships 282–4
  resource competition 174, 179
crypsis 102, *103*
Cryptochaetum spp. 427
cryptophytes 125
Cryptopygus antarcticus (Antarctic
    springtail) *81*, 81
Crytobagous spp. 260
Cuscuta sp. (dodder) 228, *229*, 237
cutworm caterpillars 257
cycles of abundance 274, *275*
  disease outbreaks 279–82
  dispersal effects 284, *285*
  Lotka–Volterra model 274, 276–7, *277*,
      278
  observational studies 274–5, 278–9,
      279, 280
Cyclotella meneghiniana 194–5, *196*
Cynodon dactylon 289
cynpipid wasps 241
cysts 159

Dalbulus spp. 248
damping-off fungi (*Pythium* sp.) 241
Daphnia magna 105, *106*
Darwin, Charles 39–41, *41*, 44, 162–3
Darwin's (Galapagos) finches 57–8, *59*
Dasypus novemcinctus (nine-banded
    armadillo) 471

data collection
    adequacy 13, 14
    planning 15, 16
data-less management 414
DDT 152, 448
    resistance 425
de Wit, C.T. 407
death rate 152, 185
    human demographic transition 405
    measurement 154, 157
    see also mortality
deciduous forest 131, 132
decision analysis 487
decision tree construction 487–8
decomposers 96, 373, 374
    detritivores 378–80, *379*
        plant matter break down 380–1, *381*,
        383
    microbivores 378–80
decomposition 378–81, 383, 393
    agricultural waste 444, 445
    ammonia release 387
    carrion 383
    definitions 378
    energy flow 375
    feces 383, *384*
    human sewage 440
    northern coniferous forest (taiga) 133
    secondary productivity 377
    temperate forest soils 132
    tropical rain forest 129
    water oxygen content impact 133, 440
deep sea vent communities *60*, 60–1, 76,
    141
defenses 100–3
    chemical 101–2, 131, 262, *264*
definitions of ecology 4–5, 8
deforestation 420, 451, 474
    large-scale experiment 28–9, *30*
Deinacrida mahoenuiensis (weta) 316
Deleatidium (mayfly) 19
Delphinium barbeyi (larkspur) 194, *195*,
    209
demographic transition *405*, 405–6, *406*
Dendrophthoe falcata 237
density-dependent processes 174–5, *176*,
    179
    carrying capacity (K) 175, 180
    compensation 105, *106*
    competition 104–5, *105*
    parasite populations 246, *247*, 282–3
    population size regulation *297*, 297,
        298
        key-factor analysis 302–3, *303*, 304,
        306
    predator populations 282
density-independent processes *106*, 106
density of population 171–2
descriptive studies 7
desert 112, 120, 131
    ephemeral "annual" life cycles 159, *160*
    polar 133

primary productivity 371
    vegetation 125, 131
desert honeysuckle (*Tidestromia
    oblongifolia*) 91
desertification 420
determinate development 153
detritivores 378–80, *379*
    aquatic 379–80, *380*
    assimilation efficiency 376
    collector–filterers 380
    plant detritus 380, *381*, 383
        mutualisms with cellulase-producing
        microflora 380–1
    shredders 379
development, environmental stimuli 80
diapause 80
diatoms
    coexistence 194–5
    competition for silicate 189–90, *190*
Dicerorhinus sumatrensis (Sumatran rhino)
    487, 487
dieldrin poisoning 446
diet width 270, 271
diphtheria 282
Diphyllobothrium latum 234
disease outbreaks 409
    biomonotonous communities 132
    contaminated water 422
    environmental conditions influence
        82–3, *83*
    monoculture farming 417–18
    population dynamics 279–82, 305
        transmission threshold 280, 281
dispersal 170–4, 185–6, 306–7
    colonizing tree species 314
    definition 171
    distance 171, *172*, 306–7
    measurement 173
    metapopulation dynamics 284, *285*,
        482
        experimental study in mites 285–6
        protected areas planning 494
    mutualists 252
    parasites 243, 246–9, 252, 267, *268*
        gut 233
        plant tissues 236–7
    pioneer plant species 313
    population abundance effects 172–4
distribution 20, 112–18
    climate change effects 61–4, *63*
    continental drift 64
    parasites/mutualists 245–6
    spatial patterns 112–18, 145
        aquatic environments 118
        large-scale climatic influences
        112–14
        local factors 114–17
    temporal patterns 118–19
        post-Pleistocene ice ages 118, *119*
disturbance
    evolutionary timescale 345
    frequency 343–4

gaps creation 309–10
    succession stages 311
    intermediate disturbance hypothesis
        343–5, *345*, 364
    predation 288
    species richness effects 343–5, 364
    streambed 134–5
diversity indexes 331, *332*, 332
Dobzhansky, Theodosius 55
dodder (*Cuscuta* sp.) 228, *229*, 237
Dolly Varden char (*Salvelinus malma*) *83*,
    83, 190–2
dolphin 446
domesticated grasses 130
dominance-controlled communities 310,
    311, 343
Doris (sea slug) 42
dormancy 80
    ephemeral/annual life cycles 159
Douglas fir 264
downy mildews 225, 229
Dreissena polymorpha (zebra mussel) 23,
    24
Dromaius novaehollandiae (emu) *66*, 66,
    67
Drosophila 182, 352
    Hawaiian Island populations 58, 60
Drosophila melanogaster 181
drought adaptations, perennial plants
    90–1
drugs 471
dry environment
    C4/CAM photosynthetic pathways 91
    plant adaptations 90–1
    root systems 92, *93*
dryfall 387
duckweed (*Lemna*) *154*
dune grass (*Vulpia fasciculata*) 104, *105*,
    106
dung 202
    see also feces
dung beetle 383, *384*, 385
Dust Bowl 418, 419
dusts, mineral nutrient sources 387

eastern hemlock (*Thuja canadensis*) 116
eastern red cedar (*Juniperus virginiana*)
    119, 313
ecological economics 471
ecological energetics 367
ecological engineers 228, 461
economic aspects 471–2
economic injury level (EIL) 423, *424*
ecosystem processes 367
    terminology 367–8
ecosystems 8
ecotourism 471, 475, 484
ectoparasites 253
    animals 229–30
        male fitness effects 230–1
        specialization 229–30, **232**, 232
    plants 229

ectotherms 85–6, 87
  production efficiency 376
El Niño phenomenon 414, 452
Elaphurus davidianus (Pere David deer)
      490, *491*
elephant 488–9, *489*, *490*
  cellulose digestion 235
Elton, Charles 5, 6, 106, 224
emigration 152, 172, 173, 282
  see also dispersal
emu (*Dromaius novaehollandiae*) 66, 66,
      67
Encarsia formosa 427
endangered species 468, *470*, *470*, 497
  conservation 6
    succession manipulation 316
endotherms 86, 87
  production efficiency 376
energy availability
  diet width 271, 272
  optimal foraging theory 269, 270
  potential evapotranspiration (PET)
      measurements 335
  species richness relationship 335, *336*,
      337
  see also solar radiation
energy flux 8, 366–78, 384
  consumer system importance 376–8,
      377
  decomposer system 375
  energy transfer efficiencies 374–6, *376*
  trophic levels 373–4, *374*
energy resources 450
energy transfer efficiencies 374–6, *376*
  assimilation efficiency 375–6
  consumption efficiency 375
  production efficiency 375, 376
Engraulis ringens (Peruvian anchoveta)
      141, *412*, 412–13, 414
Enterobius vermicularis (threadworm)
      242–3
Enteromorpha intestinalis 289
environmental conditions 76–84, 108
  disease spread relationship 82–3, *83*
  effects on organisms 77–80
    competition 83
    interactions 81–3
    population distribution study **20**, 20
    response curves 77, *78*
    sedentary organisms 84–5
  extremes 76–7, *77*, 340–1
    microorganisms 88
    species richness impact 340–2, *342*,
      363
  heterogeneity 201–3, 340, *341*, 363
  records *84*, 84–5, *85*
  seasonal cycles 84
  as stimuli 80–1
environmental heterogeneity 201–3, 213
  species richness 340, *341*, 363
Eotetranychus sexmaculatus 285–6, *287*
ephemeral environments 202, 203

epilimnion 137
Epilobium angustifolium 209
epiphytes 127, 129, 228
Equus przewalski (Przewalski's/Mongolian
      wild horse) 490, *491*
Erisiphe spp. 229
ermine 133
Erophila verna 169, *170*
Esox lucius (pike) 134
estuaries 118, 142–3
  primary productivity 368, 372
  salt water distribution (salt wedge)
      142–3
  see also wetlands
ethical aspects 471
Eucalyptus (gum tree) 121, 131
euphotic zone 139, 372
Euphydryas editha (checkerspot butterfly)
      482, *483*
European otter (*Lutra lutra*) 490
European rabbit (*Oryctolagus cuniculus*)
      249, 250
eutrophication
  sewage discharge 440, 442
  species richness reduction 337
evidence sources 10–11, 35–6
evolution 38–72
  constraints 68–9
  fossil record *see* fossil record
  interspecific competition effects 203–7,
      221
  island communities 350, 352
  natural selection 39–43
  pesticides resistance *425*, 425, 429, *430*
  timescale, species richness relationship
      345–7, 364
  within species 43
exclusion 190, 194, 213
  Competitive Exclusion Principle
      (Gause's Principle) 197, 201
  environmental heterogeneity effects
      201–2
  predation disturbance interactions 288
  successional sequence 311, 312, 313
exploitation *104*, 104
exponential growth 178, *179*, 179
extinction
  chains 483, 497
  following species introductions 475,
      476
  fossil record 359–60, *360*, 468
  global climate change 454
  habitat area associations *481*, 481
  hotspots 492
  metapopulations 307, 308, 481, 482–3,
      483
  rate estimation 468
  risk 6, 55, 469–70, *470*, *470*, 473, *479*,
      479, 497
    critical 470
  small populations 481, *482*
  vortices *479*, *480*

extreme environments 76–7, *77*, 340–1
  definitions 340–1
  microorganisms 88
  species richness impact 340–1, 363
eye spots 103

facultative parasites 241
Fagus grandifolia (American beech) 314
Falco peregrinus (peregrine falcon) 448,
      473, 490
false pink bollworm 424, 425
fat hen (*Chenopodium album*) 32–3, 159
feces 202
  decomposition 383, *384*, 385, 440
  responses in social animals 439–40
  urban disposal systems 440–2, *441*
  see also manure
fecundity 42
  age-specific fecundity schedules 163
  intraspecific competition impact 105,
      106, 106
fecundity schedules 163, 185
  annual life cycle 166
ferns 228
fertilizer application
  artificial fertilizers 418–19, 445
  organic farming 432
Ficedula hypoleuca (pied flycatcher) 258,
      259
Ficus (fig) 161, 228
field experiments 10, 26–7
field gentian (*Gentianella campestris*) 262,
      263
fig (*Ficus*) 161, 228
fire 213, 312, 387
  savanna 130
  seed germination influence 81
fish
  lakes 139
  ocean current upwellings 141
  oxygen requirements 133
    stream communities 134
  predation, experimental study 49, *50*,
      51
fish farming (aquaculture) 416
fisheries 409–14, *415*
  environmental fluctuations 414
  exploitation levels *410*
  exploitation strategies 410
    recruitment curve 410–11, *411*
  fixed effort harvesting *413*, 413
  fixed quota harvesting 411–12, *412*
  maximum sustainable yields 411
  population structure of exploited species
      414
  precautionary management approach
      414
  stock collapse 413, 414
fitness 42, 68
fixed effort harvesting *413*, 413
fixed organisms *see* sedentary organisms
fixed quota harvesting 411–12, *412*

flamingo (*Phoeniconaias minor*) 139
flax rust 249
flea 225, 229
flightless birds 64, *66*, 66–7, *67*
  fossil record 359
floodplains 136
Florida manatee (*Trichechus manatus*)
  471
flukes 225
fly parasite vectors 248
flying fox (*Pteropus* sp.) 483
food chain biomagnification of pesticides
  446, *447*
food stealing behavior 282
food webs 317–27, 328–9
  chains of extinctions 483, 498
  complexity 323–7, **326**, *326*, *327*
    stability relationship 323–7, *327*
  connectance 324, *325*
  direct/indirect interactions 317–18,
    319
  habitat fragmentation 485–6
  interaction strength 324
  keystone species 318–20
  mathematical models 324
  resilience 323, 324
  resistance 323, 324, 325
  species introduction 483–4, *484*
  species removal 317–18
    top carnivore removal experiments
      339, 339–40
  stability 323–7
    definition 323
  top-down versus bottom-up control
    320, 320–2, *321*
  trophic levels 320–2, *322*
foot and mouth disease 417
foraging behavior 267, *268*, 268–73, 291
  optimal foraging approach 269
forest
  acid rain impact *456*, 456
  clearance 420, 451, 474
  disturbance-related gaps 309–10
  energy flow through consumer/
    decomposer systems 377
  nitrate leaching from soil 445
  photosynthetic efficiency *370*, 370–1
forest catchment areas 28, *29*
  chemical budget studies 28–9, **29**
  large-scale felling experiment 28–9
forest tent caterpillars 280
fossil fuels combustion 450
  atmospheric pollution 465
    carbon dioxide 450–3, **451**, *451*,
      452, *453*
    nitrogen oxides 456
    sulfur dioxide 456
fossil record *358*, 358–60, 364
  extinctions 359–60, *360*, 468
founder-controlled communities 310–11,
  311
founder effect 61

founder populations 61
Fragaria (strawberry) *155*
fragile stability 323, 325
frost hardening 81
fruit
  nutrient content 99
  rotten 202
fruit-eating bats 129
Fucus serratus 230
fundamental niche 192, 193, 194
fungal decomposers 378
fungal parasites 225, *226*, 229

Galapagos (Darwin's) finches 57, 58, *59*
Galapagos islands 58, *59*
Galaxias 18, *19*, 20, 21
gall formation 241, *242*, 257
gannet 141
garrigue vegetation 121
Gasterophysa viridula 261, 261
gastropod defensive secretions 101
Gause's principle (Competitive Exclusion
  Principle) 197, 201, 204
Gelidium coulteri 312
gene flow, barriers for speciation 56, *56*
generalists consumers 96, 97, 271, 272
  detritivores 378
genet 153
genetic variation 42, 68
  small populations 476–7
genetically modified (GM) crops 448
  influence on biodiversity 32–5, 36
Gentianella campestris (field gentian) 262,
  263
Gentianella germanica 478, *478*, *479*
Geochelone elephantopus (giant tortoise)
  320, 490, *491*
geographical variation 43–51
  common garden experiments 44, *45*,
    46
  industrial melanism 51–3, *52*, *53*
  local adaptation 44, 46, 46–8, *47*
  plant heavy-metal tolerance 53, *54*
  reciprocal transplant experiments *45*,
    47–8, *49*
geology, local variation 115, 116
geophytes 131
Geospiza fortis 58, *373*, 373
Geospiza fuliginosa 58
Geospiza scandens 58
giant cave weta (*Gymnoplectron
  giganteum*) 473
giant panda (*Ailuropoda melanoleuca*)
  473, 473
giant tortoise (*Geochelone elephantopus*)
  320, 490, *491*
Gigartina caniculata 312, 344
Gigartina leptorhychos 312
gills 133
glaciation
  cycles 62, *62*
    tropical forest refuges 63, *64*

species distribution impact 118, *119*,
  132, 346
Glanville fritillary (*Melitaea cinxia*) 307,
  309
Glaucidium passerinum (pigmy owls) 288
glaucus-winged gull (*Larus glaucescens*)
  318
globalization of biota 24
glyphosate-resistant GM crops 448
  sugar beet 32
gold mining 461, 462, 463
gonorrhoea 243
Gordius dimorphus 243
grand Otago skink (*Leiolopis grande*) 473
grassland
  disturbance-related gaps 309–10
  energy flow through consumer/
    decomposer systems 377
  see also temperate grassland
gray mould (*Botrytis cinerea*) 241
grazers 96, 257
  savanna 130
  stream invertebrates 136
  temperate grasslands 130
grazing 291
  coexistence promotion 288–90
    pasture vegetation study 288–9,
      289
    rocky intertidal zone study 289–90,
      290
  plant fitness/abundance effects 258
  plant interspecific competition
    interactions *261*, 261
great tit (*Parus major*) 160, *162*, *215*, 215,
  272, *273*
greenhouse effect 453–5, *454*
greenhouse gases 95, 143, 453
grizzly bear (*Ursus arctos*) *484*, 484
gross primary productivity (GPP) 368
ground squirrel 283–4
growth
  environmental stimuli 80
  life cycle stages 158
  plant compensatory responses to
    herbivory 262, *263*
  trade-offs 181
guilds 195–6
Gulf Stream 114
gum tree (*Eucalyptus*) 121, 131
guppy (*Poecilia reticulata*) 49, *50*, 51, 55
gut parasites 232–5, 253
  complex life histories 233
  dispersal 233
Gymnoplectron giganteum (giant cave
  weta) 473
gypsy moth *103*

habitable sites 306–7
habitat 106
  area–species richness relationship 347,
    348, 351, 359
  disruption 474–5

habitat (*cont.*):
  fragmentation
    extinction risk *481*, 481
    nature reserve planning 493
    trophic organization changes 485–6
habitat preservation programs 488, 490
  estimation of required area 489, *490*
  protected areas 492–4
Haeckel, Ernst 4
Haematobia irritans exigua (buffalo fly)
  385
Haematopus bachmani (oystercatcher) 318
Haliaetus leucocephalus (bald eagle) *484*,
  484
handling costs/handling time 271, 272
Haplopappus venetus 265
Hawaiian Islands 58, 60
heath hen (*Tympanuchus cupido cupido*)
  481
heather (*Calluna vulgaris*) 119, *120*
heavy-metal tolerant plants 53, *54*
hedgehog 103
Heliocopris dilloni 384
Heliothis zea (cotton bollworm) 425, 429
Helminthosporium maydis (southern corn
  leaf blight) 82, *83*, 418
hemicryptophytes 125
Hemidactylus frenatus 190
hemiparasites 237
herbicide pollution 447–8
  impact on weeds/wild flowers 448–9,
  449
herbivores 96
  assimilation efficiency 376
  biomes characterization 125, *126*, 126
  consumption efficiency 375
  desert 131
  digestive processes 99–100
    digestive tract fermentation chambers
    233, *234*, 235
    mutualistic cellulose-digesting
    microorganisms 100, 234–5
  excretory products 100
  northern coniferous forest (taiga) 133
  nutrient composition of plant foods *98*,
  98
  savanna 130
  secondary productivity 373–4
  temperate grasslands 130
  tropical rain forest 128–9
herbivory
  plant compensatory responses 262, *263*
  plant defenses 100–1, *101*, 131, 262,
  264, 321
heritable variation 42, 68
herpes simplex virus 251
Herpestes edwardsii 204, *205*, 206
Herpestes javanicus 204, *205*, 206
Herpestes smithii 204, *205*, 206
herring gull (*Larus argentatus*) 57
Hesperia comma (silver-spotted skipper)
  307–9

heterotrophs 96, 372
  bacteria, secondary productivity *373*,
  373
hibernation *86*, 86, 87
high temperature 78–9
  endotherm morphological adaptations
  86
  microbial communities 88
  plant morphological adaptations 91
Hippoglossus stenolepis (Pacific halibut)
  413
Hirundo rustica (swallow) 174
historical aspects 4–6, 7
  phenological records 84
hole-dwellers 102
homologous structures 67
Hopkins' bioclimatic law 85
horse, cellulose digestion 235
host–parasite interactions 223, 240–5,
  253–4
  behavior change 242–3, *243*
  death of host 241
  defensive responses 244–5
  gall formation 241–2
  growth responses 241
  immune response 244–5, *245*
  selection pressures 55
hot spring communities 76
Hubbard Brook Experimental Forrest
  28–32, 36
Hulett, H.R. 408
human activities
  biogeochemical cycles impact 392
  community stability impact 326
  corpse disposal 438–9
  hunting, fossil record 359, 474
  riparian vegetation impact 136
  species richness impact 361
    exotic species introductions 361, *362*
  stream ecology impact 138
  temperate grasslands cultivation 130
  waste disposal 440–2, 465
human food resources 471
  food production levels 415, *416*
  monoculture farming 415–22
  natural (wild) resources 409–14, 434
human populations 400–9, 434
  age structure 401, 404–7
  demographic transition *405*, 405–6,
  406
  global carrying capacity 407–9, *408*
  numbers 401
    future predictions 404–6
    past gowth 402–3, *403*, *404*
  unequal resource distribution 402
  urban/rural distribution 401
Humboldt Current 114
humming-bird 282
hybridization 55, 56
Hydra 154
  algal symbionts 240
hydrological cycle *389*, 389–90

hydrosphere 386
hyperparasites 237
hypolimnion 139

Icerya ourchasi (cottony cushion scale)
  427
Idotea granulosa 262
immigration 152, 172, 173, 265
  see also dispersal
immune response to parasites 244–5,
  245
inbreeding depression 477, 479
indeterminate development 153
Indian meal moth (*Plodia interpunctella*)
  259, *260*
Indian mongoose (*Herpestes* sp.) 204, *205*,
  206
individual organisms 8, 18
  counting methods 153–4
  definitions 152–3
industrial melanism 51–3, *52*, *53*
industrial waste pollution 463
influenza virus 249
inoculation for biological pest control 427
insects
  fossil record *358*, 359
  galls 241, *242*
  mouthparts *99*
  parasites 225
    plants 225, *226*
  pest outbreaks 298
  pollinators 56, 225, 471
    tropical rain forest 129
  stream communities 135, *136*
  temperate forest 132
  vectors 248
integrated farming systems 428–31, 435
  soil maintenance 429
intensity of infection 245
intensive livestock management 443–4
interference competition 104
Intergovernmental Panel on Climate
  Change (IPCC) 95
intermediate disturbance hypothesis
  343–4, *345*, *346*, 364
interspecific competition 178, 188–221
  community structure effects 207–13,
  221
  Competitive Exclusion Principle
  (Gause's Principle) 197, 201
  current prevalence 216–17
  diffuse competition 195
  ecological effects 189–94, 221
    coexistence 190–1, 192, 194–5
    exclusion 190, 194
    temperature-mediated 190–2, *191*,
    193
  evolutionary effects 203–7, 221
    Canadian stickleback study *206*,
    206–7
    Indian mongoose (*Herpestes* sp.)
    study 204, *205*, 206

exotic species introductions 213–14
grazing impact interactions *261*, 261
Lotka–Volterra model 197, *198*, 198–200, *200*
neutral models 218
parasitism impact interactions 261
plant roots/shoot competition *212*, 212
practical significance 213–21
species richness influence 334
unrelated species 195
woodland tits (*Parus* sp.) study *215*, 215–16
see also resource competition
intertidal zone 141
predatory–prey relationships 289–90, *290*
see also rocky intertidal zone
intraspecific competition 103–6, 109
carrying capacity (*K*) 175, *176*, 176
compensation 105, *106*
density dependence 104–5, *105*
exploitation *104*, 104
fisheries management strategies 410
impact on vital rates 104
interference 104
natural selection 104
parasites 282–3
population effects 174–80, 186
predation effects 264–5
see also resource competition
intraspecific variation 43
geographical 43–51
intrinsic rate of natural increase (*r*) 178, 179
invertebrate communities
lakes 139
oceans
abyssal environment 141
deep sea vents 141
fossil record 358, 359
latitudinal gradients 352, *353*
streams 134–6, *135*, *136*, *137*
iodine-131 458–9
Ipomopsis aggregata 210
iron 93, 388
island biogeography 56, 57–61, 347
applications to nature reserve planning 492–3
equilibrium theory 347, 348–9, *349*, 350
evolutionary rates 350, 352
removal of introduced species 494
species–area relationship 347, *348*, 350, *351*
species–habitat diversity relationship 350, *351*, 364
iteroparous species 159
continuous breeders *161*, 161
life table construction 168
seasonal breeders 160–1, *161*
ivy 228

jellyfish, algal symbionts 240
Juncus squarrosus (rush) 82
Juniperus virginiana (eastern red cedar) 119, 313
juvenile phase *161*
semelparous species *161*, 161

K-selecting habitats 183
K species 183, 185
progeny size/numbers trade-off 183
woodland trees 183–4, *184*
k-values 300
Kalmia 116
kangaroo 235
Kates, R.W. 408
kelps 118, 142
surface commensals 228
key-factor analysis 299–304, 306
k-values determination 300–1
life tables *300*, 300, *301*, **302**, **303**
key-phase mortality *302*, 304, 306
keystone species 318–20
kiwi *66*, 66
Koebele 427
kokanee salmon (*Oncorhynchus nerka*) 483–4, *484*

laboratory experiments 10
simple systems 10–11
Lack, David 215
Lactobacillus sakei 180
Lagopus lagopus scotius (red grouse) 10, 262
lake trout 139
lakes 118, 137, 139
acid rain impact reduction 456
depth gradients in species richness 356
mineral nutrients 139
budgets 388
landscape position influence 139, *140*
physical conditions 118
saline 139, 388
stratification 118, 137, 139
larch budmoth (*Zeiraphera diniana*) 262
larch (*Larix*) 132
large-scale experiments 28–9
Larix (larch) 132
larkspur (*Delphinium barbeyi*) 194, *195*, 209
Larrea mexicana (creosote plant) 131
Larus argentatus (herring gull) 57
Larus fuscus (lesser black-backed gull) 56–7, *57*
Larus glaucescens (glaucus-winged gull) 318
Lasiommata maera 80
Lates nilotica (Nile perch) 475
latitudinal gradients
primary productivity 368, *369*
species richness 352–4, *353*, 364
lead pollution 463

leaves
defenses against herbivory *101*, 101
environmental adaptation 85
mineral nutrients uptake 387
resource utilization 89
shape/size diversity 89–90
shade leaves 90
sun leaves 90
surface phyllosphere 229
leech 257
Leiolopis grande (grand Otago skink) 473
lemming (*Lemmus*) 133, 283
Lemna (duckweed) *154*
lemur 121
Lepidodactylus lugubris 190
Lepomis macrochirus (bluegill sunfish) 272, *273*
Leptinotarsa decemlineata (Colorado potato beetle) *300*, 300, *301*, **302**, *303*
Lepus americanus (snowshoe hare) 275, 278, 279, 287
lesser black-backed gull (*Larus fuscus*) 56–7, *57*
lesser snow goose (*Chen caerulescens*) 319, 320
lichens 132, 133, 228
algal symbionts 252
life cycles 157–63, *158*, 185
annual 158, 159, *160*
continuous breeders *161*, 161
key phase mortality *302*, 304, 306
long-lived species 162–3, *161*
parasites 233, *234*, 246–9, *248*
seasonal breeders 160–1, *161*
stage-related resource allocation *158*, 158–9
see also life-history patterns; life-history trade-offs
life forms 121, *124*
life-history patterns 181–5, 186
life-history trade-offs 181–2, *182*
progeny size/numbers *183*, 183
LIFE (lower input farming and environment) 429
life tables 163–9, 185
annual life cycles *165*, 165–8, **167**
basic reproductive rate (*R*) calculation 166
cohort 163, *164*, 164–5
key-factor analysis *300*, *301*, **302**, **303**
k-values determination 300
populations with overlapping generations *168*, 168
static 163, *164*, 164–5, 168
survivorship curves 166–7, *167*
light *see* solar radiation
lignin digestion 380
limpet 141, 339
bird predation 318, *319*
Linnaeus 43

lion 272
Liriodendron tulipifera (yellow poplar) 116
LISA (low input sustainable agriculture) 429
lithosphere 386
littoral zone 139
Littorina littorea 289, *290*, 290
Littorina obtusata 262, *264*
livestock farming 416, 417
　disease outbreaks 417
lizards 322
　latitudinal gradients 352, *353*
　neutral models of community structure 218–20, *219*
local adaptation 44, *46*, 46–8, *47*
local factors 112
　soil 115
　topography 115
local speciation 43
lodgepole pine 81
logistic equation 178–9, *179*
logistic (S-shaped) growth curve *179*, 179, *180*, 180
Lolium multiflorum 93
long-term studies 9, 10, 28
　acid rain 30–2, *31*
　forest catchment area chemical budget 28–9, *29*
Lonicera japonica 154
Lotka–Volterra model 197, *198*, 198–200, *200*, 274, 276–7, *277*, *278*
Lottia digitalis 318
Lottia pelta 318
Lottia strigella 318
louse 130, 225, 229
low temperature 79
　acclimatization/acclimation *81*, 81
　endotherm morphological adaptations 86, *87*, 87
　leaf adaptations 91
　microbial communities 88
　seed germination influence 80, 81
Loxodonta africana (African elephant) 488–9, **489**, *490*
Lucilia (blowfly) 241
Lutra lutra (European otter) 490
Lygus hesperus (lygus bug) 429
Lyme disease 305
Lymnaea elodes 202–3, *203*
lynx 133
Lynx canadensis (Canada lynx) 278, *279*

magnesium 93, 388
Mahonia 247
malaria 248
　see also Plasmodium
malic acid 91
Malthus 40–1
manganese 93
mangrove swamp 118, 142

manure
　intensive livestock management waste **443**, 443
　organic farming 432
　soil quality maintenance 419, 432
maquis vegetation 121
marine protected areas 414, 475, 492
mark–recapture methods 153, 156
marsh tit (*Parus palustris*) 215
Marshallagia marshalli 106
Marsham, Robert 84
marsupial evolution 68, *69*
　fossil record 359
mast years 182
mastitis 417
mathematical models 11, 32
　value 35
matter flux 8, 366, 384–9, 393
　compartments 385
　global biogeochemical cycles 389–92
　nutrient budgets 386, 386–9
maximum sustainable yields 411, 413
　fixed effort systems 413
　fixed quota systems 411–12, *412*
　limitations of concept 411
mayfly 243
　activity levels 18, *19*, 20, 21
Mayr, Ernst 55
measles 281–2, *282*
Mediterranean vegetation 125
Melitaea cinxia (Glanville fritillary) 307, *309*
Mercurialis annua 93
mercury pollution 463
metapopulations 307–9, 328
　definition 307
　dispersal (connectedness) between subpopulations 482
　extinction risks 481, 482–3, *483*
　formation during habitat fragmentation 481–2
　population dynamics 307
　predator–prey dynamics 284, *285*, *286*
　　experimental study in mites 285–6, *287*
　　starfish–mussel observations 286–7
　theoretical aspects 308
methane 143, 442, 453, 455, 457
Miconia 127, *128*
microbivores 378–80
　assimilation efficiency 376
microorganisms
　extreme environmental conditions 88
　production efficiency 376
Micropus apus (swift) 295–6
migration 9, 87, 132, 152, 170–4, 185–6
　definition 171
millipedes 102
mimicry 102

mineral nutrients 385
　aquatic ecosystems 372, *388*, 388–9
　　lakes 139, *140*
　　oceans 139, 141
　forest catchment area long-term study 28, **29**, *30*
　　response to felling 28–9, *30*
　green plants 88, 92–4
　immobilization 378
　nutrient budgets 386, 386–9
　primary productivity 370, 371, 372
　solubility 133
　terrestrial ecosystems 386–8
　tropical rain forest 129
mineralization of chemical nutrients 378
minimum viable population (MVP) 486
mining pollution 461–3, 466
　restoration ecology 464, 494–5
Minnesota old field studies 25–8, *26*, 36, 314
mistletoe (*Phoraradendron*) 228, *229*, 237, 257
mites 225
mixed cropping 445
mixed feeders 126
modular organisms 153, *154*, *155*
　resource competition 174
modules 153
mole 102
mollusks 142
　algal symbionts 240, 252
monarch butterfly 102
Mongolian (Przewalski's) wild horse (*Equus przewalski*) 490, *491*
monkshood (*Aconitum columbianum*) 194, *195*
monoculture farming 415–22, 434
　crop production 416, *417*, 417
　disease outbreaks 417–18
　fish production (aquaculture) 416
　livestock production 416, 417
　nitrate leaching from soil 445
monophagy 96
monotremes 68
mortality
　density dependence 174–6, *176*, 179
　intraspecific competition 105, *106*
　k-values 300, 301
　key phase 302, 304, 306
　measurement 154, 157
　population dynamics *297*
　survivorship curve classification *169*, 169
　see also death rate
mosses 132, 133, 228
mouthpart specialization 97, *97*, *99*, 99
Musca vetustissima 385
mussel 141
　starfish interactions 286–7
mutual interference 282

mutualism/mutualists 225, 252, 253
 cellulase-producing microflora
  detritivores 380–1
   herbivores 100, 234–5
 coevolution 249
 definitions 225–6
 dispersal 252
 distribution within populations 245–6
mycetocytes 240
mycorrhizas 225, 236, 253
Mysis relicta (opossum shrimp) 483–4, 484
Mytilus californianus (Californian mussel) 201, *202*, 318, 339, 340
myxomatosis *180*, 249–51, *251*

natural experiments 25–6
natural selection 39–43, 70, 203–4
 adaptive peaks 53, *54*
 environmental pollution 51–3
 intraspecific competition 104
 predation pressure 49, *50*, 51
 specialization 53, *54*, 55
necrophoresis 439
nematodes 225, 233
Nesameletus ornatus (mayfly) *19*
net primary productivity (NPP) 368
neutral models 218–20, *219*
niche 68, 106–8, 109
 breadth (*n*) 333
 complementarity 210
 definition 106–7
 differentiation 197, 210, 211–13, *212*, 214
  plants versus animals 211–12
  resource partitioning/differential resource utilization 211
  woodland tits (*Parus* sp.) study *215*, 215–16
 dimensions *107*, 107, 108
 fundamental 192, 193, 194
 realized 192, 193–4, 197, 210
Nicholson, A.J. 298
Nile perch (*Lates nilotica*) 475
nine-banded armadillo (*Dasypus novemcinctus*) 471
Nippostrongylus brasiliensis 233
nitrate
 agricultural pollution 444–6, 465
 artificial fertilizers 445
 decomposition-related production 445
 leaching from soil 445
  agricultural waste **444**, 444, 445
 sewage disposal inputs to water 440
nitrogen 92, 385, 386, 387
 atmospheric deposition (pollution) 27–8
 long-term budget in forest catchment area 28
 manipulation experiments in old fields 26–7, *27*
 plant roots absorption 94

primary productivity limitation 371, 372
nitrogen cycle 390–2, *391*
nitrogen fixation 237, 239
 nitrate leaching from soil 445
nitrogen-fixing bacteria 237–40, 252, 253, 387
 host plant interactions 238–9, *239*
 regulation of numbers 246
nitrogenase 387
nitrous oxides 453, 457
 acid rain 456
 atmospheric pollution 143
northern coniferous forest (taiga) 120, 132–3
 fauna 133
 soil 133
 vegetation 132–3
northern white cedar (*Thuja occidentalis*) 116
nuclear radiation pollution 457–63, 466
 Brookhaven National Laboratory experiment 460–1, *461*
null hypothesis 12, *13*, 218
nutrient budgets *386*, 386–9
 aquatic ecosystems 388–9
 terrestrial ecosystems 386–8
nutrition in animals 98–100

oak eggar moth 100–1
oak (*Quercus robur*) 155
oak (*Quercus* sp.), gall formation 241, *242*
Obelia 154
Obelia dichotoma 230
Obelia geniculata 230
observational studies 10
ocean 118, 139, 141
 abyssal environment 141
 carbon dioxide absorption 155
 consumer/decomposer system 377–8
 deep sea vent communities *60*, 60–1, 76, 141
 depth gradients in species richness 356
 euphotic zone 139
 food chain biomagnification of pesticides 446–7, *447*
 mineral nutrients 139, 141
  nutrient budgets *388*, 388
 productivity 139, 141, 368, 372
 thermal capacity 118
ocean currents 112–13, *113*
 climatic influences 113–14
 upwellings 139, 141, 372
oil spills 450
 restoration ecology 464
old field studies 36
 nitrogen manipulation experiments 26–7, *27*
 succession following abandonment 25–6, *26*, 119, 313–14, *315*, 316–17
Onchocerca vulvulus 246

Oncorhynchus mykiss (steelhead trout) 104, *105*
Oncorhynchus nerka (kokanee salmon) 483–4, *484*
Onitis uncinatus 385
Onitis viridualus 385
opossum 102
opossum shrimp (*Mysis relicta*) 483–4, *484*
optimal diet 272–3, *273*
 diet width 270, 271, 272
optimal foraging theory 269, *270*
orchid 129, 228
Orconectes rusticus 246
organic farming **431**, 431, 432
organic matter
 lakes 139
 soil 117
 streams 136
Oryctolagus cuniculus (European rabbit) 249, 250
Oryx leucoryx (Arabian oryx) 490, *491*
ostrich (*Struthio camelus*) *66*, 66, 67
overcompensation 105, *106*
overexploitation 474
oxygen
 decomposition demand 440
 lake water stratification 137, 139
 water content 133
  streams 134
oxygen isotope cores 62
oystercatcher (*Haematopus bachmani*) 318
ozone 453
 atmospheric pollution 143, 457
 destruction of upper atmosphere layer (ozone layer) 457, *458*

P-values 12, 14
Pacific halibut (*Hippoglossus stenolepis*) 413
Pacific salmon 161–2
pandemics 249
Pangaea 359
Paracantha culta 428
Paragonimus kellicotti 246
parallel evolution 68, *69*, 70
parasites 96, 224, 232–5, 253, 257, 291
 biodiversity *224*, 224, *227*, *228*
 coevolution 249–52, 254
  gene-for-gene relationship 249
  transmissibility 251
  virulence 250–1, *251*
 definitions 225–6
 density dependence 282–3
 dispersal (transmission) 236–7, 243, 246–9, 252, 267, *268*
 distribution within populations 245–6
 facultative 241
 host fitness/abundance effects 258–60
 of humans 227
 intensity of infection 245
  mean intensity 245–6

parasites (*cont*.):
  interspecific competition interactions 261
  life cycles 246–9
    alternate hosts 247, 248
    alternative hosts 247–8
    host specialization 247
    vectors 248
  macro-/microparasites 225, 257
  of plants 226
  prevalence 245
  regulation of numbers 246, *247*
  see also disease outbreaks; ectoparasites; host–parasite interactions
parasitic flowering plants 225, 228, 237
parasitoid wasps *257*, 282
parasitoids *257*, 258
Parsi corpse disposal 438–9
Parus ater (coal tit) 215, 288
Parus caeruleus (blue tit) *215*, 215, *283*, 283
Parus cristatus (crested tit) 288
Parus major (great tit) 160, *162*, *215*, 215, 272, *273*
Parus montanus (willow tit) 215, 288
Parus palustris (marsh tit) 215
Parus sp. (woodland tits), interspecific competition study *215*, 215–16
Pasteuria ramosa 106
patches/patchiness 112, 117, 120, 306, 328
  disturbance-related gaps 309–10
  dynamics of community organization 309
  ephemeral environments 202, 203
  exploitation, optimal foraging approach 269–70, *270*
  host colonization by parasites 252
  interspecific competition effects 201–2
  predator–prey relationships 284–8, *285*, *286*
  scale 117
  small population habitat fragments 481
  successional stages following disturbance 313
  temperate forest 131
Pearl, Raymond 178
peat 132
peat bog 117
  succession (hummock–hollow cycle) 118–19, *120*
pelican 141
peppered moth (*Biston betularia*) 51–2, *52*, *53*, 55
percolating filters 442
Pere David deer (*Elaphurus davidianus*) 490, *491*
peregrine falcon (*Falco peregrinus*) 448, 473, 490
Peregrinus maidis 248
perennial life cycles 158, 160

perennial plants 25
  drought adaptation 90
  old fields succession 313
permafrost 132
Peromyscus leucopus 269, 270
pertussis (whooping cough) *282*, 282
Peruvian anchoveta (*Engraulis ringens*) 141, *412*, 412–13, 414
pest control 5–6, 423–8, 434–5
  aims 423, *424*
  biological agents 385, 426–8, 471
  chemical pesticides 424–6
    evolved resistance 429, *430*
  control action threshold (CAT) 423
  economic injury level (EIL) 423, *424*
  integrated management 428–9
    cotton pest control 429, *430*, **431**
  organic farming 432
  pest natural enemies 423, *424*, 426–7
  secondary pest outbreaks 429, *430*
  target pest resurgence 424, 429, *430*
pesticide pollution 446–9, 465
  food chain biomagnification 446–7, *447*
pH, environmental 77
  microbial specialists 88
  soil 116
  streams 134
phanerophytes 121, 125, 127
phenological observations
  records 84–5
  temperature relationship 84, *85*, 85
phlorotannins 262
Phlox drummondii 166, **167**
Phoeniconaias minor (flamingo) 139
Phoraradendron (mistletoe) 228, *229*, 237, 257
phosphorus 92, 386, *388*, 388, 389
  interspecific competition in diatoms 194–5, *196*
  primary productivity limitation 371, *372*, 372
  root absorption 94
  sewage disposal inputs to water 440
phosphorus cycle 390, *391*
photic zone 356
photoperiod *80*, 80
  reproductive activity triggering 160
  seed germination influence 81
photosynthesis 76, 384, 385
  C3 pathway 91
  C4 pathway 91
  CAM pathway 91
  plant successional sequence 313–14, **314**, *315*
  resources/limiting factors 370
    carbon dioxide 90
    solar energy *89*, 89
  solar radiation conversion efficiency *370*, 370–1
  water losses 90, 91
Phyllobius argentatus 267

phyllosphere 229, 253
Physa gyrina 202–3, *203*
phytoalexins 245
Phytophthora infestans (potato late blight) 229, 417–18
phytoplankton
  diurnal migration 174
  lakes 139
  oceans 139, 141
  primary productivity 373
  resource competition 207–8, *208*
  see also diatoms
Phytoseiulus persimilis 427
Picea critchfeldii 64
Picea (spruce) 132
pied flycatcher (*Ficedula hypoleuca*) 258, *259*
pigmy owls (*Glaucidium passerinum*) 288
pike (*Esox lucius*) 134
Pilaster ochraceus (starfish) *339*, 339, 340
pine (*Pinus*) 132
pioneer species 311
  old fields succession 313
placental mammal evolution 68, *69*
  fossil record *358*, 359
planktonic communities
  energy flow through consumer/decomposer systems *377*, 377
  experimental stratification disturbance 345, *346*
  food webs 320–1, *321*
plant macroparasites 225, 228, 237
plant resources 88–96, 108–9
  carbon dioxide 88, 90, 94–6
  minerals 88, 92–4
  radiant energy 88, 89–90
  water 88, 90–2
plants
  aquatic environments 134
  defenses against herbivory 99, 100–1, *101*, 321
    secondary chemicals 101
  food sources
    digestion 99–100
    nutrient composition *98*, 98
  fossil record *358*, 359
  as habitats 235
    surface commensals 228, *229*
  heavy-metal tolerance 53, *54*
  hypersensitive response to infection 245
  leaf surface phyllosphere 229
  roots/shoot competition *212*, 212
Plasmodiophora brassica 237
Plasmodium azurophilum 261
Plasmodium (malaria parasite) 225, 237
  life cycle *238*
  mosquito host behavior change *243*, 243
playing dead 102
Plectonema nostocorum 88

Plectrocnemia conspersa (caddis fly) 243, 267, *268*
Plegejus argus (silver-studded blue) 307, 309, *310*
Plodia interpunctella (Indian meal moth) 259, *260*
Poa annua 210, *211*
Poecilia reticulata (guppy) 49, *50*, 51, 55
polar desert 133
pollen record analysis 62, *63*
Pollicipes polymerus 318
pollution 6, 437–66
  acid rain 30–2, *31*
  agricultural 443–9
  atmospheric 449–57, 465
    nitrogen deposition 27–8
  chemical waste 442–3
  habitat disruption 474, 475
  industrial waste 463
  mining/quarrying 461–3
  nuclear radiation 457–63, 466
  prevention 464
  restoration ecology 464, 466
  selection pressure 51–3
  urban 438–43
polychaete worms 142
polyphagy 96
Polysiphonia 230
population age structure 168
population density 171–2
  resource competition 174
population dynamics 42, *43*, 293–329
  biotic/abiotic determinants 294–5, 328
  density-dependent processes 174–5, *176*, 297, *297*, 298
  density-independent processes *297*
  disease outbreaks 6, 279–82
  fluctuations 296
  intraspecific competition effects 174–80, 186
    mathematical models 178–9, *179*
  key-factor analysis 299–304, 306
    key-phase mortality 302, 304, 306
  life-history patterns 182–3
  long-term studies 9
  patterns of growth 176
    differential equations 178
    exponential growth 178, *179*
    logistic (S-shaped) growth curve *179*, 179, *180*, 180
  pest control 423, *424*, 424
  predator–prey relationships 273–88, 291
  recruitment curves 180, *181*
  small populations 480–3
  stable populations 295–6
  theoretical determinants of abundance 296–9, *297*, *299*
  see also metapopulations
population viability analysis (PVA) 488–9, **489**, 490

populations 8
  definition 152
  measurement methods 153–4, 156–7, 185
    representative samples 154
  size 152, 176
    life table data 166, 168
Populus (aspen) 132
porpoise 446
Posidonia 142
Postelsia palmaeformis (sea palm) 201, *202*
potassium 92, 386
potato late blight (*Phytophthora infestans*) 229, 417–18
potential evapotranspiration (PET) measurements 335, *336*
powdery mildews 225, 229, 249
prairie dog 103
praying mantis 243
precautionary management approach 414
predation 291
  interacting factors in prey vulnerability 261–2
  natural selection experimental study 49, *50*, 51
  prey fitness/abundance effects 258–60
  prey responses
    activity levels 18, *19*, 20
    population distribution 20
  species richness influence 334, 339–40, 363
predator-mediated coexistence 288, 339
predator–prey relationships 96, 273–88
  community structure 288–91, *292*
  crowding effects 282–4
  cycles of abundance 274, *275*
    Lotka–Volterra model 274, 276–7, *277*, *278*
    observational studies 274–5, 278–9, *279*, 280
  disease outbreaks 279–82
  patchy environments 284–8, *285*, *286*
    experimental study in mites 285–6, *287*
    starfish–mussel observations 286–7
  population dynamics 291
  selection pressures 55
predators 96, 257
  definition 257
  diet width 270, *271*
  food webs
    top carnivore removal experiments *339*, 339–40
    top-down control 320
  mutual interference 282
  optimal foraging behavior 269, *270*, 270
prevalence 245

prey
  abundance
    effects of predation 258–60
    optimal diet theory 272
  crowding effects 283
  defensive responses 262, 264
  parasite infection-related vulnerability 262
  population responses to predation 264–7
    compensation 262, 264–5
    removal of vulnerable individuals 266–7
  predation risk, optimal foraging approach 269, *270*
primary producers 372
primary productivity 368–72, 393
  community structure 21, *23*
  definition 368
  fate 372, 393
  geographical patterns 368, **369**, 370
  gross 368
  latitudinal gradients 368, *369*
  limiting factors 370–2, *371*, *372*
  net 368
  secondary productivity relationship 373, 373–4
probabilities 12–14
proboscis length 194, 209, 210
Prochloron 252
production efficiency 375, *376*
productivity
  actual evapotranspiration (AET) measure 335, *336*
  oceans 139, 141
  potential evapotranspiration (PET) measure 335, *336*
  species richness relationship 335, 337–8, 363
    latitudinal gradients 353
    paradox of enrichment 337
  tropical rain forest 127, *128*
  see also primary productivity; secondary productivity
progeny size/numbers trade-off *183*, 183
protected areas 492–4, 498
  biodiversity hotspots (centers of diversity/endemism) *492*, 492
  fragmented habitats
    area 493
    dispersal corridors 494
protozoan parasites 225
proximate causes 7
Przewalski's (Mongolian) wild horse (*Equus przewalski*) 490, *491*
Pseudomyrmex ferruginea 235
Pteropus sp. (flying fox) 483
Puccinellia spp. 319
Puccinia graminis tritici (black stem rust) 248
pursuit time 272

Pyrodictium occultum 79
Pythium sp. (damping-off fungi) 241

quarrying-related pollution 461–3, 466
Quercus robur (oak) 155
Quercus sp. (oak) 241

r-selecting habitats 183
r species 182–3, 184, 185, 461
    progeny size/numbers trade-off 183
rabbit 103
    cellulose digestion 235
    myxomatosis 180, 249–51, 251
rabbit flea (*Spilopsyllus cuniculi*) 251
rain shadow 114
rainfall 116
    atmospheric elements in rain 387
    biome ranges 116
    global climate patterns 112
    primary productivity limitation 371,
        371
    seasonality
        savanna 130
        temperate grasslands 130
    topographic influences 114, 114
Rana sylvatica (wood frog) **303**, 304
random distribution 171, 171, 172
random sampling 17
Rangifer tarandus (caribou/reindeer)
    133
rank–abundance diagrams 331, 332,
    333
Ranunculus repens (creeping buttercup)
    296
rare alleles 476
rare species 473, 497
    vulnerability to overcollection 474
raspberry beetle 98
raspberry moth 98
ratites 66
rats, economic impact 23
Raunkiaer's "life forms" 121, 124
realized niche 192, 193–4
    complementarity 210
    differentiation 197, 210
reciprocal transplant experiments 45,
    47–8, 49
records, environmental seasonal changes
    84, 84–5, 85
recruitment
    curves 180, 181
        fisheries exploitation strategies
            410–11, 411
        predation effects 265
red brome 213
red fire ant (*Solenopsis invicta*) 23, 24
red grouse (*Lagopus lagopus scotius*) 10,
    262
red oak 131
refection 235
refuges 286
    coastal communities 141

regression coefficients 302
regular (evenly spaced) distribution 171,
    171, 172
reindeer (*Rangifer tarandus*) 133
relative humidity 77
representative samples 15
    population size estimation 154
reproduction
    body size relationship 162–3, 163
    costs 181–2
    early
        impact on population growth 182
        interspecific competition 202, 203
    life cycle stages 157, 158–9
    photoperiod trigger 80, 160
    resource allocation 158, 158–9
    trade-offs 181–2, 182
reproductive success 42, 43
reptiles 350
    fossil record 358
resilience 323
resistance 323
    complex communities 324
resource competition 76, 103–6, 109
    bumble bee foraging behavior 194,
        195
    diatoms for silicate/phosphate 189–90,
        190, 194–5, 196
    K species 183–4, 184
    planktonic community structure
        diversity 207–8, 208
    plant roots/shoots 212
    seed-eating ants/rodents 195–7, 196
resource depletion zones (RDZs) 92
resource partitioning 211
    plants versus animals 211–12
resources 75–110
    food webs bottom-up control 320
    plants 88–96, 108–9
    primary productivity 370
    species richness relationship 335
    utilization 76
respiratory heat (R) 368, 374, 376, 377
restoration ecology 464, 466, 494–5
    conflicting human demands 495
rhea (*Rhea americana*) 66, 66, 67
Rhinocyllus conicus 428
Rhizobium 237, 253, 255
    life cycle 238–9, 239
rhizosphere 228–9, 253
Rhododendron 116
Rhododendron lapponicum 163
Rhodoglossum affine 312
rindepest 180
riparian vegetation 464
    impact of human activity 136
    stream community interactions 136
rivers 118
    dams 422
    drainage basin relationships 134
    floodplains/seasonal flooding 136
robust stability 323, 325

rocky intertidal zone
    disturbance impact study 344, 346
    successional sequence 311–12
    top carnivore removal experiments 339,
        339–40
rodents 257
    cellulose digestion 235
    desert 131
    northern coniferous forest (taiga) 133
    seed-eaters 337, 338
        ant competition/coexistence 195–7,
            196
Rodolia cardinalis (vedalia) 427
root hairs 92, 92
root systems 92, 93
roots
    branching 94, 94
    mineral nutrient uptake 92, 93–4, 94,
        387, 388
    mycorrhizas 236
    rhizosphere 228–9
rose periwinkle (*Catharanthus roseus*)
    471
rotenone 432
roundworm (*Ascaris*) 42, 232
rumen microbial community 100, 234,
    235
Rumex crispus 261, 261
Rumex obtusifolius 261, 261
rush (*Juncus squarrosus*) 82
rush moth (*Coleophora alticolella*) 82
rusts 225, 247, 248

Sagina procumbens 93
saguro cactus 79, 80
Sahara mustard 214
saline lakes 139, 388
salinity, estuaries 142–3
salinization 420, 422
Salix alba 471
Salix cinerea (willow) 180
Salix cordata (sand-dune willow) 258,
    259
Salmo trutta see brown trout
salmonids, interspecific competition
    190–1
Salvelinus leucomaenis (white-spotted
    char) 83, 83, 190–2
Salvelinus malma (Dolly Varden char) 83,
    83, 190–2
Salvinia molesta 260
sampling 14–15, 16
    criteria 16–17
    direction of effort 17
    population size estimation 154
    random 17
    species richness estimation 331
    stratified random 17
    subgroups 17
sand dune 10, 125
    annual life cycles 159
sand-dune willow (*Salix cordata*) 258, 259

sand shrimp (*Crangon septemspinosa*)
107, *108*
sandy soil 91
sapphire rockcress (*Arabis fecunda*) 44,
46
Sapraxis grandiflora 158
savanna 120, 130
rainfall 130
vegetation 130
Saxifraga bronchialis 155
scale 8–10, 35
biological 8
geographical (distribution) patterns
112–18
patches 117
spatial 8–9
temporal 9–10
scansorial (climbing) mammals *126*,
126
schistosomiasis 422
schooling behavior 51
scientific rigor 11, 36
sea anemone 141
sea grass 142
sea lion 141
sea otter 177
sea palm (*Postelsia palmaeformis*) 201,
202
sea slug (*Doris*) 42
searching costs/searching time 271,
272
seasonal breeders 160–1, *161*
seasonal environments
primary productivity 371
records *84*, 84–5, *85*
species richness 343, *344*
seasonal flooding 136
seasonal population size variation 176
seasonal responses 80
seaweed 142
surface commensals 228, *230*
secondary productivity 368
decomposer system 377
energy transfer efficiencies 374–6
primary productivity relationship *373*,
373–4
sedentary organisms 84–5
indeterminate development 153
resource partitioning constraints
211–12
resource requirements 88–9
seed banks 32, 33, 34, 159, 169
seed-eaters 195–7, *196*, 257, 337, *338*
seeds
defenses against herbivory 101
dispersal 172, 173
dormancy 90
ephemeral/annual life cycles 159
germination 80–1
nutrient content 99
semelparous species 159, *161*, 161
sessile organisms *see* sedentary organisms

sewage disposal 439–42, *441*, 465
primary treatment 441
secondary treatment
activated sludge 442
percolating filters 442
tertiary treatment 442
sexual signaling 56
shade leaves/species 90
Shannon (Shannon–Weaver) diversity
index (*H*) 332, *332*
sharks 474
shredders 136
detritivores 379
significance testing 12, *13*
silicate, diatom interspecific competition
189–90, *190*, 194–5, *196*
silt deposition 422
silver birch (*Betula pendula*) 174, *175*
silver mining 461
silver-spotted skipper (*Hesperia comma*)
307–9
silver-studded blue (*Plegejus argus*) 307,
309, *310*
simulation modeling 488–9
sit-and-wait predators 267, *268*
skeleton weed (*Chondrilla juncea*) 212,
212
skylark (*Alauda arvensis*) 32–3, *33*, *34*
sleepy lizard (*Tiliqua rugosa*) 226
sludge disposal 442
small ground mammals *126*, 126
small populations
dynamics 480–3, 497
demographic uncertainty 480
environmental uncertainty 480–1
spatial uncertainty 481
extinction risk 481, *482*, 497
genetic problems 476–8, *478*, 497
small watershed technique 28
smuts 225
snowshoe hare (*Lepus americanus*) 275,
278, *279*, 287
soil 6
acidic 116
calcareous 116
degradation/erosion 418–20, 434
contour-based cultivation 419–20,
420
forest clear-felling 420
formation/persistence 418
local variation 115–16, 117
northern coniferous forest (taiga)
133
nutrient element loss 387
organic matter 117
integrated farming systems 429
maintenance 418–19
pH 116
primary productivity 371
rhizosphere 228–9
temperate forest 132
tropical rain forest 129–30

soil water 91
root hair absorption *92*, 92
solar radiation 385
global climate patterns 112, *113*
lakes 137
oceans 139
photosynthesis/primary productivity
89–90, 370
aquatic environments 372
conversion efficiency 370, 370–1
streams 136
Solenopsis invicta (red fire ant) 23, *24*
southern corn leaf blight
(*Helminthosporium maydis*) 82,
83, 418
Spartina 142
spatial distribution patterns 112–18, 145,
170–1, *171*
large-scale climatic influences 112–14
small-scale 114–17
spatial scale 8–9
specialist consumers 96, 97, 271, 272
tropical rain forest 129
specialization 53, *54*, 55
coevolution 55
constraints 97
ectoparasites of mammals/birds
229–30, *232*, 232
extinction risk 55
plant solar radiation utilization 90
speciation 55–61, 70
barriers to gene flow 56
clines 56–7
island populations 57–61
role of isolation 56, *56*
species 43
definition 55–7
estimation of numbers *468*, 469
extinction risk 469–70, *470*, 470
species introductions 361, 362
biodiversity impact 475–6
biological pest control 385, 427
brown trout (*Salmo trutta*) studies
18–23
community structure impact 21, *22*
economic impact 23–4, *24*, 24
food web impact 483–4, *484*
interspecific competition 213–14
removal programs 494
species management plans 486–90, *487*
decision analysis 487
population viability analysis (PVA)
488–9, **489**, 490
simulation modeling 488–9
species richness 330–65
centers of diversity/endemism 492
climate variation effects 343, *344*,
363
disturbance effects 343–5, 364
intermediate disturbance hypothesis
343–4, *345*, *346*, 364
diversity indexes 331, 332

species richness (*cont.*):
  environmental harshness 340–2, *342*, 363
  environmental heterogeneity 340, *341*, 363
  estimation methods 331
  evolutionary timescales 345–7, 364
  food web stability 323–4, *325*, **326**, *326*
  fossil record *358*, 358–60, *360*, 364
  founder-controlled communities 310–11
  general correlations 360–1
  gradients 347, 364
    altitude 354–6, *355*
    depth *356*, 356
    latitude 352–4, *353*
  habitat area relationship 347, *348*, 350, *351*, 359, 493
  human activities impact 361
  interspecific competition effects 334
    phytoplankton study 207–8, *208*
  island biogeography 347, *348*, 350, *351*, 364
  niche breadth/overlap models 333–4, *334*
  parasite communities *224*, 224
  predation effects 334–5
    intensity 334, 339–40, 363
  productivity relationship 335, 337–8, 363
    paradox of enrichment 337
  rank–abundance diagrams 331, 332, *333*
  spatially varying determinants 335–42
  succession stages 311, *312*, 356–7, *357*, 364
  temporally varying determinants 342–7
  see also biodiversity
Spergula arvensis 159
Sphagnum 118, 119, *120*
spiders 257, 322
Spilopsyllus cuniculi (rabbit flea) 251
Spirorbis 228, *230*
Spodoptera exigua (beet army worm) 429
sponges, algal symbionts 240
spores 88, 159
spotted alfalfa aphid (*Therioaphis trifolii*) 423
spruce budworm (*Choristoneura fumiferana*) 132
spruce (*Picea*) 132
standard errors 14–15
standing crop 367
starfish–mussel interactions 286–7
starfish (*Pilaster ochraceus*) *339*, 339, 340
static life tables 163, *164*, 164–5, 168
statistical significance 12–13
statistics 11–17, 36

steelhead trout (*Oncorhynchus mykiss*) 104, *105*
Stellaria media 210, *211*
stomata 90, 91, 94
  desert plants 131
stonefly 243
stratification 118, 137
stratified random sampling 17
strawberry (*Fragaria*) 155
streamflow 387, 388
streamlining 133–4
streams 118, 134–6
  bed disturbance 134–5, 344–5
  drainage basin relationship 134
  energy flow through consumer/ decomposer systems *377*, 377
  food webs 322, 322
  human acitivity impact 138
  invertebrate communities 134–6, *135*, *136*, *137*
  nutrient element input into aquatic systems 387, 388
  pH 134
  riparian vegetation influences 136
  temperature 134
strontium-90 459
Struthio camelus (ostrich) 66, 66, 67
Stylophora pistillata 246
subgroup sampling 17
subpopulations 284, 307
  see also metapopulations
subterranean clover (*Trifolium subterraneum*) *212*, 212
succession 9, 118–19, 312–17, 328
  animal species 314, 316
  climax community 311, *312*, 317
  dominance-controlled community dynamics 311–12, *312*
  early colonization stage 311, *312*
  manipulation for endangered species conservation 316
  midsuccessional stage 311, *312*
  old field studies following abandonment 25–6, *26*, 119, 313–14, *314*, *315*, 316–17
  peat bog (hummock–hollow cycle) 118–19, *120*
  plant photosynthetic activity 313–14, *314*, *315*
  post-Pleistocene ice ages 118, *119*
  rocky intertidal zone studies 311–12
    disturbance frequency effects 344
  seed dispersal capacity 314
  species richness 356–7, *357*, 364
succulents 91, 131
sugar maple (*Acer saccharum*) 119, 131, 313–14
sulfur 92
sulfur cycle *391*, 392
sulfur dioxide pollution 456
Sumatran rhino (*Dicerorhinus sumatrensis*) *487*, 487

sun leaves/species 90
survivorship
  cohort life tables 163, 166
  curves 166–7, *167*
    classification *169*, 169
  static life tables 163
sustainability 399–436
  definition 400
swallow (*Hirundo rustica*) 174
swallowtail butterfly *103*
swift (*Micropus apus*) 295–6
swine fever 417
Sylvilagus brasiliensis 249
Symbiodinium 240, 252
symbiosis 224, 253
  intracellular organisms 240
Synchitrium endobioticum 237
Synedra ulna 189–90, *190*, 192
syphilis 243

Taenia saginata 233
taiga *see* northern coniferous forest
tannins 101
Tansley, A.G. 5
tapeworms 225, 232, 257
  complex life histories 233, *234*, 247
target pest resurgence 429, *430*
tectonic plate movements, continental drift 64, *65*
Tectus niloticus (trochus) 414
temperate forest 120, 131–2
  fauna 132
  soil 132
  vegetation 131
temperate grassland 120, 130
  cultivation for agriculture 130
temperature
  aquatic environments 118, 134
  biome ranges *116*
  influence on organism interactions 82
    competition 83, *191*, 199–2
  lake stratification 137, 139
  metabolic effects of increase 77–8, *78*
  niche definition *107*, 107, 108
  oceans 118
  primary productivity 370, *371*, 371
  regulation in animals 85–6
  response to extremes 77, *78*, 78–80, *79*
    duration/frequency of events 79–80
  seasonal variation
    animal responses 86–7
    phenological correlations *84*, 84, *85*, 85
  streams 134
  water oxygen content 133
temporary ponds 202, 203
termite cellulose digestion 235
  gut flora 8, *9*, *235*
terrestrial biomes *see* biomes
territorial behavior 282, 284
Therioaphis trifolii (spotted alfalfa aphid) 423

thermocline 137
thermophiles 79, 88, 141
therophytes 125
Thiobacillus ferroxidans 88
Thomson's gazelle *266*
threadworm (*Enterobius vermicularis*) 242–3
threat displays *103*, 103
Three Mile Island 460
three-spined stickleback (*Gasterosteus aculeatus*) *206*, 206–7
thrush (*Turdus*) 448
Thuja canadensis (eastern hemlock) 116
Thuja occidentalis (northern white cedar) 116
ticks 225, 226
    parasite vectors 248–9
    Lyme disease 305
tides 141
Tidestromia oblongifolia (desert honeysuckle) 91
Tiliqua rugosa (sleepy lizard) 226
timescale 9–10
    climate change 64
tinamou *66*, 66
tobacco mosaic virus 257
top carnivore removal experiments *339*, 339–40
topography
    climatic influences *114*, 114
    coastal communities influence 141
    local variations 115, 120
    rainfall influences *114*, 114
toxic plant chemicals 101, 102
toxins, environmental 77
trace minerals 77
transhumance 174
translocation programs 490
transmission of parasites 236–7, 213, 246–9, 252, 267, *268*
    diseases of livestock 417
transmission threshold 280, 281
transpiration
    cooling effect 91
    water requirement 90
trapping studies 153
Trebouxia 252
tree frog 228
trees
    as habitats 228
    latitudinal gradients 352, *353*
    old fields succession 313–14
    productivity–species richness relationship 335
Trichechus manatus (Florida manatee) 471
Trichocorixa verticalis 320, *321*
trichomes *101*, 101
Trichoplusia ni (cabbage looper) 429
Trichostrongylus tenuis 262
Trifolium repens (white clover) 48, *49*
    cyanogenic forms 101, **102**

Trifolium subterraneum (subterranean clover) 212, *212*
Trilepidia adamsii 474
trochus (*Tectus niloticus*) 414
trophic levels 320–2, *322*, 372, 373
    energy flow 373–4, *374*
    pyramidal structure 373
    transfer efficiency 376
tropical land-use changes 451–2
tropical rain forest 120, 127–30
    clearing for agriculture 451, 452
    productivity 127, *128*
    refugia 63, *64*, 354
    soil/soil nutrients 129–30
    species richness
        herbivores/fruit-eaters 128–9, *129*
        insects 129
        trees 127–8
    vegetation 125
tropical species richness 6, 63, *64*, 127–9, 346–7, 353–4
Trypanosoma 258, *259*
Trypanosoma cruzi 247
tube-forming worms (*Spirorbis*) 228
tuberculosis 249, *250*, 257, 409
Tubularia crocea 155
tundra 120, 132–3
    species richness *342*, 342
Turbinaria reniformis 155
Turdus (thrush) 448
Tympanuchus cupido cupido (heath hen) 481
Typhlodromus occidentalis 285–6, *287*

Ulex europaeus 316
ultimate causes 7
Ulva spp. 312, 344
undercompensation 105, *106*
unitary organisms 153
urban waste disposal 439–43, 465
    chemical waste 442–3
    feces/urine 440–2
Uredinales (rusts) 225
    complex life cycles 247, *248*
Ursos americanus (black bear) 272
Ursos arctos (brown bear/grizzly bear) 272, *484*, 484
Ustilago tritici 243

value of ecosystem services 471
variation
    between individuals 42
    genetic 42, 68
        small populations 476–7
    intraspecific 43
        geographical 43–51
vectors 248
vedalia (*Rodolia cardinalis*) 427
vegetation
    altitudinal gradients 115, *117*
    biome characterization 121, *122*, *123*, 124–5

desert 131
intertidal zone 142
lakes 139
northern coniferous forest (taiga) 132
related herbivore distribution *125*, 125
savanna 130
soil pH 116
temperate forest 131
Venturia canescens 259–60, *260*
Verhulst, Pierre-François 178
Vibrio cholerae 232
vines 228
viral pathogens 225
    vectors 248
Viscum capitellatum 237
vital rates 104
vole 283
vulnerable species 469, **470**, *470*
Vulpia fasciculata (dune grass) 104, *105*, *106*
vultures, Parsi corpse disposal 438, 439

Wallace, Alfred Russell 39–41, *41*, 44
walleye 139
warbler finch (*Certhidea olivacea*) 58
water 384
    freezing behavior 134
    global hydrological cycle *389*, 389–90
    green plant utilization/primary productivity 88, 90–2, 370, 371
    human resource use 420–2, *421*, 434
        contamination 422, 444, 445
        supply shortages 422
        wildlife conservation impact 422
    oxygen content 133
    physical properties 133
    soil reserves 91
    thermal capacity 134
water louse (*Asellus aquaticus*) 231
water vole (*Arvicola terrestris*) 307
waves 141
weta (*Deinacrida mahoenuiensis*) 316
wetfall 387
wetlands 422, 445, 464, 485
    see also estuaries
whales 446, 474
whelk 339
White, Gilbert 295
white clover (*Trifolium repens*) 48, *49*
    cyanogenic forms 101, **102**
white-spotted char (*Salvelinus leucomaenis*) 83, *83*, 190–2
whooping cough (pertussis) 282, *282*
wild dog 266
wildebeest 130, *180*
willow (*Salix cinerea*) 180
willow tit (*Parus montanus*) 215, 288
wilting 90
wind erosion of soils 418, *419*

winter moth 101
wood frog (*Rana sylvatica*) **303**, 304
wood mouse (*Apodemus sylvaticus*) 224,
    224
woodland tits (*Parus* sp.), interspecific
    competition study *215*,
    215–16
woodpigeon (*Columba palumbus*) 264
wormwood (*Artemisia*) 131

yellow fever 248
yellow poplar (*Liriodendron tulipifera*)
    116
yellow star thistle (*Centaurea solstitalis*)
    23, *24*

zebra 130
zebra mussel (*Dreissena polymorpha*) 23,
    24

zebu 288
Zeiraphera diniana (larch budmoth) 262
zinc 93
zonation, coastal ecosystems 141, *142*
zooplankton 257
    secondary productivity *373*, 373
zoos 491
zoospores 236–7
Zostera 142